ELEMENTS OF QUEUEING THEORY

With Applications

ELEMENTS OF QUEUEING THEORY

With Applications

THOMAS L. SAATY

UNIVERSITY PROFESSOR
UNIVERSITY OF PITTSBURGH

DOVER PUBLICATIONS, INC.
NEW YORK

Published in Canada by General Publishing Company, Ltd., 30 Lesmill Road, Don Mills, Toronto, Ontario.
Published in the United Kingdom by Constable and Company, Ltd., 10 Orange Street, London WC2H 7EG.

This Dover edition, first published in 1983, is an unabridged and unaltered republication of the work originally published by McGraw-Hill Book Company, New York, Toronto, London, in 1961.

Manufactured in the United States of America
Dover Publications, Inc., 180 Varick Street, New York, N.Y. 10014

Library of Congress Cataloging in Publication Data

Saaty, Thomas L.
  Elements of queueing theory, with applications.

  Reprint. Originally published: New York : McGraw-Hill, 1961.
  Bibliography: p.
  Includes index.
  1. Queueing theory. I. Title.
T57.9.S2 1983   519.8′2   83-5203
ISBN 0-486-64553-3

To My Mother

PREFACE

A "queue" is a waiting line. Queues are commonly observed to form before ticket offices, cafeterias, and bus stops. Thus, to have a queue, one must have arrivals at a service facility where they often wait; for example, at a doctor's office to see the doctor in the order of their appointments. Telegrams are classified according to the order of their arrival with priority assigned to regular telegrams over night letters.

Queueing theory is a branch of applied mathematics utilizing concepts from the field of stochastic processes. It has been developed in an attempt to predict fluctuating demands from observational data and to enable an enterprise to provide adequate service for its customers with tolerable waiting. However, the theory also basically improves understanding of a queueing situation, enabling better control. For example, loss of customers at a grocery store because of slow check-out-counter service may be remedied by rapid service at a larger number of counters. To put this into effect does not require a theory. But to organize airport take-off and landing operations at a minimum inconvenience does, because of the many factors entering the problem and the large size of the operation. Thus, the theory provides one with predictions about waiting times, the number waiting at any time, the length of a busy period and so forth. These predictions help the manager of the enterprise to anticipate situations and to take appropriate measures to alleviate congestion. In addition it makes both manager and customer aware of a constant need for new ideas to simplify the complications of modern living.

Among the important problems arising from overpopulation (this Malthusian concept is nowadays referred to as the population explosion) are the food and housing problems, the great demand for service in a system, and the ability of people to live with this congestion. The last point is an aesthetic one; the first is an economic one. However, it is the second point that concerns us in studying queues. How should one organize congestion situations to minimize the resulting losses in time, inefficient operation, and more generally in investment of effort and capital?

The subject of queueing is not directly concerned with optimization. Rather it attempts to explore, understand, and compare various queueing situations and thus indirectly achieve optimization approximately. The fact that many people who provide public service should make greater efforts to appreciate the delicate nature and effect of queues is demonstrated by the following letter to the editor of the *Washington Post*:[1]

> It must have taken some high-level planning to close one inbound lane of Memorial Bridge [a six lane bridge] for stripe painting on one of the heaviest traffic days of the year, the day before Easter, thereby decreasing the flow by one third.   Just back up one car out of three by one car length and multiply by several thousand, the last car would be somewhere in the neighborhood of Falls Church [seven miles away].
>
> It seems to me the Great White Father in charge of painting white stripes could have painted this bridge earlier and used Saturday to paint some side street, of which I am certain he would have no trouble finding.

In passing we point out that the methods of queueing are analytic rather than synthetic.   The synthetic procedure is used to integrate ideas and give direction to analytic research on new problems.

The principal purpose of the book is to produce a general text and a summary of scattered papers and monographs on the subject of queues. Secondarily, it includes a wide bibliography and indicates some unsolved problems.   It also includes a descriptive introductory chapter most of which is aimed at the layman.   Many ideas are illustrated with examples, and a number of exercises are intended to fill in some of the omitted detail and develop results along indicated lines.

While the book is intended for graduate students with a good background in basic calculus, probability, introductory complex variable theory, and matrix theory, it can also be used by research workers who require a ready access to the subject of queues, its ideas and methods.

The contents have been divided into four parts.   The first part, which contains the first three chapters, gives basic material on queueing concepts and methods and on probability theory.   The second part contains two chapters.   It specializes in Poisson queues which, historically, comprised the first analytical treatment of the subject by A. K. Erlang of Copenhagen.   The third part, consisting of five chapters, is concerned with non-Poisson queues, which are studied mostly for cases in which statistical equilibrium has been reached.   Chapter 9 is particularly concerned with the study of waiting times.   The fourth part also contains five chapters and is concerned with the effect of various queue disciplines,

---

[1] Bridge Blunder, letter to the editor, *Washington Post*, Apr. 21, 1960, from Joseph M. Wray, Alexandria, Va.

special networks of queues, and a variety of queueing phenomena. A chapter on applications and a chapter on renewal theory are also included in this part. A brief look at some problems which require formulation precedes the bibliography.

It was felt that the varied material of Chap. 2 would be more stimulating to the reader if it were placed before the more formal presentation of the material in Chap. 3. In addition it exposes the interested reader to a variety of concepts early in the book and may increase his desire to go on.

It will be clear that sometimes we refer to the arrivals to a queue as "units," "items," "calls," and "customers." The "service time" is sometimes called the "holding time." An input source or sources of customers are populations from which customers arrive into the system.

Generally, the notation used will be clarified when it is introduced. Most notations are used often enough so that the reader should have no difficulty with them.

I am indebted to Dr. George Morgenthaler for a critical review of the first version of the manuscript, to Dr. Ryszard Syski and Dr. Eric Wolman for their helpful suggestions and proofreading, and to Dr. Leila Bram and Mr. Alexander Craw for help and comments in proofreading.

Thomas L. Saaty

CONTENTS

Preface vii

Essential Symbols xiii

Part 1.   Structure, Technique, and Basic Theory
          CHAPTER 1.   A Description of Queues   3
                  2.   Some Queueing Models.   26
                  3.   Probability, Markoff Chains and Processes, Ergodic
                       Properties of Queues   54

Part 2.   Poisson Queues
          CHAPTER 4.   The Birth-Death Process in Queueing Theory . . .   83
                  5.   The Case of Small Barriers.   135

Part 3.   Non-Poisson Queues
          CHAPTER 6.   Special Non-Poisson Cases   153
                  7.   Bulk Queues   171
                  8.   Poisson Input, Arbitrary Service Distribution, Distribu-
                       tions of the Number in the System, the Busy Period,
                       and the Number Served   193
                  9.   Poisson and Arbitrary Input, Arbitrary Service; the
                       Waiting Time for Single and Multiple Channels . .   198
                 10.   General Independent Input, Exponential or Erlangian
                       Service   219

Part 4.   Queueing Ramifications, Applications, and Renewal Theory
          CHAPTER 11. Other Queue Disciplines   231
                 12.   Queues in Tandem or Series   255
                 13.   Interesting Queueing Phenomena   269
                 14.   Applications   302
                 15.   Basic Renewal Theory   359

Bibliography   373

Index   415

ESSENTIAL SYMBOLS

X, Y, Z	Usually used to denote general random variables
S_n	Denotes the nth partial sum $X_1 + \cdots + X_n$
M	Refers to Poisson-distributed times or to exponentially distributed intervals
G	Refers to a general distribution
GI	Refers to a general distribution with independence assumptions
D	Refers to a constant distribution
$\varphi(t)$	Characteristic function of a random variable
$M(t)$	Generating function of a random variable
$E[\ ]$	Expected value
C_t	Coefficient of variation of variable t
σ	Standard deviation and σ algebra; σ_ρ is standard deviation of ρ
μ_r'	rth moment about an arbitrary point
μ_r	rth moment about the origin
κ_r	rth cumulant
$\equiv$	Definition sign
$*$	Used to denote Laplace transform
$\sim$	Used to denote Laplace-Stieltjes transform
τ	Often refers to a time interval
δ_{in}	Kronecker's delta
X_t	Stochastic process
$F(x,t)$	Cumulative distribution of the random variable X at time t
E_j	State j of a system of states
P	Matrix of transition probabilities
$p_{ij}^{(n)}$	Transition probability from state E_i to state E_j at nth transition time
$f_{ij}^{(n)}$	Probability that, starting with state E_i at time zero, the system is, for the first time, in state E_j at time n
μ_{ij}	Mean first-passage time from E_i to E_j

$p(j,t;i,s)$	Probability of being in state E_j at time t, having been in state E_i at time s
$A(t)$	Cumulative-interarrival-interval distribution
$a(t)$	Corresponding density function
$B(t)$	Cumulative-service-time distribution
$b(t)$	Corresponding density function
$\beta(s)$	Laplace-Stieltjes transform of $B(t)$
λ_n, μ_n	Parameters appearing in birth-death-process equations
q	Random variable of the number of units in queue
$P_{in}(t)$	Generally the probability that there are n units in a system (in queue and in service) at time t, given that there were i units in the system at $t = 0$
$P_n(t)$	Same as above with i suppressed; also the probability of finding exactly n channels busy at time t
p_n	Steady-state probability of having n units in the system; also used to denote discrete probability density
π_n	Steady-state probability of a single-channel system being in state E_n at times before service begins or after it has just completed service
$P(z)$	Generating function of probabilities, usually of p_n but sometimes of π_n
$P(z,t)$	Generating function for $P_n(t)$
c	Number of channels
λ	(Sometimes $1/a$) arrival rate
μ	(Sometimes $1/b$) service rate
s	Number in a batch (mostly in Chap. 7); parameter of Laplace transforms in other chapters
w_n	Random variable of the waiting time in line of the nth item
s_n	Random variable of the service time of the nth item
t_n	Random variable of the interarrival time interval of the nth item
$u_n = s_n - t_n$	
ρ	Traffic intensity or utilization factor, usually the ratio of the input rate to the service rate
L	Average number in system
L_q	Average number in queue
$W(t)$	Waiting time in queue at time t (it is a deterministic function, not a probability distribution)
$w_n(t)$	Waiting-time density function of $(n + 1)$st item

in the system; $W_n(t)$ is the corresponding cumulative distribution

$P(w,t)$ — Cumulative distribution of waiting time w in queue at time t; as $t \to \infty$ this has the limit $P(w)$ if it exists

$P(<t), P(w)$ — Cumulative distribution of waiting time in the queue in the steady state [It is a historical abuse of notation to use $P(<t)$. Thus, $P(<t)$ is the same as $P(w)$. Note that because we compute expressions of the form $P(t = 0)$ and $P(t > 0)$ we have retained old notation.]

$P(t = 0)$ — Probability of not waiting in queue

$P(t > 0)$ — Probability of any waiting in queue

$w(t)$ — Density function of the waiting time in queue, that is, $w(t) = dP(<t)/dt$

$\gamma(s)$ — Laplace-Stieltjes transform of $P(<t)$

$\gamma(s,t)$ — Laplace-Stieltjes transform of $P(w,t)$

W — Average waiting time in the system

W_q — Average waiting time in the queue

W_p — Average waiting time in queue of pth priority

$G(x)$ — Cumulative distribution of a busy period for $M/G/1$ in Chap. 8

$\Gamma(k)$ — Gamma function

$\Gamma(s)$ — Laplace-Stieltjes transform of $G(x)$

$U(u_n)$ — Cumulative distribution of u_n

$p_m(t)\, dt$ — Probability that the busy period will terminate in $(t,\ t + dt)$ for $GI/M/1$

$f_n(t)$ — Probability that, among n items waiting at any time, a given item will wait subsequently for time t when items are selected for service at random

$P_{nm}(t)$ — Probability of n first-priority and m second-priority items in the system at time t in a preemptive-priority queue

$P(n_1, \ldots, n_k; t)$ — Probability that for k channels in series there are n_j $(j = 1, \ldots, k)$ items waiting in the jth channel at time t, etc., with $p(n_1, \ldots, n_k)$ for steady state

PART 1

STRUCTURE, TECHNIQUE, AND BASIC THEORY

CHAPTERS 1 TO 3

In the first chapter of Part 1 of this book we give a qualitative structure of queues based on the developments up to the present and on extrapolations of these developments, along with many illustrations. In Chap. 2 we give an introduction to different models, the use of statistics, quantitative decision-making criteria, and Monte Carlo analysis of queues. Chapter 3 gives fundamentals of probability and stochastic processes, particularly Markoff chains and processes.

CHAPTER 1

A DESCRIPTION OF QUEUES

1-1. Introduction

In this book we present the existing theory of queues (waiting lines) within a structural framework and emphasize different mathematical models and useful methods of solution with applications of various ideas to several activities such as telephone traffic, inventory, machine repairs, dam operation, aircraft operation, and others.

In this chapter, in a painless descriptive manner, a structure for queues is presented. Chapter 2 is an introductory, technical chapter, with a variety of ideas for the interested reader. The following three chapters give ideas of the theory (mostly probability) and the mechanics (mostly mathematical analysis) involved in solving queueing problems. In fact, Chaps. 4 and 5 specialize in Poisson queues. Other parts of the structure are then studied in Chaps. 6 to 13, followed by Chap. 14 on applications. Chapter 15 is concerned with renewal theory. A comprehensive bibliography is then given.

Because of the limitation of available techniques, it is natural that the material in Chap. 1 should be more extensive in ideas than the representation provided in the rest of the book. However, progress in queueing research is rapidly providing many of the missing links toward a coherent representation of the structure outlined and even beyond.

A queue, or a waiting line, involves arriving items that wait to be served at the facility which provides the service they seek. Suppose that the facility is a check-out counter at a grocery store. If the line is long, customers may become impatient and leave, thus causing a loss in profit. The owner of the facility may decide that investment in another check-out counter is worthwhile because its cost is offset by the profits taken from the impatient customers, more of whom now remain to be served. In this manner, one introduces optimization into queueing theory. In general, unlike optimization theory in which the main concern is to maximize or minimize an objective function subject to constraints, queueing theory is mostly a mathematically descriptive theory. It attempts to formulate, interpret, and predict for the purpose of better understanding of queues and for the sake of introducing remedies.

The reader might reflect on how often and how long he is made to wait or is delayed in queues in his day-to-day activities. He waits for transportation, for food, for the doctor, in an airplane waiting to land, at the theater, etc. Then, from a practical point of view, he might roughly compute how long, on an average, he waits per week in all the congestions he meets. Extended to a total for a lifetime, the amount of time wasted in queues is appalling. If one is not properly occupied, many queueing activites amount to an appreciable depreciation of the enjoyment of life. But by anticipation and good planning the agony and waste of queues could be minimized.

That queues are inevitable and are here to stay is a corollary of the increase in population. The more individuals that there are demanding a certain type of service (whether in the theater or in city traffic), the longer the waiting line. This demand is rising with increasing population; therefore the total amount spent in queues in a lifetime is an increasing function of time. Of course, one might proportionately increase service, but the costs are frequently prohibitive, e.g., widening city streets to allow for increased traffic. Projected planning could anticipate some of these difficulties by allowing for an increase in population.

Most queueing systems involve human beings. Studying causes and remedies of queueing problems cannot be completely divorced from consideration of human factors and their influence on the problem. Nor can remedies be applied without regard to the fact that it is the people using the facilities who matter in the final analysis.

While built-in queues, or planned queues, are a well-known feature of city life, they are also indispensable in introducing order in the various phases of mass production. The presence of a queue can be a good or a bad recommendation for a public facility. For example, at times it is commonly understood that regular queues before a theater or a cinema are one sign that the show is worth seeing and that a feature which is easily accessible is not. Thus, to find a good show, look for the larger queue. On the other hand, some of the most persistent and exasperating queues are bus queues. They are sometimes but a little removed from the chaos which they are set up to avoid. Many queueing places are served by more than a single bus; consequently, when two or more buses arrive simultaneously there is a scramble and a general disintegration of the queue. If the buses become full, the rejected rider has great difficulty finding his rightful place in the queue and frequently must queue up all over again.

Waiting for the doctor on the basis of a schedule designed to accommodate both doctor and patient would be very desirable. Frequently, the patient's loss of time comprises a greater social and economic loss than that of the doctor. If it is observed that the schedule of appointments is carefully developed after some study, it could add to the prestige of the doctor. It would, at least, point to the high regard that he has for his patients' time.

There is need for both the "queuee" joining the waiting line and the "queuer" who organizes it to understand and anticipate queues and their subtleties. Often the queuee would benefit by knowing that the line in which he waits has been optimally designed to minimize his wait individually or as a member of a group, and the queuer would know better how to plan for the queue associated with his activity. The most significant queue of all is the life queue, i.e., birth-life-death cycle with life as the waiting aspect. In a sense, then, queue consciousness becomes a consciousness for planning toward better living.

Some general types of queueing phenomena to which the ideas of this book have applicability are in the fields of communications traffic (telephone, telegraph, post office), transportation traffic (air, land, sea), queueing for service (theaters, restaurants, buses, hospitals, and clinics), inventories and industrial processes (maintenance, assembly lines, machine interference), physical processes (operations of a set of dams, particles moving through a hole), epidemic processes in biology, population growth, even refereeing papers for publication, psychological flow of nervous impulses, etc.

1-2. An Illustrative Example and a Numerical Illustration

Many of the ideas arising in queueing theory can be illustrated in one important example: take-off and landing of aircraft at a metropolitan airport—an operation of interest to a large number of people who use the facilities. For emphasis, the queueing terms illustrated by this example are italicized in the next paragraphs.

The airport is assumed to have several runways (*parallel channels*) used for take-off and for landing. These runways lead to a smaller or larger number of paths ending at the terminals (channels in series or queues in tandem). After an aircraft, which arrives according to a certain *arrival distribution*, lands, it joins the queue of aircraft awaiting service (movement) on a path to the debarkation point. Thus, the *output* of one queue becomes the *input* to another. The waiting line itself is both on the ground (take-off of aircraft) and stacked in the air (landing aircraft). Both these queues have input distributions. Landing aircraft may arrive in *batches* where the members of each batch must be spaced for circling over the airport and landing in order. (In case a runway is *very* wide it is not difficult to conceive of aircraft landing in batches.) The *duration of the service operation* (landing or take-off time on a runway) is about a minute. In any case, there is a *service distribution,* and if different types of aircraft are allowed to use different runways, which may be larger for jet aircraft, for example, the service distribution may vary from one runway to another.

It is essential that an appropriate *measure of effectiveness* be chosen for the selection for landing. For example, if it is desired to *minimize the total waiting time of individuals,* it may be more desirable first to land those aircraft with the greater number of passengers.

An informal type of *priority* system is often used whereby a circling aircraft is allowed to land prior to take-off of waiting aircraft. This priority system is further extended to emergency cases where a late-arriving aircraft is allowed to land first for urgent reasons. Frequently jet aircraft, because of limited fuel capacity, are given landing priority. Sometimes, by the nature of the holding pattern, an arriving aircraft, having joined the queue of stacked aircraft waiting to land, is chosen at random for landing (a form of service priority). Thus the aircraft which is nearest to a point where it can leave the pattern will be given instructions to land. Between being given the priority for landing and a "clear to land" instruction, an aircraft moves from the stack to the "landing aid." The time used is known as the "approach time." The "landing time" is spent in the landing operation until the aircraft turns off the runway.

A circling aircraft may be in a quasi emergency with others that are in actual emergency and decide to *join a shorter queue* at a nearby airport and land there. An arriving aircraft may decide not to stack and goes (*balks*) to another airport. This aircraft is then said to be "lost" to the airport—as distinguished from being delayed; or, after joining the queue and waiting longer than desirable, it leaves (*reneges*) for the neighboring airport. A landing aircraft may be considered to *cycle* when it joins the queue of aircraft waiting to take off, again becoming an input item into the system. If a landing aircraft has information on the size of the stacked queue in a neighboring airport, it might join that queue; if it has information on yet another airport, it might go there (a rare situation). This moving back and forth when there are several lines is called jockeying (*queue-selection rule*).

An airport may be temporarily shut down and arriving aircraft diverted to another airport if the number of stacked aircraft reaches a prescribed size. The service operation may be speeded up by building "turn offs" which make it possible for an aircraft to turn off the main runway at high speed.

A fundamental problem in airport operations is that of communications. When the input both on land and in the air is large, the airport control must communicate rapidly with the aircraft and obtain a response. How many operators and *communication channels* there must be to handle various congestion situations which might arise is an important communications problem. Here one must decide on an *optimum number of channels* to service items arriving by a given distribution. One may even compare the *cost of an additional channel* with the cost of increased service at the existing channels.

The problem of having adequate *waiting space* for the queue is important. For example, an essential aspect in airport design must be adequate ground taxi-way for the aircraft ready for take-off.

For many queueing problems it is enough to know the input distribution, the queue discipline (e.g., random, ordered, or priority selection for service), and the service-time distribution to determine the desired measures of effectiveness. In other queueing problems one must have additional information. For example, in the case of balking (reneging) one must determine the probability that an arriving unit would balk on (after) arrival and hence abandon the queue before (after) joining it.

From a theoretical standpoint, queues may be regarded in terms of flow through a network connecting service points in series and in parallel, as the situation may be. The flow is influenced by various phenomena which can delay it, cause it to overflow, etc.

Consider the following simplified situation of a single queue as shown in Table 1-1. For customer A the first row gives his arrival time and the second row the elapsed time between the arrival of the preceding customer and customer A. The third row gives his service time, and the fourth row is the sum of the service time and waiting time of the previous customer, minus the interarrival time of the customer whose waiting time is being computed. Thus customer A would experience a wait equal to the waiting time plus service time of his predecessor minus the arrival-time interval of A. If the result is zero or negative the waiting time is zero.

TABLE 1-1. SAMPLE QUEUEING DATA

Chronological time	0	2	6	11	12	19	22	26	36	38	45	47	49	52	61
Between-arrival times	0	2	4	5	1	7	3	4	10	2	7	2	2	3	9
Service time	5	7	1	9	2	4	4	3	1	2	5	4	1	2	1
Waiting time	0	3	6	2	10	5	6	6	0	0	0	3	5	3	0

A considerably larger sample would be required for a statistically valid study of an actual operation, but some important queue data can now be computed. In the above example, of the total number of customers, 10 have waited. The average waiting time of those who waited is $^{49}\!/_{10}$, whereas the average waiting time for everyone is $^{49}\!/_{15}$.

The total idle time of the channel may also be computed. The channel is idle and waiting for customer A if the interval between the arrival of A and his predecessor exceeds the total wait in queue and in service of the predecessor. Thus the total idle time is equal to the sum of the differences between the arrival interval of the present unit and the waiting time plus service time of the previous unit whenever the difference is positive. The fraction of time during which the channel is idle is the ratio of the previous quantity and the total time of operation.

If the sample were large enough, the frequency of a single unit waiting, that of two, etc., could be computed by counting the frequency of occurrence of single-unit waits, groups of two, etc. This gives the probability of occurrence of these groups. A question which we shall ask the reader to answer is: What use can one make of this probability?

Another useful quantity is the probability of a given number waiting at any time. Note, for example, that the fourth arrival waited with the third arrival for one time unit and was then left to wait with the fifth arrival for one time unit. This gives two groups of two units waiting in line.

The reader is urged to construct a diagram of horizontal parallel lines, each corresponding to a customer and with lengths corresponding to the duration of wait of each customer, over a base line which is divided according to chronological time. Each line must begin with the arrival time of the customer and terminate with the time in which he enters service. In this manner the number waiting and the duration of wait of this number can be measured. For example, the line corresponding to the second customer extends from the second to the fifth time unit, and that of the third customer extends from the sixth to the twelfth time unit. There is no overlap between the two lines. However, the line corresponding to the fourth customer extends from the eleventh to the thirteenth time unit and overlaps by one time unit the line corresponding to the previous customer. It overlaps the line corresponding to the subsequent customer by an equal amount, etc. Thus, to obtain the frequency of waiting of a group of two at any time, one takes the ratio of the total number of time units in which a group of two waited and divides by the total time.

Repetition of the above experiment with new data and table gives rise to new situations. With sufficient repetitions for a practical case in which measurements are taken, one can compute the probability of a given number waiting at a given time; this is different from the previous probability, which is given at any time. It is obtained by counting for the many runs of the experiment the frequency of occurrence of a single wait, groups of two, etc., at the prescribed time.

Thus we have three types of quantities to compute for the number waiting. They are (1) the frequency of occurrence of a given group of items waiting together (i.e., how often they occur), (2) the frequency of occurrence of a group at any time, and (3) the frequency of occurrence of a group at a prescribed time.

In passing, we note that a queueing situation must be studied over the period in which meaningful action is required with regard to congestion. For example, at a restaurant congestions usually occur at noon and in the evening. Sometimes it is of little use to study the two together because of the different intensities in traffic and because of the difference in its fluctuation in the two periods. If the congestion were independent of the time of day (i.e., homogeneous in time), matters would be simpler. However, caution is required in properly examining and sorting the periods in which congestion occurs in a given operation.

As a final remark, note that one can also obtain the arrival and service rates from the above data. If the data were more extensive, the *distribution* of arrival times and service times could be found.

1-3. Varieties of Queues

We divide a queueing operation into four parts: the input, the waiting line, the service facility, and the output. With each of these is associated a set of alternative assumptions concerning the queueing process, some of which have been the object of research in the field, as indicated in the historical background. Other assumptions lead to as yet unsolved queueing problems which require investigation. We give a general description of various queueing possibilites with some repetition of ideas given in the previous section.

1. *Types of Arrival and Service-time Distributions*

Arrivals into a queue (which may have an initial number waiting before the operation starts) occur by assumption according to a certain frequency distribution, and so do the intervals between arrivals. These intervals may be independently distributed for many application purposes or may be dependent, as, for example, in the case of flow leaving a traffic light. The same remarks apply to times of entry to service and to service times.

2. *Initial-input Variations*

The initial number in a system when an operation begins may be given by a distribution because it is different for each complete run of the operation (e.g., from day to day). The input to a queue may be from a limited or an unlimited population which may also consist of several categories (populations) of customers, each of which may arrive by a different distribution, singly or in batches, and may queue in a prescribed order. The input distribution may depend on the output distribution, as in a hospital where patients are admitted if there are vacant beds.

3. *Customer Behavior*

a. Balking. Customer behavior can vary. Arriving customers may balk (i.e., not join the queue) because of the length of the existing queue, or simply because they have to wait at all, and are consequently lost. Sometimes they are lost because they have no opportunity to wait, as is the case with a busy telephone signal (lost calls) although they can reinitiate a call. It is also possible to hold such a call, delaying it until a trunk becomes free. There may be a single line in which to wait before going into service, or there may be several lines, as in banks or in supermarkets. An item may join the nearest line independently of its length. If arrivals are designed to occur at constant intervals, they can, for example, still occur sooner or later by a distribution about the arrival point as a mean.

b. Influence of Incomplete Information. For many problems a decision may be required as to which line of a multiple-queue operation to join when information about only a few is immediately available—a case of incomplete information. In congested traffic the lack of knowledge as to which is the best route to follow without trying them all is also "incomplete information."

c. Customer Adaptation to Queue Conditions. On the basis of experience, passengers may learn to travel earlier or later to avoid intolerable queueing, and such measures, when adequately studied, may even relieve congestion. For example, ships approaching the Suez Canal can be notified to slow down until the queue at Port Said is reduced to an acceptable size. An item may join a large waiting line at closing hours for fear that a short line which it encounters may be closed suddenly—a familiar experience. But there are situations in which an item which arrives before another must go into service before the following one. There are cases in which each service facility has its own specialty and consequently its own queue, as a stamp-sale counter and money-order or special registry counter at the Post Office.

d. Collusion, Jockeying, and Reneging. Several customers may be in collusion whereby only one person waits in line while the rest are then free to attend to other things. Some may even arrange to take turns waiting. Units may jockey from one line to another, as in a bank. A customer may lose patience and leave the line, i.e., renege.

4. *Queue and Channel Variations*

a. Full or Limited Availability. Service channels may be available to any unit waiting in a system (full availability) or may be available only to some waiting units. Other units are blocked and must wait until a channel that can provide the required service becomes available. In telephone link systems, whether one obtains a free connection depends on whether a free inlet to the next waiting line can be combined with a free outlet.   The idea itself shows the need for economy in setting up possible combinations.   This is particularly true for some long-distance calls which may have to pass through more than one switching center.

   b. Service Procedures or Discipline.   While in line customers may be chosen for service by allocation to the channels in an ordered first-come–first-served manner or at random, they may be assigned priorities with errors committed when initially it is not clear which priority to assign, or the priority assignment may change in time.   Priorities may be pre-empted if higher priorities arrive, or they may be allowed to finish the service.   Finally, items may be chosen for service on a last-come–first-served basis.   We assume throughout the book that, once a channel that is able to provide service to a waiting unit becomes free, the unit immediately enters service without loss of time.

   c. Pooling Queues.   There are various ways of pooling queues. Some result in a shorter average wait, particularly when the dispersion of service time by a server before which a separate queue was formed is high.

   d. Series-Parallel Facilities.   A service facility may consist of several channels in parallel, some of which may be in series with other channels, or several parallel channels may all lead to one or more channels in series.   In the case of channels in series, a queue may or may not be permitted before each channel.   In a supermarket, an arriving customer serves himself immediately on arrival and thus the number of service channels (though not the number of check-out counters) varies with the number of arrivals.   All customers queue before the check-out counters for a second service.   A service channel can have various service distributions to use on different-category customers.   Idle servers may be asked to perform other tasks when the queue is empty; for example, in machine repair, ancillary work is undertaken.   This frequently depends on the amount of service required, the capacity of the server, and the size of the queue.   The channel itself may move as a production-line belt with a server to provide the work on units spaced along the belt.

   e. Specialized Service Channels.   Some of the service channels may specialize while others remain general, as in certain air traffic facilities where some counters are there to serve passengers whose departure time is within a prescribed interval.   Thus the arriving customer's demand for special service varies, depending on the length of the interval between his arrival and the departure time of the airplane.   Parallel channels may all cooperate to serve the many needs of each customer.   Customers may cycle by returning to the waiting line for additional service.

   f. Queue Interference.   Two queues may interfere with one another, as is the case if a single traffic lane on a part of a highway must be used by cars coming from either direction. Cars on one side must wait for a car coming in the opposite direction to pass. If more cars arrive from the same direction before a car goes through, generally they also attempt to pass, most likely in batches. This could also be the case at the Suez Canal, even though it is not the case at present. Ships pass according to a schedule.

5. *The Output of a Queue*

The output of a queue may also be of importance, particularly when it forms an input to another queue in series with the first one. Arrival and service distributions may depend on each other. For example, correla-

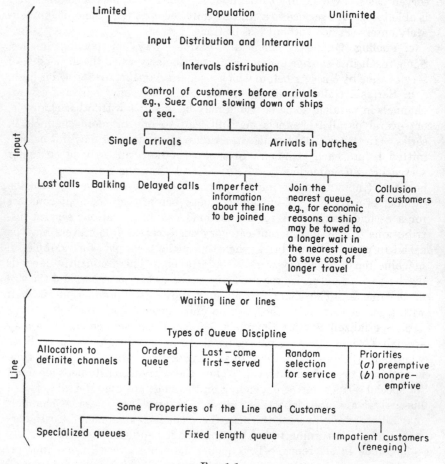

Fig. 1-1

tion exists when service might influence arrivals and conversely.   Thus, in a set of competing identical enterprises, regarded as a set of parallel channels, that enterprise which offers special and perhaps quicker service obtains a greater influx of population.

These are some examples of various situations to which queueing theory may be applicable.   Later we shall examine many of these ideas.

Figure 1-1 is an abbreviation of many of the foregoing ideas.   It attempts to show the various possibilities that influence arrivals, the times of arrivals, etc., the queue and the different types of queue discipline, the service channels, and finally the output.   Breakdown of queues is also indicated as an antithesis to the operation of queues, since not all queueing activities are in continuous healthy operation.   The manner in

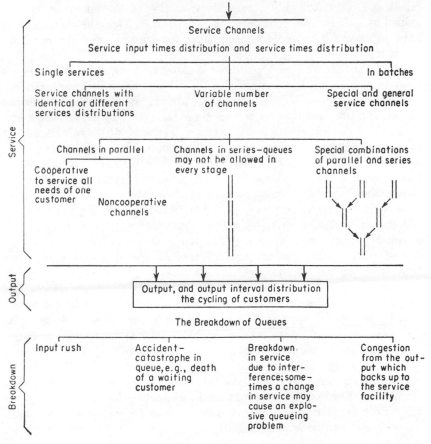

FIG. 1-1   (*Continued*)

which these concepts are related is by no means rigid. The reader might make his own attempt at relating them.

1-4. Elementary Descriptive Analysis

In the first example of the next chapter we consider arrivals and service at constant intervals. Although we consider all arrival intervals equal and all service intervals equal, it is clear that this assumption is not necessary to obtain the number waiting and their waiting time.

In general, however, no such designed regularity exists in queues. Customers may arrive at random so that one cannot, in advance, be sure of the exact arrival times. This is also true of the service times. As customers wait, there is a probability at any time that a certain number is in the line because the arrival and service times are now given probabilistically; therefore only a general evaluation of the operation can be gained by a theoretical consideration. From this, a specific customer can have an idea of his average waiting time, for example, and the variation about it.

It is clear that with each customer is associated a variety of chance, or random variables. For example, for the nth customer we have a random variable which describes the time of his arrival, another which describes his waiting time, another his service time, etc. With a set of customers, for each of the situations described, e.g., arrivals, we have a family of random variables. Each random variable may have a different probability distribution (a characteristic of the nth customer) for each point of time. Because of their dependence on time and because there can be, at least in theory, an infinite number of them, these families of random variables form stochastic processes. Thus a stochastic process is a family of random variables that depend on parameters. We have a single parameter here, which is time. The study of queueing theory is concerned with finding distributions of various combinations (sums and differences) of these random variables when enough information is given about them. The main problem is then (1) to relate the variables correctly in order to describe a queueing problem, (2) to derive the associated distributions and determine their actual form by statistical measurement, (3) to use the distributions for deriving useful measures, and (4) to apply these measures to the improvement of an operation.

An interesting phenomenon of queueing theory is that of the dependence of the various quantities involved on the time at which the system is examined. For example, the number in a supermarket check-out queue will be different 10 minutes after store opening from that after 2 hours. At each time there is a probability distribution for the length of the queue or for the waiting time to assume any prescribed value. Thus the resulting probability distribution will vary with time.

With the nth customer, then, there is a random variable that assumes different values of waiting time with probabilities that depend on when the queue is looked at. In time the system moves away from the initial number at zero time which conditions the patterns (distributions) of the queue. Later, the system may be rid of this effect (or of the effect that there was nothing in line at the start) and operates with the influx and efflux of customers establishing, ultimately, patterns that are independent of the initial number present. When the initial effect is "worn off" one might expect to find the system with the same general pattern of probabilities at one time as at any other time. Thus the pattern itself stops changing predictably with time, and therefore the probabilities acquire an equilibrium or a steady-state form.

Equilibrium may never be attained if, for example, the service time is longer than the arrival time. Thus there is always the question with queues as to whether these equilibrium probabilities actually exist, i.e., under what conditions the distributions associated with the queue become independent of time. This is a question which will be of main concern in some cases. In practice, many operations attain this equilibrium status (approximately) and are therefore studied for equilibrium solutions of the waiting-time distributions and the number in the system, for example. Note that equilibrium means that the probabilities are independent of time but not that the system becomes deterministic. The queue continues to fluctuate but the distributions describing it are fixed in time. One can compute averages, deviations, etc.

An illustration of the idea of a steady state may be given by an air-conditioning unit with which we have had experience during the hot season. This instrument is very old and is, therefore, very reluctant to operate. The initial load is a shock to it. It starts with shrill noises, vibration, and rattling. It gives the impression of recuperating after a while, but there is a reluctance to move away from the initial state of inertia. After a long while it begins to warm up, the noises subside, and the vibrations are damped.

The air conditioner may never settle into a steady state of operation but may continue in a transient state under the influence of the initial condition. In hot weather, when it must work harder, this may be the best that one could expect. Such may be the case with some queues; depending on their input and service distributions and their initial state, a steady state may never be reached.

Frequently it is easier and more practicable to assume that the behavior of a queue is independent of time, and the problem is then studied in the steady state.   Steady-state solutions are not always adequate, and it is often desirable to have time-dependent solutions.

Because of the likely similarities between customers, their arrival distributions and their service distributions are often assumed to be identical and not to depend on whether a customer arrived nth or mth in time. Many problems assume that in a small time interval it is not likely to have more than one customer arrive.   These various assumptions are made depending on the practicality of the situation and how difficult it is to find an answer.

As an illustration of relating some of the variables, we note that the waiting time (when there is a line) of the $(n + 1)$st customer, w_{n+1}, equals the waiting time of the nth customer, w_n, plus his service time s_n, minus the arrival-time interval of the $(n + 1)$st customer t_n—the arrival time of the first customer does not matter since he immediately enters service; this gives

$$w_{n+1} = w_n + s_n - t_n.$$

Now each of s_n and t_n is given by a probability distribution and hence so are w_n and w_{n+1}.   One must determine these unknown distributions.   It is therefore essential to learn some of the elementary mechanics used in probability theory in order to apply them here.

Note that, if there is no waiting room, arriving customers who cannot obtain service will be lost.   Note also that the service channels may be complicated in structure, such as being combined in series and parallel. The problem becomes still more interesting and useful.

An example of a useful quantity to compute is the waiting time.   Now, if the service is first-come–first-served, the waiting time in line of the kth customer in line is the sum of the service times of the $k - 1$ items in front of him and the remaining service time of the item in service.   Each of these is a chance variable/taking on values with probability distributions.   The problem then is to determine the probability that the sum of these variables will take on any specified value.

The waiting-time distribution is obtained by taking the probability of n items in the queue and multiplying it by the waiting-time distribution of the nth item and summing over all possible values of n.   The average waiting time may then be obtained from the waiting-time distribution by averaging over time.   These ideas are illustrated in Chap. 3.   In passing, we note that, if customers are chosen at random, the waiting distribution of the item which arrived after $n - 1$ others had joined the line will be different from what we gave above for the ordered queue.

In the case of analytical answers, there still remains a problem of how to apply the results to a practical situation.   For example, one problem may be to estimate the arrival rate and the service rate for a practical situation, using the most likely values for these parameters in the equations. How should the actual measurements be made with a minimum of error? If the equations, when available, are not solvable analytically, one might attempt numerical solutions.

We have, so far, discussed the analytical approach, i.e., when it is possible to describe a queueing situation analytically and obtain also by analytical methods useful expressions, such as the average waiting time and the average length of the queue, from which many useful measures are derivable. Frequently it happens that a queueing problem is complicated and cannot be easily described by analytical equations which can be manipulated to yield useful answers. The alternative method is to simulate the system. This method is used much as was illustrated by the numerical example, and numerical values for the waiting time, etc., are directly obtained. The experiment must be repeated sufficiently often to obtain large samples and a variety of answers, which are then taken together in some manner to obtain a value for what is desired. This is a very useful method in practice whenever complicated problems require immediate answers. However, there are pitfalls in the method, and great care is required to obtain correct simulation with enough samples and properly to combine the results to obtain an answer.

1-5. Measures of Facility Effectiveness and Some Remedies

Our study of queues is conducted through analysis of the various queueing phenomena encountered. This direction of research to solve queueing problems lies between a complacent, resigned attitude toward congestion and the morphological approach. In the latter approach one examines specific queues arising in practice and finds solutions by canvassing all conceivable courses of action, no matter how outlandish; discarding the most unfeasible; and then, in Sherlock Holmes fashion, adopting what remains, no matter how unconventional.

The latter approach is recommended for practical purposes. However, it tends to be speculative. It requires our analytic approach to assess the relative improvement produced by the new ideas. The two methods must be combined for effective remedy of difficult queueing problems.

The theory of queues has proved to be of abundant application value, as can be seen from the variety of types of problems to which it has been successfully applied. This is partly due to the fact that one has a large number of useful measures of facility effectiveness to use. In application, care is required to select the appropriate measure to be used and consequently the model needed to describe the operation from which such a measure is derivable.

Some malfunctioning queues receive little cure because of lack of better measures required for their improvement. For example, one may decide to control the input of a queue instead of speeding up service—or to do both—thus restricting the capacity. In certain instances, because of periods of heavy input congestion, what may be required is adequate waiting room for arriving units rather than faster service.

The question of how well a theoretical model fits a practical queueing operation, or parts of it (if it can be divided into parts each of which satisfies equilibrium conditions, for example), is mostly determined by what measure one needs to take. For many practical purposes one requires indices of performance which enable comparisons. One such comparison is the effect on the waiting-time distribution of the type of service used. Another is the effect of the queue discipline on the waiting time.

Whatever approach one may use to study such a problem, one must decide on those measures that can be applied to the operation in order to make a decision. For example, a facility manager must decide whether to increase the waiting space or the number of service facilities by comparing the cost of facility increase with business lost due to departing customers. The average waiting time of customers who must wait may be too great, and consequently the average waiting time of all customers may be an inadequate measure. An enterprise owner may use these quantities to determine the amount of space needed for waiting customers and how to make their wait pleasant and perhaps shorter by increasing the number of channels, providing faster service or other things to do, etc.

A measure of effectiveness may be of use to a customer in deciding whether to leave a queue or not join it, if its average length or its waiting time is too long.

Some important measures for studies of a system in equilibrium, i.e., when time does not affect the operation any longer, or in a transient, i.e., time-dependent, state are the input rate; the service rate; the traffic intensity, i.e., the ratio of the mean service time to the mean interval length between successive arrivals; the probability of the number of units in the queue or in the system being a given number n; the average number in the queue (or in the system, i.e., including the average number in service); the waiting-time distribution in the system or in the waiting line; and its average and variance. When dealing with perishable goods awaiting processing and with impatient customers, the probability of waiting longer than a given period is important; also important are the probability of not waiting, i.e., the proportion of items not delayed (the proportion of items delayed is obtained by subtracting this from unity); the probability that there will be someone waiting; the average number of waiting individuals, for those who are delayed and must actually wait; the average length of a busy period; and the average number of items being served. The ratio of mean queueing time to mean service time is particularly important to a customer requiring little service. It is also useful to calculate the probability that exactly $0 \leq n \leq c$ channels are occupied; the average number of idle channels; the coefficient of loss (being idle) for the channels; the operative efficiency, i.e., the proportion of time a service channel is busy serving customers; the average number of units not in the queueing system applied to the finite-population case; the probability of a lost call; etc.

Cost analyses are essential to many queue decisions. It may be necessary to know the cost of the entire operation; of an additional channel vs. increased service at a channel; the cost of the queue discipline chosen; that of having customers wait a certain length of time; the cost of maintaining a large number of channels vs. using the space for other purposes; the cost of maintaining different queue lengths and of different-size waiting rooms; the cost of controlling the input distribution (e.g., spacing the arrival of aircraft in time and hence comparing the cost of operating with a given service distribution under the hypothesis of a random input vs. a constant input). One may also desire to adopt measures of utilization of the facility and measures of efficiency of operation or of output per unit time. The cost of losing customers and the cost of customers' waiting time are also important.

A service facility may break down for several reasons. For example, the input rush may create a great pressure in the system, requiring the shutting down of operation. The service facility may break down, as when a storm causes temporary delays in the telephone system. A radical change in service may discourage customers completely and the entire operation ceases. A sudden mishap in a bank (e.g., the shooting of a clerk by a bandit) can cause the closing of the entire operation. A congestion at the end of service, either because of served customers who do not depart or for other reasons, may result in stopping the operation.

Quick remedies for queueing inefficiencies can be brought about by reducing the service time or the variation in the service time, by having additional service available for rush periods, and by controlling the arrival distribution. Gains are made, for example, by using regular input control instead of random arrivals, which as we shall see later lead to larger queueing time for a given service time. It is clear that the latter step can be effected only in part. Customers will continue to arrive with variation about the regular interval. Smoothing out arrival irregularities which persist is also desirable. It is important that arrivals be controlled so that the traffic intensity (ratio of arrival rate to service rate) remains less than unity. Staggering of working hours to regularize input and use of the facilities over an extended period would tend to alleviate heavy congestion at peak loads.

It will be shown later how expressions for various queue measures are obtained from the models used.   There are instances in which there is no really satisfactory solution of a queueing problem, except to replace the operation by a more efficient one.   An example of a cure in some facilities is for the customer to serve himself.   Elevated roads and superhighways relieve traffic congestion, which can continue to present difficulties, even if some bottlenecks are better controlled.

A telephone can be used for making appointments and reservations, thus eliminating most of the time spent waiting for the facilities.

We now make the observation that, while one may solve a queueing problem by increasing the number of operators to provide service, one creates a new queue, namely, that of the operators themselves who now must wait for the arrival of customers who demand their service.   In this sense, a new congestion is created (when the channels are idle) to take care of another—perhaps less desirable—congestion.

1-6. Historical Background

1. *Telephony*

It is interesting and perhaps of particular importance to note that the origin of queueing theory is to be found in telephone-network congestion problems.   It is also true that this field continues to present challenging problems to the many capable investigators working in the field throughout the world.   In presenting a brief sketch of the history of developments in queueing theory it would be well to describe some basic ideas in telephony of special interest to investigators.

We immediately observe two basic situations.   When all devices in a group are engaged, the group is said to be blocked.   If a call arises when the group is blocked, it may be either lost or delayed.   The first case is characteristic of a loss system and the second of a waiting system (mixed cases are also possible).

a. Full Availability.   A full-availability group is a set of devices (telephones, trunks, lines, etc.) that are all accessible to a group of sources (telephone subscribers). In Fig. 1-2 the capital letter indicates a group of sources, and the small letters indicate

$$A. \quad \underline{a} \quad \underline{b} \quad \underline{c}$$

FIG. 1-2. Full availability.

destinations.   Note that, if a caller from A wishes to make a connection, he can be connected through any of the free lines a, b, c, etc.   If all lines are engaged, then the caller either waits or his call is lost; i.e., he receives a busy signal.   In either case the probability of a call waiting a certain time or getting lost is of interest.

If there is a limited source of N subscribers of whom n are occupied making calls, Erlang assumed that the probability of a new call being initiated in the time interval $(t, t + dt)$ is $(N - n)\lambda \, dt$, where λ is the call rate per unit time of a caller.   If no consideration is given to the number of occupied independent callers, then the probability of a new independent call being initiated is $\lambda \, dt$, leading to the negative exponential distribution $\lambda e^{-\lambda t}$ for the distribution of intervals between successive calls.   This gives a Poisson distribution $(\lambda t)^n e^{-\lambda t}/n!$ (shown later) for the probability of exactly n calls during time t with a mean of λt calls. This distribution has been empirically verified for periods of heavy traffic and found satisfactory for the distribution of call inputs. Similarly, if $1/\mu$ is the average duration of a call, the probability of a call terminating during $(t, t + dt)$ is $\mu\, dt$, which is independent of the length of the call and of other calls. This in turn implies that call durations are distributed according to $\mu e^{-\mu t}$.

Erlang [80][1] developed telephone-traffic theory, using these assumptions, and obtained results for the probability of different numbers of calls waiting and for the waiting time when the system is in equilibrium and the probability of loss for the loss system. He published his paper in 1917 in Danish and later in English, German, and French. He assumes Poisson inputs from unlimited sources and either exponential or constant holding times.

For a delay system in the constant-service-time case he gives the results for one, two, and three service channels and for an arbitrary number of channels in the exponential-service-time case.

Erlang's work stimulated other work in the field by T. C. Fry, as indicated in his excellent book [236], by E. C. Molina (particularly in his 1927 paper), and by G. F. O'Dell [530]. Examination of the earlier papers leads to a conclusion also reached by Syski [681] that early works were concerned with proving or disproving Erlang's results. Syski calls it the "Erlang-O'Dell" period. O'Dell gave a summary of work in congestion theory prior to 1920. F. Pollaczek developed the well-known formula for a single channel with Poisson input and arbitrary holding time (Pollaczek-Khintchine formula); he also studied the constant-holding-time case for ordered-queue discipline and for allocation by subqueues in front of each server. C. D. Crommelin made an equilibrium study of the delayed-calls, constant-holding-time, several-service-channel case. His result coincides with the few cases studied by Erlang. Berkeley and also Schouten and Giltay have checked the results experimentally.

More recently F. Pollaczek [590] has studied analytically the general holding-time case with general input for multiple channels.

Palm studied the effect of varying the traffic intensity. Statistical computations of moments of waiting times and of the number of lost calls, etc., have been made by L. Kosten. Wilkinson and W. S. Hayward have studied measurement reliability in scanning the telephone lines to determine the number of busy lines.

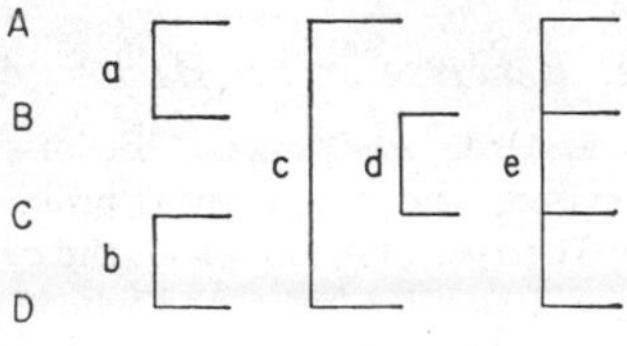

b. Limited Availability. In a limited-availability system only a few of the devices are available to the sources. Figure 1-3 shows a typical but sophisticated limited-availability system (also known as grading); for example, the line a is available only to calls from A and B, e is available to all, and c is

FIG. 1-3. Limited availability.

available to A and D. Here again questions of loss (mainly) and delay (difficult) have been studied.

Erlang, by assuming that new calls search for a fixed subset in a large number of subsets (called the ideal grading), derived an expression for the probability of a blocking of the system. Rigorous derivation of this result has been obtained by Brockmeyer. Various other investigations have been made depending on the types of connections made within the system.

---

[1] Numbers in brackets refer to the numbered Bibliography at the end of the book.

c. Link Systems. A more recent mathematical idea is that of a link system. In a link system a set of sources may have limited access to a set of destinations, as Fig. 1-4 shows for a two-stage link system.

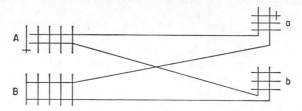

FIG. 1-4. Link system.

Note the dependence of the distribution of calls arriving at a, for example, on the distribution of calls initiated at A. The problem is: Given the distribution of calls, what is the probability that a given call would not have access to a free outlet? Note that correlations exist between the distribution functions of calls on all various stages. Jacobaeus, Jensen, and Fortet solved this by assuming independence among the distribution functions, with some dependence considered by the last two. The above problem with complete dependence remains unsolved, but Elldin has made promising attempts.

d. Common Control. Next we consider a lost-call system in which all calls arrive at a central control before going into service. If the control is busy, a call is lost even though a service channel may be free. The problem has been solved for the state probability of a call getting lost, with Poisson arrivals and exponential services for all channels, with exponential service for the common control (by Vaulot and Leroy), or with constant service time for the common control (by Fortet), and with arbitrary service time for common control (by Pollaczek). Syski has solved this problem with delayed instead of lost calls for one channel as well as for an infinite number.

In his recent book, Syski [684] pursues these ideas at great length in chapters on loss systems, waiting systems, limited availability, link systems, and interconnected exchanges.

2. Remarks on the History of General Queues

a. Ordered Queues. Feller's book, which appeared in 1950 [202] and was later revised, studies congestion problems. From a theoretical standpoint, D. G. Kendall's 1951 paper [366], in which the concept of the imbedded Markoff chain so much in use now was introduced and which was followed by a classification paper in 1953 [367], stimulated considerable interest. (It should be added that the regeneration-points technique used was first introduced by Palm.) The first transient solution of the birth and death equation with constant coefficients was given by A. B. Clarke in 1952 [107] and published later. This solution was obtained in subsequent years by various methods, e.g., Ledermann and Reuter [426], N. T. J. Bailey [19], P. M. Morse [507], and Champernowne [97]. S. Karlin and McGregor [348] have given a method for solving the problem with coefficients as functions of the number in the queue. T. L. Saaty [645] has given a method of solving many-server time-dependent Poisson queues by Laplace transforms. A. B. Clarke [110] and later Luchak studied the problem in which the coefficients are functions of time. In 1952 Lindley [442] gave an expression for the average waiting time in equilibrium for arbitrary input and holding time for a single channel and for an ordered queue (first come first served).

b. Random Selection for Service. S. D. Mellor [482] attempted in 1942 to obtain the distribution of delays for random selection for service. In 1946 Vaulot [733] gave a correct formulation of the problem, for Poisson input, exponential holding time, and multiple channels, later solved by Pollaczek [574] and by Palm [551]. In 1953 Riordan gave a method for computing the delay distribution, and curves were developed by Wilkinson [764]. In 1959, P. J. Burke [90] studied the problem of random selection for service for a single channel with Poisson input and constant holding time.

c. Priorities. In 1954, A. Cobham published the first paper on nonpreemptive priorities for Poisson input, exponential holding time, single and multiple channels. He gave expressions for the expected waiting time of an item belonging to a given priority. J. Holley simplified Cobham's work. T. Phipps generalized Cobham's results to a continuous number of priorities with application to machine repairs. P. Morse [508] gave the steady-state probability generating function for two priorities with different exponential holding times. In 1957, H. Kesten and J. T. Runnenburg considered the Cobham problem in elaborate detail, giving the Laplace-Stieltjes transforms and the first two moments of the waiting time for a general-service queue in equilibrium. Recently J. R. Jackson has studied queues with dynamic priorities [317]. J. Y. Barry and F. F. Stephan studied the problem of a preemptive-priority system with two priorities and Poisson input and exponential holding time. H. White and L. S. Christie studied the same problem for two channels, giving expressions for the expected times in the system for any priority and the generating functions for the delay distribution. They also studied similarities between this system and that of queueing with breakdown.

R. G. Miller, Jr. [486], in an extensive paper, produced several of the preceding results for two priorities, using different exponential services, giving Laplace-Stieltjes transforms and generating functions for new cases.

Heathcote has made the most recent and comprehensive contribution to preemptive-priority queues, studying several priorities and several channels in the transient case.

d. Balking, Reneging and Impatience, Cycling and Jockeying. An individual might decide on a queue length (which may be infinite) which he would accept, refusing to join a larger one. Various situations call for the acceptance of different tolerable queue lengths, the totality of such yielding a balking distribution. In 1957, F. A. Haight studied this problem for a single queue in equilibrium with Poisson input and exponential holding time for various balking distributions. He gives the mean number in the queue requiring that a customer remain for service, once he joins.

P. D. Finch [213] has studied a similar problem where the length can assume a finite maximum but for a general input distribution. He gives the steady-state queue-length distribution, having proved its existence.

In earlier papers D. Y. Barrer [26] studied the problem of impatient customers who, after a fixed amount of waiting (deterministic impatience)—in a single-channel queue in equilibrium, with Poisson input and exponential service time, in which they are selected at random for service—impatiently leave. He also studied the problem in a first-come–first-served queue [27] for a case in which customers leave the queue after a fixed wait. He derives the queue-length distribution. P. D. Finch has studied the above problem for an arbitrary input and several service channels. He has also studied the problem of indeterministic impatience with Poisson input, exponential service time, and several channels, giving the queue-length distribution. Haight has combined reneging (impatience) with balking. C. Palm investigated the case of early departures from a queue, and Cohen generalized this problem by taking account of repeated calls.

The process of cycling of customers in a queueing system was investigated by E. Koenigsberg in 1958. H. Glazer has examined the problem of customers jockeying back and forth between several queues forming before a facility, such as in a bank.

e. **Queues in Series.** In 1954, G. G. O'Brien studied the problem of two queues (two phases in equilibrium, with Poisson input and exponential holding time) in series (tandem), giving an expression for the length distribution and for the expected waiting time. In the same year R. R. P. Jackson [318] studied the problem for a limited and an unlimited input for two and three phases, giving queue-length distributions. G. C. Hunt has studied the two-phase problem, allowing no queues, finite queues, and infinite queues, and computed, for example, the expected number in the system. A. W. Marshall and E. Reich studied waiting times for queues in tandem. Burke [89], Reich [616], Cohen, and recently Finch have studied the output from a queue and established independently that the output from a Poissonian queue is also Poissonian.

f. **Non-Poisson Queues.** In a non-Poissonian queue either the input or the service distribution or both are non-Poisson. Khintchine gave an elegant derivation for the expected number waiting and the expected waiting time for an ordered queue in equilibrium, with Poisson input and arbitrary holding time (Pollaczek-Khintchine formula). Lindley, as already indicated, gave an integral equation expressing the expected waiting time with arbitrary input and arbitrary holding time. To supplement the early works of Crommelin, who computed the delays for the equilibrium case of a multiple-channel queue with Poisson input and constant holding time, J. Everett [195] gave an alternative method for obtaining these results, using a process of successive approximation. Following D. G. Kendall's imbedded-Markoff-chain theory, N. T. J. Bailey [20] and Downton [162, 163] studied a bulk-service queue (several items are serviced in a batch) in equilibrium, obtaining expressions for the expected numbers waiting and the expected waiting time, for the chi-square service-time distribution. Downton allowed the batch size to increase indefinitely. More recently R. G. Miller, Jr. [487], approached the subject of bulk arrival and bulk service from a renewal-theory standpoint and was able to obtain results not obtained by the imbedded-Markoff-chain theory. He treated a more realistic case in which periods in which no service occurs are permitted and accounted for. He also gave the expected waiting time for busy periods. T. Meisling [480], approaching the subject in a novel way from a discrete standpoint, uses the imbedded Markoff chain and gives the expected number waiting and the expected waiting time in equilibrium for a binomial input, arbitrary holding time, for an ordered single-channel queue, also treating the special cases of a geometric progression and a constant-holding-time distribution. He then obtained well-known results by passing to the limit. Pollaczek has also used this approach.

R. R. P. Jackson and D. G. Nickols have treated the case of a single-channel ordered queue in equilibrium with chi-squared input and exponential holding times, obtaining the steady-state probability distribution and the waiting-time distribution of those customers who have to wait, and specialized the results for an exponential and constant interarrival distribution.

In his 1957 book [590] Pollaczek studied extensively the single-channel queue for various types of input, service times, and queue disciplines with particular reference to air traffic problems.

L. Takács [689] and later V. E. Beneš [42] developed and studied an integrodifferential equation for the waiting-time distribution in the transient state for an ordered queue with Poisson input whose parameter depends on time and for arbitrary-holding-time distribution. A similar approach was recently used by R. Descamps. Solutions in the steady state are given in terms of (e.g., Laplace) transforms. W. L. Smith [661] has shown that, if the service-time distribution is K_n, defined below, so is the waiting-time distribution with fairly general arrival-time distribution. From this it follows that, if the service time is exponentially distributed, so is the waiting time, whatever the arrival-time distribution. B. Gnedenko has studied waiting times with reneging and with interrupted service. J. Kiefer and J. Wolfowitz [380] showed for the case of an ordered queue with arbitrary input and service distributions and multiple channels that the distribution of waiting times does not exist in the steady state if the traffic intensity is not less than unity. Under special circumstances it could exist when equality holds. A general expression for the waiting-time distribution is also given. Lindley's result is the special case of this for a single channel.

Volberg [742], following Pollaczek's work, developed the waiting-time distribution in the transient and the steady state for a Poisson-input, arbitrary-service-time, multiple-channel queue in which arriving units are each assigned sequentially to the different channels and further arriving units are again assigned in the sequential order.

D. M. G. Wishart has examined waiting times for the systems $GI/E_k/1$ and $D/E_k/1$ (see Kendall's classification below).

There are many other interesting queueing situations which we study in the text but have not attempted to include in the historical account.

We briefly mention the classification of queues due to D. G. Kendall. If M refers to the fact that the distribution of occurrences in a fixed time is Poisson (and hence that the distribution of the intervals is exponential), then $M/M/c$ refers to the c-channel case with Poisson input and exponential service time. The symbol D is used for the constant distribution, K_n for the chi square with even degrees of freedom, M_s for service in batches, E_n for the Erlangian with mean independent of n, GI for a general independent distribution (e.g., input intervals may be independently and identically distributed by a general distribution), and G refers to a general distribution without requiring independence. Thus $D/K_n/1$ refers to a constant-input, chi-square service-time distribution in the single-channel case, etc.

PROBLEMS

1. Estimate and compute for a period of 40 years the congestion delays that you encounter in daily activity. From a material standpoint, compute the loss in creative or other activity which you could have developed in this time. (It is realized that such an assessment may be difficult, but it is worth trying.)

2. Give an example of a queueing activity (as we did for aircraft) which incorporates most of the ideas given. Insert at least two new queueing situations not mentioned in this chapter.

3. Use a set of random numbers to simulate (samples of size 100) a queue as illustrated in the text, and compute the different quantities mentioned.

4. Everyone is familiar with the chaos resulting from waiting lines before the many parallel service counters of a supermarket on a busy day. The lines before each counter extend way back to the grocery aisles, blocking traffic. Much space is wasted between these queues. One may wait a long time in a short line because of a slow clerk. Find a first-come–first-served scheme of serving all customers forming a single line, not wasting space or time in unloading groceries on the check-out counter. Suggest your scheme to your neighborhood store.

CHAPTER 2

SOME QUEUEING MODELS

2-1. Introduction

In this chapter we shall treat (with some elaboration) basic approaches to queueing problems. The purpose is to aid the reader in following the descriptive approach of the first chapter by the more analytical aspects. However, we shall not be satisfied with formulas alone. We shall indicate some of the statistical problems involved to give the reader early in the book a broader outlook on the types of elementary mathematical and scientific details needed to analyze real queueing problems.

We therefore divide the chapter basically into three parts. The first part deals with model building and solution. We begin our study of this part with a deterministic queueing problem which has not received sufficient attention in the past despite the fact that, because of its simplicity, it can help to clarify a large number of ideas encountered in queues. We shall treat some of the ideas, leaving others as exercises.

This example is then followed by an interlude which is the beginning of a transition from the deterministic to the probabilistic aspect of queues. The latter is the very element on which the development of queueing theory is based. It is followed by a discussion of the Poisson process which has many useful properties for queueing theory and requires some discussion. Based on the property of the Poisson process (and its intimate, the exponential distribution), a second queueing model is derived in the time-dependent case. Various properties of the Poisson process are obtained as an indication of how to derive some of the properties of queueing processes.

A similar analysis is applied to study the equilibrium, i.e., time-independent, solution of a single-channel queue with Poisson-input ordered service and exponential service times. Yet another example is given deriving the well-known Pollaczek-Khintchine formula in the steady state, i.e., without resort to the transient solution. By examining this equilibrium solution, it is possible to obtain useful results which are then tested for sensitivity to parameters.

The second part deals with the testing of a hypothesis about an observed distribution and with parameter estimation from observation values. Some decision problems are briefly discussed.

The last part deals with a Monte Carlo example of a simple queueing problem. It is followed by numerical methods of solving some queueing equations when a solution in closed form is not readily obtainable for applications.

2-2. A Deterministic Queue

We start by studying an elementary queueing problem. In the following example we take as fixed the arrival times and the service times of customers. It then becomes possible to study the evolution of the waiting line. Our concern is to find out how many wait in the queue and what the total waiting time is.

Suppose that a queueing operation is organized so that arrivals occur at regular intervals and are serviced regularly. The problem is deterministic and can be worked out in an elementary manner.

If customers arrive in a single channel at regular time intervals of length a (i.e., arrival rate $1/a$) and are served at regular time intervals of length b (i.e., service rate $1/b$), then if $b < a$, that is, if $b/a < 1$, no one will wait. If $b/a > 1$, the number of waiting customers will increase indefinitely. If $b = a$, no one will wait if the operation begins with an empty line. Otherwise the queue will maintain a constant length.

We suppose that $b/a < 1$ and the operation begins with a line of i customers (note that $i \geq 2$, for if $i = 1$ that customer will finish service before the first arrival and no one waits). All i customers will be served by the end of a time interval of length ib. However, by then $[ib/a]$ customers would have arrived and would have to wait. The brackets indicate the largest integer which is less than ib/a since customers do not arrive in fractions. The last arriving customer waits for the service facility to complete its service.

The service time of these customers who, of course, will first wait in the line is $b[ib/a]$, and again during this time $\left[\dfrac{ib}{a} + \left[\dfrac{ib}{a} \right] \left(\dfrac{b}{a} - 1 \right) \right]$ customers will have arrived. Note that this number must be less than $[ib/a]$ since $b < a$, etc., until a point is reached when arriving customers need not wait.

To illustrate, we assume that the service time b is the unit of time; let a be three time units long. Suppose that $i = 50$. The service time of the i waiting customers is 50 time units. During this time $[5\%] = 16$ units arrive and will require 16 time units for service during which six units will arrive. During the service time of the latter, two units will arrive. A unit will arrive at the termination of service of the last one of the two units and hence will not have to wait. The total number of

units who have waited in line is 74. This includes the first of the initial
i customers who immediately enters service at the start of operation.

If we denote by $P_n(t)$ the probability that there are n units waiting at
time t, then clearly

$$P_n(0) = \begin{cases} 1 & \text{for } n = i, \\ 0 & \text{for } n \neq i. \end{cases}$$

These two possibilities are summarized by one symbol $P_n(0) = \delta_{in}$,
which is called Kronecker's delta. It is equal to 1 if $n = i$ and 0 if $n \neq i$.
In this case we can state with certainty what the number is. Everything
is determined exactly; hence $P_n(t)$ is either 0 or 1 at any value of time.
But this will not be the case for the nondeterministic examples to be
considered in the following sections.

It is clear that the length of the line itself is zero after a lapse of time
in which every item which has waited has also been served. Let A
new arrivals have to wait in line. Then, certainly, the $(A + 1)$st
arrival occurs *after* the service time of $i + A$ customers; i.e., $(i + A)b$
$\leq a(A + 1)$, or $ib - a \leq A(a - b)$. The number A is the smallest such
integer, so that $A = [(ib - a)/(a - b)] + 1$. Therefore, the length of
time required to serve all items which have waited is

$$b\left(i + \left[\frac{ib - a}{a - b}\right] + 1\right) = b\left[\frac{ia - b}{a - b}\right].$$

Again, this includes the first of the initial i customers.

Note that, the closer b is to a, the more customers will have to wait
until, after a long lapse of time, the queue disappears. By viewing the
situation slightly differently, we shall be able to write more useful and
readily computable expressions. For the remainder of the discussion we
impose the conditions that b divides a and that a divides the time t.

Suppose that we start the operation by admitting one of the initially
waiting customers into service; if he finishes service and nothing arrives,
another is admitted. If a customer arrives during his service or that of a
subsequent initial customer, the arriving unit will be admitted into service once the facility becomes vacant. During the time of this service it
is possible that an item will arrive. This depends on the magnitude of
$a - b$. The arriving customer will enter service once it becomes vacant,
and all customers who may have arrived would also be serviced. When
no arrivals are waiting, another of the original i customers is admitted into
service; additional arrivals are again served once the facility is empty,
and so on until all i customers go into service. When the last of the
initial customers goes into service, there will be no one waiting. If
anyone arrives during this service time he waits and is then served, etc.,
until arriving customers do not have to wait.

We have looked at the operation in this manner to make clear the following analysis. Consider the increments of service time gained by the fact that $a > b$. Thus, there is a gain of $a - b$ time units from each of the arriving customers. Theoretically, since the service time is b, $b/(a - b)$ customers must arrive to contribute enough slack time to service one of the $i - 1$ waiting customers who would be there after the operation begins. Note that the first one has already gone into service. Thus, in all, $(i - 1)b/(a - b)$ arrivals are needed to vacate the line.

Therefore the total number of customers who have waited at all in line after the operation began is

$$i - 1 + \frac{(i - 1)b}{a - b} = \frac{(i - 1)a}{a - b}, \qquad (2\text{-}1)$$

and if we include the initial customer in service we have

$$\frac{(i - 1)a}{a - b} + 1 = \frac{ia - b}{a - b}. \qquad (2\text{-}2)$$

Hence the total time until the service facility first becomes idle is

$$T \equiv \left(\frac{ia - b}{a - b}\right) b. \qquad (2\text{-}3)$$

Thus if a customer arrives at time $t < T$, then t/a customers would have arrived before him, bringing the total of those who waited in line to $t/a + i - 1$, and t/b customers would have been served during time t. Note that there is no loss in generality by writing t as if it were continuous as long as one keeps in mind that t at which an arrival occurs must be an integral multiple of a. Hence at time t there would be

$$\frac{t}{a} + (i - 1) - \frac{t}{b} = t\left(\frac{1}{a} - \frac{1}{b}\right) + (i - 1) \qquad (2\text{-}4)$$

customers in line ahead of him. (Note that if $b > a$ the number will grow in proportion to t.) However, there would be $t(1/a - 1/b) + i$ customers in the system ahead of him.

The waiting time spent in the line by a customer arriving at the first multiple of a past t is

$$W(t) = \begin{cases} 0, & T - a \le t, \\ \left(t\dfrac{b - a}{ab} + i\right) b - a & 0 < t \le T - a, \\ (k - 1)b, & t = 0, \end{cases} \qquad (2\text{-}5)$$

where $W(0)$ is the time in the line for the kth member of the initial group of i customers. This transient solution, which depends on time and on the initial number, reaches the steady state after time T, for then the system acquires a pattern independent of time and of the initial number i.

The total time spent in the system by a customer arriving at the first multiple of a past t is

$$W(t) + b \qquad\qquad (2\text{-}6)$$

where we have included the time spent in service.

Suppose now that the service time and the arrival times are altered during the period T so that the new quantities are b_1 and a_1, respectively. Assume that this is done at time $t_0 < T$. The item in service will finish service by the b service-time regime. Then the number in the queue and the unit in service become the initial input of the new queue. This establishes a new T which we denote by T_1, and we have a new waiting-time expression. The value of T_1 will depend on the choice of a_1 and b_1; hence we could have $T_1 < T$. Thus $T_1 = T$ if, and only if, $a_1 = a$ and $b_1 = b$, as seen by examining the form of T.

If there are c service channels and the initial number is $i > c$ with the same service time at all channels, then, assuming $b/ca < 1$, to empty this line requires increments of length $a - b/c$, etc. One may assign priorities to arriving customers whose arrival times are different and study the problem using both preemptive (i.e., the lower priority is returned to the line if a higher priority arrives) and nonpreemptive systems. Thus, even though the structure of this queueing problem is basically simple, a variety of ideas may be tested on it. However, ideas can also become complicated with this simple problem as the reader will readily verify by examining two priorities with different arrival and different service times.

Exercise 1: What influence would arrivals have on the number waiting and on the waiting time if the first arrival occurs at time $a - \varepsilon$, the second at $2a + \varepsilon$, the third at $3a - \varepsilon$, etc., alternately occurring early and late, where $0 < \varepsilon < a$?

Exercise 2: Let $i = 7$, $b = 2$, and $a = 4$. Compute $W(t)$, the time in the system, and the average waiting time of a customer chosen at random from all those who wait.

Exercise 3: Derive expressions for waiting times in the line and in the system for two nonpreemptive priorities with arrival intervals a_1 and a_2 for the higher and lower priorities, respectively, and with common service time b, where $b/a_1 + b/a_2 < 1$ and where $a_1 < a_2$, starting with i_1 higher priorities and i_2 lower priorities at $t = 0$.

2-3. Probability Models

Normally one does not have prior information on a customer's arrival time and length of service. In order to make an analysis of queue behavior, assumptions are made based on a study of customer habits and needs. The fact that the use of probabilistic reasoning makes this possible is an indication of the power of the concept of probability.

In the deterministic example, it was known that customers would arrive exactly at known times and receive service of exact length. Now consider the service times. In general, it is not known how long an arriving customer's service time might be. But from his previous habits, it is possible to observe the relative frequency with which he requires service of a given length, i.e., the ratio of the number of times he requires this length of service to the total number of times in which he requires any service. Or, stated differently, there is a probability that his service time will last up to a given length. This probability is the sum of all the frequencies of lengths varying up to the prescribed length. For example, if the service times are discrete and their lengths are 1, 2, 3, . . . time units and the probability (relative frequency) of their occurrence is $p_1, p_2, \ldots$, respectively, then, if P_n is the probability that the customer will require service of duration no longer than n, we have

$$P_n = p_1 + \cdots + p_n \tag{2-7}$$

and hence
$$p_n = P_n - P_{n-1}.$$

Among the useful quantities computed from these probabilities are the mean and variance. The mean is obtained by multiplying each value n by its probability of occurrence and summing over n. Thus, for example, in rolling an unbiased die, the probability of each face is $\frac{1}{6}$, and the mean value is given by

$$1 \times \tfrac{1}{6} + 2 \times \tfrac{1}{6} + 3 \times \tfrac{1}{6} + 4 \times \tfrac{1}{6} + 5 \times \tfrac{1}{6} + 6 \times \tfrac{1}{6} = 3.5.$$

Hence over a large number of rolls the average value of the numbers turned up is 3.5. In the case of a waiting line, a similar computation gives the expected size of the queue. In addition, we can compute the dispersion about this value as a measure of the fluctuaton in the size of the queue. In the case of the die, to obtain the dispersion we first calculate the variance. The square root of the variance is the standard deviation, which is a measure of dispersion; i.e., as the number of throws is increased, one may compute the probability that the average value obtained lies within a standard deviation from 3.5. We have for the variance

$$(1 - 3.5)^2 \tfrac{1}{6} + (2 - 3.5)^2 \tfrac{1}{6} + (3 - 3.5)^2 \tfrac{1}{6} + (4 - 3.5)^2 \tfrac{1}{6}$$
$$+ (5 - 3.5)^2 \tfrac{1}{6} + (6 - 3.5)^2 \tfrac{1}{6}$$

which is approximately equal to 2.92, and the standard deviation is $\sqrt{2.92}$. The differences are squared since the main concern is the fluctuation about the mean, regardless of whether the number falls below or above the mean value. Each of these differences is weighted by the probability of occurrence of the value considered.

In the mean and the variance we have the first moment about the origin and the second moment about the mean, respectively. Of course, we may compute other moments about the origin and the mean. For example, the third moment about the origin is the sum of the cubes of the values, each multiplied by its probability of occurrence. Returning to p_n we must have $\sum_{n=0}^{\infty} p_n = 1$, since this gives the sum of the probabilities that any one of the durations will occur.

If the time is continuous, then instead of P_n we use $F(x)$ to denote the probability that the duration will not be longer than a time x; that is, $F(x) = \text{Prob } (X \leq x)$, where X is a chance variable which can take on any of the possible values of service-time duration. If $f(x) \, dx$ is the corresponding frequency function (i.e., the probability of occurrence of any value x lying in an interval of length dx), since x is continuous, we write, by analogy with the discrete case, for the cumulative probability distribution

$$F(x) = \int_0^x f(y) \, dy, \qquad (2\text{-}8)$$

and if we can differentiate $F(x)$ we may write

$$f(x) = \frac{dF(x)}{dx}. \qquad (2\text{-}9)$$

If each of n customers has different habits, we have several chance variables $X_1, \ldots, X_n$ with corresponding cumulative distributions $F_1(x_1), \ldots, F_n(x_n)$. If their habits are independent (in a sense to be defined more precisely later), we can write for the probability distribution function of their joint behavior

$$F(x_1, \ldots, x_n) = F_1(x_1)F_2(x_2) \cdots F_n(x_n) = \prod_{i=1}^{n} F_i(x_i). \quad (2\text{-}10)$$

Once the joint distribution function is known, everything else can be determined for that aspect described by the distribution.

Now it may be that the habits of a customer are different at different times. Thus he may tend to demand longer service in the evening (e.g., at a restaurant) than at noon, for example. At each time, the probability of duration of service is described by a probability distribution that is now a function of time. The durations themselves are described by a random or chance variable X_t that depends on time. Here we have

$$\text{Prob } (X_t \leq x) = F(x;t)$$

as the probability distribution function at time t. For each value of t, X_t has a probability distribution. Thus for a range of values of t one has a family of random variables. As previously indicated, a family of random variables that depends on a parameter defines a stochastic process. If there is no reason to discriminate among habits of different customers at the same time, the same stochastic process may be used to describe the times of all those demanding service. Otherwise, each customer may have a stochastic process $X(k,t)$ which describes his identity (the kth to enter service) and time of entry into service.

Again, if X, for example, were a chance variable that assumed different weather conditions (with a probability of occurrence of each) at different times and in different places, we would have a stochastic variable that depends on four parameters, i.e., time and the three space coordinates. In queueing theory, time is usually the only parameter involved. An example of a stochastic process is the Poisson process which we shall study in the next section. It is represented by

$$P_n(t) = \frac{(\lambda t)^n e^{-\lambda t}}{n!}. \tag{2-11}$$

where, for purposes of Sec. 2-4, $P_n(t)$ is the probability that n customers arrive in an interval of length t. We shall investigate several properties of this useful process, introducing various concepts of value in subsequent parts of the book.

2-4. The Poisson Process

The model developed here will be concerned with the distribution of arrivals under applicable assumptions. The development examines the properties of the Poisson process. By making explicit assumptions in keeping with these Poisson properties, the Poisson process is then derived as an arrival distribution.

From $(\lambda t)^n e^{-\lambda t}/n!$ we note that, during a time interval t, the probability of no arrivals is $e^{-\lambda t}$ and that of a single arrival is $\lambda t e^{-\lambda t}$; hence the probability of more than one arrival is

$$1 - (e^{-\lambda t} + \lambda t e^{-\lambda t}) = 1 - \left\{ \left[1 - \lambda t + \frac{(\lambda t)^2}{2!} - \cdots \right] \right.$$
$$\left. + \lambda t \left[1 - \lambda t + \frac{(\lambda t)^2}{2!} - \cdots \right] \right\}$$
$$= \frac{(\lambda t)^2}{2} + \cdots = O(t^2), \tag{2-12}$$

a function which behaves as t^2.

Thus if t is small, terms with t^2 are negligible compared with terms without t or with the first power of t. Hence for small t the probability of more than one arrival is negligible. This is a very desirable property for many practical applications, thus making the Poisson process useful. The probability of at least one arrival during t is given by

$$1 - e^{-\lambda t} = \lambda t + O(t^2).$$

The probability of no arrivals is $e^{-\lambda t} = 1 - \lambda t + O(t^2)$. In both of these expressions we may ignore the last quantity on the right if we consider only small values of t.

At the risk of some repetition but in conformity with standard methods of procedure, let us assume these properties, i.e., that the probability of a single arrival during a small time interval Δt is $\lambda \, \Delta t$ and that of more than a single arrival during Δt is negligible; then we can derive the Poisson distribution, which, of course, has these properties.

We let $P_n(t)$ be the probability of n items having arrived by time t. We first observe that $0 \le P_n(t) \le 1$ since the $P_n(t)$ are probabilities. Besides, $\sum_{n=0}^{\infty} P_n(t) = 1$ because either nothing arrived or something must have arrived by time t. To see what happens during the subsequent small time interval Δt we write

$$P_0(t + \Delta t) = P_0(t)(1 - \lambda \, \Delta t),$$
$$P_n(t + \Delta t) = P_n(t)(1 - \lambda \, \Delta t) + P_{n-1}(t)\lambda \, \Delta t, \qquad n \ge 1. \qquad (2\text{-}13)$$

The first equation gives the probability that no arrivals have occurred by time $t + \Delta t$. This probability may be related to the state of the system at time t. Thus, by the law of compound probabilities of two events which occur independently (i.e., one takes the product of the probabilities of these events), it is equal to the probability that nothing had arrived by time t multiplied by the probability that nothing arrived during Δt. For the cases $n \ge 1$ this property (i.e., having the same number at time t and nothing arriving during Δt) also holds, but in addition there might have been $n - 1$ arrivals during time t, followed by an additional arrival during Δt. The product of these quantities yields the second term on the right. We have not mentioned the possibility of more than one arrival during the small interval Δt since it is negligible and can be shown to vanish in what follows.

On multiplying, transposing $P_n(t)$ to the left, and dividing by Δt, the system of equations becomes

$$\frac{P_n(t + \Delta t) - P_n(t)}{\Delta t} = -\lambda P_n(t) + \lambda P_{n-1}(t), \qquad n \ge 1.$$

If we take the limit as $\Delta t \to 0$, then, by definition, the left side is the derivative $P'_n(t) = dP_n(t)/dt$, and the equations become

$$P'_0(t) = -\lambda P_0(t),$$
$$P'_n(t) = -\lambda P_n(t) + \lambda P_{n-1}(t), \qquad n \geq 1. \tag{2-14}$$

These are linear differential equations with respect to t and linear first-order (one shift in subscript) difference equations with respect to n, generally called differential-difference equations.

We can solve these equations conveniently by the use of a generating function. We define such a function by

$$P(z,t) \equiv \sum_{n=0}^{\infty} P_n(t)z^n = P_0(t) + P_1(t)z + P_2(t)z^2 + \cdots. \tag{2-15}$$

It can be seen that $P_n(t)$ are obtained by differentiating $P(z,t)$ n times with respect to z, then dividing by $n!$, and putting $z = 0$. Thus, if $P(z,t)$ is known, $P_n(t)$ are easily determined in this manner.

The time origin for a specific study could be chosen anywhere, even after arrivals had actually occurred. Thus it may be that, by $t = 0$, i units had arrived. In that case, $P_n(0)$ is zero if $n \neq i$ and unity if $n = i$. Thus

$$P(z,0) = \sum_{n=0}^{\infty} P_n(0)z^n = P_i(0)z^i = z^i. \tag{2-16}$$

Note also that $P(1,t) = 1$ and

$$\frac{\partial P(z,t)}{\partial t} = \frac{\partial}{\partial t} \sum_{n=0}^{\infty} P_n(t)z^n = \sum_{n=0}^{\infty} P'_n(t)z^n. \tag{2-17}$$

We use partial differentiation because two variables are involved, that is, z and t.

Now if we multiply the system of differential-difference equations for $n \geq 1$ by z^n and the first equation by z^0 and add over n (it might be helpful if the reader substitutes several values of n and z and then adds to see what is obtained), we have for the sum on the left $\partial P(z,t)/\partial t$ and for the sum of the first terms on the right $-\lambda P(z,t)$.

The second terms on the right when summed over n give

$$\sum_{n=1}^{\infty} \lambda P_{n-1}(t)z^n = \lambda P_0(t)z + \lambda P_1(t)z^2 + \cdots. \tag{2-18}$$

If we factor λz out of the terms on the right, they may be written as $\lambda z P(z,t)$. Thus the system has been reduced to a linear differential equation in the generating function given by

$$\frac{\partial P(z,t)}{\partial t} - \lambda(z - 1)P(z,t) = 0. \tag{2-19}$$

This equation is very similar to the most elementary types of ordinary differential equations encountered in elementary calculus courses. Its solution, with z treated as a constant since it is independent of t, is given by

$$P(z,t) = Ce^{\lambda(z-1)t}. \tag{2-20}$$

This can be verified by substitution, where C might depend on z.

Suppose that at $t = 0$ nothing had arrived; then $P(z,0) = 1$ since $i = 0$. Thus $C = 1$, and we have

$$P(z,t) = e^{\lambda(z-1)t}. \tag{2-21}$$

But as we pointed out, $P_n(t)$ may be obtained by differentiation; hence

$$P_n(t) = \frac{1}{n!} \frac{\partial^n P(z,t)}{\partial z^n}\bigg|_{z=0}. \tag{2-22}$$

Thus
$$P_0(t) = e^{-\lambda t}, \tag{2-23a}$$
$$P_1(t) = \lambda t e^{-\lambda t}, \tag{2-23b}$$

and
$$P_n(t) = \frac{(\lambda t)^n e^{-\lambda t}}{n!}, \tag{2-23c}$$

which gives the desired Poisson process. By now it should be clear that t is the length of the time interval in which the events occur rather than absolute time.

The foregoing method of derivation is standard for many queueing problems. The equations may be more complicated, but the basic procedure often is the same. The reader should have no difficulty in writing for $i \neq 0$ and $n \geq i$

$$P_n(t) = \frac{(\lambda t)^{n-i} e^{-\lambda t}}{(n-i)!}, \tag{2-24}$$

since for this case $P(z,0) = z^i$, C has the value z^i, and

$$P(z,t) = z^i \sum_{n=i}^{\infty} P_n(t) z^{n-i}.$$

The reader has just encountered a typical situation that we shall frequently examine in this book, namely, computing quantities $P_n(t)$ which in other cases, instead of being the probability that n items arrive during time t, may be the probability that there are n items waiting in line and being served at time t in a queueing system or the probability that n channels are occupied if there are c channels in all, etc. Although calculating $P_n(t)$ for such situations is very useful to obtain a variety of measures of effectiveness for studying queues, it is not always possible or even desirable to study queues in terms of $P_n(t)$. Waiting-time expressions, as will be seen later, can be derived by other arguments.

Having obtained $P_n(t)$, one may wish to compute the mean number arriving by time t, that is,

$$L = \sum_{n=0}^{\infty} nP_n(t),$$ (2-25)

which for the Poisson process equals λt. In fact, the variance

$$\sum_{n=0}^{\infty} (n - L)^2 P_n(t)$$

also equals λt.

Exercise 4: Verify the values of the mean and variance for the Poisson process.

The time intervals between arrivals from a Poisson process with parameter λ have the exponential frequency distribution $\lambda e^{-\lambda t}$. To see this, note that, if an arrival has just occurred, the time to the next arrival is less than t if, and only if, there are one or more arrivals in this interval. The probability of one or more arrivals and hence the probability that the time between arrivals is less than or equal to t is

$$\sum_{n=1}^{\infty} \frac{(\lambda t)^n e^{-\lambda t}}{n!} = 1 - e^{-\lambda t}.$$ (2-26)

The quantity on the right is a cumulative distribution from which, by differentiation, one obtains the exponential density function $\lambda e^{-\lambda t}$ for the distribution of times between arrivals. *Hence when the times of arrival to a queue occur by a Poisson process, the interarrival times have a related exponential distribution.* Conversely, if, for example, service times have an exponential distribution, customers are admitted to service by a Poisson process.

Again, if $\mu \, \Delta t$ is the probability of completing service during Δt (hence $1 - \mu \, \Delta t$ is the probability of not finishing service during Δt) and if $P(t)$ is the probability that service does not terminate during time t, we have

$$P(t + \Delta t) = (1 - \mu \, \Delta t)P(t).$$ (2-27)

Thus we have

$$\frac{dP(t)}{dt} = \lim_{\Delta t \to 0} \frac{P(t + \Delta t) - P(t)}{\Delta t} = -\mu P(t).$$

This differential equation has the solution

$$P(t) = ce^{-\mu t}.$$ (2-28)

Since service must terminate sooner or later, we must have

$$\int_0^{\infty} ce^{-\mu t} \, dt = 1.$$

Hence $c = \mu$. Thus the service is described by the negative exponential distribution with density $\mu e^{-\mu t}$. Note that the exponential distribution has the property of "forgetfulness": If, for example, it is applied to service times, the exponential distribution applies for the time of completing service at any time just as it does at the start of service.

For many problems time dependence is not the typical situation to be studied. After a long period of operation the system acquires a pattern of behavior which, although describable in terms of probabilities, does not depend on time. Thus as $t \to \infty$ this situation may be expected to occur for many systems. The question of whether such $p_n = \lim_{t \to \infty} P_n(t)$ exist is a main concern of ergodic analysis where one actually looks for the probabilities p_n which describe the steady or equilibrium state. But one also notes that the requirement that the probabilities $P_n(t)$ do not change with time is the definition of the steady state.

The changing of $P_n(t)$ with respect to t is described by the derivative $P'_n(t)$, and the steady-state condition requires that $P'_n(t) = 0$. Thus essentially we have two ways of obtaining the steady-state probabilities: (1) from $P'_n(t) = 0$, which gives probabilities p_n independent of t, and (2) $\lim_{t \to \infty} P_n(t)$, which also gives the p_n independent of t. We now use this concept of an equilibrium solution to develop two queueing problems. The first is due to Erlang, and the second is the Pollaczek-Khintchine formula.

2-5a. Erlang's Model

If we assume that an operation starts with no items waiting in line, the following equations provide a representation for a Poisson input with parameter λ, exponential holding time with parameter μ, first-come–first-served, single-channel queue:

$$P_n(t + \Delta t) = P_n(t)[1 - (\lambda + \mu)\, \Delta t] + P_{n-1}(t)\lambda\, \Delta t$$
$$+ P_{n+1}(t)\mu\, \Delta t, \qquad n \geq 1, \quad (2\text{-}29)$$
$$P_0(t + \Delta t) = P_0(t)(1 - \lambda\, \Delta t) + P_1(t)\mu\, \Delta t, \qquad\qquad n = 0.$$

These equations state that the probability of n items in the system at time $t + \Delta t$ equals the probability of n items in the system at time t multiplied by the probability of no arrivals and no departures, plus the probability of $n - 1$ items in the system at time t multiplied by the probability of one arrival and no departures, plus the probability of $n + 1$ items in the system at time t multiplied by the probability of a single departure and no arrivals.

Note that the probability of no arrivals and no departures is given by $(1 - \lambda\, \Delta t)(1 - \mu\, \Delta t)$. The term involving $(\Delta t)^2$ drops out in forming the differential equation. Thus one may use $1 - (\lambda + \mu) \Delta t$. For the remaining two terms above, also note that $\lambda \Delta t(1 - \mu \Delta t) \sim \lambda \Delta t$ and $\mu \Delta t(1 - \lambda \Delta t) \sim \mu \Delta t$.

By transposing and passing to the limit with respect to Δt, these equations become

$$\frac{dP_n(t)}{dt} = -(\lambda + \mu)P_n(t) + \lambda P_{n-1}(t) + \mu P_{n+1}(t), \qquad n \geq 1,$$
$$\frac{dP_0(t)}{dt} = -\lambda P_0(t) + \mu P_1(t), \qquad n = 0. \tag{2-30}$$

As already indicated, the time-independent steady-state solution is obtained either by solving the time-dependent transient equations given above and letting $t \to \infty$ in the solution (to be shown in a later chapter) or by setting the derivatives with respect to time equal to zero and solving the resulting steady-state equations. Transient solutions are particularly useful when the traffic intensity or utilization factor $\rho = \lambda/\mu \geq 1$, since in this case no steady state occurs. Here, an expression for the expected number waiting and the expected waiting time will be derived in the steady state with $\lambda/\mu < 1$. If $\lambda/\mu \geq 1$, the number waiting would be infinite. Compare these conditions with those of the deterministic queue previously considered.

By setting the time derivatives equal to zero and eliminating time from the above equations, there result after transposing

$$(\lambda + \mu)p_n = \lambda p_{n-1} + \mu p_{n+1}, \qquad n \geq 1,$$
$$\lambda p_0 = \mu p_1, \qquad n = 0. \tag{2-31}$$

Let $\rho = \lambda/\mu$; then these equations become

$$(1 + \rho)p_n = p_{n+1} + \rho p_{n-1}, \qquad n \geq 1,$$
$$p_1 = \rho p_0, \qquad n = 0.$$

Let $n = 1$ in the first equation. Then $(1 + \rho)p_1 = p_2 + \rho p_0$. Substituting for p_1 from the second equation, this becomes $p_2 = \rho^2 p_0$.

Repetition of this process yields $p_n = \rho^n p_0$. Now $\sum_{n=0}^{\infty} p_n = 1$, since the sum gives the total probability that there are no items, one item, two items, . . . in the system. This total probability must yield certainty since it accounts for all the possible states of the system. Thus

$$\sum_{n=0}^{\infty} \rho^n p_0 = 1,$$

or

$$\sum_{n=0}^{\infty} p_0 \rho^n = p_0 \sum_{n=0}^{\infty} \rho^n = \frac{p_0}{1 - \rho} = 1.$$

or $$p_0 = 1 - \rho.$$
Hence $$p_n = \rho^n(1 - \rho),$$ (2-32)

which is a type of probability distribution known as a geometric distribution.

Now the expected number in the system is given by

$$L = \sum_{n=0}^{\infty} np_n = (1 - \rho) \sum_{n=0}^{\infty} n\rho^n = \frac{\rho}{1 - \rho},$$ (2-33)

as can be readily verified. Note that L is an expected value, and fluctuations in the number waiting can, in fact, occur. This can best be seen by calculating the variance:

$$\sum_{n=0}^{\infty} (n - L)^2 p_n = \sum_{n=0}^{\infty} n^2 p_n - \left(\frac{\rho}{1 - \rho}\right)^2.$$

But

$$\sum_{n=0}^{\infty} n^2 p_n = (1 - \rho) \sum_{n=0}^{\infty} n^2\rho^n = (1 - \rho)\rho \frac{d}{d\rho} \rho \frac{d}{d\rho} \sum_{n=0}^{\infty} \rho^n$$

$$= \frac{\rho}{1 - \rho} + \frac{2\rho^2}{(1 - \rho)^2}.$$

Therefore the variance is given by

$$\frac{\rho}{1 - \rho} + \frac{\rho^2}{(1 - \rho)^2} = L + L^2.$$ (2-34)

The expected number in the line is given by

$$L_q = \sum_{n=1}^{\infty} (n - 1)p_n = L - \rho = \frac{\rho}{1 - \rho} - \rho = \frac{\rho^2}{1 - \rho}.$$ (2-35)

Note that there are times when the channel is not occupied. The average waiting time in the line, W_q, equals L_q/λ or L/μ.

A useful device for a quick check for the possible incorrectness of a formula is dimensional analysis. Thus, for example, in the last expression above, both the expressions on the left and on the right are dimensionless.

Exercise 5: Solve (2-31) using a generating function similar to (2-15) with t suppressed.

2-5b. The Pollaczek-Khintchine Formula

Suppose now that arrivals occur at random by a Poisson process with the rate λ per unit time, to a waiting line, in statistical equilibrium.

before a single service facility. Also suppose that they are served by an arbitrary-service-time distribution at the rate of μ per unit time, first come–first served. As in the deterministic case, we assume that $\lambda/\mu < 1$. Suppose that a departing customer leaves q in line, including the one in service, whose service time is t. Let r customers arrive during this time t. If the next departing customer leaves q' customers behind, we can relate q and q' as follows:

$$q' = \max(q - 1, 0) + r = q - 1 + \delta + r, \qquad (2\text{-}36)$$

where
$$\delta(q) = \begin{cases} 0 & \text{if } q > 0, \\ 1 & \text{if } q = 0. \end{cases} \qquad (2\text{-}37)$$

Note that by introducing δ we avoid using the max expression.

It is assumed that equilibrium values for the first and second moments $E[q]$ and $E[q^2]$ of the queue exist. Note that q is treated as a random variable. Now, from the definition, $\delta^2 = \delta$ and $q(1 - \delta) = q$. Also, $E[q] = E[q']$ and $E[q^2] = E[q'^2]$, since both q and q' are assumed to have the same equilibrium distribution. Note that, because the system is in equilibrium, there is no difference between the types of queue left behind any customer; i.e., they all have the same probability distribution independent of time.

Thus, taking the expected value of the equation in the different variables, we have

$$E[q'] = E[q] - E[1] + E[\delta] + E[r], \qquad (2\text{-}38)$$

from which we have

$$E[\delta] = 1 - E[r]. \qquad (2\text{-}39)$$

But during a service time of length t we have

$$E[r] = \sum_{r=0}^{\infty} r \, \frac{(\lambda t)^r}{r!} \, e^{-\lambda t} = \lambda t, \qquad (2\text{-}40)$$

$$E[r^2] = \sum_{r=0}^{\infty} r^2 \, \frac{(\lambda t)^r}{r!} \, e^{-\lambda t} = (\lambda t)^2 + \lambda t. \qquad (2\text{-}41)$$

Thus, on taking expectations with respect to the service time t, one has $E[r] = \lambda/\mu \equiv \rho$. For example, if the service distribution is exponential, we have

$$E[r] = \mu \int_0^{\infty} (\lambda t) e^{-\mu t} \, dt = \frac{\lambda}{\mu} \equiv \rho, \qquad (2\text{-}42)$$

since the average value of the service-time distribution is $1/\mu$.

Note that $E[r]$ is a number which is unaffected by taking averages. This gives $E[\delta] = 1 - \rho$. Now the probability of r arrivals is independent of q, the length of the queue, and of δ, which assumes values that depend only on q which is independent of r. Consequently, if we take the expected value over the variables r, q, and δ, we may take the expected value of r and of q separately wherever we encounter their product. This is also true for r and δ.

Also, averaging r^2 over time,

$$E[r^2] = \lambda^2 \text{ var } (t) + \rho^2 + \rho. \tag{2-43}$$

Again, by squaring the equation relating q and q' and using the facts that $\delta^2 = \delta$, $\delta q \equiv 0$,

$$q'^2 = q^2 - 2q(1 - r) + (r - 1)^2 + \delta(2r - 1). \tag{2-44}$$

Hence, because of equilibrium,

$$0 = E[q'^2] - E[q^2] = 2E[q]E[r - 1] + E[(r - 1)^2] + E[\delta]E[2r - 1],$$

or, simplifying and using the foregoing relations, we have the Pollaczek-Khintchine formula:

$$\begin{aligned} E[q] &= \frac{E[(r - 1)^2] + E[\delta]E[2r - 1]}{2E[1 - r]} \\ &= \frac{E[r^2] - 2E[r] + 1 + E[\delta](2E[r] - 1)}{2(1 - E[r])} \\ &= \frac{\lambda^2 \text{ var } (t) + \rho^2 + \rho - 2\rho + 1 + (1 - \rho)(2\rho - 1)}{2(1 - \rho)} \\ &= \rho + \frac{\rho^2 + \lambda^2 \text{ var } (t)}{2(1 - \rho)}. \end{aligned} \tag{2-45}$$

Thus, once we know the variance of the service time t from its given distribution, the average number in the system is determined. It is important to observe that the foregoing average is taken over instants just *following* departures and is not the time average value of the number in the system. In fact, if $E_t(q)$ is the time average, all we can say without further argument (see Chap. 7) is that $E[q] \leq E_t(q) < E[q] + 1$.

Note the fact, which holds in general (i.e., even if one has several channels in parallel), that the average number in the system equals the sum of the average number of busy channels (in this case it is ρ, the traffic intensity) and the average number in line.

Now to obtain the average waiting time we argue as follows: If we write $E[w]$ for the average waiting time in the queue (not including service), $\lambda[E(w) + 1/\mu]$ is the expected number of arrivals during the total waiting plus service time of one customer, i.e., of his stay in the system. But this must be just the number in the system immediately after his departure, namely $E[q]$. Thus[1]

$$W_q \equiv E[w] = \frac{\rho^2 + \lambda^2 \text{ var } (t)}{2\lambda(1 - \rho)} = \frac{L_q}{\lambda}. \tag{2-46}$$

---

[1] The name Pollaczek-Khintchine should actually be given to (2-46) throughout. This change will be made in revised editions of the book.

A useful measure to introduce at this point is the ratio of the average waiting time of a customer to his service time. Naturally one would not wish to wait a very long time when he requires service of short duration (unless, of course, he is at the Pearly Gates). We have for this measure

$$\mu E[w] = \frac{\rho}{2(1 - \rho)} (1 + C_t^2), \tag{2-47}$$

where C_t is the coefficient of variation of the service time defined by

$$C_t^2 = \frac{\text{var } (t)}{E^2[t]} = \frac{\text{var } (t)}{1/\mu^2}. \tag{2-48}$$

Clearly, the above measure $\mu E[w]$ depends on ρ and on C_t^2. Now any change in the input rate λ or in the service rate μ is reflected in their ratio ρ. Therefore to test the effect, i.e., the sensitivity to changes of values of ρ from a given average value $\bar{\rho}$ (say the present value) cf ρ, we expand the right side in a Taylor's series about the point $\bar{\rho}$ for small variations about it. We have

$$f(x) = f(a) + \frac{f'(a)}{1!} (x - a) + \frac{f''(a)}{2!} (x - a)^2 + \cdots$$
$$+ \frac{f^{(n)}(a)}{n!} (x - a)^n + \cdots,$$

where the number of primes indicates the order of the derivative to be taken.

When applied to the right side of the above formula, this gives

$$\mu E[w] = \frac{\bar{\rho}}{2(1 - \bar{\rho})} (1 + C_t^2) + \frac{1 + C_t^2}{2(1 - \bar{\rho})^2} (\rho - \bar{\rho})$$
$$+ \frac{1 + C_t^2}{2!(1 - \bar{\rho})^3} (\rho - \bar{\rho})^2 + \cdots. \tag{2-49}$$

If we take the average with respect to ρ, then the middle term becomes zero since $E[\rho - \bar{\rho}] = \bar{\rho} - \bar{\rho}$, and up to second-degree terms we have

$$\frac{\bar{\rho}}{2(1 - \bar{\rho})} (1 + C_t^2) \left[1 + \frac{\sigma_\rho^2}{\bar{\rho}(1 - \bar{\rho})^2} \right], \tag{2-50}$$

where $\sigma_\rho^2 = E[(\rho - \bar{\rho})^2]$.

Cox [138] uses the values $\bar{\rho} = 0.6$, $\sigma_\rho = 0.1$ to show that the last factor in parentheses yields 1.10 and hence a 10 per cent increase in waiting time. This analysis is useful when one can consider a queueing operation as a set of operations each in the steady state and each with its parameters. Then one can determine how wide a range of fluctuation in parameter values to allow (before shifting to a new steady-state solution) for a tolerated change in the waiting-time to service-time measure.

2-6. Testing a Hypothesis

In collecting data for estimating the distribution of the number of arrivals or of holding times, the number of arrivals and the intervals of time in service are recorded as a function of time. These data may then be grouped in time intervals, after which statistical tests for independence of the means in successive time intervals may be made, e.g., by the variance-ratio test and other appropriate methods found in standard texts on statistics. Should there prove to be lack of independence, the data must be subdivided into time intervals within which a steady state is a reasonable approximation. A non-steady-state situation arises, for example, in a restaurant operation before and after the meal hours.

With the data, or portion of the data which appears to be homogeneous, a further grouping is made and a frequency histogram constructed, taking 10 to 20 subgroups. The appearance of this histogram will suggest certain standard distributions such as the normal, Poisson, or exponential, or the distribution of a sum of independently and exponentially distributed random variables.

The parameters of the hypothesized distribution are then estimated from the data. The observed and model distributions are compared, using the chi-square test. When slight but statistically significant deviations are found, a decision as to whether or not the model should be altered can be made only with a knowledge of the objectives of the particular problem.

Another consideration must be borne in mind when the data are not very numerous. Parameters estimated from such data may vary quite markedly from the time-population values, even though a chi-square test shows no significant deviation from the model.

The considerations of the preceding two paragraphs are less important if the model can be tested in the actual operation. Ingenuity may sometimes be required to verify the model experimentally, such as closing one lane of a multilane highway in order to verify the applicability of the model and subsequently to predict the effect of opening an additional lane.

In exceptional circumstances where no experimental verification is possible, greater care should be taken in fitting the available data with the model. The problem of the effect of changes in parameter values on the model may perhaps be approached analytically by varying the parameters of the distribution of arrivals and holding times over a range corresponding to the expected uncertainty and observing the effects of the changes upon the predictions, as we indicated in the previous sensitivity analysis.

Suppose that it is desired to test the hypothesis that the arrivals to a waiting line are Poisson-distributed with a mean arrival rate per minute

equal to 0.4. The intervals for counting the number of arrivals are fixed
at 2 minutes. In each such interval, one counts the total number of
arrivals. Then one counts the frequency of intervals in which a given
number of arrivals occurs. This gives the first two columns of Table 2-1.

TABLE 2-1. COMPARISON OF ARRIVAL DATA WITH POISSON DISTRIBUTION

Number of arrivals n	Number of 2-minute intervals observed with n	Value from Poisson distribution	Number of 2-minute intervals expected with n
0	21	.449	26.9
1	23	.359	21.5
2	10	.144	8.6
3	4		
4	1		
5	0 } 6	.048	2.9
6	0		
7	1		
	60		

When a number in the frequency column is small, e.g., less than five,
it is grouped with others to increase the sample size; e.g., the frequencies
for three to seven arrivals are grouped to yield a total frequency of 6.
Using the Poisson distribution

$$P_n(t) = \frac{(.4t)^n e^{-.4t}}{n!},\qquad (2\text{-}51)$$

with $t = 2$, one computes the third column for values of n from the first
column. This gives the probabilities for the indicated arrivals. By
multiplying each probability by the total frequency 60, one has the
number of 2-minute intervals expected to contain the indicated number
of arrivals n. Finally to judge whether the Poisson process is an acceptable fit statistically, one forms the chi-square expression

$$\chi^2 = \sum \frac{(\text{number observed} - \text{number expected})^2}{\text{number expected}},\qquad (2\text{-}52)$$

where the differences are taken between the numbers in the second and
last columns, squared, and divided by the last-column number. The
sum is then taken for all corresponding pairs. This gives

$$\frac{(26.9 - 21)^2}{26.9} + \frac{(21.5 - 23)^2}{21.5} + \frac{(8.6 - 10)^2}{8.6} + \frac{(2.9 - 6)^2}{2.9} = 4.94.$$

Entering the chi-square table with $k - 1 = 3$ degrees of freedom (four
groups are considered, and the degrees of freedom are this number reduced
by unity), one finds that this number is less than that indicated at the

highest significance levels. Hence, one may conclude that, if the hypothesis that the mean of the Poisson distribution is 0.4 is accepted, the observed data do not differ significantly from values obtained from the "fitted" Poisson distribution. If the computed value exceeds the value from the table, the fit of the curve is rejected at that significance level.

2-7. Parameter Estimation in Queueing Theory

A. B. Clarke [111] has derived a maximum-likelihood (see below) procedure for estimating λ and μ, the input and service parameters, respectively, in a single-channel Poisson-input and exponential-holding-time, first-come–first-served queue in equilibrium. In general, to estimate λ, one observes the operation for a time of length s; if n persons arrive, then one has $\lambda = n/s$.

In estimating μ, the average number of customers which the channel can serve per unit time, it is fair to count only time when there are customers available to be served or in service. This is called the channel "busy time." If m persons are served and the total busy time of the service channel is τ, then $\hat{\mu} = m/\tau$, which is the best one can do when $\rho = \lambda/\mu \geq 1$. For the case $\rho < 1$ it is possible to use the initial queue size i to improve the above estimates for the queue in equilibrium.

If one has all the relevant distributions of the quantities involved, one may obtain the sampling distribution as products of these (assuming that the samples are independent); the likelihood function is formed as the product of the sampling distributions. It is then differentiated with respect to the parameters to obtain the maxima.

Let $m > i$, i being the initial queue size, and suppose that the busy time of the channel is observed until a preassigned value τ is reached; let T be the time at which the mth departure occurs. Let the random variables X_j, Y_j, Z_j give the jth arrival time, the jth departure time, and the busy time of the channel up to the jth departure (all vanishing for $j \leq 0$), respectively. Then one has recursively

$$Y_j = \max (Y_{j-1}, X_{j-i}) + Z_j - Z_{j-1}. \tag{2-53}$$

This gives the departure time of the jth departing customer as his arrival time X_{j-i} plus his service time if he does not have to wait, or as the departure time of the $(j - 1)$st customer plus the service time of the jth customer if the latter must wait. Thus, once the initial number i and the X_j and Z_j are known, the entire queueing process can be completely described. Clearly, X_j is independent of Z_j.

Assume that the initial queue size i is distributed according to the geometric distribution $(1 - \rho)\rho^i$ ($i = 0, 1, 2, \ldots$), which is the equilibrium solution of the queue size.   From the assumptions of the problem, the sequence X_j $(j = 1, 2, \ldots)$ is a Poisson process with parameter λ, and Z_j is a Poisson process with parameter μ.   Both processes are independent of i.

Note that m, the number of persons served during time τ, is a function of the Z_j only, since m is the maximum index that yields $Z_m \leq \tau$. Therefore m is independent of i and is distributed according to

$$e^{-\mu\tau}\frac{(\mu\tau)^m}{m!}, \qquad m = 0, 1, 2, \ldots . \tag{2-54}$$

The conditional distribution of Z_j $(j = 1, \ldots, m)$, given m, is independent of μ and λ.   It is a random subdivision of the interval τ into $m + 1$ parts.   Thus we know the distribution of X_j, Z_j, i, and m.   We now obtain the distribution of samples of size m.

Note that, since the initial number in the queue is i, the arrival times X_j are applicable for $m - i$ numbers of the sample.   The probability density for an arrival at X_1 is $\lambda e^{-\lambda X_1}$.   That of an arrival at X_2 is $\lambda e^{-\lambda(X_2 - X_1)}$, and so on to the $(m - i)$th arrival.   The product of these is $\lambda^{m-i}e^{-\lambda X_{m-i}}$, which is independent of the Z_j.

Consider that n customers have been observed to arrive in the time up to T during which m customers have been served.   Between X_{m-i} and the time of the mth departure at time $Y_m = T$ there must be $n - m + i$ arrivals to bring the total arrivals to n.   The probability of this number of arrivals during the interval (X_{m-i}, T) is given by

$$e^{-\lambda(T - X_{m-i})}\frac{[\lambda(T - X_{m-i})]^{n-m+i}}{(n - m + i)!}, \qquad \begin{array}{l} n - m + i = 0, 1, \ldots, \\ n + i > 0. \end{array} \tag{2-55}$$

Multiplying the distributions for the arrivals, the distribution for the initial number i, the distribution for the Z_j (the latter does not depend on μ and λ), and the distribution for m, we have for the likelihood function

$$K\left(1 - \frac{\lambda}{\mu}\right)\mu^{m-i}\lambda^{n+i}e^{-\mu\tau-\lambda T},$$

where K is a function which does not depend on λ or μ and whose form does not matter since it disappears on equating the derivatives with respect to λ and μ, respectively, to zero.   Thus one obtains the maximum-likelihood estimates of λ and μ from the two equations obtained by inserting the observed i, m, and n and equating the respective derivatives with respect to λ and μ to zero:

$$\begin{aligned} \hat{\lambda} &= (\hat{\mu} - \hat{\lambda})(n + i - \hat{\lambda}T), \\ \hat{\lambda} &= (\hat{\lambda} - \hat{\mu})(m - i - \hat{\mu}\tau). \end{aligned} \tag{2-56}$$

Also, by substituting $\hat{\lambda} = \hat{\mu}\hat{\rho}$ and eliminating $\hat{\mu}$, one obtains a quadratic equation from which $\hat{\rho}$, an estimate of ρ, is obtained. Exactly one solution of this equation lies between zero and unity. It is therefore the unique estimate $\hat{\rho}$.

Exercise 6: Obtain this quadratic and the unique estimate $\hat{\rho}$.

2-8. A Decision Process (a Method for Testing Fault Frequency in the Service Channels)

This section has nothing to do with the theory of queues per se, but the method presented may be useful for determining when the service facility must be repaired or reconditioned.

A service channel, if automatic, as in telephony, requires occasional inspection for faults before difficulties of a greater order of magnitude develop. By considering maintenance costs and fault-finding costs at different fault frequencies, a limit of action α_0 is determined. If samples show that the fault frequency is greater than α_0, the service channel is then corrected; otherwise no action is taken. Wrong conclusions from the samples may cause loss.

If n is the number of observations, x the observed number of faults, and α the fault frequency in the population and if X, the number of faults, is a stochastic variable with a Poisson distribution (a reasonable assumption if the probability for a fault at each observation is small and constant and if the number of observations is large), then

$$\text{Prob } (X \leq x) = \sum_{k=0}^{x} \frac{(n\alpha)^k}{k!} e^{-n\alpha} \equiv F_\alpha(x). \tag{2-57}$$

For a given n one tests the hypothesis $\alpha = \alpha_0$ against the hypothesis $\alpha \neq \alpha_0$. The null hypothesis is $\alpha = \alpha_0$. The alternative hypothesis is $\alpha < \alpha_0$ or $\alpha > \alpha_0$. Given n and α_0, one determines x_1 and x_2 such that

$$F_{\alpha_0}(x_1) \leq \varepsilon_1,$$
$$1 - F_{\alpha_0}(x_2) \leq \varepsilon_2,$$

where $\varepsilon_1 + \varepsilon_2$ is the level of significance chosen. If $x_1 < x \leq x_2$, the null hypothesis is accepted; otherwise it is rejected. In the latter case there is always a risk $\leq (\varepsilon_1 + \varepsilon_2)$ of making a wrong decision; i.e., the hypothesis is rejected when it is true.

By successive sampling one has different values of n with corresponding values of x_1 and x_2. An upper limit to the number of observations must often be set.

It is also possible to develop expressions for optimum decision values by studying different risk levels [179].

Exercise 7: Quantify the symbols occurring above for an interpretation of the ideas.

2-9. Simulation and Monte Carlo Application

The Monte Carlo method has frequently been applied to queueing problems to obtain numerical solutions. Application of the method may be illustrated by an example for obtaining a single-channel first-come–first-served waiting-time result.

A plot of the cumulative-input and service-time distributions is made either from data or from formulas. Then a column of two-digit random numbers is formed. As decimals, they are used as ordinates in the figure for the input distribution, and the corresponding abscissas are then obtained. The latter provide a column of the lengths of the time intervals between arrivals. Similarly, two adjacent columns are included to provide service times. In a fifth column the waiting time for every item is calculated as the sum of the holding time and waiting time of the previous item minus the delay in the arrival of the new item. If this difference is negative, the waiting time is zero.

The average total wait is calculated in a straightforward manner. Note that, for the steady-state results, $\lambda/\mu < 1$, which must be taken into consideration when using the method. The accuracy of the method improves roughly as $\sqrt{N}$, where N is the number of trials. This will now be illustrated by a numerical example.

In a production-line operation it was calculated that speeding up operation would bring items to the inspection line at the constant rate of one every 5 minutes. The present inspection-service times in minutes, when grouped and fitted with a curve, were observed to be exponentially distributed according to

$$b(t) = .5e^{-.5t}.$$

The Monte Carlo method was applied to this problem as illustrated in Table 2-2 to calculate the average waiting times for possible increase in service capacity caused by the increase in production rate.

Obviously, the input is fixed at constant intervals, and any experiment must use this information as it is. On the other hand, the service times for each item must be selected (at random) without any discrimination; i.e., one service time is as likely as any other. The cumulative-service-time distribution was plotted, and the columns in the table were generated as described above. This was carried out for 100 arrivals, and the average waiting time was calculated to be .25 minute, which did not warrant additional inspection service to maintain a smooth operation.

Exercise 8: Repeat the above experiment, interchanging the input and the service distributions and their parameters; i.e., use an exponential-interarrival-interval distribution of $.2e^{-.2t}$ and a constant holding interval of length 2. Compute the average waiting time. Compare with the value obtained from the Pollaczek-Khintchine formula.

TABLE 2-2. A MONTE CARLO ILLUSTRATION

Intervals between consecutive arrivals	Random numbers taken as decimals $(1 - e^{-.5t})$	The corresponding service time t obtained from the figure, using the decimals	The waiting time
5	.10	0.20	0
5	.22	0.50	0
5	.24	0.55	0
5	.42	1.05	0
5	.37	0.90	0
5	.77	2.90	0
5	.99	9.00	0
5	.96	6.30	4.00
5	.89	4.50	5.30
5	.85	3.80	4.80
5	.28	0.65	3.60
5	.63	2.00	0
5	.09	0.20	0
. . .	. . .	. . .	. . .

Note: There is rounding off in decimals.

2-10. Power Series Expansions

We give a procedure to illustrate numerical calculation to obtain solutions of queueing problems. At this point it is not essential to understand the significance of the equations used. By way of example, we shall consider a typical queueing problem and shall show how to find series solution expansions for the probabilities $P_n(t)$, using the initial queue length i and the given parameters λ and μ.

Consider the problem (with two absorbing barriers at the origin and at N, that is, the process stops once it reaches these points) described by

$$P_0'(t) = \mu P_1(t),$$
$$P_1'(t) = -(\lambda + \mu)P_1(t) + 2\mu P_2(t),$$
$$P_n'(t) = (n - 1)\lambda P_{n-1}(t) - n(\lambda + \mu)P_n(t) + (n + 1)\mu P_{n+1}(t),$$
$$P_{N-1}'(t) = (N - 2)\lambda P_{N-2}(t) - (N - 1)(\lambda + \mu)P_{N-1}(t),$$
$$P_N'(t) = (N - 1)\lambda P_{N-1}(t),$$

with the initial condition $P_n(0) = \delta_{in}$.

Let $i = 2$ and $N = 5$ and assume that

$$P_n(t) = P_n(0) + P_n'(0)t + P_n''(0)\frac{t^2}{2!} + \cdots.$$

We shall now calculate the coefficients $P_n^{(k)}(0)$ in terms of the constants λ and μ. We have $P_2(0) = 1$, $P_n(0) = 0$, $n \neq 2$, and on starting with the third equation we have

$$P_2'(0) = \lambda P_1(0) - 2(\lambda + \mu)P_2(0) + 3\mu P_3(0)$$
and hence
$$P_2'(0) = -2(\lambda + \mu),$$
since
$$P_1(0) = P_3(0) = 0.$$

Also, from the equation defining $P_2(t)$, we have

$$P_2''(0) = \lambda P_1'(0) - 2(\lambda + \mu)P_2'(0) + 3\mu P_3'(0).$$
But
$$P_1'(0) = -(\lambda + \mu)P_1(0) + 2\mu P_2(0),$$

which on using the initial condition equals 2μ, and

$$P_3'(0) = 2\lambda P_2(0) - 3(\lambda + \mu)P_3(0) + 4\mu P_4(0),$$

which equals 2λ; therefore

$$P_2''(0) = 2\mu\lambda + 4(\lambda + \mu)^2 + 6\mu\lambda.$$

Again
$$P_2'''(0) = \lambda P_1''(0) - 2(\lambda + \mu)P_2''(0) + 3\mu P_3''(0),$$
where
$$P_1''(0) = -(\lambda + \mu)P_1'(0) + 2\mu P_2'(0) = -6\mu(\lambda + \mu)$$
and
$$P_3''(0) = 2\lambda P_2'(0) - 3(\lambda + \mu)P_3'(0) + 4\mu P_4'(0).$$
But
$$P_4'(0) = 3\lambda P_3(0) - 4(\lambda + \mu)P_4(0) = 0;$$
hence
$$P_3''(0) = -10\lambda(\lambda + \mu).$$

We finally have

$$P_2'''(0) = \lambda[-6\mu(\lambda + \mu)] - 2(\lambda + \mu)[8\mu\lambda + 4(\lambda + \mu)^2]$$
$$+ 3\mu[-10\lambda(\lambda + \mu)].$$

One may similarly compute $P_2^{IV}(0)$, etc. Thus

$$P_2(t) = 1 - 2(\lambda + \mu)t + [8\mu\lambda + 4(\lambda + \mu)^2]\frac{t^2}{2!} + \cdots.$$

We also have

$$P_1(t) = P_1(0) + P_1'(0)t + P_1''(0)\frac{t^2}{2!} + \cdots,$$

which, on substitution from the above calculation, becomes

$$P_1(t) = 2\mu t + [-6\mu(\lambda + \mu)]\frac{t^2}{2!} + \cdots.$$

Since $P_0'(t) = \mu P_1(t)$,

$$P_0(t) = 2\mu^2 \frac{t^2}{2!} - 6\mu^2(\lambda + \mu) \frac{t^3}{3!} + \cdots.$$

Also $$P_3(t) = 2\lambda t - 10\lambda(\lambda + \mu) \frac{t^2}{2!} + \cdots,$$

etc., for $P_4(t)$ and $P_5(t)$.

As an exercise the reader should compute $P_4(t)$ as indicated. Because of convergence problems, these expansions are valid only near $t = 0$. It is possible, using difference methods, to compute values of $P_n(t)$ away from the origin. To do this we divide the time axis into intervals of unit length, for example. To compute the probabilities at a time period of a given length, we start with $t = 0$ and obtain the value for $P_n(0)$. To compute $P_n(1)$ we use the difference approximation which determines P_n at time $t + \Delta t$.

$$P_n'(t) \sim \frac{P_n(t + \Delta t) - P_n(t - \Delta t)}{2 \Delta t} \tag{2-58}$$

in the original system, the nth equation of which becomes

$$P_n(t + \Delta t) = P_n(t - \Delta t) + 2 \Delta t[(n - 1)\lambda P_{n-1}(t) - n(\lambda + \mu)P_n(t) \\ + (n + 1)\mu P_{n+1}(t)]. \tag{2-59}$$

Thus one starts by putting $t = 0$ and $\Delta t = 1$. For the next sum one puts $t = 1$ and $\Delta t = 1$, etc. Note that the probabilities are zero for negative values of the argument except that $P_2(-1) = 1$.

In these calculations one encounters two types of errors: (1) rounding errors in ignoring the remainder term, using the series approximation, and (2) truncation errors in approximating the derivative by the first differences. It is not the purpose here to go more deeply into this matter; it may be continued by the interested reader through numerical-analysis techniques which have become sophisticated and highly useful. It is clear, however, that something analogous to the above procedure can be used.

Remark: G. D. Camp [94] has discussed various methods for putting bounds on solutions of queueing problems. This is another interesting idea which is useful when it is clear what type of answer is required. Knowledge of the behavior of a solution with upper and lower bounds on its values constitutes valuable information. Both these ideas will be encountered in the text.

PROBLEMS

1. Using the Pollaczek-Khintchine formula, derive the expected waiting number in the system and in line and the corresponding waiting times for the exponential-service-time case.

2. Solve the system

$$qp_1 + qp_2 = p_1,$$
$$pp_{j-1} + qp_{j+1} = p_j, \qquad j = 2, \ldots, c-1,$$
$$pp_{c-1} + pp_c = p_c,$$

where

$$\sum_{j=1}^{c} p_j = 1$$

and

$$p + q = 1.$$

Obtain

$$p_j = \frac{1 - p/q}{1 - (p/q)^c} \left(\frac{p}{q}\right)^{j-1} \qquad \text{if } p \neq q,$$

$$p_j = \frac{1}{c} \qquad \text{if } p = q.$$

3. Using a table of random numbers, a Poisson-input distribution with an arrival rate of one per unit time (i.e., exponential interarrival distribution), and a constant service rate of two per unit time, with a single channel and an ordered queue, simulate the operation for 100 times, obtaining the average waiting time and the average idle time for the channel. Compare the answer with that obtained from the Pollaczek-Khintchine formula.

4. To derive the exponential distribution, consider an urn with a large number of balls of diameter d of which B are black balls and A are white balls. Given that a black-ball has been drawn, the probability that the distance x between its center and the center of the next black ball to be drawn will be not more than rd is given by

$$\text{Prob } (x \leq rd) = 1 - (1 - b)^r, \qquad b = \frac{B}{A + B}.$$

If the cumulative probability density function of x is $F_s(x)$, let $d = 1/s$ and $sb = \lambda$; as $b \to 0$ and $s \to \infty$, so that λ and x remain unchanged, then

$$F_s(x) = 1 - \left(1 - \frac{\lambda}{s}\right)^{sx} \to 1 - e^{-\lambda x}.$$

Interpret λ as the expected frequency of events per unit interval of distance.

CHAPTER 3

PROBABILITY, MARKOFF CHAINS AND PROCESSES, ERGODIC PROPERTIES OF QUEUES

3-1. Introduction

To help the reader, this chapter gives in outline form various ideas from probability and from stochastic processes. With the possible exception of theorems on the ergodicity of queues and examples of waiting-time computations, most of the material of this book does not depend on what is said here. The reader will profit by examining introductory texts on probability for the development of the ideas in addition to what is presented below. A teacher might provide guidelines for greater elaboration. Most of the equations derived in various parts of the book, however, will not require a great deal of familiarity with probability.

Experience with a few special cases of the birth-death process such as we demonstrated in Erlang's model in the previous chapter should give adequate insight when combined with a good understanding of the queueing system under study. The idea of a Markoff chain and a Markoff process is fundamental and useful. Although Markovian processes are encountered in queueing theory, one must not conclude that all queueing processes are Markovian. This property, which requires that the history of the randomly evolving system has no effect on its future development, makes the problem of analyzing a queue simpler because of the manageable functional equations to which it leads. But non-Markovian properties have also been investigated.

The outline that we give here briefly examines the notions of an event, a Boolean algebra, a σ algebra, a measure space, and probability. The concepts of conditional probability and independence are also introduced. This is followed by the definition of a random variable. Useful ideas from probability are then briefly given. The reader is urged to expand and obtain proofs and examples of the facts given, which, because of space limitations, are omitted. Formulas of several useful distributions are also given. Then the notion of a stochastic process is again introduced, after which Markoff chains and Markoff processes are discussed, together with Feller's classification of states and theorems related to this subject.

The ideas are then applied to queues. Finally, a few examples of a stochastic process are given. A very useful one among these is the birth-death process, which we treat in the following chapter, because of its many applications to queues.

3-2. Elements of Probability

Probability theory provides quantitative description of occurrence of events. To understand the basic concepts, one must begin with the events themselves. Events may be abstractly regarded as point sets. They may be combined to yield other events. For two events a and b, $a \cup b$ is used to denote an event which occurs if, and only if, at least one of a or b occurs, and $a \cap b$ is an event whose occurrence requires the occurrence of both a and b. The event a' is complementary to a; it occurs if, and only if, a does not occur. The sure event is denoted by e and the null event by 0.

The only reason for defining an algebraic structure here is that (1) an event can be a combination of other subevents, and (2) one requires rules for manipulating combinations of events defined by "or" and "and."

A system of elements (in this case events) 0, a, b, . . . , e in which the operations a', $a \cup b$, $a \cap b$ are defined is a Boolean algebra if the following relations are satisfied:

$$
\begin{aligned}
0 \cap a &= 0 & 0 \cup a &= a \\
e \cap a &= a & e \cup a &= e \\
a \cap b &= b \cap a & a \cup b &= b \cup a \\
(a \cap b) \cap c &= a \cap (b \cap c) & (a \cup b) \cup c &= a \cup (b \cup c) \\
a \cap (b \cup c) &= (a \cap b) \cup (a \cap c) & a \cup (b \cap c) &= (a \cup b) \cap (a \cup c) \\
0' &= e & e' &= 0 \qquad (a')' = a \\
(a \cap b)' &= a' \cup b' & (a \cup b)' &= a' \cap b' \\
a \cap a' &= 0 & a \cup a' &= e
\end{aligned}
$$

This definition is adequate for finite events. For an infinite number of events we require the notion of a σ algebra. One first defines $a \subset b$, that is, a is contained in b or b implies a (e.g., the collection of elements $\{1,2,3\}$ is contained in $\{1,2,3,4\}$). In addition to the above relations, a σ algebra requires that in any set which contains an infinite sequence of events $\{a_n\}$ there be a smallest one given by $a \equiv a_1 \cup a_2 \cup \cdots$. Thus when we include this "countable-additivity" property in the definition of a Boolean algebra we have a Boolean σ algebra.

We also have a Boolean algebra if the following holds: $a \cap a = a$; $a \cup a = a$; $a \subset a$; $a \subset b$ and $b \subset a$ imply $a = b$; $a \subset b$ and $b \subset c$ imply $a \subset c$; and the three conditions $a \subset b$, $a \cap b = a$, $a \cup b = b$ are equivalent.

Probability is a numerically valued function $P(a)$ of the elements of a Boolean σ algebra. This function must represent the proportion of times an event is expected to occur. Thus $0 \leq P(a) \leq 1$, and one must have $P(a) = 0$ for $a = O$ and $P(a) = 1$ for $a = e$. Also

$$P(a \cup b) = P(a) + P(b)$$

if $a \cap b = O$, and, by induction,

$$P(a_1 \cup \cdots \cup a_n) = P(a_1) + \cdots + P(a_n)$$

for $a_i \cap a_j = O$ when $i \neq j$. If $\{a_n\}$ is an infinite set, then we require countable additivity to hold for a probability function, i.e.,

$$P(a_1 \cup a_2 \cup \cdots) = P(a_1) + P(a_2) + \cdots,$$

where $a_i \cap a_j = O$, $i \neq j$.

The foregoing properties also apply to a general measure function with minor modification (e.g., it need not lie between zero and unity). Thus numerical probability is a measure function; i.e., it is a finite, nonnegative, and countably additive function of elements of a Boolean σ algebra, and in addition $P(a) = 1$ is equivalent to $a = e$.

As Halmos points out, if B is any Boolean σ algebra and P a probability measure on B, there exists a measure space Ω (the triple: sure event e, σ algebra B, and probability measure) such that the system B is abstractly identical with an algebra of subsets of Ω reduced by identification according to sets of measure zero, and the value of P for any event is identical with the values of the measure for the corresponding subsets of Ω. Thus measure and probability are identical, and the theory of measure and integration carries over. In probability theory the measure of the whole space is finite and scaled to unity.

We now define conditional probability and the independence of events. We define

$$P_b(a) \equiv \frac{P(a \cap b)}{P(b)}, \tag{3-1}$$

where $P_b(a)$ indicates the probability of occurrence of a given that b has occurred. Of course, if the occurrence of b does not affect the occurrence of a we have

$$P_b(a) = P(a).$$

Hence a and b are independent if, and only if,

$$P(a \cap b) = P(a)P(b). \tag{3-2}$$

The following are useful formulas.

1. The law of addition of probabilities:

$$P(a_1 \cup \cdots \cup a_n) = \sum_{i=1}^{n} P(a_i) - \sum_i \sum_j P(a_i \cap a_j)$$

$$+ \sum_i \sum_j \sum_k P(a_i \cap a_j \cap a_k) - \cdots \pm P(a_1 \cap a_2 \cap \cdots \cap a_n).$$

2. The law of total probability:

$$P(a \cap b) = \sum_i P(b_i)P(a|b_i),$$

where $P(a|b_i)$ is the conditional probability of occurrence of a, given that b_i has occurred with probability $P(b_i)$, and $b = \bigcup_i b_i$, where the b_i are independent.

A random variable is a measurable function defined on a measure space with total measure equal to unity.[1] A function $x(\omega)$ defined on Ω is said to be measurable if, and only if, the ω sets for which $\alpha \leq x(\omega) \leq \beta$ belong to the basic σ algebra B of Ω.

The distribution function $F(x)$ of a random variable X is a function of a real variable defined for each real value x to be the probability that $X \leq x$, that is, $F(x) = \text{Prob } (X \leq x)$. The values x may be discrete or continuous. We shall treat only the continuous case since we have already given a brief description of the discrete case. The reader should apply the ideas of the discrete case to calculating $F(x)$ for the case of casting two dice, for example.

The function $F(x)$ (which is continuous on the right) is nondecreasing and tends to zero as $x \to -\infty$ and to unity as $x \to \infty$. The density function $f(x)$ of a random variable X is defined by

$$f(x) = \frac{dF(x)}{dx} \tag{3-3}$$

which exists whenever $F(x)$ is absolutely continuous. In that case,

$$F(x) = \int_{-\infty}^{x} f(y) \, dy.$$

We define the characteristic function of $F(x)$ by the Riemann-Stieltjes integral:

$$\varphi(t) \equiv E(e^{itx}) = \int_{-\infty}^{\infty} e^{itx} \, dF(x), \qquad i = \sqrt{-1}, \tag{3-4}$$

which exists for real t but need not exist for imaginary t. When the

[1] P. R. Halmos, The Foundations of Probability, *Am. Math. Monthly*, vol. 51, pp. 493–510, 1944.

moment-generating function exists it is given by

$$M(t) = \int_{-\infty}^{\infty} e^{tx}\, dF(x),$$ (3-5)

which is the Laplace-Stieltjes transform of $F(x)$. We have $M(t) = \varphi(it)$ when the left member exists.

The characteristic function and the moment-generating function are important tools to compute moments of distributions, to assist in finding the distribution of sums of random variables, and to study limits of sequences of distributions [see (3-8) and Prob. 3]. This will be exemplified.

The characteristic (moment-generating) function of the sum of independently distributed random variables is the product of their characteristic (moment-generating) functions. Alternatively, if $Z = X + Y$, where X and Y are independently distributed according to probability functions $F(x)$ and $G(y)$, respectively, the probability function of Z is given by

$$H(z) = \int_{-\infty}^{\infty}\int_{-\infty}^{z-y} dF(x)\, dG(y) = \int_{-\infty}^{\infty} F(x)\, \Big|_{-\infty}^{z-y}\, dG(y)$$

$$= \int_{-\infty}^{\infty} F(z-y)\, dG(y).$$ (3-6)

If the corresponding density functions exist, we have

$$h(z) = \int_{-\infty}^{\infty} f(z-y)g(y)\, dy.$$ (3-7)

If the distributions are defined only for positive values of the variables and are otherwise zero, then

$$H(z) = \int_{0}^{\infty} F(z-y)\, dG(y) = \int_{0}^{z} F(z-y)\, dG(y).$$ (3-8)

One can obtain the density function by differentiation with respect to z, provided that it exists. [Note that the integral

$$\int_{0}^{x} f(x-y)g(y)\, dy$$ (3-9)

is called the convolution of f and g.] The foregoing procedure can be generalized to the sum of several variables, as we shall see in a later chapter. In the discrete case, one can introduce generating functions to obtain similar results.

From the characteristic function one can recover the distribution $F(x)$ as indicated by Doob [161]:

$$\frac{1}{2}\,[F(x+0) - F(x-0)] - \frac{1}{2}\,[F(0+0) - F(0-0)]$$

$$= \frac{1}{2\pi} \int_{-\infty}^{\infty} \varphi(t)\, \frac{1 - e^{-itx}}{it}\, dt,$$ (3-10)

where, for example, $x + 0$ indicates that the argument tends to x through values larger than x. The moments (the first of which is the mean μ_1) about the origin are given by the nth derivative of φ at the origin:

$$\mu_n = E(X^n) = (-i)^n \varphi^{(n)}(0), \qquad n = 0, 1, \ldots .$$

In addition to the moments, the cumulants κ_r constitute another set of useful parameters. They are defined by the identity

$$\exp\left(\kappa_1 t + \frac{\kappa_2 t^2}{2!} + \cdots + \frac{\kappa_r t^r}{r!} + \cdots\right) = 1 + \mu_1' t + \frac{\mu_2' t^2}{2!}$$
$$+ \cdots + \frac{\mu_r' t^r}{r!} + \cdots,$$

where $\mu_r' \equiv E[(X - a)^r]$ is the rth moment about an arbitrary point a. If we replace t by it, the right side is the series expansion of $\varphi(t)$. Thus μ_r' is the coefficient of $(it)^r/r!$ in $\varphi(t)$, whereas κ_r is the coefficient of $(it)^r/r!$ in the expansion of $\log \varphi(t)$, that is,

$$\kappa_r = (-i)^r \frac{d^r}{dt^r} \log \varphi(t) \Big|_{t=0}.$$

Thus
$$\mu_1' = \kappa_1, \qquad \mu_2' = \kappa_2 + \kappa_1^2.$$

For moments about the mean μ_1 we have

$$\kappa_1 = 0, \qquad \mu_2 = \kappa_2.$$

For a Poisson distribution whose characteristic function is $\exp[\lambda(e^{it} - 1)]$, for example, all the cumulants are equal to λ.

We illustrate some of the foregoing ideas with an example and then proceed to apply the results of the example to the calculation of waiting times for a Poisson queue.

Example 1: Show that the Poisson process may be derived by considering the distribution of the sum of independently and identically distributed variables t_i $(i = 1, \ldots, n)$ with corresponding exponential distributions $\mu e^{-\mu t_i}$.

The moment-generating function of this exponential density is

$$M(\theta; t_i) = \int_0^\infty e^{\theta t_i} \mu e^{-\mu t_i}\, dt_i = \frac{\mu}{\mu - \theta}.$$

If we consider each t_i as the service time of the ith item at a single-channel queue with ordered service, we are asking for the waiting time in the line of the nth item, given that there are n items in the line and one is just beginning service.

To obtain the distribution of the sum of the variables we take the product of their moment-generating functions and find the density function corresponding to the result. The product is $(1 - \theta/\mu)^{-n}$; hence the density is

$$\mu \frac{(\mu t)^{n-1} e^{-\mu t}}{(n - 1)!}, \qquad t \geq 0.$$

Exercise: Show that the mean of this distribution is n/μ and its variance is n/μ^2.

The cumulative distribution with the above density function gives the probability that the nth item waits for a time less than t. It is desired to compute the probability $P_n(\tau)$ to serve exactly n items during τ. Let $\tau = t + (\tau - t)$. The probability of n items being served during time t is derived above and the probability of serving none during time $\tau - t$ is $e^{-\mu(\tau-t)}$, since the probability of being served during time x is $1 - e^{-\mu x}$ and that of not being served during x is $1 - (1 - e^{-\mu x}) = e^{-\mu x}$. Thus

$$P_n(\tau) = \int_0^\tau \mu \frac{(\mu t)^{n-1}}{(n - 1)!} e^{-\mu t} e^{-\mu(\tau-t)} \, dt = \frac{(\mu \tau)^n}{n!} e^{-\mu \tau},$$

which is the Poisson process.

Example 2: Compute the waiting times in line and in the system for Erlang's model. As an application of the foregoing ideas, in Example 1, we compute the waiting-time distribution for Erlang's model directly. We compute the waiting-time distribution in the line and in the system as follows:

An arriving item is $(n + 1)$st in the system if it finds n items ahead of it. As we have seen, the probability of this event is $\rho^n(1 - \rho)$. The waiting time of the arriving unit is a random variable. It is the sum of the service-time random variables of the n items ahead. We know that all the service-time distributions are identical exponential distributions with parameter μ. Because of the "forgetfulness" property of the exponential distribution, we also know that the remaining service time of the item in service satisfies the same law as the service time of the unserved items.

From the analysis of the previous paragraphs, we have for the waiting-time (in queue) density function of the $(n + 1)$st item

$$w_n(t) \equiv \frac{\mu(\mu t)^{n-1} e^{-\mu t}}{(n - 1)!}. \tag{3-11}$$

Since the item considered is arbitrary and may occupy any position in line, to obtain the probability of waiting between time t and $t + dt$ we multiply the probability of n items in the system by the waiting time of the $(n + 1)$st item and sum over $n \geq 1$ (why?). This gives for $t > 0$

$$w(t) \equiv \sum_{n=1}^{\infty} \rho^n(1 - \rho)\mu \frac{(\mu t)^{n-1} e^{-\mu t}}{(n - 1)!} = \mu\rho(1 - \rho)e^{-(1-\rho)\mu t}. \tag{3-12}$$

a. The average waiting time in line is obtained by multiplying the expression on the right by t and integrating from zero to infinity, yielding

$$W_q = \frac{\rho}{\mu(1 - \rho)}. \tag{3-13}$$

b. The cumulative distribution $P(\leq t)$ which gives the probability of waiting no more than time t is obtained as follows:

$$P(\leq t) = 1 - P(> t) = 1 - \int_t^\infty w(x)\, dx = 1 - \rho e^{-(1-\rho)\mu t}. \tag{3-14}$$

Note that $w(0)$ has the value $(1 - \rho)\delta(t)$, where $\delta(t)$ is the Dirac delta function, which is zero everywhere except for an infinite spike at the origin, with the result that its integral over the infinite range is unity.

Exercise: Show that the cumulative distribution for waiting in the system rather than in the line is given by

$$1 - e^{-(1-\rho)\mu t}.$$

Find the average waiting time (i.e., stay in system).

Example 3: Compute the number in the system and the waiting times for limited waiting room for Erlang's model. If the number in the system may not exceed N (i.e., the waiting room is limited), the equations of Erlang's model apply up to $n = N$ with slight modification in the last equation of the model. One easily obtains the solution of that case by adjusting the solution for $N = \infty$.

$$p_n = \frac{p_0(\lambda/\mu)^n}{p_0 \displaystyle\sum_{n=0}^{N} (\lambda/\mu)^n} = \frac{1 - \lambda/\mu}{1 - (\lambda/\mu)^{N+1}} \left(\frac{\lambda}{\mu}\right)^n, \qquad n = 0, \ldots, N. \tag{3-15}$$

One can also write the solution for $\lambda = \mu$, using L'Hospital's rule. As $N \to \infty$, one obtains the probabilities for an unrestricted queue length with $\lambda < \mu$.

For the waiting-time density we have

$$w(t) = \sum_{n=0}^{N} p_n w_n(t).$$

The average waiting time (in queue) for $\lambda < \mu$ then is given by

$$W_q = \sum_{n=0}^{N} p_n \int_0^\infty t w_n(t)\, dt = \frac{1}{\mu} \sum_{n=1}^{N} n p_n$$

$$= \frac{\rho[1 - (N + 1)\rho^N + N\rho^{N+1}]}{\mu(1 - \rho^{N+1})(1 - \rho)}, \qquad \rho = \frac{\lambda}{\mu}, \tag{3-16}$$

which also gives the well-known result for the average waiting time when no restriction on the waiting room is imposed, that is, $N \to \infty$. Note that (3-16) equals L/μ.

Exercise: Verify this expression for W_q and evaluate it as $N \to \infty$. Obtain $P(\leq t)$ from $w(t)$.

Example 4: A slight modification of the relation given in the first chapter between the waiting times of the nth and $(n + 1)$st units, to include the possibility of no waiting, gives

$$w_{n+1} = \max (w_n + s_n - t_n, 0).$$

We wish to calculate the waiting-time distribution $P(\leq t)$ of a unit, in the steady state. Let us suppose that $t > 0$; that is, the $(n + 1)$st customer will wait. We seek

$$\text{Prob } (w_{n+1} < w) = \text{Prob } (w_n + s_n - t_n < w)$$
$$= \text{Prob } (w_n < w - s_n + t_n).$$

We denote by $P_{n+1}(w)$ and $P_n(w)$ the distributions of waiting times in queue of the $(n + 1)$st and nth unit. Note in the expression on the right that the variables t_n and s_n are independently distributed according to the distribution functions $A(x)$ and $B(y)$. In addition, the waiting time of the nth unit is independent of its own service time and of the interarrival interval of the $(n + 1)$st unit.

Accordingly, the probability of the events that the waiting time w_n satisfies the above inequality, that the interarrival time t_n and the service time s_n have lengths $x \leq t_n < x + dx$, $y \leq s_n < y + dy$, equals

$$P_n(w + x - y) \, dA(x) \, dB(y).$$

By summing all these probabilities we have [591]

$$P_{n+1}(w) = \int_0^\infty \int_0^\infty P_n(w + x - y) \, dA(x) \, dB(y).$$

This formula is also obtained by considering the distribution of the sum of three independently distributed random variables. But in the steady state all units have the same waiting-time distribution $P(w)$, as will be discussed in Chap. 9; hence

$$P(w) = \int_0^\infty \int_0^\infty P(w + x - y) \, dA(x) \, dB(y).$$

We now let the service time be of constant length equal to unity. Then

$$B(y) = \begin{cases} 0, & y < 1, \\ 1, & y > 1, \end{cases}$$
$$B(1 + 0) - B(1 - 0) = 1,$$

and
$$P(w) = \int_0^\infty P(w + x - 1) \, dA(x),$$

where
$$P(w) = 0 \quad \text{for } w < 0.$$

If $w < 1$, then, since P vanishes for negative argument, we have

$$P(w) = \int_{1-w}^{\infty} P(w + x - 1) \, dA(x).$$

If we write (integrating by parts)

$$\gamma(s) = \int_0^{\infty} e^{-sw} \, dP(w) = s \int_0^{\infty} e^{-sw} P(w) \, dw$$

for the Laplace-Stieltjes transform of $P(w)$, put $A(x) = 1 - e^{-\lambda x}$, and substitute for $P(w)$, recalling the last remark above, we have

$$\gamma(s) = s \int_0^1 e^{-sw} \, dw \int_{1-w}^{\infty} \cdots + s \int_1^{\infty} e^{-sw} \, dw \int_0^{\infty} \cdots,$$

which, from the definition of the transform, yields an expression that simplifies to

$$\gamma(s) = \frac{se^{-\lambda}\gamma(\lambda)}{s - \lambda + \lambda e^{-s}}.$$

Let $s \to 0$; then $\gamma(s) \to 1$ and $e^{-\lambda}\gamma(\lambda) \to 1 - \lambda$. Thus finally

$$\gamma(s) = \frac{(1 - \lambda)s}{s - \lambda + \lambda e^{-s}}.$$

The steady-state assumption on the traffic intensity requires $\lambda < 1$.

The inverse transform may be obtained from the formula

$$P(w) = \frac{1}{2\pi i} \int_C e^{st} \gamma(s) \frac{ds}{s},$$

where C extends from $-i\infty$ to $i\infty$, bypassing the origin by a small semicircle on the right. By substituting for $\gamma(s)$, moving the integration path to

$$R(s) = \lambda + \delta > \lambda,$$

which is permissible since $\gamma(s)$ is regular for $R(s) \geq 0$, and developing the denominator in geometric series, one has

$$P(w) = (1 - \lambda) \sum_{n=0}^{[t]} \frac{\lambda^n (n - t)^n}{n!} e^{\lambda(t-n)},$$

where $[t]$ is the largest integer less than t. The residue for the terms indicated by the sum is obtained at the pole of the integrand $s = \lambda$, whereas the integral of the rest of the series, where $t - [t] - 1 < 0$, is zero, as is known by extending the path of integration to a semicircle in the right half plane and allowing its radius to become infinite (see Chap. 9 for a description of a similar problem).

Suppose that we have several random variables X_i $(i = 1, \ldots, N_t)$ that are independently, identically, and continuously distributed according to $F(x)$. Also suppose that N_t is also a random variable; i.e., the number of variables itself is a random variable. We ask for the distribution of their sum $S_{N_t} = X_1 + \cdots + X_{N_t}$. Some reflection will show the following to hold (called a compound distribution) if N_t is independent of the X_i:

$$\text{Prob } (S_{N_t} \leq x) = \sum_{n=0}^{\infty} \text{Prob } (N_t = n) \text{ Prob } (S_n \leq x).$$

Taking the moment-generating function of the above equation, we have for this function

$$\sum_{n=0}^{\infty} \text{Prob } (N_t = n) \left[\int_{-\infty}^{\infty} e^{-\theta x} \, dF(x) \right]^n.$$

We take the nth power of the moment-generating function since we have a convolution of n variables. But the result is the generating function $P(z)$ of the discrete probabilities Prob $(N_t = n)$ except that, instead of using z as in Chap. 2, we have used

$$z = \int_{-\infty}^{\infty} e^{-\theta x} \, dF(x) \equiv M(\theta).$$

Hence the moment-generating function of Prob $(S_{N_t} \leq x)$ is given by $P[M(\theta)]$. The mean is obtained by differentiating with respect to θ and putting $\theta = 0$. Thus

$$E(S_{N_t}) = \frac{dP}{dM} \bigg|_{M=1} (-1) \frac{dM}{d\theta} \bigg|_{\theta=0} = E(N_t)E(X),$$

where X is any of the X_i since they all have the same mean. The derivation of the variance is left as an exercise to the reader. He should obtain

$$\text{var } (S_{N_t}) = \text{var } (N_t)E^2(X) + \text{var } (X)E(N_t).$$

These ideas will be used in renewal theory.

3-3. Some Useful Distributions

We give in Table 3-1 a list of distributions and some of their corresponding characteristic functions. For a better understanding of some of these distributions, the reader is referred to standard texts on probability.

Pearson Type-III distribution is a general type of distribution obtained by Pearson in an attempt to classify distributions. By properly specializing its parameters, one obtains known distributions belonging to that type. The gamma, chi-square, Erlangian, and exponential distributions are of this type.

TABLE 3-1

Name of distribution	Expression for density function	Characteristic function
Uniform or rectangular	$\dfrac{1}{b-a}$ $a < x < b$	$\dfrac{e^{ibt} - e^{iat}}{i(b-a)t}$
Binomial.............	$\dfrac{n!}{(n-k)!k!} \, q^{n-x}p^x$ $p + q = 1$	$(q + pe^{it})^n$
Hypergeometric........	$\dfrac{\dbinom{N_1}{k}\dbinom{N_2}{n-k}}{\dbinom{N}{n}}$ $(N_1 + N_2 = N)$	
Poisson..............	$\dfrac{\mu^k e^{-\mu}}{k!}$	$e^{\mu(e^{it}-1)}$
Normal..............	$\dfrac{1}{\sigma\sqrt{2\pi}} e^{-[(y-\mu)^2/2\sigma^2]}$	$e^{i\mu t - \sigma^2 t^2/2}$
log Normal...........	$\dfrac{1}{\sigma\sqrt{2\pi}}\dfrac{1}{y-a} \exp\left\{-\dfrac{1}{2\sigma^2}[\log(y-a)-\mu]^2\right\}$ where $a \leq y$	
Pearson Type III......	$A(y-\mu)^{a-1}e^{-b(y-\mu)}$	
Gamma..............	$\dfrac{b^a y^{a-1}e^{-by}}{\Gamma(a)}$	$\left(1 - \dfrac{it}{b}\right)^{-a}$
Chi-square...........	$\dfrac{1}{2^{n/2}\Gamma(n/2)} y^{(n/2)-1}e^{-y/2}$	
Erlangian (the time necessary to observe k phenomena)	$\dfrac{1}{(k-1)!}(k\mu)^k y^{k-1}e^{-k\mu y}$	
Exponential..........	$\mu e^{-\mu y}$	
Pascal or negative binomial	$\dbinom{n+k-1}{k} p^k q^n$	$\left(\dfrac{q}{1 - pe^{it}}\right)^n$

Chi square is given by

$$\chi^2 = X_1^2 + \cdots + X_k^2, \qquad (3\text{-}17)$$

where the X_i are independently distributed according to the normal distribution with zero mean and unit variance. The number k is called the degrees of freedom of the distribution. The Erlangian distribution is a gamma distribution with $2k$ degrees of freedom.

The presence of the parameter k in the Erlangian distribution, which, as we have seen (in modified form) earlier in the chapter, is obtainable as the distribution of the sum of identically exponentially distributed random variables each with mean $1/k\mu$, is of important practical use. First we note that it yields the exponential distribution for $k = 1$ and a constant distribution as $k \to \infty$, that is, from randomness to certainty.

Many important distributions occurring in real problems lie between the two cases and may be approximated by the Erlangian distribution on a proper choice of value of k. Hence, if queueing models are solved with the Erlangian distribution, the results can be applied to those cases where the choice of k results in a good fit.

A modified Erlangian input (to a single channel) in which k is replaced by c and $k\mu$ by λ may be regarded as a filtered Poisson input with parameter λ to c parallel channels, each with its queue before it. The first arrival joins the first-channel queue, the second arrival joins the second-channel queue, etc., the cth arrival joins the cth-channel queue and then the $(c + 1)$st arrival joins the first-channel queue, the $(c + 2)$nd the second-channel queue, etc., in that order. The input to each channel would be modified Erlangian of the above form. This idea will be encountered later on in the book.

As an exercise the reader should attempt to compute the first and second moments of these distributions. Then compute the variance $\sigma^2 =$ (second moment) $-$ (mean)$^2 = \mu_2 - \mu_1^2$, where σ is the standard deviation.

For many purposes it is useful to know that the sample mean from an arbitrary distribution approaches normality in the sense of the central-limit theorem. Specifically, if a population has a finite variance σ^2 and mean μ, the distribution of the sample mean approaches the normal distribution with variance σ^2/n and mean μ as the sample size n increases.

3-4. Stochastic Process

A process is a system which, when evolving in time, is continually subjected to chance effects. We shall generally be concerned with a system that occupies a number of states $E_1, E_2, \ldots, E_m$, at the instants of time $t_1, \ldots, t_n$. At any of these time instants there is a given probability that the system will pass from the state it is in, e.g., state E_i, to any other state.

We have already indicated that a stochastic process is defined by a family of random variables $\{X_t\}$ depending on a parameter t. We also indicated that the number of parameters in the subscript can be more than one; however, we restricted our attention to one parameter, namely, time. If the time is considered in discrete units, such as is the case with the tossing of a coin, we write $X_1, X_2, \ldots$. Each X_i has a probability distribution which may itself be discrete or continuous.

Coin tossing is an example of a finite or countable family of random variables, each with two outcomes. In tossing a coin we have two possible outcomes at the time of a toss, and each of the two outcomes has a probability of occurrence. For example, Prob $(X_i = 1) = p$ gives the probability of a head, which we label by the number 1. The probability of a tail, which we denote by -1, is Prob $(X_i = -1) = q$. We must also have $p + q = 1$, since one of the outcomes must occur.

Note that the probabilities do not change in time. They are the same for all i. This is an illustration of a *stationary process* in which probabilities are independent of time (i.e., of the trials).

One may be interested[1] here in the number of heads obtained in n throws for large n. When this number is divided by n, and n is allowed to tend to infinity, one has

$$\lim_{n \to \infty} \frac{X_1 + \cdots + X_n}{n} = \lim_{n \to \infty} \frac{S_n}{n}.$$

The law of large numbers asserts that

$$\text{Prob}\left\{ \left| \frac{S_n}{n} - p \right| < \varepsilon \right\} \to 1,$$

as n increases, for arbitrary $\varepsilon > 0$. That is, the probability itself improves with increased trials.

Another interesting question is to ask how fast $S_n/n - p$ tends to zero. The answer is given by the law of the iterated logarithm:

$$\text{Prob}\left\{ \limsup_{n \to \infty} \frac{\sqrt{n}(S_n/n)}{\sqrt{2pq \log \log n}} = 1 \right\} = 1.$$

We also have from the central-limit theorem that for large n, $(S_n - np)/\sqrt{npq}$ is approximately normally distributed.

The number of possible outcomes of X_t for each t need not be finite, as in the above example, or even countable. It may have a continuum of outcomes such as might be the case in randomly selecting real numbers. Any number may be chosen when X_t is a continuous random variable.

Finally, the index t may assume continuous values, with X_t having continuous or discrete distributions at each value. The Poisson process is an illustration of the discrete-distribution case where the parameter itself can assume any value. The number of telephone subscribers initiating a call at any time has a discrete probability distribution which at any time t of the day assigns a probability to the event of one subscriber initiating a call, two subscribers, etc. These quantities are determined experimentally by noting over many days at a fixed time how many subscribers have initiated a call at that time and computing the frequency. Even though it may not be practically meaningful, one

[1] J. L. Doob, What Is a Stochastic Process, *Am. Math. Monthly*, vol. 49, pp. 648–653, 1942.

frequently allows t to assume values from $-\infty$ to $+\infty$ and then obtains the required answer on appropriate interpretation of the results.

Suppose now that t assumes discrete values and the experiment has a set of finite outcomes called states, denoted by E_i ($i = 1, \ldots, m$). The coin-tossing problem is an example in which $m = 2$. We attempted to answer various interesting questions for that case. But there was no dependence from one run of the experiment to another; hence the probability of occurrence of an outcome was independent of whether it previously occurred or not.

Suppose now that, with Markoff, we introduce a weak type of dependence which gives the transition probability p_{ij} of being in state E_j, on the nth trial of the experiment, given that E_i ($i = 1, \ldots, m$) has occurred at the previous trial. We must also specify probabilities $p_i(0)$ ($i = 1, \ldots, m$) to obtain the ith outcome on the first trial, i.e., at time zero. The resulting situation is more interesting. Note that the probabilities p_{ij} describe transitions $E_i \rightarrow E_j$ from one trial to the immediately following one.

We could complicate matters by introducing probabilities for transitions among nonconsecutive pairs of trials; i.e., future events may depend on the occurrence of events in some or all previous trials. There are many actual physical systems that require the latter description. In fact, one may require probabilities that must describe outcomes of several trials at a time.

In the simple case above, the initial probabilities a_i and the transition probabilities p_{ij} (which can be arranged in a square matrix P, called a stochastic matrix) define a Markoff chain (which is finite if the number of outcomes is finite).

The matrix P is written as

$$P = \begin{bmatrix} p_{11} & p_{12} & \cdots \\ p_{21} & p_{22} & \cdots \\ \cdots & \cdots & \cdots \end{bmatrix}.$$

In any case, the sum of each row of the matrix is unity, since from each state one must be able to make a transition in the next trial. The transition from a trial to the next does not preclude the possibility that the same state might occur again. If the sums of the elements in each column are also unity, the matrix is called doubly stochastic.

Note that, instead of occurring directly in consecutive trials, transitions from one state to another may occur by going through intermediate states in several trials. We use superscripts to denote the number of steps for the transitions to occur. We have

$$p_{ij}^{(1)} = p_{ij}, \qquad p_{ij}^{(2)} = \sum_k p_{ik} p_{kj}, \ldots, \qquad p_{ij}^{(n)} = \sum_k p_{ik} p_{kj}^{(n-1)}. \quad (3\text{-}18)$$

That is, the probabilities for being in any state at the $(n - 1)$st transition time are used together with the transition probabilities p_{ij} to obtain the probability of a transition leading from E_i to E_j in the nth transition time. Thus at any time the probability of the system being in a state depends only on the knowledge of the state of the system at the immediately preceding time. This is the essential property of Markoff chains and Markoff processes.

In terms of powers of the transition matrix, the last equation may be written as $P^n = PP^{n-1}$. In fact, one may write a more general relation of the form $P^{m+n} = P^m P^n$ for which representative elements satisfy $p_{ij}^{(n+m)} = \sum_k p_{ik}^{(m)} p_{kj}^{(n)}$.

Generally, one is interested in the probability

$$P_j(n) = \sum_{i=0}^{\infty} P_i(0) p_{ij}^{(n)}.$$

This is the probability of being in state j at time n regardless of which state one reached in the previous time. Now as $n \to \infty$ it is desirable to have an asymptotically stationary property; i.e., one has a limiting vector on the left and a limiting transition matrix in which the probabilities are independent of the state i on the right. This property is reasonable to expect on intuitive grounds. This type of property (the ergodic property) is useful for long trends in which the probabilities become independent of time. For many queueing problems such a result is of greater application value than one obtains from time-dependent probabilities.

We first illustrate the ideas with a finite Markoff chain with the transition matrix

$$P = \begin{bmatrix} \frac{1}{6} & 0 & \frac{1}{6} & 0 & \frac{2}{3} \\ 0 & \frac{1}{4} & 0 & 0 & \frac{3}{4} \\ \frac{1}{4} & \frac{3}{8} & \frac{1}{8} & \frac{1}{4} & 0 \\ 0 & 0 & 0 & 1 & 0 \\ 0 & \frac{1}{3} & 0 & 0 & \frac{2}{3} \end{bmatrix}.$$

A set of states in which there is a nonzero probability for every pair of states of a transition from one state to another is an equivalence class. We have three such classes here. They consist of (1) the first and third states, (2) the second and the fifth states, and (3) the fourth state by itself. For example, $p_{25} = \frac{3}{4}$, and $p_{52} = \frac{1}{3}$; hence one can stay in state 2 or 5 or go from state 2 to state 5 and return. No other state can be entered.

We note that it is possible to go from the first of these equivalence classes to the second and from the first to the third. Once a transition occurs to the second or third class, it is not possible to return to the first class. The first class is therefore unstable. It is called a *transient equivalence class*, and its states are transient states. The second and third equivalence classes are in equilibrium; i.e., once one of these classes is reached, it is not possible to leave for other classes. In addition to being closed classes, in that transitions from them to other classes are impossible (i.e., the p_{ij} are zero for i in the class and j not in it), the last two are called *ergodic equivalence* classes, and their states are called ergodic states. The third class, which consists of a single ergodic state, is called an absorbing state. A Markoff chain in which every ergodic state is absorbing is itself an *absorbing Markoff chain*. In such a chain, transitions in one or more steps are possible from any state to an absorbing state.

Note that, in general, the submatrix corresponding to transitions among an ergodic class is itself a stochastic matrix; i.e., the sum of p_{ij} in a row is unity. In general, a stochastic matrix is called *regular* if for some power of the matrix all its elements are positive. Thus if the submatrix corresponding to an ergodic class is regular (and thus its states are called regular), the class itself is called regular. Otherwise the ergodic equivalence class is *periodic* (with periodic states).

If P is the stochastic matrix of a regular chain, then for the nth power of P we have $\lim_{n \to \infty} P^n = T$ (we shall see in Chap. 5 how to compute functions of matrices), and T has identical positive elements in each column. Thus all row vectors of T are identical. If V is such a row vector, one can show that $\lim_{n \to \infty} W P^n = V$ for any probability vector W (a vector whose elements lie between 0 and 1 and whose sum is unity). The reader may attempt to verify this, using the definition of V. From this property we have the independence of the probability of being in a given state from the initial vector of a_i.

One also has $VP = V$, obtained from the definition of T on multiplying on the right by P. We have the fact that the average time to return to a state in a regular Markoff chain, having been there previously, is equal to the reciprocal of the limiting probability of being in that state.

In the case of absorbing states, the probability of reaching such a state, the average time for this to occur, and the average number of times the process will be in each transient state before absorption are of interest, and here the answer may depend on the initial probabilities a_i.

Note that in the foregoing example there were two classes of closed states contained in the chain. If there are no such closed classes in a Markoff chain, except of course for the entire class of states, the chain is then called irreducible. Otherwise it is called decomposable.

Let $f_{ij}^{(n)}$ denote the probability that, starting with state E_i at time zero, the system is for the first time in state E_j at time n. We then have

$$f_{ij}^{(n)} = p_{ij}^{(n)} - \sum_{k=1}^{n-1} f_{ij}^{(k)} p_{jj}^{(n-k)},$$

$$f_{ij}^{(1)} = p_{ij}.$$

Define the probability that the system will ever reach E_j by

$$F_{ij} = \sum_{n=1}^{\infty} f_{ij}^{(n)}.$$

Thus if E_j is the initial state, F_{jj} gives the probability of ever returning to E_j. To continue in this line, we define the mean first-passage time from E_i to E_j by

$$\mu_{ij} = \sum_{n=1}^{\infty} n f_{ij}^{(n)},$$

where μ_{ij} is measured in the units of basic trial time.

Feller's [202] classification of states for a Markoff chain with a denumerable number of states is as follows:

A state is recurrent if $F_{jj} = 1$ (i.e., return to it is certain).

A recurrent state is null if $\mu_{jj} = \infty$.

A state is transient if $F_{jj} < 1$ (i.e., return to it is uncertain).

A state is periodic with period t if return is possible only in $t, 2t, \ldots$ steps, where $t > 1$ is the greatest integer with this property and hence $p_{jj}^{(n)} = 0$ if n is not an integral multiple of t.

A recurrent, nonnull, nonperiodic (aperiodic) state is called an ergodic state.

Note that a finite Markoff chain can contain no null state, and it is impossible that all its states are transient. Thus if M is the number of states, $p_{ij}^{(n)} \geq 1/M$ for at least one j, since the sum of the probabilities over j is unity and there are M of them. Hence it is impossible that $p_{ij}^{(n)} \to 0$ as $n \to \infty$ for all j.

Feller gives theorems which characterize different types of states. We shall give some of the pertinent ideas from Foster's 1953 paper and also his application to queues where we give a proof.

We are interested in $\lim\limits_{n \to \infty} p_{ij}^{(n)}$.

Theorem 3-1: An irreducible aperiodic Markoff chain (called the system) is ergodic if there exists a nonnull solution of

$$\sum_{i=0}^{\infty} x_i p_{ij} = x_j, \qquad j = 0, 1, 2, \ldots.$$

Theorem 3-2: The system is ergodic if

$$\sum_{j=0}^{\infty} p_{ij} y_j \leq y_i^{-1}, \qquad i \neq 0,$$

has a nonnegative solution with $\sum_{j=0}^{\infty} p_{0j} y_j < \infty$.

Theorem 3-3: If the system is ergodic, the (finite) mean passage times d_j from the jth to the zero state satisfy

$$\sum_{j=1}^{\infty} p_{ij} d_j = d_i - 1$$

and

$$\sum_{j=1}^{\infty} p_{0j} d_j < \infty.$$

Theorem 3-4: The system is transient if, and only if, there exists a bounded nonconstant solution of the equations

$$\sum_{j=0}^{\infty} p_{ij} y_j = y_i.$$

Theorem 3-5: The system is recurrent if there exists a solution $\{y_i\}$ of

$$\sum_{j=0}^{\infty} p_{ij} y_j \leq y_i, \qquad i \neq 0,$$

such that $y_i \to \infty$ as $i \to \infty$.

Theorem 3-6: The system is transient if, and only if, there exists a bounded solution $\{y_i\}$ of the last set of inequalities such that $y_i < y_0$ for some i.

Theorem 3-7: If $\{p_n\}$ $(n = 0, 1, \ldots)$ is a probability distribution with $p_0 > 0$, then $\sum_{n=0}^{\infty} z^n p_n = z$ has a root ξ in $0 < \xi < 1$ if, and only if,

$$\sum_{n=1}^{\infty} n p_n > 1.$$

In the single-channel queueing system $M/G/1$ with Poisson input whose parameter is λ and arbitrary-service-time distribution is $B(t)$, let k_n be the probability of n new arrivals during a single service time. We have (see Chap. 7 if more discussion is desired)

$$k_n = \frac{1}{n!} \int_0^{\infty} e^{-\lambda t} (\lambda t)^n \, dB(t), \qquad n = 0, 1, \ldots . \qquad (3\text{-}19)$$

Note that the transition matrix is given by

$$[p_{ij}] = \begin{bmatrix} k_0 & k_1 & k_2 & \cdots \\ k_0 & k_1 & k_2 & \cdots \\ 0 & k_0 & k_1 & \cdots \\ 0 & 0 & k_0 & \cdots \\ \cdot & \cdot & \cdot & \cdot & \cdot & \cdot & \cdot & \cdot & \cdot & \cdot \end{bmatrix}.$$

Theorem 3-8: The system $M/G/1$ is ergodic if, and only if, the following holds:

$$\rho \equiv \sum_{n=1}^{\infty} n k_n < 1.$$

It is recurrent if, and only if, $\rho \le 1$.

Proof: If $\rho < 1$, define $y_j = j/(1 - \rho)$; then y_j satisfies Theorem 3-2.

If, on the other hand, the system is ergodic, it can be seen from the transition matrix that μ_{ij}, the mean first-passage time from state E_i to state E_j, satisfies

$$\mu_{i,i-1} = \mu_{10}, \qquad i \ne 0.$$

The state E_0 can only be reached from E_i via E_{i-1}.   Thus

$$\mu_{i0} = \mu_{i,i-1} + \mu_{i-1,0},$$

and by induction

$$\mu_{i0} = i\mu_{10}.$$

Then by Theorem 3-3 above

$$\mu_{10} \sum_{j=1}^{\infty} j p_{ij} = \mu_{10} - 1.$$

Therefore

$$\rho\mu_{10} = \mu_{10} - 1$$

and

$$\rho = 1 - \frac{1}{\mu_{10}} < 1.$$

If $\rho \le 1$, let $y_j = j$; then $y_j \to \infty$ as $j \to \infty$ and Theorem 3-5 is satisfied. Therefore the system is recurrent.   On the other hand, if $\rho > 1$, we apply Theorem 3-7 to $\sum_{n=0}^{\infty} k_n z^n = z$ and define $y_j = \xi^j$, which satisfies the condition of Theorem 3-4 and $y_j \to 0$ as $j \to \infty$, with $y_0 = 1$; by that theorem the system is transient.

Note that $\lim_{n \to \infty} p_{ij}^{(n)} = 0$ for $\rho \ge 1$ and for all i and j.   A system is non-dissipative if the row sums of the limiting probabilities are unity.   Otherwise it is dissipative.

For the system $GI/M/1$ one may also write the transition matrix and prove a similar theorem.   Note that a queueing process must be defined in terms of the Markoff chain having one-step transitions p_{ij} in order to apply the above ideas.   It is more difficult to treat the queueing process in continuous time.

If the transition probabilities p_{ij} are made to depend on a continuous parameter, the time parameter t, then we have a Markoff process which is an example of a continuous stochastic process.   The Poisson process and Erlang's model are other examples.   The ideas are analogous to the foregoing discussion of a Markoff chain.

One writes $p(j,t;i,s)$ for the transition probability to state j at time t, given that the system (set of outcomes) was in state i at time s.   Of course there are systems occurring in statistical mechanics in which information from all previous states has an effect on the outcome of a new state.

A one-dimensional Markoff process is described by the Chapman-Kolmogorov equation (a generalization of the equations describing Markoff chains):

$$p(j,t;i,s) = \sum_{k=0}^{\infty} p(j,t;k,u)p(k,u;i,s),\tag{3-20}$$

with $s < u < t$ and $i, j = 0, 1, 2, \ldots$.   This equation gives rise to two sets of equations:

1. The forward equations which are obtained by taking the derivative with respect to t and then replacing u by t
2. The backward equations which are obtained by taking the derivative with respect to s and then replacing u by s

Thus, for example, if transitions to state i at time t can occur only from the two neighboring states $i - 1$ and $i + 1$, as is the case in a birth-and-death process, the backward equations are given by

$$p_s(j,t;i,s) = p(j,t;i,s)p_s(i,s;i,s) + p(j,t;\, i - 1,\, s)p_s(i - 1,\, s;\, i,s) \\ + p(j,t;\, i + 1,\, s)p_s(i + 1,\, s;\, i,s), \quad (3\text{-}21)$$

where p_s is the derivative of p with respect to s.

If the transitions are of form $E_i \rightarrow E_{i+1}$ only, as in the case of a Poisson process, we have a pure birth process ($E_i \rightarrow E_{i-1}$ only, a pure death process).

In formulating the forward equations one considers the ways of reaching a state; for the backward equations one considers the ways of getting out of a state.   In general, a solution of one of these two sets, with initial conditions, also satisfies the other set.   Sometimes it is easier to solve the backward system than the forward system.

Example 5: As an illustration, we now show that the Poisson process, which is a Markoff process, satisfies the Chapman-Kolmogorov equation. On using the appropriate form of the Poisson process,

$$p(j,t;k,u) = e^{-\lambda(t-u)} \frac{[\lambda(t-u)]^{j-k}}{(j-k)!},$$

$$p(k,u;i,s) = e^{-\lambda(u-s)} \frac{[\lambda(u-s)]^{k-i}}{(k-i)!},$$

and substituting in the Chapman-Kolmogorov equation we have

$$e^{-\lambda(t-s)} \sum_{k=i}^{j} \frac{[\lambda(t-u)]^{j-k}}{(j-k)!} \frac{[\lambda(u-s)]^{k-i}}{(k-i)!}$$

$$= e^{-\lambda(t-s)} [\lambda(t-u)]^{j} [\lambda(u-s)]^{-i} \sum_{i}^{j} \left(\frac{u-s}{t-u}\right)^{k} \frac{1}{(j-k)!(k-i)!}.$$

Let $m = k - i$; then $k = m + i$, and we have

$$\sum_{m=0}^{j-i} \left(\frac{u-s}{t-u}\right)^{m+i} \frac{1}{(j-m-i)!m!}$$

$$= \frac{[(u-s)/(t-u)]^{i}}{(j-i)!} \sum_{m=0}^{j-i} \left(\frac{u-s}{t-u}\right)^{m} \frac{(j-i)!}{(j-i-m)!m!}$$

$$= \frac{1}{(j-i)!} \left(\frac{u-s}{t-u}\right)^{i} \left(1+\frac{u-s}{t-u}\right)^{j-i} = \frac{1}{(j-i)!} \left(\frac{u-s}{t-u}\right)^{i} \left(\frac{t-s}{t-u}\right)^{j-i}.$$

The first expression then becomes

$$\frac{e^{-\lambda(t-s)}}{(j-i)!} \frac{[\lambda(t-u)]^{j}}{[\lambda(u-s)]^{i}} \left(\frac{u-s}{t-u}\right)^{i} \left(\frac{t-s}{t-u}\right)^{j-i} = \frac{e^{-\lambda(t-s)}}{(j-i)!} [\lambda(t-s)]^{j-i}, \quad (3\text{-}22)$$

which is the desired result.

We finally note that the Poisson process is a stationary process; it depends only on the length of the time interval rather than on its beginning (i.e., its position on the time axis). Thus the Poisson process is a pure birth, a stationary (i.e., the transitions are stationary), and a Markovian process.

Since we shall usually be concerned with initial conditions (i.e., initial state E_i) at $t = 0$, we avoid the above cumbersome notation by writing $P_{ij}(t)$ for the probability of being in state E_j at time t, having been in state E_i at time zero. We have a transition matrix $P(t) = [P_{ij}(t)]$ such that

$$0 \leq P_{ij}(t), \qquad \sum_{j=0}^{\infty} P_{ij}(t) = 1, \qquad \lim_{t \to 0} P_{ij}(t) = \delta_{ij},$$

$$P_{ij}(s + t) = \sum_{k=0}^{\infty} P_{ik}(s)P_{kj}(t), \qquad 0 < s, t < \infty.$$

The case in which $\sum_{0}^{\infty} P_{ij}(t) < 1$ is called the degenerate case.

If we denote by $P_i(0)$ the probability that the system was originally in state i, then $\sum_{i=0}^{\infty} P_i(0) = 1$, and the probability that the system is in state j after time t is given by

$$P_j(t) = \sum_{i=0}^{\infty} P_{ij}(t)P_i(0), \qquad j = 0, 1, \ldots . \qquad (3\text{-}23)$$

With this notation, the coefficients of the matrix $P(t)$ satisfy the forward Chapman-Kolmogorov equation:

$$\frac{dP_{ij}(t)}{dt} = \sum_{k=0}^{\infty} P_{ik}(t)a_{kj}, \qquad i, j = 0, 1, 2, \ldots , \qquad (3\text{-}24)$$

where $A \equiv (a_{ij})$ is the matrix of transition probability constants ($a_{ij} \, dt$ is the probability of a transition from E_i to E_j during dt, and, hence, A is known as the infinitesimal transition matrix) such that

$$a_{ij} \geq 0, \qquad i \neq j, \qquad a_{ii} \leq 0, \qquad \sum_{j=0}^{\infty} a_{ij} \leq 0,$$

and the backward equation is obtained by using the transpose of the matrix A. In matrix notation these equations are $P'(t) = P(t)A$ and $P'(t) = AP(t)$ with $P(0) = I$. When either equation is known to hold, one has $P'_{ij}(0) = a_{ij}$.

Hille [302] and G. E. H. Reuter [626] have examined the solution of the above problem, using the theory of semigroups of linear bounded operators.

For the birth-death process the matrix A has the special form

$$a_{i,i+1} = \lambda_i, \qquad a_{ii} = -(\lambda_i + \mu_i), \qquad a_{i,i-1} = \mu_i, \qquad (3\text{-}25)$$

and $a_{ij} = 0$ if $|i - j| > 1$, where $\lambda_i > 0$ for $i \geq 0$ and $\mu_i > 0$ for $i \geq 1$ and $\mu_0 \geq 0$.

For a more complete discussion of the birth-death process in forward and backward equations, see the next chapter.

In conclusion, we note that a Markoff process has a simple probability structure which is best described by the fact that the future of the system depends only on the present. Information about the past history has no effect on the future. It is this property that has made Markoff processes useful in queueing theory.

PROBLEMS

1. Show by using moment-generating functions and by convolution that the distribution of a sum of independently and identically distributed variables

$$S_n = X_1 + \cdots + X_n$$

with common distribution $\mu e^{-\mu t}$ is given by

$$\frac{\mu(\mu t)^{n-1}e^{-\mu t}}{(n - 1)!} \quad \text{for } t \geq 0.$$

2. Solve Prob. 1 for two variables with the same form of density function but with different means μ_1 and μ_2.

3. Show that the Laplace transform of the convolution of two functions is equal to the product of the transforms of the functions. Do the same for two sequences.

4. When one assumes the exponential-service-time distribution one has $\mu e^{-\mu t}\, dt$ for the probability that a service will have a duration lying in the interval $(t,\ t + dt)$. Suppose that we have c channels with different exponential service at each one and corresponding parameters μ_n, and let there be an infinite population waiting for service. Each time a customer is admitted to service, he goes to one of the channels, and all the others become inoperative; i.e., people are served one at a time and by only one of the channels. Let a_n be the probability that at the instant an item enters service it goes to the nth channel. Show that the distribution of time which separates two customers who enter service is

$$\sum_{n=1}^{c} a_n \mu_n e^{-\mu_n t}, \qquad \sum_{n=1}^{c} a_n = 1,$$

which is known as the hyperexponential distribution. Show that its mean is

$$m = \sum_{n=1}^{c} \frac{a_n}{\mu_n}$$

and its variance is

$$\sigma^2 = 2 \sum_{n=1}^{c} \frac{a_n}{\mu_n^2} - \left(\sum_{n=1}^{c} \frac{a_n}{\mu_n} \right)^2.$$

Note that $\sigma/m \geq 1$ always holds and that, for fixed m, one can find nonzero values of a_n and μ_n $(n = 1, \ldots, c)$ such that σ can be made as large as desired.

Palm has divided distributions into two classes. He considers $1 - F(t)$ instead of the cumulative distribution $F(t)$. If m is the mean computed in the usual way for $1 - F(t)$ and if σ^2 is its variance, we define

$$\varepsilon \equiv \frac{2}{m^2} \int_0^\infty t[1 - F(t)]\,dt = 1 + \frac{\sigma^2}{m^2},$$

and we have $\varepsilon \geq 1$ always. When $1 \leq \varepsilon < 2$ one has a steep distribution, whereas for $2 < \varepsilon$ a distribution is flat. The exponential distribution falls on the boundary $\varepsilon = 2$. The constant-service-time distribution where $\varepsilon = 1$ is an example of a steep distribution. The hyperexponential distribution can be shown to be useful in representing (fitting) any completely monotone function. Such functions constitute an important subclass of flat distributions. A completely monotone function $f(t)$ has the following alternation in sign on its pth derivative:

$$(-1)^p f^{(p)}(t) \geq 0.$$

It has the form

$$\int_0^\infty e^{-xt}\,dH(x),$$

where $H(x)$ is bounded and nondecreasing for $x \geq 0$, $H(0) = 0$. In fact, every function which has this form is completely monotone. For such functions used as holding-time distributions the mean residual holding-time duration, given by

$$\int_0^\infty \frac{f(x + t)\,dx}{f(t)},$$

increases with time. On the other hand, it decreases as time increases when $f(t)$ is a steep distribution. Illustrate each idea by an example.

5. Consider the logarithm of the likelihood function of a sample of n observations $(x_1, \ldots, x_n)$ taken from an Erlangian distribution:

$$\log \prod_{i=1}^n \frac{\lambda^k}{(k-1)!} x_i^{k-1} e^{-\lambda x_i} = nk \log \lambda - n \log \Gamma(k) - \lambda \sum_1^n x_i + (k - 1) \sum_1^n \log x_i.$$

Find the maximum-likelihood estimators of λ and k. The estimator of k may be left in implicit form.

6. By referring to Foster's work, prove that $GI/M/1$ is ergodic if, and only if, $\rho < 1$.

7. Write the transition matrix of four states with two equivalence classes, one of which is ergodic.

8. Give an example of a decomposable Markoff chain.

9. Find the mean and variance of the distributions listed in the chapter.

10. What specialization of parameters in the Pearson Type-III distribution yields the gamma, the Erlangian, and the exponential distributions?

11. Give an example of a process which is non-Markovian.

12. Suppose that unoccupied taxis arrive at a taxi stand by a Poisson process with parameter a. Also suppose that customers arrive at the stand to take a taxi by another Poisson process with parameter b. Suppose that at $t = 0$ there are neither customers nor taxis waiting and at time t there are X customers and Y taxis. Let $Z = X - Y$ indicate the lengths of the line of customers. Show that the mean value of Z is $(b - a)t$ and that its variance is $(a + b)t$. Compute the characteristic function of Z and show that all cumulants of even order are equal to $(a + b)t$ and all those of odd order are equal to $(b - a)t$.

If $b > a$, by examining the standard deviation of Z, show that the number of customers becomes infinite with time with probability approaching unity. Also examine the case $a = b$. As $t \to \infty$, in neither case is a stable regime reached.

13. In a single-channel queue with arbitrary input and service distributions [138] let the arrival rate be λ and let the service rate be μ. Let $\rho = \lambda/\mu$ be the traffic density, and consider the problem in the transient state where $\rho > 1$. Suppose that the operation starts with an initial number i waiting in line at time $t = 0$. During t, roughly μt items would be served. However, during this time λt arrivals occur. The net result is $\lambda t - \mu t = \mu t(\rho - 1)$. Thus the expected number in the line at time t is $i + \mu t(\rho - 1)$, which coincides with the result for the deterministic queue studied in Chap. 2. Prove that the variance of the number in the queue at time t is the sum of the arrival-distribution variance at time t, that is, $\sigma_a^2(t)$, and the service-distribution variance, that is, $\mu t C^2$, where C^2 is the coefficient of variation of the service distribution.

If the initial population has a distribution, then its mean replaces i in the expression for the expected number in the line at time t and its variance is added to the above variance. If $w(t)$ is the waiting-time density in the queue at time t and if $L_q(t)$ is the queue-length density at time t, show that

$$E[w(t)|L_q(t)] \sim \frac{L_q(t)}{\mu},$$

$$\sigma^2[w(t)|L_q(t)] \sim \frac{L_q(t)C^2}{\mu^2},$$

where the vertical line may be replaced by the word "given." Thus show that

$$W_q = E[w(t)] \sim \frac{i}{\mu} + t(\rho - 1),$$

$$\sigma^2[w(t)] = E[L_q(t)]\sigma^2[w(t)|L_q(t)] + \sigma^2[L_q(t)]E[w(t)|L_q(t)] \sim \frac{iC^2}{\mu^2} + \frac{t}{\mu}\rho C^2 + \frac{\sigma_a^2(t)}{\mu^2}.$$

If arrivals are uniformly distributed over $[0,T]$ with density λ, then the total queueing time of all customers joining the queue in $[0,T]$ is approximately

$$\lambda \int_0^T E[w(t)]\, dt = i\rho T + \frac{\rho(\rho - 1)\mu T^2}{2}.$$

14. A random variable is infinitely divisible if for every natural number n it can be represented as the sum of n identically distributed random variables. (See Gnedenko and Kolmogorov, "Limit Distributions for Sums of Independent Random Variables," Addison-Wesley Publishing Company, Reading, Mass., 1954.) A distribution function is infinitely divisible if and only if its characteristic function is the nth power of some characteristic function (depending on n) for every natural number n with the qualifications that the last characteristic function is continuous and has unit value at the origin. It can be determined uniquely as the nth root of the characteristic function of the infinitely divisible distribution function.

Show that a random variable is infinitely divisible if it is distributed according to (1) the normal law, (2) the Poisson law, and (3) the gamma distribution. Also show by means of these examples the fact that the characteristic function of an infinitely divisible law never vanishes.

Prove that the distribution function of the sum of a finite number of independent infinitely divisible random variables is infinitely divisible.

15. Note that the distribution of a sum of independently and identically distributed random variables is a tighter distribution (i.e., the greater the amount of information, the smaller is the region of uncertainty). By considering means and variances, etc., verify this observation.

POISSON QUEUES

CHAPTERS 4 AND 5

In the two chapters of Part 2 single- and multiple-channel queues with ordered-service discipline and with Poisson input and exponential service times are investigated. Cases in which the parameters depend on the number in queue and on the time are treated in Chap. 4. Chapter 5 emphasizes the use of matrix operator methods in queueing theory. We begin with Poisson queues because of their historical significance. Their study provides a reasonably good start with queueing problems and with techniques used in their formulation and solution. Steady-state solutions are treated in a general manner in the text, and a large number of special cases are left as problems at the end of Chap. 4.

CHAPTER 4

THE BIRTH-DEATH PROCESS IN QUEUEING THEORY

4-1. Introduction

Historically, Poisson queues (both input and service are Poisson-distributed) were initially observed to form in telephone systems, where calls originated by a Poisson process and the duration of calls was experimentally verified to have an exponential distribution. The idea that in a small time interval the chances are small that more than a single arrival occurs, together with the "forgetfulness" property of the exponential distribution, has wide applicability. The operations of arrival and unloading of ships at a large port have also been found reasonably to satisfy these assumptions. In any case, Poisson queues have been among the first queues to be investigated. As there is a more adequate theory for them, we shall start by investigating the birth-death equations which contain the Poisson assumptions. As we have already observed in Chap. 3, the birth-death equations are established in terms of the probability of a given number in the queue at time t. From this probability we can obtain by further reasoning the distribution of waiting times.

It is now clear that the equations of the birth-death process, of which Erlang's model is a special case, play an important role in queueing theory. On specializing the coefficients, one obtains appropriate descriptions of queueing systems with Poisson input and exponential service times.

In this section we give the fundamental equations and in subsequent sections examine solutions of various queueing problems, the first of which is the single-channel problem. The steady-state solutions will be tabulated as an exercise.

1. *The Birth-Death Equations*

The "forward" birth-death equations are given by

$$P'_{in}(t) = \frac{dP_{in}(t)}{dt} = -(\lambda_n + \mu_n)P_{in}(t) + \lambda_{n-1}P_{i,n-1}(t) + \mu_{n+1}P_{i,n+1}(t),$$

$$t \geq 0, \quad n > 0, \quad i = 0, 1, 2, \ldots,$$

$$P'_{i0}(t) = \frac{dP_{i0}(t)}{dt} = -\lambda_0 P_{i0}(t) + \mu_1 P_{i1}(t), \quad n = 0,$$

$$\quad (4\text{-}1)$$

$$i = 0, 1, 2, \ldots,$$

$P_{in}(t)$ is the probability of a transition from state E_i at $t = 0$ to state E_n by time t (we have adhered to a common usage of the subscripts). Note that

$$P_{in}(0) = \delta_{in} = \begin{cases} 0, & i \neq n, \\ 1, & i = n, \end{cases}$$

where δ_{in} is the Kronecker symbol.

When this theory is applied to queues, the system is said to be in state E_n if there are n units in it at a given time. The probability of a transition from state E_n to state E_{n+1} during the time interval $(t, t + dt)$ is given by $\lambda_n \, dt + o(dt)$, and the probability of a transition from E_n to E_{n-1} during such an interval is $\mu_n \, dt + o(dt)$. Note the relation between these assumptions and those of a Poisson process and of the exponential distribution. If $\mu_n = 0$, the equations become those of the Poisson process. If $\lambda_n = \lambda$, $\mu_n = \mu$, we have seen how the equations describe a single-channel queue.

With the forward equations are associated the "backward" equations, given by

$$P'_{in}(t) = -(\lambda_i + \mu_i)P_{in}(t) + \lambda_i P_{i+1,n}(t) + \mu_i P_{i-1,n}(t), \qquad n > 0, t \geq 0,$$

$$P'_{0n}(t) = -\lambda_0 P_{0n}(t) + \lambda_0 P_{1,n}(t).$$

(4-2)

A solution of the forward equations (which satisfies the initial conditions) also satisfies the backward equations. One has a choice as to which of the two systems to solve.

It is appropriate to introduce the notions of absorbing and reflecting barriers again at this point. Note in the backward equations, for example, that if $\lambda_0 = 0$ it is impossible to go from state E_0 to state E_1. Thus, once the system reaches E_0 it stays there; hence E_0 is an *absorbing state*. On the other hand, if $\lambda_0 \neq 0$, E_0 is a *reflecting state* and return to E_1 is possible. It may be that the number of states is finite. In that case, the last state E_N may itself be absorbing or reflecting. In any case, one need not start with the state E_0. Any other state could be the lower absorbing or reflecting state of the system.

It is important and interesting to discuss the existence and uniqueness problem. To do this, we give a summary of some results from an excellent paper by G. E. H. Reuter [626], who has studied the problem of the existence of unique solutions of the birth-death-process equations.

If $\sum_{k=0}^{\infty} a_{ik} = 0$, the matrix A referred to in Chap. 3 is called conservative. Define recursively

$$f_{ij}^0(t) \equiv 0,$$

$$f_{ij}^{n+1}(t) \equiv \delta_{ij}e^{-a_i t} + e^{-a_i t}\int_0^t \Big[\sum_{k \neq i} a_{ik} f_{kj}^n(u) \Big] e^{a_i u} \, du,$$

where $a_i \equiv -a_{ii}$ (≥ 0). The f_{ij}^n increase with n and tend to finite limits:

$$f_{ij}(t) \equiv \lim_{n \to \infty} f_{ij}^n(t).$$

They also satisfy

$$f_{ij}^{n+1}(t) = \delta_{ij} e^{-a_i t} + e^{-a_i t} \int_0^t \Big[\sum_{k \neq j} f_{ik}^n(u) a_{kj} \Big] e^{a_j u} \, du.$$

The limit functions f_{ij} define a process $F = \{f_{ij}(t)\}$ such that $f'_{ij}(0) = a_{ij}$. Both forward and backward equations hold for F. Any process $P_{ij}(t)$ with $P'_{ij}(0) = a_{ij}$ is called an A process. One appropriately calls F the minimal A process since it can be shown that $f_{ij}(t) \leq P_{ij}(t)$ for any A process. An A process is called honest if $\sum_k P_{ik}(t) = 1$; otherwise it is dishonest. The solution of both the forward and backward equations is always provided by the minimal A-process F. Necessary and sufficient conditions for F to be the only solution of these two sets of equations are given in Reuter's paper. When A is conservative, every A process (in particular, every solution of the forward equations) is a solution of the backward equations.

If we specialize the coefficients a_{ij} as indicated in Eqs. (3-25) to obtain the birth-death-process equations, we have the following:

Theorem 4-1: If A with coefficients defined for the birth-death process is conservative and if

$$R = \sum_{n=1}^{\infty} \left(\frac{1}{\lambda_n} + \frac{\mu_n}{\lambda_n \lambda_{n-1}} + \cdots + \frac{\mu_n \cdots \mu_2}{\lambda_n \cdots \lambda_2 \lambda_1} \right),$$

$$S = \sum_{n=1}^{\infty} \left(\frac{1}{\mu_{n+1}} + \frac{\lambda_n}{\mu_{n+1} \mu_n} + \cdots + \frac{\lambda_n \cdots \lambda_1}{\mu_{n+1} \cdots \mu_2 \mu_1} \right),$$

$$T = \sum_{n=1}^{\infty} \left(\frac{\mu_n \cdots \mu_2}{\lambda_n \cdots \lambda_2 \lambda_1} + \frac{\lambda_n \cdots \lambda_1}{\mu_{n+1} \cdots \mu_2 \mu_1} \right),$$

then:

1. If $R = \infty$, there is exactly one A process; it is honest and satisfies the forward equations.

2. If $R < \infty$ and $S = \infty$, there are infinitely many A processes. Only one of these satisfies the forward equations, but it is dishonest.

3. If $R < \infty$ and $S < \infty$, equivalently if $T < \infty$, there are infinitely many A processes satisfying the forward equations. Exactly one of these is honest.

As an application of this theorem, it is easy to see that for the c server queue (see later) condition 1 holds. Note that $\lambda_0 = 0$ and A is conservative.   Also note from the series defining R that it majorizes term by term the series

$$\sum_{n=1}^{\infty} \frac{1}{\lambda_n} = \frac{1}{\lambda}\left(1 + \frac{1}{2} + \cdots + \frac{1}{n} + \cdots + \frac{1}{c-1} + \frac{1}{c} + \frac{1}{c} + \cdots\right)$$

which diverges; hence $R = \infty$.   Thus there is a unique solution of the problem which is honest and satisfies the forward equations which we have used.   Note that for Erlang's model we have $c = 1$.

Karlin and McGregor [349, 350] also give a necessary and sufficient condition that the birth-death-process equations have a unique solution $P_{in}(t) \geq 0$ satisfying the initial conditions, and

$$\sum_{n=0}^{\infty} P_{in}(t) \leq 1.$$

This condition is

$$\sum_{n=0}^{\infty} \left(\pi_n + \frac{1}{\lambda_n \pi_n}\right) = \infty,$$

where $$\pi_0 = 1, \qquad \pi_n = \frac{\lambda_0 \lambda_1 \cdots \lambda_{n-1}}{\mu_1 \mu_2 \cdots \mu_n}, \qquad n \geq 1.$$

Remark: We have the integral representation

$$P_{in}(t) = \pi_n \int_0^{\infty} e^{-xt} R_i(x) R_n(x)\, d\psi(x).$$

Here $R_0(x) = 1,$
$$-xR_0(x) = -(\lambda_0 + \mu_0)R_0(x) + \lambda_0 R_1(x),$$
$$-xR_n(x) = \mu_n R_{n-1}(x) - (\lambda_n + \mu_n)R_n(x) + \lambda_n R_{n+1}(x), \qquad n \geq 1,$$

and ψ is a positive regular measure on $0 \leq x < \infty$, satisfying

$$\int_0^{\infty} R_i(x) R_n(x)\, d\psi(x) = \frac{\delta_{in}}{\pi_n}, \qquad i, n = 0, 1, 2, \ldots.$$

Such a ψ is called a solution of the moment problem.   If we write

$$\rho = \frac{1}{\displaystyle\sum_{n=0}^{\infty} \pi_n},$$

then, since $R_n(0) = 1,$

$$\lim_{t \to \infty} P_{in}(t) = \rho \pi_n \equiv p_n.$$

ρ is the mass of ψ at $x = 0$. The constant ρ is interpreted as zero if the series diverges. We now have the classical ergodic theorem for the behavior of $P_{in}(t)$ as $t \to \infty$ as a consequence of the integral representation. After a proper classification of states, it is found that the system is ergodic if, and only if, $\Sigma\pi_n < \infty$, $\Sigma 1/\lambda_n\pi_n = \infty$, etc. For applications of queueing theory the above conditions are assured.

For many cases a statistical-mechanics type of argument is used instead of the classification-of-states approach, and equilibrium, when assumed to exist, requires that $dP_{in}(t)/dt$ be zero (i.e., the probabilities do not vary with time and hence become independent of time). It then becomes possible to solve a set of difference equations for the stationary (i.e., time-independent and often called equilibrium) solution.

Thus the steady-state probabilities (when the above conditions for the existence of the steady state are satisfied) are given by solving the above equations on setting the time derivative equal to zero or on using the result of the above argument. We have

$$p_n = p_0 \prod_{i=1}^{n} \frac{\lambda_{i-1}}{\mu_i},$$

where p_0 is determined from the condition $\sum_{n=0}^{\infty} p_n = 1$. As an exercise the reader should derive the foregoing steady-state solution, recursively, in preparation for the problems that are given at the end of this chapter.

We are now at the point where the reader should recognize a special case of p_n developed in Chap. 2, that is, $p_n = p_0(\lambda/\mu)$,^{n} $p_0 = 1 - \lambda/\mu$.

We now proceed to obtain transient solutions and other ramifications of queueing situations by a variety of methods intended to illustrate ways of solving such problems. Because of the nature of the birth-death process, the ideas of this chapter apply to queues with Poisson inputs and exponential service times.

The analysis will often carry us into differential equations, difference equations, Laplace transforms, and complex-variable theory. An important fact in the latter field to which we shall have frequent recourse and which we urge the reader to remember is the following theorem due to Rouché:

Theorem 4-2: If $f(z)$ and $g(z)$ are analytic inside and on a closed contour C and if $|g(z)| < |f(z)|$ on C, then $f(z)$ and $f(z) + g(z)$ have the same number of zeros inside C.

A proof of this theorem is found in the well-known work of Titchmarsh on the subject.

4-2. Case 1: The Single-channel Solution (Erlang's Model) $\lambda_n = \lambda$, $\mu_n = \mu$

This specialization of parameters yields equations describing a queue with a single server, a much-studied process.

Several methods have been used to solve this problem since A. B. Clarke gave his time-dependent solution in 1953. However, we shall give a simple step-by-step method of obtaining the solution by refining some existing procedures.

Our method proceeds along lines similar to those followed by Bailey until the Laplace transform of the generating function is explicitly obtained.

1. *The Equations*

By suppressing the subscript i to simplify the notation and putting $\lambda_n = \lambda$, $\mu_n = \mu$, one has the following single-channel Poisson-input, exponential-holding-time equations giving the probability that there are n items in the system at time t, given that there were i items in it at time $t = 0$:

$$\frac{dP_n}{dt} = -(\lambda + \mu)P_n(t) + \lambda P_{n-1}(t) + \mu P_{n+1}(t), \qquad n \geq 1, \qquad (4\text{-}3)$$

$$\frac{dP_0}{dt} = -\lambda P_0(t) + \mu P_1(t).$$

2. *The Solution*

Define the generating function of $P_n(t)$ by

$$P(z,t) = \sum_{n=0}^{\infty} P_n(t)z^n \qquad (4\text{-}4)$$

which must converge within the unit circle $|z| = 1$.

On multiplying each equation in $P_n(t)$ by z^{n+1} and summing over n, one has

$$z \frac{\partial P(z,t)}{\partial t} = (1 - z)[(\mu - \lambda z)P(z,t) - \mu P_0(t)]. \qquad (4\text{-}5)$$

Note that the initial condition becomes $P(z,0) = z^i$, since $P_n(0) = 0$ except when $n = i$, in which case $P_i(0) = 1$.

Applying the Laplace transform to this linear, first-order partial differential equation and solving for the transform of $P(z,t)$, one has

$$P^*(z,s) = \frac{z^{i+1} - \mu(1 - z)P_0^*(s)}{sz - (1 - z)(\mu - \lambda z)}, \qquad (4\text{-}6)$$

where, for example,

$$f^*(s) = \int_0^\infty e^{-st}f(t)\, dt.$$

Note that

$$\int_0^\infty e^{-st} \frac{\partial P}{\partial t}\, dt = e^{-st}P\Big]_0^\infty + s \int_0^\infty e^{-st}P\, dt = -z^i + sP^*. \quad (4\text{-}7)$$

In the next few pages a more convenient expression for $P^*(z,s)$ will be found, and the coefficient of z^n in the power series expansion of $P^*(z,s)$, $P_n^*(s)$, is recognized as the transform of $P_n(t)$. Taking the inverse Laplace transform gives the transient solution $P_n(t)$ sought.

Since $P^*(z,s)$ converges inside and on the unit circle for Re $(s) > 0$, the zeros of the denominator of Eq. (4-6) inside and on $|z| = 1$ must coincide with corresponding zeros of the numerator. But the zeros $\alpha_k(s)$ of the denominator are obtained by setting it equal to zero and solving a quadratic in z:

$$\alpha_k(s) = \frac{\lambda + \mu + s \pm [(\lambda + \mu + s)^2 - 4\lambda\mu]^{1/2}}{2\lambda}, \quad k = 1, 2. \quad (4\text{-}8)$$

Here the value of the square root with positive real part is taken, that is, $\alpha_k(s)$ has Re $(s) > 0$. $\alpha_1(s)$ has the positive sign before the radical.

Also by Rouché's theorem the denominator of P^* has a single zero within the disk $|z| < 1$. Here

$$f(z) = (\lambda + \mu + s)z, \qquad g(z) = -(\lambda z^2 + \mu),$$
$$|f(z)| = |\lambda + \mu + s| > |\lambda + \mu|, \qquad |g(z)| = |\lambda + \mu|,$$

and therefore $|g(z)| < |f(z)|$ on $|z| = 1$.

Since both $f(z)$ and $g(z)$ are analytic inside and on $|z| = 1$, $f(z)$ and $f(z) + g(z)$ have the same number of zeros. This zero must be $\alpha_2(s)$ since $|\alpha_2(s)| < |\alpha_1(s)|$ and there is no zero on $|z| = 1$. Note that

$$\alpha_1 + \alpha_2 = \frac{\lambda + \mu + s}{\lambda}, \qquad \alpha_1\alpha_2 = \frac{\mu}{\lambda}, \qquad s = -\lambda(1 - \alpha_1)(1 - \alpha_2).$$

The numerator of (4-6) must vanish for $z = \alpha_2(s)$; otherwise, for this value $P^*(z,s)$ would not exist, as it must. This leads to

$$P_0^*(s) = \frac{\alpha_2^{i+1}}{\mu(1 - \alpha_2)}, \quad (4\text{-}9)$$

which when substituted above gives

$$P^*(z,s) = \frac{z^{i+1} - [(1 - z)\alpha_2^{i+1}/(1 - \alpha_2)]}{-\lambda(z - \alpha_1)(z - \alpha_2)}. \quad (4\text{-}10)$$

When the numerator is multiplied through by $1 - \alpha_2$, it simplifies and factors. We have

$$P^*(z,s) = \frac{\begin{array}{c}(z - \alpha_2)(z^i + \alpha_2 z^{i-1} + \cdots + \alpha_2^i) \\ - z\alpha_2(z - \alpha_2)(z^{i-1} + \alpha_2 z^{i-2} + \cdots + \alpha_2^{i-1})\end{array}}{\lambda\alpha_1(z - \alpha_2)(1 - z/\alpha_1)(1 - \alpha_2)}. \quad (4\text{-}11)$$

Canceling $z - \alpha_2$ in numerator and denominator, subtracting and adding α_2^{i+1}, which enables factoring $1 - \alpha_2$ out of the numerator, and then writing $(1 - z/\alpha_1)^{-1} = \sum_{k=0}^{\infty} (z/\alpha_1)^k$ yield

$$P^*(z,s) = \frac{1}{\lambda\alpha_1}(z^i + \alpha_2 z^{i-1} + \cdots + \alpha_2^i)\sum_{k=0}^{\infty}\left(\frac{z}{\alpha_1}\right)^k$$

$$+ \frac{\alpha_2^{i+1}}{\lambda\alpha_1(1 - \alpha_2)}\sum_{k=0}^{\infty}\left(\frac{z}{\alpha_1}\right)^k. \quad (4\text{-}12)$$

Note that $|z/\alpha_1| < 1$.

Now $P_n^*(s)$ is the coefficient of z^n in the Laplace transform of (4-4). It is clear that the contribution to this coefficient of the second term in Eq. (4-12) (i.e., the coefficient of z^n in that term) is

$$\frac{\alpha_2^{i+1}}{\lambda\alpha_1^{n+1}(1 - \alpha_2)} = \frac{\alpha_2^{i+1}}{\lambda\alpha_1^{n+1}}(1 + \alpha_2 + \alpha_2^2 + \cdots)$$

$$= \frac{1}{\lambda}\left(\frac{\lambda}{\mu}\right)^{n+1}\sum_{k=n+i+2}^{\infty}\left(\frac{\mu}{\lambda}\right)^k\frac{1}{\alpha_1^k}, \quad (4\text{-}13)$$

the fact that $|\alpha_2| < 1$ and that $\alpha_1\alpha_2 = \mu/\lambda$ having been used.

The latter fact gives rise to powers of μ/λ by multiplying each α_2 raised to a power in the numerator by α_1 raised to the same power and therefore also multiplying the denominator by this power of α_1. The first term on the right of $P^*(z,s)$ yields the remaining coefficients of z^n on multiplying the left factor by the appropriate power of z in the series and adding the coefficients of z^n.

For example, it is clear that, if $n \geq i$, all the terms of $(1/\lambda\alpha_1)(z^i + \alpha_2 z^{i-1} + \cdots + \alpha_2^i)$ will contribute to this coefficient. Thus z^i is multiplied by z^{n-i} in the series, and the coefficient in this case is $1/\alpha_1^{n-i}$ to be multiplied by the outside factor $1/\lambda\alpha_1$. To this is added the coefficient resulting from multiplying z^{i-1} by z^{n-i+1}, which is $(1/\lambda\alpha_1)(\alpha_2/\alpha_1^{n-i+1})$, and so on.

In general, the contribution to this coefficient from multiplying z^{i-m} by z^{n-i+m} is

$$\frac{1}{\lambda\alpha_1}\frac{\alpha_2^m}{\alpha_1^{n-i+m}} = \frac{1}{\lambda\alpha_1}\frac{\alpha_2^m\alpha_1^m}{\alpha_1^{n-i+2m}} = \frac{(\mu/\lambda)^m}{\lambda\alpha_1^{n-i+2m+1}}. \quad (4\text{-}14)$$

Hence, for $n \geq i$, we have

$$P_n^*(s) = \frac{1}{\lambda}\left[\frac{1}{\alpha_1^{n-i+1}} + \frac{\mu/\lambda}{\alpha_1^{n-i+3}} + \frac{(\mu/\lambda)^2}{\alpha_1^{n-i+5}} + \cdots + \frac{(\mu/\lambda)^i}{\alpha_1^{n+i+1}} \right.$$
$$\left. + \left(\frac{\lambda}{\mu}\right)^{n+1} \sum_{k=n+i+2}^{\infty} \frac{(\mu/\lambda)^k}{\alpha_1^k}\right]. \quad (4\text{-}15)$$

Now $P_n(t)$ is the inverse transform of (4-15) given by

$$P_n(t) = \frac{1}{2\pi i}\int_{c-i\infty}^{c+i\infty} e^{st}P_n^*(s)\,ds. \quad (4\text{-}16)$$

To obtain the inverse transform, we first use the theorem that, if $f^*(s)$ is the Laplace transform of $f(t)$, then $f^*(s + a)$ is the Laplace transform of $e^{-at}f(t)$, a fact which easily follows from the definitions. Also, the inverse transform may be applied to each term of $P_n^*(s)$, that is, distributed over the entire right side.

Exercise: Prove the last two statements above.

Note from Eq. (4-8) that, in α_1, s has the constant $\lambda + \mu$ added to it; consequently in the case of the inverse transform of every term in (4-15) we apply the above theorem to functions whose Laplace transform is of the form $(\alpha_1)^{-\nu}$. Now $[(s + \sqrt{s^2 - 4\lambda\mu})/2\lambda]^{-\nu}$ is the Laplace transform of

$$(2\lambda)^\nu \nu(2\sqrt{\lambda\mu})^{-\nu}t^{-1}I_\nu(2\sqrt{\lambda\mu}\,t) = \nu\left(\sqrt{\frac{\lambda}{\mu}}\right)^\nu t^{-1}I_\nu(2\sqrt{\lambda\mu}\,t). \quad (4\text{-}17)$$

Here $I_\nu(z)$ is the modified Bessel function of the first kind given by

$$I_\nu(z) = \sum_{k=0}^{\infty} \frac{(z/2)^{\nu+2k}}{k!\Gamma(\nu + k + 1)}, \qquad I_\nu(z) = i^{-\nu}J_\nu(iz), \quad (4\text{-}18)$$

$$I_\nu(z) = \frac{z^\nu}{2^\nu \nu!} \quad \text{as } z \to 0, \qquad I_\nu(z) = \frac{e^z}{\sqrt{2\pi z}} \quad \text{as } z \to \infty. \quad (4\text{-}19)$$

Finally,

$$P_n(t) = \frac{e^{-(\lambda+\mu)t}}{\lambda}\left[\left(\sqrt{\frac{\lambda}{\mu}}\right)^{n-i+1}(n - i + 1)t^{-1}I_{n-i+1}(2\sqrt{\lambda\mu}\,t)\right.$$
$$+ \frac{\mu}{\lambda}\left(\sqrt{\frac{\lambda}{\mu}}\right)^{n-i+3}(n - i + 3)t^{-1}I_{n-i+3}(2\sqrt{\lambda\mu}\,t) + \cdots$$
$$+ \left(\frac{\mu}{\lambda}\right)^i\left(\sqrt{\frac{\lambda}{\mu}}\right)^{n+i+1}(n + i + 1)t^{-1}I_{n+i+1}(2\sqrt{\lambda\mu}\,t)$$
$$\left. + \left(\frac{\lambda}{\mu}\right)^{n+1}\sum_{k=n+i+2}^{\infty}\left(\sqrt{\frac{\mu}{\lambda}}\right)^k kt^{-1}I_k(2\sqrt{\lambda\mu}\,t)\right]. \quad (4\text{-}20)$$

Substituting the well-known relation

$$\frac{2\nu}{z} I_\nu(z) = I_{\nu-1}(z) - I_{\nu+1}(z), \tag{4-21}$$

one has

$$
\begin{aligned}
P_n(t) = \frac{e^{-(\lambda+\mu)t}}{\lambda} &\left\{ \left(\sqrt{\frac{\lambda}{\mu}}\right)^{n-i+1} \sqrt{\lambda\mu}\, [I_{n-i}(2\sqrt{\lambda\mu}\, t) - I_{n-i+2}(2\sqrt{\lambda\mu}\, t)] \right. \\
&+ \left(\sqrt{\frac{\lambda}{\mu}}\right)^{n-i+1} \sqrt{\lambda\mu}\, [I_{n-i+2}(2\sqrt{\lambda\mu}\, t) - I_{n-i+4}(2\sqrt{\lambda\mu}\, t)] \\
&+ \cdots + \left(\sqrt{\frac{\lambda}{\mu}}\right)^{n-i+1} \sqrt{\lambda\mu}\, [I_{n+i}(2\sqrt{\lambda\mu}\, t) \\
&\quad - I_{n+i+2}(2\sqrt{\lambda\mu}\, t)] \\
&+ \left(\frac{\lambda}{\mu}\right)^{n+1} \sum_{k=n+i+2}^{\infty} \left(\sqrt{\frac{\mu}{\lambda}}\right)^k \sqrt{\lambda\mu}\, [I_{k-1}(2\sqrt{\lambda\mu}\, t) \\
&\left. \qquad\qquad\qquad\qquad - I_{k+1}(2\sqrt{\lambda\mu}\, t)] \right\}. \tag{4-22}
\end{aligned}
$$

But

$$
\begin{aligned}
\left(\frac{\lambda}{\mu}\right)^{n+1} &\sum_{k=n+i+2}^{\infty} \left(\sqrt{\frac{\mu}{\lambda}}\right)^k [I_{k-1}(2\sqrt{\lambda\mu}\, t) - I_{k+1}(2\sqrt{\lambda\mu}\, t)] \\
= \left(\frac{\lambda}{\mu}\right)^{n+1} &\left[\left(\sqrt{\frac{\mu}{\lambda}}\right)^{n+i+2} I_{n+i+1}(2\sqrt{\lambda\mu}\, t) \right. \\
&+ \sqrt{\frac{\mu}{\lambda}} \sum_{k=n+i+2}^{\infty} \left(\sqrt{\frac{\mu}{\lambda}}\right)^k I_k(2\sqrt{\lambda\mu}\, t) + \left(\sqrt{\frac{\mu}{\lambda}}\right)^{n+i+1} I_{n+i+2}(2\sqrt{\lambda\mu}\, t) \\
&\left. \qquad\qquad - \sqrt{\frac{\lambda}{\mu}} \sum_{k=n+i+2}^{\infty} \left(\sqrt{\frac{\mu}{\lambda}}\right)^k I_k(2\sqrt{\lambda\mu}\, t) \right] \\
= \left(\sqrt{\frac{\lambda}{\mu}}\right)^{n-i} &I_{n+i+1}(2\sqrt{\lambda\mu}\, t) + \left(\sqrt{\frac{\lambda}{\mu}}\right)^{n-i+1} I_{n+i+2}(2\sqrt{\lambda\mu}\, t) \\
&+ \left(1 - \frac{\lambda}{\mu}\right)\left(\frac{\lambda}{\mu}\right)^n \sqrt{\frac{\lambda}{\mu}} \sum_{k=n+i+2}^{\infty} \left(\sqrt{\frac{\mu}{\lambda}}\right)^k I_k(2\sqrt{\lambda\mu}\, t); \tag{4-23}
\end{aligned}
$$

therefore the second expression of the first term of (4-22) cancels with the first expression of the second term, and so on, and one has for $n \geq i$

$$
P_n(t) = e^{-(\lambda+\mu)t} \left[\left(\sqrt{\frac{\mu}{\lambda}} \right)^{i-n} I_{n-i}(2\sqrt{\lambda\mu}\,t) \right.
$$
$$
+ \left(\sqrt{\frac{\mu}{\lambda}} \right)^{i-n+1} I_{n+i+1}(2\sqrt{\lambda\mu}\,t) + \left(1 - \frac{\lambda}{\mu} \right) \left(\frac{\lambda}{\mu} \right)^n
$$
$$
\left. \sum_{k=n+i+2}^{\infty} \left(\sqrt{\frac{\mu}{\lambda}} \right)^k I_k(2\sqrt{\lambda\mu}\,t) \right]. \quad (4\text{-}24)
$$

It will now be shown that this is also the solution for $n < i$.

Note from the first expression on the right in Eq. (4-12) that to obtain the coefficient of z^n one commences the multiplication of the series $\sum_{k=0}^{\infty} (z/\alpha)^k$ by z^n in the outside factor $(1/\lambda\alpha_1)(z^i + \alpha_2 z^{i-1} + \cdots + \alpha_2^i)$, using all the terms in z to the right of it. Everything from then on remains the same as above. Hence cancellation takes place in the same manner. However, we must show that the first term for the case $n < i$ coincides with the first term of (4-24).

Now the first contribution to the coefficient of z^n in this case is obtained by multiplying the coefficient of z^n in the outside factor by the first term of the series, which is unity. This gives

$$
\frac{\alpha_2^{i-n}}{\lambda\alpha_1} = \frac{(\mu/\lambda)^{i-n}}{\lambda\alpha_1^{i-n+1}}, \quad (4\text{-}25)
$$

which has the inverse transform

$$
\frac{e^{-(\lambda+\mu)t}}{\lambda} \left[\left(\frac{\mu}{\lambda} \right)^{i-n} \left(\sqrt{\frac{\lambda}{\mu}} \right)^{i-n+1} (i-n+1)t^{-1}I_{i-n+1}(2\sqrt{\lambda\mu}\,t) \right]
$$
$$
= \frac{e^{-(\lambda+\mu)t}}{\lambda} \left\{ \left(\sqrt{\frac{\lambda}{\mu}} \right)^{n-i+1} \sqrt{\lambda\mu}\,[I_{i-n}(2\sqrt{\lambda\mu}\,t) - I_{i-n+2}(2\sqrt{\lambda\mu}\,t)] \right\}. \quad (4\text{-}26)
$$

It is also known that if n is an integer $I_{-n}(z) = I_n(z)$, and since the second expression is canceled, as outlined above, the first terms coincide and the solution is complete in detail.

3. The Sum Is Unity

As an exercise in interchanging summations, we show that the sum of the probabilities is unity.

$$\sum_{n=0}^{\infty} P_n(t) = e^{-(\lambda+\mu)t} \Bigg[\sum_{m=-i}^{\infty} \left(\sqrt{\frac{\lambda}{\mu}}\right)^m I_m(2\sqrt{\lambda\mu}\,t) + \left(\sqrt{\frac{\lambda}{\mu}}\right)^{-2(i+1)}$$

$$\sum_{m=i+1}^{\infty} \left(\sqrt{\frac{\lambda}{\mu}}\right)^m I_m(2\sqrt{\lambda\mu}\,t) + \left(1-\frac{\lambda}{\mu}\right) \sum_{n=0}^{\infty} \sum_{k=n+i+2}^{\infty} \left(\frac{\lambda}{\mu}\right)^n$$

$$\left(\sqrt{\frac{\mu}{\lambda}}\right)^k I_k(2\sqrt{\lambda\mu}\,t) \Bigg]. \quad (4\text{-}27)$$

On interchanging summation and summing with respect to n, the last expression becomes

$$\left(1-\frac{\lambda}{\mu}\right) \sum_{k=i+2}^{\infty} \sum_{n=0}^{k-i-2} \left(\frac{\lambda}{\mu}\right)^n \left(\sqrt{\frac{\mu}{\lambda}}\right)^k I_k(2\sqrt{\lambda\mu}\,t)$$

$$= \sum_{k=i+2}^{\infty} \left[1 - \left(\frac{\lambda}{\mu}\right)^{k-i-1}\right] \left(\sqrt{\frac{\mu}{\lambda}}\right)^k I_k(2\sqrt{\lambda\mu}\,t)$$

$$= \sum_{m=i+2}^{\infty} \left(\sqrt{\frac{\mu}{\lambda}}\right)^m I_m(2\sqrt{\lambda\mu}\,t) - \left(\sqrt{\frac{\lambda}{\mu}}\right)^{-2(i+1)}$$

$$\sum_{m=i+2}^{\infty} \left(\sqrt{\frac{\lambda}{\mu}}\right)^m I_m(2\sqrt{\lambda\mu}\,t). \quad (4\text{-}28)$$

Combining the first expression on the right side of (4-28) with the first expression inside the brackets in (4-27), using the fact that $I_n(z) = I_{-n}(z)$, and then adding the second expression of (4-28) to the second expression in the brackets in (4-27), one has

$$\sum_{m=0}^{\infty} P_m(t) = e^{-(\lambda+\mu)t} \Bigg[\sum_{m=-i}^{\infty} \left(\sqrt{\frac{\lambda}{\mu}}\right)^m I_m(2\sqrt{\lambda\mu}\,t)$$

$$+ \sum_{m=-\infty}^{-(i+2)} \left(\sqrt{\frac{\lambda}{\mu}}\right)^m I_m(2\sqrt{\lambda\mu}\,t)$$

$$+ \left(\sqrt{\frac{\lambda}{\mu}}\right)^{-(i+1)} I_{-(i+1)}(2\sqrt{\lambda\mu}\,t) \Bigg]$$

$$= e^{-(\lambda+\mu)t} \sum_{m=-\infty}^{\infty} \left(\sqrt{\frac{\lambda}{\mu}}\right)^m I_m(2\sqrt{\lambda\mu}\,t)$$

$$= e^{-(\lambda+\mu)t} \exp\left[\sqrt{\lambda\mu}\left(\sqrt{\frac{\lambda}{\mu}}+\sqrt{\frac{\mu}{\lambda}}\right)t\right] = 1. \quad (4\text{-}29)$$

We have used the fact that

$$e^{(x/2)(y+1/y)} = \sum_{n=-\infty}^{\infty} y^n I_n(x) = \sum_{n=0}^{\infty} y^n I_n(x) + \sum_{n=1}^{\infty} y^{-n} I_n(x), \quad (4\text{-}30)$$

since $I_{-n}(x) = I_n(x)$ for integral n.

4. The Steady-state Result

As an exercise in using asymptotic expressions, we show that the above solution for $\lambda < \mu$ tends to a steady-state limit as $t \to \infty$. We shall simplify the computation by assuming that $i = 0$. For fixed λ and μ as $t \to \infty$ one has

$$I_n(2\sqrt{\lambda\mu}\,t) \sim \frac{\exp(2\sqrt{\lambda\mu}\,t)}{\sqrt{2\pi(2\sqrt{\lambda\mu})t}}, \quad (4\text{-}31)$$

which is independent of n.

When this expression is substituted for I_n in the first two terms in brackets in (4-24) and the latter multiplied by the outside factor, each of the resulting two expressions is readily seen to approach 0 as $t \to \infty$, the fact that $\lambda + \mu > 2\sqrt{\lambda\mu}$ having been used. The latter fact follows from the assumption $\lambda/\mu < 1$. As for the limiting behavior of the last factor of (4-24), note that

$$\sum_{k=n+2}^{\infty} \left(\sqrt{\frac{\mu}{\lambda}}\right)^k I_k(2\sqrt{\lambda\mu}\,t) = \sum_{k=0}^{\infty} \left(\sqrt{\frac{\mu}{\lambda}}\right)^k I_k(2\sqrt{\lambda\mu}\,t)$$
$$- \sum_{k=0}^{n+1} \left(\sqrt{\frac{\mu}{\lambda}}\right)^k I_k(2\sqrt{\lambda\mu}\,t). \quad (4\text{-}32)$$

When multiplied by $e^{-(\lambda+\mu)t}$, the finite sum on the right tends to zero as $t \to \infty$, since each term tends to zero separately. The series will now be examined.

When multiplied by $e^{-(\lambda+\mu)t}$, the last sum on the right of (4-30) with $y = \sqrt{\mu/\lambda}$ and $x = 2\sqrt{\lambda\mu}\,t$ tends to zero as $t \to \infty$ when the asymptotic expression of $I_n(x)$ is used together with the fact that $\lambda/\mu < 1$. On the other hand, the first sum on the right of (4-30) is the desired series of (4-32) under study. Hence, finally,

$$e^{-(\lambda+\mu)t} \sum_{n=0}^{\infty} \left(\sqrt{\frac{\mu}{\lambda}}\right)^k I_k(2\sqrt{\lambda\mu}\,t)$$
$$= e^{-(\lambda+\mu)t} \exp\left[\frac{1}{2}t(2\sqrt{\lambda\mu})\left(\sqrt{\frac{\mu}{\lambda}} + \sqrt{\frac{\lambda}{\mu}}\right)\right] = 1. \quad (4\text{-}33)$$

Therefore $\qquad P_n(t) \to \left(\frac{\lambda}{\mu}\right)^n \left(1 - \frac{\lambda}{\mu}\right) \qquad$ as $t \to \infty$. $\qquad (4\text{-}34)$

5. The Probability That Not Less than a Given Number Is in the System

The reader may show that the probability that the number in the system is greater than or equal to j is given by

$$\sum_{n=j}^{\infty} P_n(t) = e^{(-\lambda+\mu)t} \left[\sum_{n=j-i}^{\infty} \left(\sqrt{\frac{\lambda}{\mu}}\right)^n I_n(2\sqrt{\lambda\mu}\,t) \right.$$

$$\left. + \sum_{n=j+i+1}^{\infty} \left(\sqrt{\frac{\mu}{\lambda}}\right)^{n-2j} I_n(2\sqrt{\lambda\mu t}) \right]. \quad (4\text{-}35)$$

6. Another Solution Method

Conolly [126] has solved Case 1 by applying the Laplace transform to the system of equations rather than to the equation for the sum $P(z,t)$, as we have done, and has introduced the initial condition in the equation with $P'_i(t)$, where i is the initial state at $t = 0$. The resulting infinite system of difference equations is linear. Its general solution may be written as

$$P_n^*(s) = A\alpha_1^n + B\alpha_2^n, \quad (4\text{-}36)$$

where A and B are independent of n.

From the condition $\sum_{n=0}^{\infty} P_n(t) = 1$, which implies that

$$\sum_{n=0}^{\infty} P_n^*(s) = \frac{1}{s}, \quad (4\text{-}37)$$

we conclude that the last infinite series would diverge since $|\alpha_1| > 1$, unless a distinction is made between values of $n \leq i$ and those of $n \geq i$. We use the above expression of $P_n^*(s)$ for $n \leq i$ and use $P_n^*(s) = C\alpha_2^n$ for $n \geq i$, where C is independent of n. The two expressions must coincide for $n = i$. This gives one relation among A, B, C. Another relation is obtained from the first equation $(s + \lambda)P_0^*(s) = \mu P_1^*(s)$. The third relation is obtained from the equation for $n = i$ in which the proper solution is substituted, recalling that $n = i$ is the cutoff point. This determines the solution for $n \leq i$, $n \geq i$.

7. Random-walk Solution Method

Case 1 has also been solved by Champernowne [97], using random-walk methods.

Let q_t denote the queue length at time t, λ the Poisson-input parameter, and b_t the numbers of new arrivals during $(0,t)$. Thus

$$\text{Prob}\,(b_t = b) = e^{-\lambda t}\frac{(\lambda t)^b}{b!} \equiv g_t(b). \quad (4\text{-}38)$$

Let the service rate be the unit of time so that $\mu = 1$. Since the service epochs are Poisson-distributed, if a_t, the number of customers served during time $(0,t)$, is a Poisson variate of parameter t, then

$$\text{Prob }(a_t = a) = e^{-t}\frac{t^a}{a!} \equiv h_t(a). \tag{4-39}$$

Let

$$c_t = b_t - a_t \tag{4-40}$$

and ν_t be the most negative value of $c_{t'}$ in $0 \leq t' \leq t$. Then

$$\begin{aligned} q_t &= q_0 + c_t &&\text{if } \nu_t + q_0 \geq 0, \\ q_t &= -\nu_t + c_t &&\text{if } \nu_t + q_0 < 0. \end{aligned} \tag{4-41}$$

The problem is to determine $f_t(q) \equiv \text{Prob }(q_t = q)$.

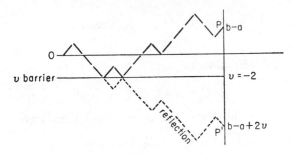

FIG. 4-1

One must first determine $J_t(\nu|a,b)$, the distribution function for ν_t, given $a_t = a$, $b_t = b$. All paths a_T and b_T in $0 \leq T \leq t$ such that $a_T = a$, $b_T = b$ have equal probability density. Thus $J_t(\nu|a,b)$ is the proportion of those $\binom{a+b}{a}$ arrangements with b forward steps and a backward steps which reach the ν barrier. There are $\binom{a+b}{a+\nu}$ such arrangements when $0 \geq \nu \geq -a$. It follows that

$$J_t(\nu|a,b) = \frac{a!b!}{(a+\nu)!(b-\nu)!}, \qquad 0 \geq \nu \geq -a, \tag{4-42}$$

$$J_t(\nu|a,b) = \begin{cases} 1, & 0 \leq \nu, \\ 0, & \nu < -a. \end{cases} \tag{4-43}$$

We illustrate with the case $\nu = -2$, $a = 7$, and $b = 9$, as in Fig. 4-1. All arrangements lead from O to P; all arrangements which reach the ν barrier have modified arrangements which lead from O to P', where P' is the reflection of P in the ν barrier. Moreover, every set of steps from

O to P' is the modified arrangement from one distinct arrangement leading from O to P.

But to get from O to P' in $a + b$ steps involves $b - \nu$ forward and $a + \nu$ backward steps. Therefore, there are $\binom{a + b}{a + \nu}$ such arrangements. In the specific example there are $\binom{16}{5}$ paths from O to P' in 16 steps. Hence there are $\binom{16}{5}$ paths from O to P which ever fall to -2; this bears the proportion $\binom{16}{5} \big/ \binom{16}{7}$ to the number of all paths of 16 steps from O to P.

Let $F_t(q|a,b)$ be the probability that $q_t \geq q$, given $a_t = a, b_t = b$, and let $F_t(q)$ be the unconditional probability that $q_t \geq q$; then

$$F_t(q) = \sum_{a=0}^{\infty} \sum_{b=0}^{\infty} h_t(a) g_t(b) F_t(q|a,b), \qquad (4\text{-}44)$$

and the desired distribution is

$$f_t(q) = F_t(q) - F_t(q + 1). \qquad (4\text{-}45)$$

From Eqs. (4-40) and (4-41) and $a_t = a$, $b_t = b$, we have

$$F_t(q|a,b) = 1 \qquad \text{for } q \leq q_0 + b - a$$

and, from (4-41) and (4-43),

$$F_t(q|a,b) = J_t(-a - 1|a,b) = 0 \qquad \text{for } q > q_0 + b - a \text{ and } q > b.$$

From (4-42) and $a_t = a$, $b_t = b$,

$$F_t(q|a,b) = J_t(b - a - q|a,b)$$
$$= \frac{a!\,b!}{(a + q)!\,(b - q)!} \qquad \text{for } b \geq q > q_0 + b - a.$$

We then have, for example, for the case $q \geq q_0$

$$F_t(q) = e^{-(1+\lambda)t} \left[\sum_{a=0}^{\infty} \sum_{b=a+q-q_0}^{\infty} \frac{\lambda^a t^{a+b}}{a!\,b!} + \sum_{b=q}^{\infty} \sum_{a=b-q+q_0+1}^{\infty} \frac{\lambda^b t^{a+b}}{(a + q)!\,(b - q)!} \right]$$

$$= e^{-(1+\lambda)t} \left[\sum_{c=q-q_0}^{\infty} \lambda^{c/2} I_c(2\lambda^{1/2}t) + \sum_{c=q-q_0-1}^{\infty} \lambda^{q-c/2} I_c(2\lambda^{1/2}t) \right], \qquad (4\text{-}46)$$

and $f_t(q)$ is obtained using (4-45), which gives the well-known solution (4-24). One similarly obtains the solution for the case $q < q_0$.

8. Remark

For the distribution of the length of a busy period, see Prob. 15 at the end of the chapter.

4-3. Case 2: $\lambda_n = \lambda$, $\mu_n = n\mu$ (The Case of an Infinite Number of Channels)

Obviously no queue forms in this case since an arrival immediately enters service. The equations are given by (again suppressing the subscript i for simplicity)

$$\frac{dP_n(t)}{dt} = -(\lambda + n\mu)P_n(t) + \lambda P_{n-1}(t) + (n + 1)\mu P_{n+1}(t), \qquad n \geq 1,$$

$$\frac{dP_0(t)}{dt} = -\lambda P_0(t) + \mu P_1(t). \tag{4-47}$$

Define the generating function of $P_n(t)$ by

$$P(z,t) = \sum_{n=0}^{\infty} P_n(t)z^n,$$

etc. As before, we have the following partial differential equation in the generating function:

$$\frac{\partial P}{\partial t} = (1 - z)\left(-\lambda P + \mu \frac{\partial P}{\partial z}\right), \tag{4-48}$$

or, in standard form,

$$\frac{\partial P}{\partial t} - (1 - z)\mu \frac{\partial P}{\partial z} = -\lambda(1 - z)P.$$

We give the solution, using ordinary methods.

This has the associated Lagrange equations

$$\frac{dt}{1} = \frac{dz}{-(1 - z)\mu} = \frac{dP}{-\lambda(1 - z)P}, \tag{4-49}$$

where the coefficient under dt is that of $\partial P/\partial t$, under dz is that of $\partial P/\partial z$, and under dP is the right side of the equation.

By Lagrange's method we use the first equation to obtain a one-parameter family of surfaces $u(z,t,P) = c_1$ and the second equation for another, $v(z,t,P) = c_2$. The general solution is an arbitrary function f of u and v such that $f(u,v) = 0$. One can solve for v and obtain it as a function of u, that is, $v = g(u)$. The form of g can then be determined from the given initial conditions.

For our example,

$$\frac{dt}{1} = \frac{dz}{-(1-z)\mu}$$

yields $u(z,t,P) \equiv (1-z)e^{-\mu t} = c_1.$

Also $$\frac{dz}{-(1-z)\mu} = \frac{dP}{-\lambda(1-z)P}$$

yields $v \equiv Pe^{-(\lambda/\mu)z} = c_2.$

Hence the general solution is given by

$$P = e^{(\lambda/\mu)z}g[(1-z)e^{-\mu t}].$$

Now when $t = 0$, $P = z^i$, since all the $P_n(0)$ are zero, except that $P_i(0) = 1$ because of the fact that the initial state is E_i. This gives

$$z^i = e^{(\lambda/\mu)z}g(1-z).$$

Let $y = 1 - z$; then

$$g(y) = e^{-(\lambda/\mu)(1-y)}(1-y)^i.$$

But for arbitrary t, the argument of g is $(1-z)e^{-\mu t}$, and in that case we must replace y in the right side of the above equation by this argument to obtain $g[(1-z)e^{-\mu t}]$.

When this is substituted in the general solution, we obtain

$$P(z,t) = e^{(\lambda/\mu)z} \exp\left\{-\frac{\lambda}{\mu}[1-(1-z)e^{-\mu t}]\right\}[1-(1-z)e^{-\mu t}]^i$$

$$= \left\{\exp\left[-\frac{\lambda}{\mu}(1-z)(1-e^{-\mu t})\right]\right\}[1-(1-z)e^{-\mu t}]^i. \quad (4\text{-}50)$$

Now $$P_n(t) = \frac{1}{n!}\frac{\partial^n P(z,t)}{\partial z^n}\bigg|_{z=0}.$$

Note that $P(z,t)$ is the product of two factors. If we denote the first by $f(z)$ and the second by $g(z)$, we may apply Leibnitz differentiation formula

$$\frac{d^n f(z)g(z)}{dz^n} = \sum_{k=0}^{n}\binom{n}{k}f^{(n-k)}(z)g^{(k)}(z), \quad (4\text{-}51)$$

where the superscripts in parentheses indicate the order of the derivative of the functions with respect to z.

For $n < i$ we have

$$P_n(t) = \frac{\exp[(-\lambda/\mu)(1-e^{-\mu t})]}{n}$$

$$\sum_{k=0}^{n}\binom{n}{k}\frac{i!}{(i-k)!}\left(\frac{\lambda}{\mu}\right)^{n-k}(1-e^{-\mu t})^{n+i-2k}e^{-\mu kt}. \quad (4\text{-}52)$$

One may similarly obtain expressions for $n > i$. (For the expected number in the system, see Prob. 13.)

If the initial number has a random, i.e., Poisson, distribution with mean equal to m, then one must compute

$$\sum_{i=0}^{\infty} \frac{m^i e^{-m}}{i!} P_{in}(t)$$

to obtain probabilities that do not depend on the initial number but on its mean. However, one can start by obtaining a new generating function

$$\sum_{i=0}^{\infty} \frac{m^i e^{-m}}{i!} P(z,t).$$

Exercise: Compute a new generating function for this case.

4-4. Case 3: $\lambda_n = \lambda(t)$, $\mu_n = \mu(t)$ **(Time-dependent Parameters)**

The single-channel problem starting with i in the system at $t = 0$ but with parameters depending on time has been investigated by A. B. Clarke [110]. In analyzing a queueing problem satisfying these assumptions, it was observed that the input parameter was a periodic function of time. This section deals with such a case. The analysis is involved and is given in outline form here. The reader is urged to solve the similar case given in Prob. 24. We start by introducing a time unit given by

$$\tau \equiv \tau(t) = \int_0^t \mu(s) \, ds. \tag{4-53}$$

Then, using τ as the time unit to simplify calculations, we put

$$\rho(\tau) = \frac{\lambda(t)}{\mu(t)}, \tag{4-54}$$

$$R(\tau) \equiv \frac{\int_0^t \lambda(s) \, ds}{\int_0^t \mu(s) \, ds} = \frac{1}{\tau} \int_0^\tau \rho(s) \, ds, \tag{4-55}$$

$$\frac{d}{d\tau} P_0(\tau) = -\rho(\tau) P_0(\tau) + P_1(\tau),$$

$$\frac{d}{d\tau} P_n(\tau) = -[1 + \rho(\tau)] P_n(\tau) + \rho(\tau) P_{n-1}(\tau) + P_{n+1}(\tau), \qquad n > 0, \tag{4-56}$$

$$P_n(0) = \delta_{in}.$$

Let $\qquad Q_n(\tau) = e^{\tau[1+R(\tau)]} P_n(\tau), \qquad n = 0, 1, \ldots ,$

and the above system of equations becomes

$$\frac{dQ_0(\tau)}{d\tau} = Q_0(\tau) + Q_1(\tau),$$

$$\frac{dQ_n(\tau)}{d\tau} = \rho(\tau)Q_{n-1}(\tau) + Q_{n+1}(\tau), \qquad n > 0. \tag{4-57}$$

$Q_{in}(0) = \delta_{in}$ as before. (Note again that we have suppressed the subscript i which indicates the initial state, in order to simplify notation.)

Use of the generating function

$$Q(z,\tau) = \sum_{n=0}^{\infty} Q_n(\tau) \frac{(z-\tau)^n}{n!} \tag{4-58}$$

reduces the above system to the hyperbolic partial differential equation known as the telegrapher's equation:

$$\frac{\partial^2 Q}{\partial\tau\,\partial z} = \rho(\tau)Q, \tag{4-59}$$

with the boundary conditions

$$\frac{\partial Q(z,\tau)}{\partial\tau}\bigg|_{z=\tau} = Q(\tau,\tau)$$

obtained by differentiating $Q(z,\tau)$ with respect to τ, putting $z = \tau$, and then using the equation in $Q_0'(\tau)$. Also $Q(z,0) = z^i/i!$, obtained from the initial state of the system condition. Let $f(\tau) = \partial Q(0,\tau)/\partial\tau$.

The above differential equation has the (modified Bessel function) Riemann function

$$I_0(2\{[R(\tau)\tau - R(\sigma)\sigma](z - \zeta)\}^{\frac{1}{2}}), \tag{4-60}$$

which on integration gives

$$Q(z,\tau) = A_i(0,\tau,z) + \int_0^\tau A_0(\sigma,\tau,z)f_i(\sigma)\,d\sigma. \tag{4-61}$$

Here

$$A_n(\sigma,\tau,z) = z^{n/2}[R(\tau)\tau - R(\sigma)\sigma]^{-n/2}I_n(2\{[R(\tau)\tau - R(\sigma)\sigma]z\}^{\frac{1}{2}}),$$
$$n = 0, \pm 1, \ldots,$$

$$\frac{\partial}{\partial z}A_n(\sigma,\tau,z) = A_{n-1}(\sigma,\tau,z),$$

$$\frac{\partial}{\partial\tau}A_n(\sigma,\tau,z) = \rho(\tau)A_{n+1}(\sigma,\tau,z), \tag{4-62}$$

$$A_n(0,0,0) = \delta_{0n},$$

$$A_{-n}(\tau,\tau,\tau) = \begin{cases} 0, & n > 0, \\ 1, & n = 0. \end{cases}$$

Let
$$B_n(\sigma,\tau) = A_n(\sigma,\tau,\tau) - \rho(\tau)A_{n+1}(\sigma,\tau,\tau).$$

Then
$$f_i(\tau) = B_i(0,\tau) + \int_0^\tau B_0(\sigma,\tau)f_i(\sigma)\,d\sigma, \qquad (4\text{-}63)$$

which when solved can be used to obtain the explicit solution of our problem, i.e.,

$$P_n(\tau) = \exp\{-\tau[1 + R(\tau)]\}[A_{i-n}(0,\tau,\tau) + \int_0^\tau A_{-n}(\sigma,\tau,\tau)f_i(\sigma)\,d\sigma]. \quad (4\text{-}64)$$

When $R(\tau) \equiv \rho$ is a constant, that is, $\mu(\tau) = \rho\lambda(\tau)$, it can be shown that

$$P_{in}(\tau) = e^{-\tau(1+\rho)}[A_{i-n}(0,\tau,\tau) + \rho^n A_{n+i+1}(0,\tau,\tau)$$
$$+ (1 - \rho)\rho^n \sum_{k=n+i+2}^\infty A_k(0,\tau,\tau)]. \quad (4\text{-}65)$$

Note that

$$P_{in}(\tau) = \rho^{-i}P_{0,i+n}(\tau) + e^{-\tau(1+\rho)}[A_{i-n}(0,\tau,\tau) - \rho^n A_{i+n}(0,\tau,\tau)], \quad (4\text{-}66)$$

thus enabling the computation of $P_{in}(\tau)$ from tables giving $P_{0n}(\tau)$.

Differentiation of the mean and variance,

$$L(\tau) = \sum_{n=0}^\infty nP_n(\tau),$$

$$\sigma^2(\tau) = \sum_{n=0}^\infty [n - L(\tau)]^2 P_n(\tau),$$

and using the system of differential-difference equations in τ and the fact

that $\sum_{n=0}^\infty P_n(t) = 1$ yield

$$\frac{dL(\tau)}{d\tau} = \rho(\tau) - 1 + P_0(\tau),$$

or
$$L(\tau) = [R(\tau) - 1]\tau + \int_0^\tau P_0(\sigma)\,d\sigma + i, \qquad (4\text{-}67)$$

$$\frac{d\sigma^2(\tau)}{d\tau} = 2\{\rho(\tau) + [\rho(\tau) - 1]L(\tau)\} - [1 + 2L(\tau)]\frac{dL(\tau)}{d\tau}, \qquad (4\text{-}68)$$

or $\sigma^2(\tau) = 2\int_0^\tau \{\rho(\sigma) + [\rho(\sigma) - 1]L(\sigma)\}\,d\sigma - L(\tau)$
$$- [L(\tau)]^2 + i + i^2. \quad (4\text{-}69)$$

Both mean and variance are $O(\tau)$ as $\tau \to \infty$ if $\rho(\tau)$ is bounded for all τ.

If $\rho > 1$ is a constant, and using the fact that

$$\int_0^\infty e^{-\tau(1+\rho)}A_n(0,\tau,\tau)\,d\tau = \frac{1}{\rho^n(\rho - 1)},$$

assuming $\tau = \int_0^t \mu(s)\,ds \to \infty$ as $\tau \to \infty$, one has

$$\int_0^\infty P_0(\sigma)\,d\sigma = \frac{1}{\rho^i(\rho - 1)} \quad \text{or} \quad \int_0^\tau P_0(\sigma)\,d\sigma$$

$$= \frac{1}{\rho^i(\rho - 1)} + o(1), \quad (4\text{-}70)$$

and
$$L(\tau) = (\rho - 1)\tau + \frac{1}{\rho^i(\rho - 1)} + i + o(1), \quad (4\text{-}71)$$

$$\sigma(\tau) = \sqrt{\tau(\rho + 1)} + o(\sqrt{\tau}). \quad (4\text{-}72)$$

If $\rho = 1$, the solution gives

$$P_0(\tau) = e^{-2\tau}[I_i(2\tau) + I_{i+1}(2\tau)] \sim \frac{1}{\sqrt{\pi\tau}}\left[1 + O\left(\frac{1}{\tau}\right)\right] \quad (4\text{-}73)$$

for large τ, and consequently

$$L(\tau) = \int_0^\tau P_0(\sigma)\,d\sigma + i = 2\sqrt{\tau/\pi} + O(1), \quad (4\text{-}74)$$

$$\sigma(\tau) = \sqrt{2\tau\left(1 - \frac{2}{\pi}\right)} + O(1). \quad (4\text{-}75)$$

4-5. Case 4 : $\lambda_n = \lambda(t), \mu_n = n\mu$ (A Time-dependent Input Parameter with an Infinite Number of Channels)

For simplicity, we take the mean service time as the time unit; hence $\mu = 1$. $P_n(t)$ is the probability that at time t we find exactly n channels busy. It satisfies the following equation:

$$P_n'(t) = -[\lambda(t) + n]P_n(t) + (n + 1)P_{n+1}(t) + \lambda(t)P_{n-1}(t), \quad (4\text{-}76)$$

etc., for $P_0'(t)$ [744].

If we write

$$P(z,t) = \sum_{n=0}^\infty P_n(t)z^n,$$

we have

$$\frac{\partial P}{\partial t} + (z - 1)\frac{\partial P}{\partial z} = \lambda(t)(z - 1)P \quad (4\text{-}77)$$

as a first-order linear partial differential equation in the generating function.

If one assumes that $\lambda(t) = \lambda$ is a constant, then

$$\lim_{t \to \infty} P(z,t) = e^{\lambda(z-1)}. \quad (4\text{-}78)$$

From this one has

$$p_n = \lim_{t \to \infty} P_n(t) = \frac{\lambda^n}{n!}e^{-\lambda}, \cdot$$

coinciding with the case of random traffic; i.e., traffic conditions converge to randomness independently of the initial state.

If we assume random traffic at $t = 0$, we have for $\lambda = \lambda(0)$

$$P(z,0) = e^{\lambda(z-1)}, \tag{4-79}$$

and the final solution using these conditions is given by

$$P_n(t) = \frac{[\Lambda(t)]^n}{n!} e^{-\Lambda(t)}, \qquad \Lambda(t) = e^{-t}\left[\lambda + \int_0^t \lambda(t)e^t \, dt\right]. \tag{4-80}$$

Exercise: Fill in the intermediate steps.

Note that in the special case, where λ is constant and the number of channels is N, we have for the steady state

$$\sum_{n=0}^{N} p_n = 1 \qquad \text{and} \qquad p_n = \frac{\lambda^n e^{-\lambda}/n!}{e^{-\lambda}\sum_{j=1}^{N} \lambda^j/j!} = \frac{\lambda^n/n!}{\sum_{j=1}^{N} \lambda^j/j!}, \tag{4-81}$$

which is called Erlang's distribution.

More generally we may replace λ by $\lambda/\mu \equiv \rho$.

4-6. Case 5: Phase-type Service

1. *The Number in the System*

Consider the case of a single-channel queue with Poisson arrivals with parameter λ, and suppose that service is accomplished in phases, the times of each of which are exponentially distributed with identical distributions and with parameter μ. Let c_j be the probability that an arrival will demand that j phases of service all be completed before another unit is admitted to service. If the initial state is i phases in the system at $t = 0$ and if $P_n(t)$ is the probability of n phases in the system at time t, then

$$P_0'(t) = -\lambda P_0(t) + \mu P_1(t),$$

$$P_n'(t) = -(\lambda + \mu)P_n(t) + \mu P_{n+1}(t) + \lambda \sum_{j=1}^{n} c_j P_{n-j}(t), \qquad n \geq 1. \tag{4-82}$$

Introducing the generating function

$$P(z,t) = \sum_{n=0}^{\infty} P_n(t)z^n$$

with

$$P(z,0) = z^i,$$

we have [455]

$$z\frac{\partial P}{\partial t} = \left[\mu - (\lambda + \mu)z + \lambda z \sum_{j=1}^{\infty} c_j z^j\right]P + \mu(z - 1)P_0(t).$$

The Laplace transform of $P(z,t)$ is

$$P^*(z,s) = \frac{z^{i+1} + \mu(z-1)P_0^*(s)}{(\lambda + \mu + s)z - \mu - \lambda z \sum_{j=1}^{\infty} c_j z^j}. \tag{4-83}$$

Note that if $c_1 = 1$, $c_j = 0$, $j \neq 1$, the problem reduces to the first case.

As demonstrated previously, using Rouché's theorem, it is clear that the denominator has a single simple zero within the unit circle which we denote by ξ, at which the numerator also vanishes. This enables computing $P_0(t)$ and subsequently $P_n(t)$, either from the original system or from $P^*(z,s)$. We give a method of determining $P_0(t)$ which we use to compute the number of phases in the system.

Let

$$\rho = \frac{\lambda}{\mu},$$

$$\zeta = \frac{\mu}{\lambda + \mu + s},$$

$$w = \frac{\lambda}{\lambda + \mu + s},$$

$$g(z) = z \sum_{j=1}^{\infty} c_j z^j,$$

and substitute $\xi^m = F(\xi)$, where m is an arbitrary integer and $F(\xi)$ is an analytic function of ξ given by Lagrange's expansion (see Prob. 25) [759]:

$$F(\xi) = F(\zeta) + \sum_{n=1}^{\infty} \frac{w^n}{n!} \frac{\partial^{n-1}}{\partial \zeta^{n-1}} \{F'(\zeta)[g(\zeta)]^n\}. \tag{4-84}$$

Finally, let

$$\left(\sum_{j=1}^{\infty} c_j \zeta^{j-1}\right)^n \equiv \sum_{j=0}^{\infty} b_{nj} \zeta^j; \tag{4-85}$$

then one has

$$\xi^m = \zeta^m + m \sum_{n=1}^{\infty} \frac{\rho^n}{n!} \sum_{j=0}^{\infty} \frac{(j+2n+m-1)!}{(j+n+m)!} b_{nj} \zeta^{j+2n+m}. \tag{4-86}$$

It is desired to compute the inverse Laplace transform of this expression and use the result to compute the inverse transform of $P_0^*(s)$, which, when expanded in powers of ξ, requires computing the inverse transform of all such powers. Therefore it suffices to write the inverse transform of the right side of the above equation and use it to write $P_0(t)$. We have for the former

$$\frac{m}{t}\left[\frac{(\mu t)^m}{m!} + \sum_{n=1}^{\infty} \frac{\rho^n}{n!} \sum_{j=0}^{\infty} b_{nj} \frac{(\mu t)^{j+2n+m}}{(j+n+m)!}\right] e^{-(1+\rho)\mu t} \qquad (4\text{-}87)$$

and

$$P_0(t) = \frac{1}{\mu t} \sum_{m=i+1}^{\infty} m\left[\frac{(\mu t)^m}{m!} + \sum_{n=1}^{\infty} \frac{(\rho \mu t)^n}{n!} \sum_{j=0}^{\infty} b_{nj} \frac{(\mu t)^{j+n+m}}{(j+n+m)!}\right] e^{-(1+\rho)\mu t}.$$

$$(4\text{-}88)$$

We also have for the expected number of phases in the system

$$L(t) = \frac{\partial P}{\partial z}\Big|_{z=1} = i + \left(\lambda \sum_{j=1}^{\infty} j c_j - \mu\right) t + \mu \int_0^t P_0(x)\, dx, \quad (4\text{-}89)$$

from which the expected waiting time at time t is given by $L(t)/\lambda$, etc., for the variance.

For

$$c_j = \begin{cases} 1, & j = k, \\ 0, & j \neq k, \end{cases}$$

the service time is Pearson Type III (see also Erlangian distribution) and

$$b_{n,n(k-1)} = 1, \qquad b_{nj} = 0, \qquad j \neq n(k-1).$$

Then Luchak gives for this case

$$P_0(t) = 2 \sum_{m=i+1}^{\infty} m \rho^{-[(m-1)/(k+1)]} \frac{I_m^k(r)}{r} e^{-(1+\rho)\mu t}, \qquad (4\text{-}90)$$

where

$$r = 2\rho^{1/(1+k)} \mu t, \qquad I_m^k(r) = \sum_{n=0}^{\infty} \frac{(r/2)^{m+n(k+1)}}{n!(m+nk)!}.$$

Note that, if $k = 1$, $I_m^k(r)$, whose properties have been studied by Luchak, is the modified Bessel function of the first kind. The reader might attempt to write $P_0(t)$ for $c_k = \alpha$, $c_{k+p} = \beta$, $c_j = 0$ otherwise, $\alpha + \beta = 1$, where it follows that

$$b_{n,qp+n(k-1)} = \binom{n}{q} \beta^q \alpha^{n-q}, \qquad q = 0, 1, \ldots, n.$$

2. The Busy-period Distribution

A busy period begins with the arrival of a customer at an idle channel (i.e., the initial number i waiting at $t = 0$ equals 1) and ends when the channel next becomes idle. This gives rise to a system in which E_0 is an absorbing state; i.e., once this state is reached, the operation stops.

Thus for the distribution of the length of a busy period we compute dP_0/dt. This rate of change of the probability of being in E_0 with respect to time is the desired distribution of the length of a busy period. It is obtained from the system

$$P'_n(t) = \mu P_{n+1}(t) - (\lambda + \mu)P_n(t) + \lambda \sum_{j=1}^{n-1} c_j P_{n-j}(t), \qquad n \geq 2,$$

$$P'_1(t) = \mu P_2(t) - (\lambda + \mu)P_1(t), \qquad\qquad\qquad\qquad (4\text{-}91)$$

$$P'_0(t) = \mu P_1(t).$$

Our generating function is $P(z,t) = \sum_{n=1}^{\infty} P_n(t)z^n$. We similarly obtain

$$P^*(z,s) = \frac{z^{i+1} - \mu z P_1^*(s)}{(\lambda + \mu + s)z - \mu - \lambda z \displaystyle\sum_{j=1}^{\infty} c_j z^j}. \qquad (4\text{-}92)$$

In this case $P_1^*(s) = \xi^i/\mu$, where ξ is the zero considered above.

$$\frac{dP_0}{dt} = \mu P_1(t) = \frac{i}{t}\left[\frac{(\mu t)^i}{i!} + \sum_{n=1}^{\infty} \frac{(\rho t \mu)^n}{n!} \sum_{j=0}^{\infty} b_{nj} \frac{(\mu t)^{j+n+i}}{(j+n+i)!} e^{-(1+\rho)\mu t} \right],$$

which becomes, for
$$c_j = \begin{cases} 1, & j = k, \\ 0, & j \neq k, \end{cases}$$

$$\frac{dP_0(t)}{dt} = 2i\mu\rho^{-[(i-1)/(k+1)]} \frac{I_i^k(r)}{r} e^{-(1+\rho)\mu t}. \qquad (4\text{-}93)$$

Let $i = 1$, and this gives the distribution of the length of a busy period.

3. *The Waiting-time Distribution*

Let the phase traffic intensity or utilization factor be given by

$$\rho = \frac{\lambda}{\mu} \sum_{j=1}^{\infty} j c_j.$$

If $w(x,t)\,dx$ is the probability that a unit which arrives at time t will obtain service in $(x, x + dx)$, then, if there are n phases of service to be completed for units arriving prior to t, the waiting-time distribution of the given unit is the sum of n identical exponential variables multiplied by the probability $P_n(t)$ of their presence, summed over n. Thus

$$w(x,t)\,dx = \sum_{n=1}^{\infty} P_n(t) e^{-\mu x} \frac{(\mu x)^{n-1}}{(n-1)!} \mu\,dx + P_0(t)\delta(x)\,dx, \qquad (4\text{-}94)$$

where $\delta(x)$ is the Dirac delta function.

Note that the service-time distribution for a unit (including all its phases) is

$$b(\tau)\,d\tau = \sum_{j=1}^{\infty} c_j e^{-\mu\tau} \frac{(\mu\tau)^{j-1}}{(j-1)!}\, \mu\, d\tau, \tag{4-95}$$

since there is a probability c_j for each unit to require j phases of service. We denote the Laplace transform of $b(\tau)$ by $b^*(s)$.

By introducing a generating function for (4-82), using $z = \mu/(\mu + s)$ as the generating-function variable, where s is the parameter of the Laplace transform $w^*(s,t)$ of (4-94), we have

$$\frac{\partial w^*}{\partial t} + \{\lambda[1 - b^*(s)] - s\}w^* = -sP_0(t), \tag{4-96}$$

which in the steady state gives

$$w^*(s) = \frac{p_0}{1 - \lambda/s[1 - b^*(s)]}, \qquad p_0 = 1 - \rho.$$

If each unit requires k phases of service, then $c_j = \delta_{jk}$, the Kronecker symbol, and the service time for each unit is a Pearson Type-III distribution:

$$b(\tau)\,d\tau = e^{-\mu\tau} \frac{(\mu\tau)^{k-1}}{(k-1)!}\, \mu\, d\tau, \tag{4-97}$$

with $b^*(s) = [\mu/(\mu + s)]^k$. Let $\rho = \lambda k/\mu$. Then

$$w^*(s) = \frac{1 - \rho}{1 - \lambda/s\{1 - [\mu/(\mu + s)]^k\}}. \tag{4-98}$$

But from the Laplace transform it is possible to obtain the moments μ_i about the origin by expanding in a power series in s, that is,

$$w^*(s) = \int_0^{\infty} e^{-sx} w(x)\, dx = 1 - s\mu_1 + \frac{s^2}{2!} \mu_2 - \cdots. \tag{4-99}$$

Hence, if we apply the ideas to (4-98), we have, on using the mean service time as the time unit and introducing the parameter a to be specialized for known distributions,

$$W \equiv \mu_1 = \frac{a + 1}{2a} \frac{\rho}{1 - \rho}, \tag{4-100}$$

$$\sigma^2 = \frac{a + 1}{a} \frac{\rho}{1 - \rho} \left(\frac{a + 1}{4a} \frac{\rho}{1 - \rho} + \frac{a + 2}{3a} \right).$$

We have the exponential service time by taking $a = 1$, $k = 1$ and the constant service time by taking $a = k = \infty$.

Exercise: The reader may now determine (4-98) and so forth for

$$c_j = \frac{1 - b}{b} b^j, \qquad j = 1, 2, \ldots,$$

and for

$$c_j = \frac{a^j e^{-a}}{(1 - e^{-a})j!}, \qquad j = 1, 2, \ldots,$$

and compare the results with those given by Gaver [247].

Exercise: Show that the Laplace transform of a Dirac delta function is unity. (See Sec. 4-8.)

4-7. Case 6: Multiple Channels in Parallel, $\lambda_n = \lambda$, $\mu_n = n\mu$, $n \leq c$, $\mu_n = c\mu$, $c \leq n$

1. *The Equations and Solution*

We assume that there are i units waiting in a system (which consists of c parallel channels) at $t = 0$. A single queue is formed and units enter the facilities when they are vacant on a first-come–first-served basis. The forward equations describing the problem are the following:

$$\frac{dP_0(t)}{dt} = -\lambda P_0(t) + \mu P_1(t),$$

$$\frac{dP_n(t)}{dt} = -(\lambda + n\mu)P_n(t) + \lambda P_{n-1}(t) + (n + 1)\mu P_{n+1}(t),$$

$$1 \leq n < c, \quad (4\text{-}101)$$

$$\frac{dP_n(t)}{dt} = -(\lambda + c\mu)P_n(t) + \lambda P_{n-1}(t) + c\mu P_{n+1}(t), \qquad c \leq n,$$

with initial conditions $P_n(0) = \delta_{in}$. The service rate is proportional to n (the number in the queue) if $n < c$ and to c if $n \geq c$. We rewrite the equations in the form

$$P_0'(t) = -(\lambda + c\mu)P_0(t) + c\mu P_0(t) + c\mu P_1(t) - (c - 1)\mu P_1(t),$$

$$P_n'(t) = -(\lambda + c\mu)P_n(t) + (c - n)\mu P_n(t) + \lambda P_{n-1}(t)$$
$$+ c\mu P_{n+1}(t) - (c - n - 1)\mu P_{n+1}(t), \quad 1 \leq n < c, \quad (4\text{-}102)$$

$$P_n'(t) = -(\lambda + c\mu)P_n(t) + \lambda P_{n-1}(t) + c\mu P_{n+1}(t), \qquad c \leq n.$$

Define the generating function

$$P(z,t) \equiv \sum_{n=0}^{\infty} P_n(t)z^n, \qquad P(z,0) = z^i, \qquad (4\text{-}103)$$

then multiply the second and third equations by z^n, and sum over the appropriate ranges of n including the first equation. This yields, for

$P \equiv P(z,t)$,

$$\frac{\partial P}{\partial t} = -(\lambda + c\mu)P + \lambda z P + c\mu \frac{P - P_0(t)}{z} + c\mu P_0(t)$$
$$- (c - 1)\mu(1 - z)P_1(t) - (c - 2)\mu z(1 - z)P_2(t) - \cdots$$
$$- (c - n)\mu z^{n-1}(1 - z)P_n(t) - \cdots - \mu z^{c-2}(1 - z)P_{c-1}(t). \quad (4\text{-}104)$$

Taking the Laplace transform, we have, after grouping terms and solving for $P^*(z,s)$,

$$P^*(z,s) = \frac{z^{i+1} - \mu(1 - z) \sum_{n=0}^{c-1} (c - n)z^n P_n^*(s)}{sz - (1 - z)(c\mu - \lambda z)}. \quad (4\text{-}105)$$

On and inside the unit circle $|z| = 1$, we apply Rouché's theorem to the denominator of the expression on the right whose two zeros are

$$\alpha_k = \frac{\lambda + c\mu + s \pm \sqrt{(\lambda + c\mu + s)^2 - 4c\mu\lambda}}{2\lambda}, \quad k = 1, 2, \quad (4\text{-}106)$$

where α_1 has the positive sign before the radical.

Arguing, in the usual manner, that since P^* must exist inside and on $|z| = 1$, and since $|\alpha_2| < 1$, the numerator must have $z - \alpha_2$ as a factor, i.e., vanish at α_2, we have

$$\sum_{n=0}^{c-1} (c - n)\alpha_2^n P_n^* = \frac{\alpha_2^{i+1}}{\mu(1 - \alpha_2)}. \quad (4\text{-}107)$$

In order to determine P_n^* ($0 \leq n \leq c - 1$) and thus be able to determine P_n^* for $n \geq c$ explicitly, we require c equations involving the c unknowns P_n^*, $0 \leq n \leq c - 1$. Now (4-107) is one such equation, and the first $c - 1$ equations of (4-101) are the remaining. Note that without (4-107) we could not use c equations, for then it would not be possible to determine the last one among them. Once we have determined P_n^* by taking inverse transforms, $P_n(t)$ can be determined from the inversion formula

$$P_n(t) = \frac{1}{2\pi i} \int_{a-i\infty}^{a+i\infty} e^{st} P_n^*(s) \, ds.$$

These equations are

$$P_0'(t) = -\lambda P_0(t) + \mu P_1(t),$$
$$P_n'(t) = -(\lambda + n\mu)P_n(t) + \lambda P_{n-1}(t) + (n + 1)\mu P_{n+1}(t),$$
$$1 \leq n \leq c - 2. \quad (4\text{-}108)$$

Note that the second equation is valid only when $c \geq 3$. As shown in the problems, the case $c = 2$ may be treated separately. As we have seen, the case $c = 1$ requires (4-107) alone, from which P_0^* is determined. Our object now is to obtain explicit expressions for the Laplace transforms of $P_n(t)$, $0 \leq n \leq c - 1$, using (4-107) and (4-108). Using (4-108), we express $P_n^*(s)$, $0 \leq n \leq c - 2$, in terms of $P_{c-1}^*(s)$ and then use (4-107) to determine the latter.

We introduce the generating function

$$Q(z,t) = \sum_{n=0}^{c-2} P_n(t)z^n, \qquad (4-109)$$

and we assume that $i > c - 2$ at $t = 0$. The ensuing argument can also be applied for the case $i \leq c$, as indicated below. Multiplying the second equation of (4-108) by z^n, summing over n, and adding the first equation yield

$$\frac{\partial Q}{\partial t} = -\lambda Q - \mu z \frac{\partial Q}{\partial z} + \mu \frac{\partial Q}{\partial z} + \lambda z Q - \lambda z^{c-1} P_{c-2} + \mu(c - 1)z^{c-2} P_{c-1},$$

or, in the standard form of a first-order linear partial differential equation,

$$-\frac{\partial Q}{\partial t} + \mu(1 - z)\frac{\partial Q}{\partial z} = \lambda(1 - z)Q + \lambda z^{c-1} P_{c-2} - \mu(c - 1)z^{c-2} P_{c-1}. \qquad (4-110)$$

The associated equations are

$$\frac{dt}{-1} = \frac{dz}{\mu(1 - z)} = \frac{dQ}{\lambda(1 - z)Q + \lambda z^{c-1} P_{c-2} - \mu(c - 1)z^{c-2} P_{c-1}}. \qquad (4-111)$$

The first equation gives

$$c_1 e^{\mu t} = 1 - z; \qquad c_1 = (1 - z)e^{-\mu t}; \qquad z = 1 - c_1 e^{\mu t}. \qquad (4-112)$$

Using this value of z in the third member and taking the equation in the first and third members, we have

$$\frac{\partial Q}{\partial t} + \lambda c_1 e^{\mu t} Q = \mu(c - 1)(1 - c_1 e^{\mu t})^{c-2} P_{c-1} - \lambda(1 - c_1 e^{\mu t})^{c-1} P_{c-2}, \qquad (4-113)$$

with the solution

$$Q = e^{-(\lambda/\mu)c_1 e^{\mu t}} \int_0^t [\mu(c - 1)(1 - c_1 e^{\mu x})^{c-2} P_{c-1}(x) - \lambda(1 - c_1 e^{\mu x})^{c-1} P_{c-2}(x)]e^{(\lambda/\mu)c_1 e^{\mu x}} dx. \qquad (4-114)$$

Note that the added integration constant $c_2 = f(c_1) = f[(1 - z)e^{-\mu t}]$ for some function $f(c_1)$, which was to be determined, vanishes since $Q(z,0) = 0$. This would not be the case if $i \leq c - 2$, for then

$$Q(z,0) = z^i, \qquad f(1 - z) = z^i e^{(\lambda/\mu)(1-z)}.$$

Let $y = 1 - z$; then $z = 1 - y$, $f(y) = e^{(\lambda/\mu)y}(1 - y)^i$, and

$$f[(1 - z)e^{-\mu t}] = [1 - (1 - z)e^{-\mu t}]^i \exp [(\lambda/\mu)(1 - z)e^{-\mu t}].$$

In that case, after replacing c_1 by its equal from (4-112) we would obtain

$$Q(z,t) = e^{-(\lambda/\mu)(1-z)} \int_0^t \{\mu(c-1)[1-(1-z)e^{-\mu(t-x)}]^{c-2}P_{c-1}(x)$$

$$- \lambda[1-(1-z)e^{-\mu(t-x)}]^{c-1}P_{c-2}(x)\} \exp\left[\frac{\lambda}{\mu}(1-z)e^{-\mu(t-x)}\right] dx$$

$$+ \exp\left[-\frac{\lambda}{\mu}(1-z)(1-e^{-\mu t})\right][1-(1-z)e^{-\mu t}]^i. \quad (4\text{-}115)$$

Continuing with the problem for $i > c - 2$, we suppress the second expression on the right and note that the first expression is a convolution, making it convenient to apply the Laplace transform which we are after.

From Erdélyi et al. [page 147, formula (40)] with $\nu = 1, \mu = -(c-2)$, we have for Re $(s) > 0$, on substituting $y = \mu w, dw = dy/\mu$,

$$\int_0^\infty e^{-sw}[1-(1-z)e^{-\mu w}]^{c-2} \exp\left[\frac{\lambda}{\mu}(1-z)e^{-\mu w}\right] dw$$

$$= \int_0^\infty e^{-(s/\mu)y}[1-(1-z)e^{-y}]^{c-2} \exp\left[\frac{\lambda}{\mu}(1-z)e^{-y}\right]\frac{dy}{\mu}$$

$$= \frac{1}{\mu}\frac{\Gamma(s/\mu)}{\Gamma(s/\mu+1)}\,\phi_1\left[\frac{s}{\mu}, -(c-2), 1, (1-z), \frac{\lambda}{\mu}(1-z)\right], \quad (4\text{-}116)$$

where the generalized hypergeometric function is given by

$$\phi_1(\alpha,\beta,\gamma,x,y) = \sum_{n=0}^\infty \sum_{m=0}^\infty \frac{(\alpha)_{m+n}(\beta)_n}{(\gamma)_{m+n}m!n!}\, x^m y^n,$$

$$(\alpha)_0 \equiv 1, \qquad (\alpha)_n \equiv \alpha(\alpha+1)\cdots(\alpha+n-1).$$

This gives

$$Q^*(z,s) = e^{-(\lambda/\mu)(1-z)}\left\{(c-1)P_{c-1}^*(s)\frac{\Gamma(s/\mu)}{\Gamma(s/\mu+1)}\right.$$

$$\phi_1\left[\frac{s}{\mu}, -(c-2), 1, (1-z), \frac{\lambda}{\mu}(1-z)\right] - \frac{\lambda}{\mu}P_{c-2}^*(s)\frac{\Gamma(s/\mu)}{\Gamma(s/\mu+1)}$$

$$\left.\phi_1\left[\frac{s}{\mu}, -(c-1), 1, (1-z), \frac{\lambda}{\mu}(1-z)\right]\right\}. \quad (4\text{-}117)$$

We first obtain $P_{c-2}^*(s)$ on the left by differentiation, then solve for it in terms of $P_{c-1}^*(s)$, and thus replace it in (4-117) by the expression which we readily obtain. Finally we obtain $P_n^*(s)$, $0 \le n \le c-2$, in terms of $P_{c-1}^*(s)$. Then the expression obtained is substituted in (4-107) and $P_{c-1}^*(s)$ is solved for.

We have

$$P_{c-2}^*(s) = \frac{1}{(c-2)!}\frac{\partial^{c-2}Q^*}{\partial z^{c-2}}\bigg|_{z=0}.$$

Let

$$f(z) = e^{-(\lambda/\mu)(1-z)}, \qquad g(z) = \phi_1\left[\frac{s}{\mu}, -(c-2), 1, (1-z), \frac{\lambda}{\mu}(1-z)\right],$$

and apply the Leibnitz differentiation theorem. This gives

$$\frac{1}{(c-2)!}\frac{\partial^{c-2}}{\partial z^{c-2}} f(z)g(z)\bigg|_{z=0} = \frac{1}{(c-2)!}\sum_{k=0}^{c-2}\binom{c-2}{k}\left(\frac{\lambda}{\mu}\right)^{c-k-2} e^{-(\lambda/\mu)}$$

$$\frac{d^k}{dz^k}\phi_1\left[\frac{s}{\mu}, -(c-2), 1, (1-z), \frac{\lambda}{\mu}(1-z)\right]\bigg|_{z=0}. \quad (4\text{-}118)$$

Thus

$$P_{c-2}^*(s) = \frac{\Gamma(s/\mu)}{(c-2)!\Gamma(s/\mu+1)}\left\{(c-1)P_{c-1}^*(s)\sum_{k=0}^{c-2}\binom{c-2}{k}\right.$$

$$\left(\frac{\lambda}{\mu}\right)^{c-k-2} e^{-(\lambda/\mu)}\frac{d^k}{dz^k}\phi_1\left[\frac{s}{\mu}, -(c-2), 1, (1-z), \frac{\lambda}{\mu}(1-z)\right]\bigg|_{z=0}$$

$$-\frac{\lambda}{\mu}P_{c-2}^*(s)\sum_{k=0}^{c-2}\binom{c-2}{k}\left(\frac{\lambda}{\mu}\right)^{c-k-2} e^{-(\lambda/\mu)}$$

$$\frac{d^k}{dz^k}\phi_1\left[\frac{s}{\mu}, -(c-1), 1, (1-z), \frac{\lambda}{\mu}(1-z)\right]\bigg|_{z=0}\right\}. \quad (4\text{-}119)$$

Solving for $P_{c-2}^*(s)$ yields

$$P_{c-2}^*(s)$$

$$= \cfrac{\dfrac{\Gamma(s/\mu)(c-1)}{(c-2)!\Gamma(s/\mu+1)} P_{c-1}^*(s)\sum_{k=0}^{c-2}\binom{c-2}{k}(\lambda/\mu)^{c-k-2}e^{-(\lambda/\mu)}(d^k/dz^k)}{\phi_1[(s/\mu), -(c-2), 1, (1-z), (\lambda/\mu)(1-z)]\bigg|_{z=0}} \Big/ \Bigg(1 + (\lambda/\mu)\dfrac{\Gamma(s/\mu)}{(c-2)!\Gamma(s/\mu+1)}\sum_{k=0}^{c-2}\binom{c-2}{k}(\lambda/\mu)^{c-k-2}e^{-(\lambda/\mu)}(d^k/dz^k)$$

$$\phi_1[(s/\mu), -(c-1), 1, (1-z), (\lambda/\mu)(1-z)]\bigg|_{z=0}\Bigg)$$

$$(4\text{-}120)$$

We also have for $n \leq c - 2$

$$P_n^*(s) = \frac{\Gamma(s/\mu)}{n!\Gamma(s/\mu+1)}\left\{(c-1)P_{c-1}^*(s)\sum_{k=0}^{n}\binom{n}{k}\left(\frac{\lambda}{\mu}\right)^{n-k} e^{-(\lambda/\mu)}\frac{d^k}{dz^k}\right.$$

$$\phi_1\left[\frac{s}{\mu}, -(c-2), 1, (1-z), \frac{\lambda}{\mu}(1-z)\right]\bigg|_{z=0} - \frac{\lambda}{\mu}P_{c-2}^*(s)\sum_{k=0}^{n}\binom{n}{k}\left(\frac{\lambda}{\mu}\right)^{n-k}$$

$$e^{-(\lambda/\mu)}\frac{d^k}{dz^k}\phi_1\left[\frac{s}{\mu}, -(c-1), 1, (1-z), \frac{\lambda}{\mu}(1-z)\right]\bigg|_{z=0}\right\}. \quad (4\text{-}121)$$

Thus

$$P^*_{c-1}(s) = \frac{\alpha_2^{i+1}}{1 - \alpha_2} \div \left(\alpha_2^{c-1} + \sum_{n=0}^{c-3} (c - n)\alpha_2^n \frac{\Gamma(s/\mu)}{n!\Gamma(s/\mu + 1)} \right\{ (c - 1)$$

$$\sum_{k=0}^{n} \binom{n}{k} \left(\frac{\lambda}{\mu}\right)^{n-k} e^{-(\lambda/\mu)} \frac{d^k}{dz^k} \phi_1 \left[\frac{s}{\mu}, -(c - 2), 1, (1 - z), \frac{\lambda}{\mu}(1 - z) \right] \bigg|_{z=0}$$

$$- \frac{\lambda}{\mu} \frac{P^*_{c-2}(s)}{P^*_{c-1}(s)} \sum_{k=0}^{n} \binom{n}{k} \left(\frac{\lambda}{\mu}\right)^{n-k} e^{-(\lambda/\mu)} \frac{d^k}{dz^k}$$

$$\phi_1 \left[\frac{s}{\mu}, -(c - 1), 1, (1 - z), \frac{\lambda}{\mu}(1 - z) \right] \bigg|_{z=0} \right\} + 2\alpha_2^{c-2} \frac{P^*_{c-2}(s)}{P^*_{c-1}(s)} \right). \quad (4\text{-}122)$$

This expression must now be substituted in (4-120) and (4-121) to obtain $P^*_n(s)$ for $0 \le n \le c - 2$ explicitly. Returning to (4-105), we note that the remaining $P^*_n(s)$ may be obtained as the coefficients of z^n. To do this, we write the denominator in the form $-\lambda(z - \alpha_1)(z - \alpha_2)$ and apply the Leibnitz theorem to the product $(z - \alpha_1)^{-1}(z - \alpha_2)^{-1}$. We have

$$\frac{1}{m!}\frac{d^m}{dz^m}(z - \alpha_1)^{-1}(z - \alpha_2)^{-1}\bigg|_{z=0} = \frac{1}{m!}\sum_{k=0}^{m}\binom{m}{k}(-1)^k k!(z-\alpha_1)^{-(k+1)}$$

$$(-1)^{m-k}(m - k)!(z - \alpha_2)^{-(m-k+1)}\bigg|_{z=0} = \frac{\lambda}{c\mu\alpha_2^{m-2}}\frac{1 - (\alpha_2/\alpha_1)^{m+1}}{1 - \alpha_2/\alpha_1}. \quad (4\text{-}123)$$

Hence the coefficient of z^n for $n \ge c$ is obtained by taking the sum of the coefficients resulting from multiplying the numerator of (4-105) by the appropriate power of z. In this expression we have [to avoid confusion we use the index j instead of n for summation in (4-105)]

$$P^*_n(s) = \frac{\mu}{\lambda}\left[\sum_{j=0}^{c-1} (c - j)P^*_j(s)\frac{\lambda}{c\mu\alpha_2^{n-j}}\frac{1 - (\alpha_2/\alpha_1)^{n-j+1}}{1 - \alpha_2/\alpha_1} \right.$$

$$\left. - (c - j)P^*_j(s)\frac{\lambda}{c\mu\alpha_2^{n-j-1}}\frac{1 - (\alpha_2/\alpha_1)^{n-j}}{1 - \alpha_2/\alpha_1} \right]$$

$$- \frac{1}{c\mu\alpha_2^{n-i-1}}\frac{1 - (\alpha_2/\alpha_1)^{n-i}}{1 - \alpha_2/\alpha_1}, \quad n \ge c. \quad (4\text{-}124)$$

We replace $P^*_j(s)$, $0 \le j \le c - 1$, by its expression that we determined previously.

From the definition of the Laplace transform as a moment-generating function, one can obtain the moments by expanding in series around $s = 0$.

2. The Steady State

The equilibrium solution p_n may be obtained directly from the equations by setting the derivatives equal to zero or from

$$\lim_{t \to \infty} P_n(t) = \lim_{s \to 0} sP^*_n(s).$$

We have (see Prob. 1)

$$p_n = \begin{cases} \dfrac{p_0}{n!}\left(\dfrac{\lambda}{\mu}\right)^n, & 1 \le n \le c, \\[2ex] \dfrac{p_0}{c!}\left(\dfrac{\lambda}{c\mu}\right)^n c^c, & c < n, \end{cases} \tag{4-125}$$

$$\sum_{n=0}^{\infty} p_n = 1,$$

$$p_0 = \frac{1}{\displaystyle\sum_{n=0}^{c-1} (c\rho)^n/n! + (c\rho)^c/c!(1-\rho)}, \tag{4-126}$$

where $\rho = \lambda/c\mu$. We have, for the average numbers in the system and in queue, respectively,

$$L = c_\rho + L_q, \qquad \rho = \frac{\lambda}{c\mu} \tag{4-127}$$

$$L_q = \frac{\rho(c\rho)^c}{c!(1-\rho)^2}\, p_0. \tag{4-128}$$

For the probability that there are c or more in the system and hence an arrival must wait, we have

$$P(>0) = \frac{(c\rho)^c}{c!(1-\rho)}\, p_0. \tag{4-129}$$

Finally, the probability of waiting longer than time t is

$$P(>t) = \exp\left[-c\mu t(1-\rho)\right]P(>0). \tag{4-130}$$

Exercise: Derive the last six expressions.

The expected waiting time (not including the waiting time in service) is

$$W_q = \frac{L_q}{\lambda}.$$

Example: As in the single-channel case,

$$P(<t) = \sum_{n=0}^{c-1} p_n + \sum_{n=0}^{\infty} p_{c+n}W_n(t),$$

where $W_n(t)$ is the waiting-time distribution of the $(n+1)$st item in line. It is derived as follows: The probability that an occupied channel completes service on a unit (already being served) by time t is $1 - e^{-\mu t}$. The probability that it continues to serve the unit by time t is $e^{-\mu t}$. The probability that all c channels continue to be occupied with the same units by time t is $e^{-c\mu t}$. The probability that any of the channels becomes free is $1 - e^{-c\mu t}$. This must happen n times in order that the $(n+1)$st unit in line enters service. Thus we must take the n-fold convolution of $1 - e^{-c\mu t}$. Hence the cumulative-waiting-time distribution of the $(n + 1)$st unit is given by

$$W_0(t) = 1 - e^{-c\mu t},$$

$$W_n(t) = \int_0^t W_{n-1}(x)\, d_x W_0(t - x) = 1 - e^{-c\mu t} \sum_{i=0}^{n} \frac{(c\mu t)^i}{i!},$$

as the reader should verify by induction.

Exercise: Carry out the operations to evaluate $P(<t)$. Specialize to obtain the result for $c = 1$.

3. *The Busy Period*

In general if there are c channels, it would be possible to put an absorbing barrier at any appropriate point to obtain the distribution of a busy period for any intermediate number of channels. For example, in the case of $k \leq c$ channels, we have

$$P'_{k-1}(t) = k\mu P_k(t),$$
$$P'_n(t) = -(\lambda + n\mu)P_n(t) + \lambda P_{n-1}(t)$$
$$\qquad\qquad + (n + 1)\mu P_{n+1}(t), \qquad k \leq n < c. \quad (4\text{-}131)$$

Ignoring the term $\lambda P_{k-1}(t)$ for $n = k$,

$$P'_n(t) = -(\lambda + c\mu)P_n(t) + \lambda P_{n-1} + c\mu P_{n+1}(t), \qquad c \leq n.$$

The initial number i must equal the number of busy channels studied.

4-8. Case 7: Absorbing Barriers; $\lambda_0 = 0$, λ_n, $\mu_n = 0$ for $n \geq N$ (An Epidemic Model)

Suppose that the Poisson arrival rate to a single-channel queue is proportional to the number present, that is, $\lambda_n = n\lambda$. Also, suppose that the service distribution is exponential with a service rate which is also proportional to the number present, that is, $\mu_n = n\mu$. Assume that once the number present is N, where N is very large, the operation immediately stops. Also, suppose that once the system is empty the operation stops. Thus E_N and E_0 are absorbing states. Both these conditions are expressed in the equations of the birth-death process by putting $\lambda_0 = 0$ and λ_n, $\mu_n = 0$ for $n \geq N$.

Suppose that one is interested in $Q_n(t)$, the probability distribution of the time required to reach state E_N, starting with E_n at $t = 0$, assuming, of course, that $\lambda > \mu$. The case of small values of N is treated in the next chapter [644]. The following formulation leads to the distribution of times for an initial dose of bacteria (a pathogen) of size n at time zero to grow to incubation size N. In the body of the host the birth and death rates of bacteria are assumed proportional to the number present.

We have from the definitions

$$Q_n(t) = \frac{dP_{nN}(t)}{dt}.$$ (4-132)

When this is applied to the backward birth-death equations we obtain

$$Q'_n(t) = -n(\lambda + \mu)Q_n(t) + n\mu Q_{n-1}(t)$$
$$+ n\lambda Q_{n+1}(t), \qquad n = 1, \ldots, N - 1, \quad (4\text{-}133)$$

$$Q_0(t) = Q_N(t) = 0,$$
$$Q_n(0) = \delta_{n,N-1}\lambda(N - 1).$$ (4-134)

The last condition follows from the last equation in the system, i.e.,

$$P'_{nN}(t) = (N - 1)\lambda P_{n,N-1}(t),$$ (4-135)

on putting $t = 0$ and using the definition of $Q_n(0)$.

By introducing in the usual manner the generating function

$$Q(z,t) = \sum_{n=1}^{N-1} Q_n(t)z^{n-1},$$

one obtains the first-order linear partial differential equation

$$-\frac{\partial Q}{\partial t} + [\mu z^2 - (\lambda + \mu)z + \lambda]\frac{\partial Q}{\partial z} = (\lambda + \mu - 2\mu z)Q$$
$$+ \mu N Q_{N-1}(t)z^{N-1}. \quad (4\text{-}136)$$

Note that $Q_{N-1}(t)$, which is one of the unknowns, appears in the equation. Note also that $Q_{N-1}(0) = \lambda(N - 1)$.

By direct integration of the equations one can show that

$$\int_0^\infty Q_n(t)\, dt = q_n = \frac{1 - (\mu/\lambda)^n}{1 - (\mu/\lambda)^N}.$$ (4-137)

Thus by integration one obtains a constant q_n which depends on n; on solving the resulting difference equations one obtains the value of the integral.

Now for large N, $Q_{N-1}(t)$, which starts at $\lambda(N-1)$ for $t = 0$, must drop rapidly in order that its integral not exceed unity. In fact, as $N \to \infty$, it behaves as a Dirac delta function which is zero everywhere except at the origin, where it has an infinite jump. Its integral is unity. Thus, from the definition of a Dirac delta function $\delta(x)$, one has

$$\int_0^\infty \delta(x) = 1,$$
$$\int_0^t f(x)\delta(x)\, dx = f(0),$$ (4-138)

if the integral exists for some $f(x)$. Actually we have used a right-sided delta function.

Now, solving the linear partial differential equation yields

$$Q(z,t) = -N\mu\alpha^2 \int_0^t Q_{N-1}(s)e^{-\alpha(t-s)}$$

$$\frac{\{\lambda[1 - e^{-\alpha(t-s)}] - z[\mu - \lambda e^{-\alpha(t-s)}]\}^{N-1}}{\{[\lambda - \mu e^{-\alpha(t-s)}] - \mu z[1 - e^{-\alpha(t-s)}]\}^{N+1}} \, ds$$

$$+ (N - 1)\lambda\alpha^2 e^{-\alpha t} \frac{[\lambda(1 - e^{-\alpha t}) - z(\mu - \lambda e^{-\alpha t})]^{N-2}}{[(\lambda - \mu e^{-\alpha t}) - \mu z(1 - e^{-\alpha t})]^N}, \quad (4\text{-}139)$$

where $\alpha = \lambda - \mu$. Because of our remarks on $Q_{N-1}(s)$ for large N, we are able to replace the integral by the value of the integrand at $s = 0$. It is then possible to obtain

$$Q_n(t) = \frac{1}{(n - 1)!} \frac{\partial^{n-1} Q(z,t)}{\partial z^{n-1}} \bigg|_{z=0},$$

and the derivatives of Q with respect to z are computed using the Leibnitz differentiation theorem.

This leads to rather complicated expressions which may be simplified on noting that the solution involves factors of the form $N^p e^{-p\alpha t}$ (with $p > 0$ an integer) since, if the power of N is less than that of $e^{-\alpha t}$, the result is zero and if greater it is large for large N [i.e., one can show from $Q_1(t)$ that $\alpha t \sim \log N$]. One finally obtains the following solution of the original system of equations for large N and $\beta = \mu/\lambda$ (see Prob. 21):

$$\frac{Q_n(t)}{\lambda} = (1 - \beta)^3 \frac{(1 - e^{-\alpha t})^{N-n-1}}{(1 - \beta e^{-\alpha t})^{N+n}} \sum_{k=0}^{n-1} \binom{n}{k+1}$$

$$\beta^{n-k-1}(1 - \beta)^{2k} \frac{(Ne^{-\alpha t})^{k+1}}{k!}$$

$$= (1 - \beta)\beta^n \exp\left(\frac{-\beta x}{1 - \beta}\right) x \frac{dL_n(-x)}{dx}, \quad (4\text{-}140)$$

where

$$x = \frac{(1 - \beta)^2}{\beta} Ne^{-\alpha t}$$

and $L_n(x)$ is the Laguerre polynomial. The mean and variance of $Q_n(t)$ may now be computed and normalized [since $Q_n(t)$ are conditional probabilities]. Note that

$$L_n(x) \equiv e^x \frac{d^n}{dx^n} (x^n e^{-x}).$$

Remark: The middle expression in the solution for $Q_n(t)$ involves extreme-value distributions inside the summation, of which the Fisher-Tippett distribution is well known.

The statement and solution of the several cases treated in this chapter should give to the reader who has taken the trouble to follow the analysis considerable insight and facility in handling similar queueing problems. The problems at the end of the chapter provide ample opportunity for testing his progress.

PROBLEMS

1. Verify the steady-state solutions of the following special cases of the birth-death process and justify the indicated applicability to queues in each case.

a.

Patient Customers

Number of channels

	$c = 1$	Unlimited input $c > 1$	$c = \infty$
Coefficients...	$\lambda_i = \lambda \ (i = 0, 1, \ldots)$ $\mu_i = \mu \ (i = 1, 2, \ldots)$	$\lambda_i = \lambda$ $\mu = \begin{cases} i\mu, & i \leq c \\ c\mu, & i > c \end{cases}$	$\lambda_i = \lambda$ $\mu_i = i\mu$
Solution......	$p_n = p_0 \left(\dfrac{\lambda}{\mu}\right)^n$	$p_n = p_0 \displaystyle\prod_{i=1}^{n} \dfrac{\lambda}{i\mu} = \dfrac{p_0}{n!}\left(\dfrac{\lambda}{\mu}\right)^n \quad 1 \leq n \leq c$ $p_n = p_0 \displaystyle\prod_{i=1}^{c} \dfrac{\lambda}{i\mu} \prod_{i=c+1}^{n} \dfrac{\lambda}{c\mu}$ $= \dfrac{p_0}{c!}\left(\dfrac{\lambda}{\mu}\right)^n \dfrac{1}{c^{n-c}} \quad n \geq c$	$p_n = \dfrac{p_0}{n!}\left(\dfrac{\lambda}{\mu}\right)^n$

Limited input*

Coefficients....	$\lambda_i = (m-i)\lambda \;\; (i = 0, 1, \ldots, m)$ $\mu_i = \mu \;\; (i = 1, 2, \ldots, m)$	$\lambda_i = (m-i)\lambda$ $\mu_i = \begin{cases} i\mu, & 1 \le i < c \\ c\mu, & c \le i \le m \end{cases}$	$\lambda_i = (m-i)\lambda$ $\mu_i = i\mu$

$$p_n = p_0 \prod_{i=1}^{n} \frac{(m-i+1)\lambda}{\mu}$$
$$= p_0 \left(\frac{\lambda}{\mu}\right)^n \frac{m!}{(m-n)!}$$

$$p_n = p_0 \prod_{i=1}^{n} \frac{(m-i+1)\lambda}{i\mu}$$
$$= p_0 \left(\frac{\lambda}{\mu}\right)^n \binom{m}{n} \frac{m!}{n!(m-n)!} \qquad 1 \le n < c$$

$$p_n = p_0 \prod_{i=1}^{c} \frac{(m-i+1)\lambda}{i\mu} \prod_{i=c+1}^{n} \frac{(m-i+1)\lambda}{c\mu}$$
$$= p_0 \left(\frac{\lambda}{\mu}\right)^n \binom{m}{n} \frac{n!}{c!\,c^{n-c}} \qquad c \le n \le m \quad n \ge c$$

$$p_n = p_0 \prod_{i=1}^{n} \frac{(m-i+1)\lambda}{i\mu}$$
$$= p_0 \left(\frac{\lambda}{\mu}\right)^n \binom{m}{n}$$

---

* In the limited input case, λ is a characteristic of the bids of a particular source, giving the constant probability density of a bid arriving at the service facility during the time the source is not being served or waiting for service. It is determined from the equation

$$\lambda = \mu \left[\sum_{n=0}^{c-1} n p_n + c \sum_{n=c}^{m} p_n \right].$$

b.

Impatient Customers

Number of channels

	Unlimited input		
	$c = 1$	$c > 1$	$c = \infty$
Coefficients	$\lambda_i = \lambda$ $\mu_i = \mu + c_i$	$\lambda_i = \lambda$ $\mu_i = \begin{cases} i(\mu + c_i), & i \le c \\ c\mu, & i > c \end{cases}$	$\lambda_i = \lambda$ $\mu_i = i\mu$
Solution	$p_n = p_0\lambda^n \displaystyle\prod_{i=1}^{n} \frac{1}{\mu + c_{i-1}}$	Solve for p_n	Solve for p_n

	Limited input		
	$c = 1$	$c > 1$	$c = \infty$
Coefficients	$\lambda_i = (m - i)\lambda$ $\mu_i = \mu + c_i$	$\lambda_i = (m - i)\lambda$ $\mu_i = \begin{cases} i(\mu + c_i), & i \le c \\ c(\mu + c_i), & i > c \end{cases}$	$\lambda_i = (m - i)\lambda$ $\mu_i = i(\mu + c_i)$
Solution	$p_n = p_0\lambda^n \displaystyle\prod_{i=1}^{n} \frac{m - i + 1}{\mu + c_i}$	Solve for p_n	Solve for p_n

Additional Arrivals and Impatient Customers

	Number of channels		
	Unlimited input		
	$c = 1$	$c > 1$	$c = \infty$
Coefficients.....	$\lambda_i = \lambda + d_i$	$\lambda_i = \lambda + d_i$	$\lambda_i = \lambda + d_i$
	$\mu_i = \mu + c_i$	$\mu_i = \begin{cases} i(\mu + c_i), & i \leq c \\ c\mu, & i > c \end{cases}$	$\mu_i = i(\mu + c_i)$
Solution......	$p_n = p_0 \displaystyle\prod_{i=1}^{n} \frac{\lambda + d_{i-1}}{\mu + c_i}$	Solve for p_n	Solve for p_n
	Limited input		
Coefficients.....	$\lambda_i = (m - i)(\lambda + d_i)$	$\lambda_i = (m - i)(\lambda + d_i)$	$\lambda_i = (m - i)(\lambda + d_i)$
	$\mu_i = \mu + c_i$	$\mu_i = \begin{cases} i(\mu + c_i), & i \leq c \\ c(\mu + c_i), & i > c \end{cases}$	$\mu_i = i(\mu + c_i)$
Solution......	Solve for p_n	Solve for p_n	Solve for p_n

2. In studying the following measures of effectiveness for the equilibrium solution of the c-channel problem $M/M/c$ with $\rho = \lambda/c\mu$, show that:

a. The expected number of waiting individuals in the queue is

$$\sum_{i=c}^{\infty} (i - c)p_i = p_c \frac{\rho}{(1 - \rho)^2}.$$

b. The probability that all c channels are occupied is

$$\sum_{i=c}^{\infty} p_i = p_c \frac{1}{1 - \rho}.$$

c. The probability that there will be someone waiting is

$$\sum_{i=c+1}^{\infty} p_i = p_c \frac{\rho}{1 - \rho}.$$

d. The mean number of waiting individuals for those who actually wait is

$$\frac{\displaystyle\sum_{i=c+1}^{\infty} (i - c)p_i}{\displaystyle\sum_{i=c+1}^{\infty} p_i} = \frac{1}{1 - \rho}.$$

e. The mean waiting time in the queue for all arrivals is

$$\frac{1}{\lambda} \sum_{i=c}^{\infty} (i - c)p_i = p_c \frac{1}{c\mu(1 - \rho)^2}.$$

f. The mean waiting time in queue for those who actually wait is

$$\frac{1}{1 - \rho} \frac{1}{c\mu} = \frac{1}{c\mu - \lambda}.$$

g. In the c-channel case, the probability that exactly $n \leq c$ channels are busy is given by p_n, with

$$\sum_{n=0}^{c} p_n = 1$$

from which p_0 can be determined.

3. Show that the expected waiting time of delayed customers is $W_q/P(>0)$.

4. Justify the following expressions and obtain results for the two-channel case:

a. The average number of items served is

$$\sum_{n=0}^{c-1} np_n + c \sum_{n=c}^{\infty} p_n.$$

b. The average number of idle channels is c minus the average number of items being served.

c. The coefficient of loss for the channels is the ratio of the average number of idle channels to the total number of channels c.

d. The efficiency is the ratio of the average number of items that are served to the number of channels.

e. If the input is finite with a total number $m > c$, the coefficient of loss due to service is obtained by replacing c in the efficiency by m.

f. The coefficient of loss in the m-population case due to waiting in line because all channels are occupied is obtained by dividing the average number in the waiting line by m.

g. The combined coefficient of loss is the sum of the service and waiting loss coefficients.

h. In the finite-population case the average number of customers not in the system is

$$\frac{\mu}{\lambda} \left[\sum_{n=0}^{c-1} np_n + c \sum_{n=c}^{m} p_n \right].$$

If the population consists of machines requiring service, then the efficiency of the machines is obtained by dividing this quantity by m.

5. Obtain the Poisson process from the solution of the single-channel queue $M/M/1$ by putting $\mu = 0$.

6. In a two-parallel-channel problem in the steady state with ordered queueing, Poisson input with parameter λ, and exponential service with parameter μ, compute the increase $\Delta\mu$ required in the service rate which yields the same effect as an increase in the number of channels by one (with the same service distribution); i.e., solve for $\Delta\mu$ in

$$W_q(3,\mu) = W_q(2, \mu + \Delta\mu).$$

One is now able to compare the cost of facility increase with service increase at the existing facilities.   How?

7. Obtain the solution of the Poisson process from the backward equations

$$P'_{in}(t) = -\lambda P_{in}(t) + \lambda P_{i+1,n}(t), \qquad P_{in}(0) = \delta_{in}.$$

Show that the generating function

$$P(z,t) \equiv \sum_{i=0}^{\infty} P_{in}(t)z^i$$

satisfies the equation

$$\frac{\partial P}{\partial t} + \lambda \left(1 - \frac{1}{z} \right) P = -\frac{\lambda}{z} P_{on}(t).$$

Show that the mean of $P_{in}(t)$ is given by $\lambda t + i$ and its variance by λt.

8. If X_1 and X_2 are random variables which are Poisson-distributed with parameters λ_1 and λ_2, respectively, show that $Y = X_1 - X_2$ is distributed as follows:

$$\text{Prob } (Y = x) = \text{Prob } (X_1 = X_2 + x) = \sum_{k=0}^{\infty} \text{Prob } (X_1 = k + x) \text{ Prob } (X_2 = k)$$

$$= e^{-(\lambda_1 + \lambda_2)} \left(\frac{\lambda_1}{\lambda_2} \right)^{x/2} I_x(2 \sqrt{\lambda_1 \lambda_2}).$$

Show that this has the characteristic function

$$\varphi(t) = e^{(\lambda_1+\lambda_2)(\cos t - 1) + i(\lambda_1 - \lambda_2)\sin t}, \qquad i = \sqrt{-1},$$

with mean $\lambda_1 - \lambda_2$ and variance $\lambda_1 + \lambda_2$. If we have $Y = \sum_{i=1}^{n} a_i X_i$, where a_i is either $+1$ or -1 and the X_i are Poisson-distributed with parameter λ_i, show that

$$\text{Prob } (Y = x) = \exp\left(-\sum_{j=1}^{n} \lambda_j\right)\left(\frac{\sum_k \lambda_k}{\sum_m \lambda_m}\right)^{x/2} I_x\left(2\sqrt{\sum_k \lambda_k \sum_m \lambda_m}\right),$$

where j extends over all variables, k over those with positive sign, and m over those with negative sign.

By considering the maximum of the likelihood function of n observations,

$$e^{-n(\lambda_1+\lambda_2)}\left(\frac{\lambda_1}{\lambda_2}\right)^{\Sigma x_i/2} \prod_{i=1}^{n} I_{x_i}\left(2\sqrt{\lambda_1\lambda_2}\right),$$

show that $\hat{\lambda}_1 - \hat{\lambda}_2 = \Sigma x_i/n$ is the estimate of the mean m and

$$\hat{\lambda}_1 + \hat{\lambda}_2 = \frac{\Sigma(x_i - m)^2}{n}$$

is the estimate of the variance.

9. Derive the Laplace transform of the solution of the two-parallel-channel problem with identical exponential service and show that it can be expressed as a difference of two factors, the first of which is the generating function of the single-channel problem with μ replaced by 2μ. Show that [645]

$$P_0^*(s) = \frac{\alpha_2^{i+1}}{(1-\alpha_2)[2\mu + (s+\lambda)\alpha_2]},$$

$$P_1^*(s) = \frac{(s+\lambda)\alpha_2^{i+1}}{\mu(1-\alpha_2)[2\mu + (s+\lambda)\alpha_2]}.$$

Note that the zeros α_1 and α_2 involve 2μ in this case. Verify the steady-state solution on P_0 and P_1.

10. Obtain the generating function and indicate a method for solving the two-channel problem with different exponential-service distributions whose parameters are μ_1 and μ_2, respectively. Note that the equations require taking into account the probability that an arriving customer will join either channel. Also $P_1(1,0,t)$ and $P_1(0,1,t)$ must be introduced, which gives the probability that one unit is in the system and it is in the first or the second channel, respectively. $P_1(t)$ is the sum of these.

11. Solve a problem with absorbing barriers similar to the one given in the text except that $\lambda_n = \lambda$, $\mu_n = \mu$. Use Laplace transforms on $Q(z,t)$, defined as in the last section of the chapter; since Q is a polynomial, its Laplace transform exists for all z. Thus the two zeros of the denominator of the Laplace transform $Q^*(z,s)$ must be canceled by two zeros of the numerator. This determines $Q_{N-1}^*(s)$ and $Q_1^*(s)$ in

$$Q^*(z,s) = \frac{\mu Q_{N-1}^* z^N - \lambda z^{N-1} + \lambda Q_1^*}{\mu z^2 - (\lambda + \mu + s)z + \lambda}$$

from two simultaneous equations. Substituting back and simplifying yield

$$Q^*(z,s) = \frac{\lambda}{\mu} (\alpha_1^n - \alpha_2^N)^{-1} \sum_{n=1}^{N-1} \left(\frac{\lambda}{\mu}\right)^{N-n-1} (\alpha_1^n - \alpha_2^n) z^{n-1},$$

where α_1 and α_2 are the two zeros. From this, $Q_n^*(s)$ is obtained as the coefficient of z^{n-1}. The inverse transform is then obtained from Erdélyi et al.

12. Show that the transient solution of the system (with an absorbing barrier at the origin and initial condition i at $t = 0$)

$$P_0'(t) = q\lambda P_1(t),$$
$$P_1'(t) = -\lambda P_1(t) + q\lambda P_2(t),$$
$$P_n'(t) = -\lambda P_n(t) + p\lambda P_{n-1}(t) + q\lambda P_{n-1}(t), \qquad n \geq 2,$$

where $p + q = 1$, is given by

$$P_n(t) = e^{-\lambda t} \left(\sqrt{\frac{p}{q}}\right)^{n-i} \lambda \sqrt{pq} \, [I_{n-i}(2\lambda \sqrt{pq} \, t) - I_{n+i}(2\lambda \sqrt{pq} \, t)], \qquad n > 0,$$

$$P_0(t) = \int_0^t e^{-\lambda(t-u)} \left(\sqrt{\frac{p}{q}}\right)^i \frac{i}{t-u} I_i[2\lambda \sqrt{pq} \, (t-u)] \, du.$$

Give the steady-state solution by passing to the limit.

13a. For the infinite-channel case, by computing

$$\frac{\partial P(z,t)}{\partial z}\bigg|_{z=1},$$

show that the expected number in the system is

$$L = \frac{\lambda}{\mu}(1 - e^{-\mu t}) + ie^{-\mu t}$$

and for the steady state

$$L = \frac{\lambda}{\mu}.$$

Also compute the expected number in the system from the new generating function, taking into consideration the distribution of the initial number. Note that the result can also be obtained by working on L. Intuitively, one replaces i by m.

b. Assuming that a waiting room is limited to a maximum number N, introduce an absorbing upper barrier in the equations and obtain the generating function. Derive the expected number in the system in this case. By putting $N = 0$ obtain the corresponding expression for a loss system, and then by letting $N \to \infty$ obtain the expression for infinite waiting room (a quantity just computed above).

14. Solve the birth-death equations with

$$\lambda_n = (m - n)\lambda, \qquad \mu_n = n\mu, \qquad n = 0, \ldots, m,$$

which is the limited-input case. Show that the solution of the differential equation

$$\frac{\partial P}{\partial z} = \frac{m\lambda P}{\mu + \lambda z}$$

in the generating function for the steady state is

$$P(z) = \left(\frac{\mu + \lambda z}{\mu + \lambda}\right)^m.$$

Show that
$$p_n = \binom{m}{n}\left(\frac{\lambda}{\lambda + \mu}\right)^n \left(\frac{\mu}{\lambda + \mu}\right)^{m-n}$$

is the solution of the steady-state equation.

15. The solution of the following problem leads to the distribution of the length of a busy period for a single channel. Solve the problem of an absorbing barrier at E_0 with an initial number i at $t = 0$:

$$P'_0(t) = \mu P_1(t),$$
$$P'_1(t) = -(\lambda + \mu)P_1(t) + \mu P_2(t),$$
$$P'_n(t) = -(\lambda + \mu)P_n(t) + \lambda P_{n-1}(t) + \mu P_{n+1}(t), \qquad n \geq 2.$$

Obtain for the Laplace transform of the generating function

$$P^*(z,s) = \frac{z^{i+1} - (1-z)(\mu - \lambda z)(\alpha_2^i/s)}{-\lambda(z - \alpha_1)(z - \alpha_2)},$$

where α_1 and α_2 are as in the single-channel case. Show that

$$P_n(t) = \left(\sqrt{\frac{\mu}{\lambda}}\right)^{i-n} e^{-(\lambda+\mu)t}[I_{i-n}(2\sqrt{\lambda\mu}\,t) - I_{i+n}(2\sqrt{\lambda\mu}\,t)], \qquad n \geq 1,$$

$$P'_0(t) = \frac{i(\sqrt{\mu/\lambda})^{-i}e^{-(\lambda+\mu)t}I_i(2\sqrt{\lambda\mu}\,t)}{t}.$$

A busy period begins with the arrival of a customer at an idle channel and ends when the channel next becomes idle. This gives rise to a system in which E_0 is an absorbing state; i.e., once that state is reached, the operation stops. Note that $P'_0(t)$ with $i = 1$ gives the distribution of a busy period for a single-channel queue. It is a measure of the rate in which E_0 is reached.

16. To derive the density function $Q_h(t)$ for the time t which elapses before the queue length for the first time reaches the size $h < i$ (where i is the initial number at $t = 0$), we study the following process, which ends once state E_h is reached. This system is given by

$$P'_h(t) = \mu P_{h+1}(t),$$
$$P'_{h+1}(t) = -(\lambda + \mu)P_{h+1}(t) + \mu P_{h+2}(t),$$
$$P'_n(t) = -(\lambda + \mu)P_n(t) + \lambda P_{n-1}(t) + \mu P_{n+1}(t), \qquad n \geq h + 2.$$

(This discussion can be derived from Prob. 15 by a simple translation.) Show that the differential equation in the generating function

$$P(z,t) = \sum_{n=h}^{\infty} P_n(t)z^n$$

is
$$z\frac{\partial P(z,t)}{\partial t} = (1-z)(\mu - \lambda z)[P(z,t) - P_h(t)z^h]$$

and its Laplace transform is given by

$$P^*(z,s) = \frac{z^{i+1} - z^h(1-z)(\mu - \lambda z)P_h^*(s)}{sz - (1-z)(\mu - \lambda s)}.$$

Also show that

$$P_h^*(s) = \frac{\alpha_2^{i-h}}{s}$$

and

$$Q_h^*(s) = P_h^{*\prime}(s) = \alpha_2^{i-h},$$

from which

$$Q_h(t) = \frac{(i-h)(\sqrt{\lambda/\mu})^{i-h}e^{-(\lambda+\mu)t}I_{i-h}(2\sqrt{\lambda\mu}\,t)}{t}.$$

The distribution of a busy period for a single-channel queue is obtained by putting $i = 1$, $h = 0$. Some moments of this function are

$$E(t) = \frac{i-h}{\mu-\lambda},$$

$$\text{var}\ (t) = \frac{(i-h)(\mu+\lambda)}{(\mu-\lambda)^3}.$$

17. The density function $Q_N(t)$ of the time elapsing before the queue grows from size i to size N (at which the operation stops) is obtained by solving

$$\begin{aligned}
P_0'(t) &= -\lambda P_0(t) + \mu P_1(t),\\
P_n'(t) &= -(\lambda+\mu)P_n(t) + \lambda P_{n-1}(t) + \mu P_{n+1}(t), \qquad 1 \le n \le N-2,\\
P_{N-1}'(t) &= -(\lambda+\mu)P_{N-1}(t) + \lambda P_{N-2}(t),\\
P_N'(t) &= \lambda P_{N-1}(t).
\end{aligned}$$

Let

$$P(z,t) = \sum_{n=0}^{N} P_n(t)z^n, \qquad P(z,0) = z^i.$$

Then show that

$$z\frac{\partial P}{\partial t} = (1-z)\{(\mu-\lambda z)[P(z,t) - P_N(t)z^N] - \mu P_0(t)\}$$

and

$$P^*(z,s) = \frac{z^{i+1} - (1-z)[\mu P_0^*(s) + (\mu-\lambda z)z^N P_N^*(s)]}{sz - (1-z)(\mu-\lambda z)},$$

which is a polynomial in z and exists for all values of z, including both zeros of the denominator. Hence $P_0(s)$ and $P_N(s)$ are obtained by setting the numerator equal to zero and substituting the two zeros α_1 and α_2 of the denominator (at each of which the numerator must vanish). We have

$$Q_N^*(s) = P_N^{*\prime}(s) = sP_N^*(s) = \frac{\lambda(\alpha_2^{i+1} - \alpha_1^{i+1}) - \mu(\alpha_2^i - \alpha_1^i)}{\lambda(\alpha_2^{N+1} - \alpha_1^{N+1}) - \mu(\alpha_2^N - \alpha_1^N)}.$$

$$E(t) = \begin{cases} \dfrac{N-i}{\lambda-\mu} - \mu\left[\dfrac{(\mu/\lambda)^i - (\mu/\lambda)^N}{(\lambda-\mu)^2}\right] & \text{for } \dfrac{\lambda}{\mu} < 1,\ \dfrac{\lambda}{\mu} > 1,\\[3ex] \dfrac{N-i}{2\mu}(N+i+1) & \text{for } \dfrac{\lambda}{\mu} = 1. \end{cases}$$

18. Solve the birth-death-process equation for $\lambda_n = n\lambda$, $\mu_n = n\mu$ (linear growth), with initial state E_i at $t = 0$. Note the presence of an absorbing barrier at the origin. Show that the generating function

$$P(z,t) = \sum_{n=0}^{\infty} P_n(t)z^n, \qquad P(z,0) = z^i,$$

satisfies

$$\frac{\partial P}{\partial t} = (z-1)(\lambda z - \mu)\frac{\partial P}{\partial z}.$$

Solve this linear first-order partial differential equation, obtaining

$$P(z,t) = \left\{ \frac{\mu[1 - e^{(\lambda-\mu)t}] - z[\lambda - \mu e^{(\lambda-\mu)t}]}{\mu - \lambda e^{(\lambda-\mu)t} - \lambda z[1 - e^{(\lambda-\mu)t}]} \right\}^i.$$

Show that when $i = 1$

$$P_n(t) = [1 - P_0(t)]\left[1 - \frac{\lambda - \lambda e^{(\lambda-\mu)t}}{\mu - \lambda e^{(\lambda-\mu)t}}\right]\left[\frac{\lambda - \lambda e^{(\lambda-\mu)t}}{\mu - \lambda e^{(\lambda-\mu)t}}\right]^{n-1}, \qquad n \geq 1,$$

$$P_0(t) = \frac{\mu e^{(\lambda-\mu)t} - \mu}{\lambda e^{(\lambda-\mu)t} - \mu}.$$

Find
$$\lim_{t \to \infty} P_n(t).$$

Show that

$$L = E(n) = e^{(\lambda-\mu)t}, \qquad \text{var } (n) = \frac{\lambda + \mu}{\lambda - \mu} e^{(\lambda-\mu)t}[e^{(\lambda-\mu)t} - 1].$$

Note that if $\lambda < \mu$, as $t \to \infty$, $P(z,t) \to 1$, and the population will certainly be extinct. With $i = 1$, $\mu = 0$,

$$P_n(t) = e^{-\lambda t}(1 - e^{-\lambda t})^{n-1}, \qquad n \geq 1,$$

which tends to zero as $t \to \infty$, giving the limiting probability of extinction.

19. In the birth-death equations let $\lambda_n = \lambda$, $\mu_n = a + b(n - 1)$, where a and b are constants. Show that the generating function satisfies

$$\frac{\partial P}{\partial t} = (1 - z)\left[(a - \lambda z)\frac{\partial P}{\partial t} + bz\frac{\partial^2 P}{\partial z^2}\right].$$

When $a = 0$ and λ and b are positive, as $t \to \infty$ the probability of extinction is zero. Also show that

$$E(n) = \frac{\lambda/b}{1 - e^{-\lambda/b}}.$$

20. For the system $M/M/1$, the autocorrelation function for the queue length $n(t)$ by the ergodic theorem (which equates space and time averages for an ergodic process) is given by

$$\psi(t) = \lim \frac{1}{T} \int_0^T n(x)n(x + t)\, dx$$

$$= \sum_{i=0}^{\infty} p_i\left[i \sum_{n=0}^{\infty} nP_{in}(t)\right]$$

$$= \frac{\lambda^2}{(\mu - \lambda)^2} + (\mu - \lambda)\frac{\lambda\mu}{\pi}\int_0^{2\pi} \sin^2\theta\, \frac{e^{-wt}}{w^3}\, d\theta,$$

where
$$w = \mu + \lambda - 2\sqrt{\lambda\mu}\cos\theta.$$

Show that for $t = 0$ the middle expression must give the second moment of the number in the queue. Using Clarke's approximation for the second sum of the middle expression for large t, obtain an approximate expression for $\psi(t)$. Show that, as $t \to \infty$, this equals L^2, where L is the average length of the line.

The cosine transform of $\psi - L^2$, the time-dependent part of which is an even function of t, is

$$w(f) = 4\int_0^\infty [\psi(t) - L^2]\cos(2\pi ft)\, dt = 4\lambda\mu\frac{\mu - \lambda}{\pi}\int_0^{2\pi} \frac{\sin^2\theta\, d\theta}{w^2(w^2 + 4\pi^2 f^2)}.$$

Show that
$$w(f) \to \begin{cases} \dfrac{4\lambda\mu(\mu + \lambda)}{(\mu - \lambda)^4} & \text{when } f \to 0, \\[2mm] \dfrac{4\mu}{(2\pi f)^2} & \text{when } f \to \infty. \end{cases}$$

$w(f)$ is the mean square of the frequency spectrum of the fluctuations of queue length away from the average length. We have, using $E[\cdot]$ for expected values,

$$w(f) = \lim \frac{2}{T} E[|S(f)|^2],$$

where

$$S(f) = \int_0^T [n(t) - L]e^{-2\pi i f t}\, dt.$$

By integrating first over f and then over θ, show that

$$\int_0^\infty w(f)\, df = E[(n - L)^2] = \frac{\lambda\mu}{(\mu - \lambda)^2}.$$

P. Morse [507] points out that, as $\lambda \to \mu$, the low-frequency components of $n(t) - L$ increase rapidly, while the high-frequency components are almost unchanged. Thus, as saturation is approached, $n(t)$ returns more slowly to its average value after displacement. Justify the rough approximations

$$\psi(t) \sim \frac{\lambda^2}{(\mu - \lambda)^2} + \frac{\lambda\mu}{(\mu - \lambda)^2} \exp\left[\frac{-(\mu - \lambda)^2 t}{\lambda}\right]$$

and

$$w(f) \sim \frac{4\lambda^2\mu}{(\mu - \lambda)^4} + \lambda^2(2\pi f)^2.$$

If one defines relaxation time for the stochastic fluctuations of the queue about L as the mean time required for the queue to go from its mean-square deviation away from L back to $1/e$ of this deviation, using the approximation for $\psi(t)$, show that the relaxation time is roughly $2\lambda/(\mu - \lambda)^2$, which is more sensitive to saturation (i.e., as $\lambda \to \mu$) than L.

21. In Eq. (4-140) we make the following simplification, which is in accordance with our assumptions on products of powers of N and $e^{-\alpha t}$.

$$\begin{aligned}
\log \frac{(1 - e^{-\alpha t})^{N-n-1}}{(1 - \beta e^{-\alpha t})^{N+n}} &= (N - n - 1) \log(1 - e^{-\alpha t}) - (N + n) \log(1 - \beta e^{-\alpha t}) \\
&= (N - n - 1)\left(-e^{-\alpha t} - \frac{e^{-2\alpha t}}{2} - \cdots\right) \\
&\qquad - (N + n)\left(-\beta e^{-\alpha t} - \frac{\beta^2 e^{-2\alpha t}}{2} - \cdots\right) \\
&= -N(1 - \beta)e^{-\alpha t}.
\end{aligned}$$

All other products are negligible.

Let

$$me^{-y} = N(1 - \beta)e^{-\alpha t},$$

$$t = \frac{1}{\alpha}[y + \log N + \log(1 - \beta) - \log m].$$

Then

$$e^{-m\alpha t} = \frac{m^m e^{-my}}{N^m(1 - \beta)^m}.$$

Let $m = k + 1$ and, substituting in (4-140), obtain

$$\sum_{k=0}^{n-1} \binom{n}{k+1}(1 - \beta)^{k+2}\beta^{n-(k+1)}\frac{(k + 1)^{k+1}}{k!} e^{-(k+1)y - (k+1)e^{-y}}.$$

The last two factors comprise the well-known extreme-value distribution

$$T_m(y) = \frac{m^m}{(m-1)!} e^{-my - me^{-y}}$$

(the Fisher-Tippett distribution for $m = 1$), whose moments are known:

$$\int_{-\infty}^{\infty} T_m(y) \, dy = 1,$$

$$\int_{-\infty}^{\infty} y T_m(y) \, dy = \log m + \gamma - \sum_{j=1}^{m-1} \frac{1}{j},$$

$$\int_{-\infty}^{\infty} y^2 T_m(y) \, dy = \sum_{j=m}^{\infty} \frac{1}{j^2} + \left(\log m + \gamma - \sum_{j=1}^{m-1} \frac{1}{j}\right)^2,$$

where γ is Euler's constant, $0.5772156649 \cdots$. For $m = 1$, the last sums in the first and second moments are ignored. Verify all the foregoing expressions.

22. According to B. Epstein, extreme-value distributions arise in sampling as follows: If a sample of n independent observations $(x_1, \ldots, x_n)$ is taken from a cumulative distribution function $F(x)$—with density $f(x)$—and if

$$y_n = \min (x_1, \ldots, x_n),$$
$$z_n = \max (x_1, \ldots, x_n),$$

then

$$\text{Prob } (y_n > y) = \text{Prob } (x_1 > y, x_2 > y, \ldots, x_n > y)$$
$$= [1 - F(y)]^n,$$

and

$$G_n(y) = \text{Prob } (y_n \leq y) = 1 - [1 - F(y)]^n,$$
$$g_n(y) = G_n'(y) = nf(y)[1 - F(y)]^{n-1}.$$

Also

$$H_n(z) = \text{Prob } (z_n \leq z) = \text{Prob } (x_1 \leq z, \ldots, x_n \leq z) = [F(z)]^n,$$
$$h_n(z) = H_n'(z) = nf(z)[F(z)]^{n-1}.$$

Thus if $F(x) = 1 - e^{-\lambda x}$, $x \geq 0$,

$$g_n(y) = n\lambda e^{-n\lambda y}, \qquad y \geq 0 \text{ and zero otherwise,}$$
$$h_n(y) = n\lambda e^{-\lambda z}(1 - e^{-\lambda z})^{n-1}, \qquad z \geq 0 \text{ and zero otherwise.}$$

For large n we can define the random variable $\eta_n = nF(y_n)$, and thus for $0 \leq u \leq n$

$$\Gamma_n(u) \equiv \text{Prob } (\eta_n \leq u) = \text{Prob } [nF(y_n) \leq u] = \text{Prob } \left[y_n \leq F^{-1}\left(\frac{u}{n}\right) \right]$$

$$= 1 - \left\{ 1 - F\left[F^{-1}\left(\frac{u}{n}\right) \right] \right\}^n = 1 - \left(1 - \frac{u}{n}\right)^n \to 1 - e^{-u} \equiv \Gamma(u), \qquad u \geq 0,$$

as $n \to \infty$. Thus the density function also converges, i.e.,

$$\gamma(u) \equiv \lim_{n \to \infty} \gamma_n(u) = e^{-u}, \qquad u \geq 0.$$

Also

$$y_n \to y \equiv F^{-1}\left(\frac{\eta}{n}\right).$$

Show that, if

$$F(x) = \begin{cases} 0, & x < 0, \\ x^\alpha, & 0 \leq x \leq 1, \\ 1, & x > 1, \end{cases}$$

then for large n the distribution of y_n is

$$\Gamma_n(y) = 1 - e^{-ny^\alpha}, \qquad y \geq 0 \text{ and zero otherwise.}$$

In the case of z_n for large n we use

$$\xi_n = n[1 - F(z_n)].$$

For $0 \leq v \leq n$

$$\Lambda_n(v) \equiv \text{Prob } (\xi_n \leq v) = \text{Prob} \left[F(z_n) \geq 1 - \frac{v}{n} \right]$$

$$= \text{Prob} \left[z_n \geq F^{-1}\left(1 - \frac{v}{n}\right) \right] = 1 - H_n\left[F^{-1}\left(1 - \frac{v}{n}\right) \right]$$

$$= 1 - \left\{ F\left[F^{-1}\left(1 - \frac{v}{n}\right) \right] \right\}^n = 1 - \left(1 - \frac{v}{n}\right)^n \to 1 - e^{-v}, \qquad v \geq 0,$$

as $n \to \infty$. Thus the random variable ξ_n converges in distribution to the random variable ξ, and the density function of the latter is

$$\lambda(v) = \lim_{n \to \infty} \lambda_n(v) = e^{-v}, \qquad v \geq 0.$$

Hence $z_n \to z$, where

$$z = F^{-1}\left(1 - \frac{\xi}{n}\right).$$

If $F(x) = 1 - e^{-\lambda x}$ for $x \geq 0$ and zero otherwise, then $\xi_n = ne^{-\lambda z_n}$ and

$$z_n \sim \frac{\log n}{\lambda} - \frac{\log \xi_n}{\lambda}.$$

Then $\qquad \Lambda_n(z) = e^{-ne^{-\lambda z}} \qquad$ for $z \geq 0$ and zero otherwise,

and $\qquad \lambda_n(z) = n\lambda e^{-\lambda z - ne^{-\lambda z}} \qquad$ for $z \geq 0$ and zero otherwise,

which is a Type-I distribution of largest values.

The modal value of $h_n(z)$ is obtained by solving $h_n'(z) = 0$.

The following are among the uses of the theory of extreme values: (1) It gives a scale for plotting the cumulative distribution of observed breakdown values linearly, i.e., $\log \log [1/\Lambda(z)]$ vs. z or $\log z$, and (2) it indicates, for example, that larger specimens are more susceptible to the operating forces. Show that, if $f(x) = 1/x^2$ for $x \leq -1$ and zero otherwise, $\gamma_n(y) = (n/y^2)e^{n/y}$ for $y < 0$ and zero otherwise, which is a Type-II asymptotic distribution of smallest values. If

$$F(x) = \begin{cases} 0, & x < 0, \\ \dfrac{x}{A}, & 0 \leq x \leq A, \\ 1, & x > A, \end{cases}$$

show that

$$h_n(z) = \frac{nz^{n-1}}{A^n}, \qquad 0 \leq z \leq A \text{ and zero otherwise.}$$

23. Using Reuter's theorem, prove the existence and uniqueness of solutions for the single-server and for the infinite-number-of-servers queueing problems.

24. Solve the birth-death equations with λ_n replaced by $\lambda(t)$ and μ_n by $n\mu(t)$ and also with $n\lambda(t)$ and $n\mu(t)$, respectively. Introduce a Poisson distribution for the initial number and compute the probabilities.

25. Let $\xi = \zeta + wg(\xi)$. By differentiating with respect to w and then with respect to ζ, the two independent variables, show that $\dfrac{\partial \xi}{\partial w} = g(\xi)\dfrac{\partial \xi}{\partial \zeta}$.

Let $u \equiv F(\xi)$ and assume that $\xi = f(u)$ and hence $g(\xi) = g[f(u)] \equiv \phi(u)$. Then

$$\frac{\partial u}{\partial w} = \frac{du}{d\xi}\frac{\partial \xi}{\partial w}, \qquad \frac{\partial u}{\partial \zeta} = \frac{du}{d\xi}\frac{\partial \xi}{\partial \zeta}.$$

Show that

$$\frac{\partial u}{\partial w} = g(\xi)\frac{\partial u}{\partial \zeta}. \tag{1}$$

Now

$$\frac{\partial}{\partial w}\left[\phi(u)\frac{\partial u}{\partial \zeta}\right] = \frac{\partial}{\partial \zeta}\left[\phi(u)\frac{\partial u}{\partial w}\right]$$

since each side is equal to

$$\phi'(u)\frac{\partial u}{\partial w}\frac{\partial u}{\partial \zeta} + \phi(u)\frac{\partial^2 u}{\partial w\,\partial \zeta}.$$

It follows from this that

$$\frac{\partial^2 u}{\partial w^2} = \frac{\partial}{\partial w}\left[g(\xi)\frac{\partial u}{\partial \zeta}\right] = \frac{\partial}{\partial \zeta}\left[g(\xi)\frac{\partial u}{\partial w}\right] = \frac{\partial}{\partial \zeta}\left\{[g(\xi)]^2\frac{\partial u}{\partial \zeta}\right\},$$

having used (1) on the right.

By induction we have

$$\frac{\partial^n u}{\partial w^n} = \frac{\partial^{n-1}}{\partial \zeta^{n-1}}\left\{[g(\xi)]^n\frac{\partial u}{\partial \zeta}\right\}.$$

Derive Lagrange's formula given in the text by expanding u in terms of a series of powers of w, valid for small values of w; i.e.,

$$u = F(\xi) = \sum_{n=0}^{\infty}\frac{w^n}{n!}\frac{\partial^n F}{\partial w^n}\bigg|_{w=0}.$$

CHAPTER 5

THE CASE OF SMALL BARRIERS

5-1. Introduction

Matrix methods have been used to solve the birth-death equations. Because the equations are linear as differential and as difference equations, it is possible to give the equations in matrix notation and write the solution at least formally. In order to obtain the solution explicitly, one must compute the characteristic roots, or eigenvalues, of the matrix of coefficients. The material of this chapter has two objectives. The first is to alert the reader to the possible use of matrix methods in queueing theory, and the second is to provide examples of how to use these methods. Together with the material of the previous chapter, there are a variety of tools from analysis and from algebra that apply, to some extent, to the mathematical ideas of queueing theory, which require some mastery.

A discussion of the solution of the infinite system of birth-death equations is given at the end of the chapter in outline form. It, at least, requires the early parts of the chapter as background.

We begin by giving explicit solutions to the birth-death model with absorbing barriers for those cases of the upper barrier for which it is possible to develop the solution. The problem is essentially related to the solvability of equations. In the process we give an outline of the theory required [234].

The reader without an adequate background may follow the computations formally, accepting as much of the arguments as possible and then resorting to the references for more complete explanations if needed.

5-2. Equations

We give the equations of the birth and death process with reflecting and with absorbing barriers. Recall that a reflecting state is a state from which transitions are possible only in one direction, either to a higher state or to a lower state. Thus it acts as a reflecting barrier. An absorbing state is a state from which, when once reached by a system, it is not possible to move out. It is a trap state or an absorbing barrier. In each case considered there are two barriers, one at E_0 and the other at E_N, which are both reflecting in the first case and absorbing in the second case.

Reflecting barriers:

$$P_0'(t) = -\lambda_0 P_0(t) + \mu_1 P_1(t),$$
$$P_n'(t) = -(\lambda_n + \mu_n)P_n(t) + \lambda_{n-1}P_{n-1}(t) + \mu_{n+1}P_{n+1}(t),$$
$$n = 1, \ldots, N-1, \qquad (5\text{-}1)$$
$$P_N'(t) = -\mu_N P_N(t) + \lambda_{N-1}P_{N-1}(t).$$

Absorbing barriers [obtained by putting $\lambda_0 = 0$, $\mu_N = 0$ in (5-1)]:

$$P_0'(t) = \mu_1 P_1(t),$$
$$P_n'(t) = -(\lambda_n + \mu_n)P_n(t) + \lambda_{n-1}P_{n-1}(t) + \mu_{n+1}P_{n+1}(t),$$
$$n = 1, \ldots, N-1, \qquad (5\text{-}2)$$
$$P_N'(t) = \lambda_{N-1}P_{N-1}(t).$$

If the system is in state E_i at $t = 0$ [for convenience we have omitted in the above equations the subscript i in $P_{in}(t)$ which shows the dependence on the initial condition] in both (5-1) and (5-2) we must naturally have

$$P_{in}(0) = \delta_{in} \equiv \begin{cases} 0, & n = i, \\ 1, & n \neq i, \end{cases}$$

where δ_{in} is the Kronecker symbol.

Exercise: Write the birth-death equations for a reflecting barrier at E_0 and an absorbing barrier at E_N, and vice versa.

5-3. Method of Solution

In the matrix notation the above systems may be represented by equations of the form $\bar{P}'(t) = B\bar{P}$, as will be explained in the next paragraphs. We have used $\bar{P}(t)$, the transpose of $P(t)$, and B, the transpose of the birth-death coefficient matrix A. We do our calculations using B. Of course the analysis can also be done using $P'(t) = P(t)A$. Note that, for the solution of

$$\frac{d\bar{P}(t)}{dt} = B\bar{P}(t),$$

we may formally write

$$\bar{P}(t) = e^{Bt}C,$$

where the matrix C is determined from the initial conditions, i.e., by putting $t = 0$. It immediately follows that C is the identity matrix;

hence $\bar{P}(t) = e^{Bt}$. These ideas apply to all the cases studied below. For a finite n by n matrix B the solution may be written as follows:

$$P(t) = e^{Bt} = \sum_{j=0}^{\infty} \frac{1}{j!} (Bt)^j,$$

but in the case of an infinite matrix the series does not always converge, especially when the elements become unbounded.

In the case of absorbing and reflecting barriers at both ends, B is a finite n by n matrix with constant elements. Note that for absorbing and reflecting barriers the appropriate coefficients of B are set equal to zero, i.e., λ_0, λ_N, and μ_N.

The matrix B considered here is the matrix of coefficients of an Nth section; i.e., we take the first $N + 1$ of the birth-death-process equations given in (4-1), cutting off a term from the last equation in order to have a finite matrix. It is given by

$$B = \begin{bmatrix} -\lambda_0 & \mu_1 & 0 & 0 & \cdots & 0 \\ \lambda_0 & -(\lambda_1 + \mu_1) & \mu_2 & 0 & \cdots & 0 \\ 0 & \lambda_1 & -(\lambda_2 + \mu_2) & \mu_3 & \cdots & 0 \\ 0 & 0 & \lambda_2 & -(\lambda_3 + \mu_3) & \cdots & 0 \\ \cdots & \cdots & \cdots & & \cdots & \cdots \\ 0 & 0 & 0 & 0 & \cdots & \mu_N \\ 0 & 0 & 0 & 0 & \cdots & -(\lambda_N + \mu_N) \end{bmatrix}.$$

On putting $\lambda_N = 0$, B becomes the coefficient matrix of (5-1). By setting λ_0, λ_N, $\mu_N = 0$, B becomes the coefficient matrix of (5-2). Also

$$\bar{P}(t) \equiv \begin{bmatrix} P_{00}(t) & P_{10}(t) & \cdots & P_{N0}(t) \\ P_{01}(t) & P_{11}(t) & \cdots & P_{N1}(t) \\ \cdots & \cdots & \cdots & \cdots \\ P_{0N}(t) & P_{1N}(t) & \cdots & P_{NN}(t) \end{bmatrix},$$

and $\bar{P}'(t)$ is defined by taking the derivative of each element in $\bar{P}(t)$.

One has the following spectral form from the calculus of bounded operators applied to an entire function $f(B)$ of a bounded operator B or from Sylvester's well-known theorem in this field [234]:

$$f(B) = \sum_{i=1}^{k} \sum_{m=0}^{m_i - 1} \frac{(B - \alpha_i I)^m}{m!} f^{(m)}(\alpha_i) Z(\alpha_i). \tag{5-3}$$

Here k is the number of distinct characteristic roots of the matrix B, m_i is the multiplicity of the ith root α_i of the characteristic equation $|\alpha I - B| = 0$, $f^{(m)}(\alpha_i)$ is the mth-order formal derivative of f evaluated at α_i, and $Z(\alpha_i)$ are complete orthogonal idempotent matrices of the matrix B; that is, they have the properties

$$\sum_{i=1}^{k} Z(\alpha_i) = I, \qquad Z(\alpha_i)Z(\alpha_j) = 0 \qquad (i \neq j), \qquad Z^2(\alpha_i) = Z(\alpha_i),$$

where I and 0 are the identity and null matrices, respectively. For a more detailed account the reader is urged to go on.

When the characteristic roots are all distinct, one has for an nth-order matrix B

$$f(B) = \sum_{i=1}^{n} f(\alpha_i)Z(\alpha_i), \tag{5-4}$$

where
$$Z(\alpha_i) = \frac{\displaystyle\prod_{j \neq i} (\alpha_j I - B)}{\displaystyle\prod_{j \neq i} (\alpha_j - \alpha_i)}. \tag{5-5}$$

If $f(B) = e^{Bt}$ and the characteristic roots of B are distinct, we have the spectral resolution of $f(B)$ given by

$$e^{Bt} = \sum_{i=1}^{n} e^{\alpha_i t} Z(\alpha_i). \tag{5-6}$$

The case of multiple characteristic roots is derived from the confluent form of Sylvester's theorem. If we write for brevity

$$f(B) = \sum_{i=1}^{k} T(\alpha_i), \tag{5-7}$$

where k is the number of distinct roots, then

$$T(\alpha_i) = f(\alpha_i)Z_{m_i-1}(\alpha_i) + f'(\alpha_i)Z_{m_i-2}(\alpha_i) + \frac{f''(\alpha_i)}{2!} Z_{m_i-3}(\alpha_i) + \cdots . \tag{5-8}$$

Here m_i refers to the multiplicity of the root α_i and

$$Z_{m_i}(\alpha_i) = \frac{1}{m_i!} \frac{d^{m_i}}{d\alpha^{m_i}} \frac{F(\alpha)}{\Delta_i(\alpha)} \bigg|_{\alpha = \alpha_i}, \tag{5-9}$$

where
$$F^{(m)}(\alpha_i) = m!(-1)^{n-m-1}(\alpha_i I - B)^{m_i-m-1} \prod_{j \neq i} (\alpha_j I - B), \tag{5-10}$$

giving the mth-order derivative of F and

$$\Delta_i(\alpha) = \prod_{\substack{j \neq i \\ n}} (\alpha - \alpha_j)$$
$$\prod_{\substack{j=1 \\ j \neq i}} (\alpha - \alpha_j). \tag{5-11}$$

We now apply these ideas to the case of a special birth and death process with absorbing barriers at E_0 and E_2 since, in this case, it is possible to compute explicitly the characteristic roots of the matrix. We have carried this through for the case of absorbing barriers at E_0 and E_3, E_0 and E_4, and E_0 and E_5.

We first study cases in which $\lambda_n = n\lambda$, $\mu_n = n\mu$. Let $N = 2$. Consider first the solution of the absorbing-barrier case for which we therefore have, with the above assumption,

$$B = \begin{bmatrix} 0 & \mu & 0 \\ 0 & -(\lambda + \mu) & 0 \\ 0 & \lambda & 0 \end{bmatrix}$$

and

$$\alpha I - B = \begin{bmatrix} \alpha & -\mu & 0 \\ 0 & \alpha + \lambda + \mu & 0 \\ 0 & -\lambda & \alpha \end{bmatrix}.$$

A word of explanation as to the physical significance of this problem may be desirable before we proceed to the solution. In a single-channel queue this corresponds to Poisson arrivals, whose arrival rate depends on the number already present, and to exponential service with service rate also depending on the number in the queue. However, once the number reaches a prescribed size N, the operation stops completely. This is also the case once the queue becomes empty. If parasites in the body of a host obey the above birth (arrival) and death (service) distributions, then N might correspond to the number of parasites which makes the host ill, and zero corresponds to the case in which the parasites are finally defeated and the body is cured. Their birth and death rates are proportional to the number present. Note that we have treated this problem for large N at the end of the previous chapter. Here, however, we treat cases for small values of N.

Now the characteristic equation $|\alpha I - B| = 0$ obtained by equating the determinant to zero gives

$$\alpha^2[\alpha + (\lambda + \mu)] = 0.$$

The characteristic roots of B are $\alpha_1 = \alpha_2 = 0$ and $\alpha_3 = -(\lambda + \mu)$.

From (5-7) we have

$$e^{Bt} = f(B) = \sum_{i=1}^{k} T(\alpha_i) = T(0) + T[-(\lambda + \mu)]. \tag{5-12}$$

The multiplicity of the first root is $m_i = 2$. Also $f(0) = e^{0t} = 1$,

$$f'(0) = te^{0t} = t;$$

thus we obtain from (5-8)

$$T(0) = f(0)Z_1(0) + f'(0)Z_0(0) + \frac{f''(0)}{2!} Z_{-1}(0) + \cdots$$
$$= Z_1(0) + tZ_0(0). \quad (5\text{-}13)$$

The derivatives of f are with respect to B as a variable. We have from (5-9)

$$Z_1(0) = \frac{d^{(1)}}{d\alpha^{(1)}} \frac{F(\alpha)}{\Delta_1(\alpha)} \bigg|_{\alpha=0} = \frac{\Delta_1(0)F'(0) - F(0)\Delta_1'(0)}{[\Delta_1(0)]^2} \quad (5\text{-}14)$$

and

$$Z_0(0) = \frac{d^{(0)}}{d\alpha^{(0)}} \frac{F(\alpha)}{\Delta_1(\alpha)} \bigg|_{\alpha=0} = \frac{F^{(0)}(0)}{\Delta_1(0)}. \quad (5\text{-}15)$$

From (5-10) we have

$$F^{(1)}(0) = 1!(-1)^1(0I - B)^0 \prod_{i \neq j} (\alpha_j I - B)$$
$$= (-1)[-(\lambda + \mu)I - B] \quad (5\text{-}16)$$

and

$$F^{(0)}(0) = (-1)^2(0I - B)[-(\lambda + \mu)I - B]$$
$$= (-B)[-(\lambda + \mu)I - B]. \quad (5\text{-}17)$$

Finally, from (5-11) we have

$$\Delta_1(0) = [0 - (-\lambda - \mu)] = \lambda + \mu,$$
$$\Delta_1'(0) = 1.$$

Now we may compute $T(0)$, which is given by

$$T(0) = \frac{-(\lambda + \mu)[-(\lambda + \mu)I - B] - (-B)[-(\lambda + \mu)I - B]}{(\lambda + \mu)^2}$$
$$+ \frac{t(-B)[-(\lambda + \mu)I - B]}{\lambda + \mu}. \quad (5\text{-}18)$$

But

$$-(\lambda + \mu)I - B = \begin{bmatrix} -(\lambda + \mu) & -\mu & 0 \\ 0 & 0 & 0 \\ 0 & -\lambda & -(\lambda + \mu) \end{bmatrix}, \quad (5\text{-}19)$$

$$0I - B = -B = \begin{bmatrix} 0 & -\mu & 0 \\ 0 & \lambda + \mu & 0 \\ 0 & -\lambda & 0 \end{bmatrix}, \quad (5\text{-}20)$$

and

$$(-B)[-(\lambda + \mu)I - B] = 0. \quad (5\text{-}21)$$

Therefore

$$T(0) = \frac{1}{\lambda + \mu} \begin{bmatrix} \lambda + \mu & \mu & 0 \\ 0 & 0 & 0 \\ 0 & \lambda & \lambda + \mu \end{bmatrix} \quad (5\text{-}22)$$

In the same manner, $T[-(\lambda + \mu)]$ is computed. Since the multiplicity of the root is unity in this case and $f(-\lambda - \mu) = e^{-(\lambda+\mu)t}$, we have

$$T[-(\lambda + \mu)] = e^{-(\lambda+\mu)t}Z_0[-(\lambda + \mu)], \tag{5-23}$$

where

$$Z_0[-(\lambda + \mu)] = \frac{d^{(0)}}{d\alpha^{(0)}} \frac{F[-(\lambda + \mu)]}{\Delta_2[-(\lambda + \mu)]} = \frac{F^{(0)}[-(\lambda + \mu)]}{\Delta_2[-(\lambda + \mu)]}, \tag{5-24}$$

$$F^{(0)}[-(\lambda + \mu)] = (-B)^2, \tag{5-25}$$

and

$$\Delta_2[-(\lambda + \mu)] = [-(\lambda + \mu) - 0]^2 = (\lambda + \mu)^2. \tag{5-26}$$

Thus

$$T[-(\lambda + \mu)] = \frac{e^{-(\lambda+\mu)t}}{(\lambda + \mu)^2}
\begin{bmatrix}
0 & -\mu(\lambda + \mu) & 0 \\
0 & (\lambda + \mu)^2 & 0 \\
0 & -\lambda(\lambda + \mu) & 0
\end{bmatrix}$$

$$= \frac{e^{-(\lambda+\mu)t}}{\lambda + \mu}
\begin{bmatrix}
0 & -\mu & 0 \\
0 & \lambda + \mu & 0 \\
0 & -\lambda & 0
\end{bmatrix}. \tag{5-27}$$

Finally,

$$e^{Bt} =
\begin{bmatrix}
P_{00}(t) & P_{10}(t) & P_{20}(t) \\
P_{01}(t) & P_{11}(t) & P_{21}(t) \\
P_{02}(t) & P_{12}(t) & P_{22}(t)
\end{bmatrix}$$

$$= \frac{1}{\lambda + \mu}
\begin{bmatrix}
\lambda + \mu & -\mu e^{-(\lambda+\mu)t} + \mu & 0 \\
0 & e^{-(\lambda+\mu)t}(\lambda + \mu) & 0 \\
0 & -\lambda e^{-(\lambda+\mu)t} + \lambda & \lambda + \mu
\end{bmatrix}. \tag{5-28}$$

We have, for example, $P_{00}(t) = 1$, $P_{01}(t) = 0$,

$$P_{12}(t) = \frac{\lambda - \lambda e^{-(\lambda+\mu)t}}{\lambda + \mu},$$

etc.   Note that the sum of elements in each column is unity.   Also putting $t = 0$ yields the identity matrix for the initial conditions, as it should.

For the absorbing-barrier case with $N = 3$ (still using $\lambda_n = n\lambda$, $\mu_n = n\mu$) we have the characteristic roots $\alpha_1 = \alpha_2 = 0$,

$$\alpha_3 = \frac{-(3\lambda + 3\mu) + \sqrt{\lambda^2 + 10\lambda\mu + \mu^2}}{2} \quad \text{and}$$

$$\alpha_4 = \frac{-(3\lambda + 3\mu) - \sqrt{\lambda^2 + 10\lambda\mu + \mu^2}}{2}. \tag{5-29}$$

These are solutions of the characteristic equation

$$\alpha^4 + (3\lambda + 3\mu)\alpha^3 + 2(\lambda^2 + \lambda\mu + \mu^2)\alpha^2 = 0.$$

Now to obtain e^{Bt} we proceed in the same manner as for $N = 2$.

We have

$$e^{Bt} = \sum_{i=1}^{k} T(\alpha_i) = T(0) + T(\alpha_3) + T(\alpha_4), \tag{5-30}$$

$$T(0) = Z_1(0) + tZ_0(0), \tag{5-31}$$

$$Z_1(0) = \frac{\Delta_1(0)F'(0) - F(0)\Delta_1'(0)}{[\Delta_1(0)]^2}, \qquad Z_0(0) = \frac{F(0)}{\Delta_1(0)}, \tag{5-32}$$

$$F'(0) = f(\alpha_3)f(\alpha_4), \qquad F(0) = -f(0)f(\alpha_3)f(\alpha_4), \tag{5-33}$$

$$\Delta_1(0) = \alpha_3\alpha_4, \qquad \Delta_1'(0) = -(\alpha_3 + \alpha_4) \tag{5-34}$$

$$f(\alpha_i) \equiv (\alpha_i I - B).$$

Then

$$T(0) = \frac{\alpha_3\alpha_4 f(\alpha_3)f(\alpha_4) - (\alpha_3 + \alpha_4)f(0)f(\alpha_3)f(\alpha_4) - \alpha_3\alpha_4 t f(0)f(\alpha_3)f(\alpha_4)}{\alpha_3^2\alpha_4^2},$$

$$\tag{5-35}$$

Also

$$T(\alpha_3) = e^{\alpha_3 t}Z_0(\alpha_3), \tag{5-36}$$

$$Z_0(\alpha_3) = \frac{F(\alpha_3)}{\Delta_3(\alpha_3)}, \tag{5-37}$$

$$F(\alpha_3) = -[f(0)]^2 f(\alpha_4), \tag{5-38}$$

$$\Delta_3(\alpha_3) = \alpha_3^3 - \alpha_3^2\alpha_4. \tag{5-39}$$

Hence

$$T(\alpha_3) = -e^{\alpha_3 t}\frac{[f(0)]^2 f(\alpha_4)}{\alpha_3^2(\alpha_3 - \alpha_4)}. \tag{5-40}$$

By interchanging α_3 and α_4 in $T(\alpha_3)$ one easily obtains $T(\alpha_4)$. Then

$$T(\alpha_4) = -e^{\alpha_4 t}\frac{[f(0)]^2 f(\alpha_3)}{\alpha_4^2(\alpha_4 - \alpha_3)}. \tag{5-41}$$

Finally, if we define

$$D = \frac{1}{\alpha_3^2\alpha_4^2(\alpha_3 - \alpha_4)}, \tag{5-42}$$

we have

$$P_{00}(t) = [\alpha_3^2\alpha_4^2(\alpha_3 - \alpha_4)]D, \tag{5-43}$$

$P_{01}(t)$, $P_{02}(t)$, $P_{03}(t)$ all are 0,

$$P_{10}(t) = \{-\alpha_3\alpha_4(\alpha_3 - \alpha_4)\mu(\lambda + \mu + \alpha_3 + \alpha_4)$$
$$+ e^{\alpha_3 t}\alpha_4^2\mu[(\lambda + \mu)(\lambda + \mu + \alpha_4) + 2\lambda\mu]$$
$$- e^{\alpha_4 t}\alpha_3^2\mu[(\lambda + \mu)(\lambda + \mu + \alpha_3) + 2\lambda\mu]\}D,$$

$$P_{11}(t) = \{-e^{\alpha_3 t}\alpha_4^2[2\lambda\mu(4\lambda + 4\mu + \alpha_4) + (\lambda + \mu)^2(\lambda + \mu + \alpha_4)]$$
$$+ e^{\alpha_4 t}\alpha_3^2[2\lambda\mu(4\lambda + 4\mu + \alpha_3) + (\lambda + \mu)^2(\lambda + \mu + \alpha_3)]\}D,$$

$$P_{12}(t) = \{e^{\alpha_3 t}\alpha_4^2\lambda[(\lambda + \mu)(7\lambda + 7\mu + 3\alpha_4) + 2\lambda\mu]$$
$$- e^{\alpha_4 t}\alpha_3^2\lambda[(\lambda + \mu)(7\lambda + 7\mu + 3\alpha_3) + 2\lambda\mu]\}D,$$

$$P_{13}(t) = [\alpha_3\alpha_4(\alpha_3 - \alpha_4)2\lambda^2 - e^{\alpha_3 t}\alpha_4^2 2\lambda^2(3\lambda + 3\mu + \alpha_4)$$
$$+ e^{\alpha_4 t}\alpha_3^2 2\lambda^2(3\lambda + 3\mu + \alpha_3)]D,$$

$$P_{20}(t) = [\alpha_3\alpha_4(\alpha_3 - \alpha_4)2\mu^2 - e^{\alpha_3 t}\alpha_4^2 2\mu^2(3\lambda + 3\mu + \alpha_4)$$
$$+ e^{\alpha_4 t}\alpha_3^2 2\mu^2(3\lambda + 3\mu + \alpha_3)]D,$$

$$P_{21}(t) = \{e^{\alpha_3 t}\alpha_4^2 2\mu[(\lambda + \mu)(7\lambda + 7\mu + 3\alpha_4) + 2\lambda\mu]$$
$$- e^{\alpha_4 t}\alpha_3^2 2\mu[(\lambda + \mu)(7\lambda + 7\mu + 3\alpha_3) + 2\lambda\mu]\}D,$$

$$P_{22}(t) = \{-e^{\alpha_3 t}\alpha_4^2[2\lambda\mu(5\lambda + 5\mu + \alpha_4) + (2\lambda + 2\mu + \alpha_4)(2\lambda + 2\mu)^2]$$
$$+ e^{\alpha_4 t}\alpha_3^2[2\lambda\mu(5\lambda + 5\mu + \alpha_3) + (2\lambda + 2\mu + \alpha_3)(2\lambda + 2\mu)^2]\}D,$$

$$P_{23}(t) = \{-\alpha_3\alpha_4(\alpha_3 - \alpha_4)2\lambda(2\lambda + 2\mu + \alpha_3 + \alpha_4)$$
$$+ e^{\alpha_3 t}\alpha_4^2 2\lambda[(2\lambda + 2\mu)(2\lambda + 2\mu + \alpha_4) + 2\lambda\mu]$$
$$- e^{\alpha_4 t}\alpha_3^2 2\lambda[(2\lambda + 2\mu)(2\lambda + 2\mu + \alpha_3) + 2\lambda\mu]\}D,$$

$P_{30}(t)$, $P_{31}(t)$, and $P_{32}(t)$ are all equal to zero, and

$$P_{33}(t) = [\alpha_3^2\alpha_4^2(\alpha_3 - \alpha_4)]D.$$

5-4. General Case

For the general case we solve the problem for the Nth section of the coefficient matrix B for $N = 2$. The corresponding reflecting- and absorbing-barrier solutions are easily obtained from this case.

The characteristic equation $|\alpha I - B| = 0$ for $N = 2$ is

$$\alpha^3 + (\lambda_0 + \lambda_1 + \lambda_2 + \mu_1 + \mu_2)\alpha^2$$
$$+ (\lambda_0\lambda_1 + \lambda_0\lambda_2 + \lambda_0\mu_2 + \lambda_1\lambda_2 + \lambda_2\mu_1 + \mu_1\mu_2)\alpha + \lambda_0\lambda_1\lambda_2 = 0. \quad (5-44)$$

To solve this cubic put

$$\alpha = \beta - \frac{b}{3}.$$

Then the original equation becomes

$$\beta^3 + p\beta + q = 0,$$

where

$$p = c - \frac{b^2}{3}$$

and

$$q = d - \frac{bc}{3} + \frac{2b^3}{27}.$$

Thus

$$p = \tfrac{1}{3}(\lambda_0\lambda_1 + \lambda_0\lambda_2 + \lambda_0\mu_2 + \lambda_1\lambda_2 + \lambda_2\mu_1 + \mu_1\mu_2 - 2\lambda_0\mu_1 - \lambda_1^2 - 2\lambda_1\mu_1$$
$$- 2\lambda_1\mu_2 - \lambda_2^2 - 2\lambda_2\mu_2 - \mu_1^2 - \mu_2^2 - \lambda_0^2) \quad (5-45)$$

and

$$q = \tfrac{1}{27}[2(\lambda_1 + \mu_1)^3 + 2(\lambda_2 + \mu_2)^3 - \lambda_0^2(3\lambda_2 + 3\mu_2 + 3\lambda_1 - 2\lambda_0 - 6\mu_1)$$
$$- 3\lambda_1^2(\lambda_2 + \lambda_0 - 2\mu_2) - 3\lambda_2^2(\lambda_0 + \lambda_1 + \mu_1)$$
$$- 3\mu_1^2(\lambda_2 + \mu_2 - 2\lambda_0) - 3\mu_2^2(\lambda_0 - 2\lambda_1 + \mu_1)$$
$$- 3\lambda_0(-4\lambda_1\lambda_2 + 2\lambda_2\mu_1 + 2\mu_1\mu_2 + 2\lambda_1\mu_2 + 2\lambda_2\mu_2 - \lambda_1\mu_1)$$
$$- 3\lambda_1(2\lambda_2\mu_1 - \mu_1\mu_2 + \lambda_2\mu_2) - 6\lambda_2\mu_1\mu_2]. \quad (5-46)$$

Let $n = \sqrt{-3/4p}$ and

$$\cos 3\theta = -\frac{q}{2}\left(-\frac{27}{p^3}\right)^{1/2},$$

from which

$$\theta = \frac{1}{3}\cos^{-1}\left[-\frac{q}{2}\left(-\frac{27}{p^3}\right)^{1/2}\right].$$

By a well-known method of solving cubic equations, we now have for the three roots of the above reduced cubic

$$\beta_1 = \frac{1}{n} \cos \theta, \qquad \beta_2 = \frac{1}{n} \cos (\theta + 120), \qquad \text{and} \qquad \beta_3 = \frac{1}{n} \cos (\theta + 240).$$

The desired roots are

$$\alpha_1 = \beta_1 - \frac{\lambda_0 + \lambda_1 + \lambda_2 + \mu_1 + \mu_2}{3},$$

$$\alpha_2 = \beta_2 - \frac{\lambda_0 + \lambda_1 + \lambda_2 + \mu_1 + \mu_2}{3},$$

$$\alpha_3 = \beta_3 - \frac{\lambda_0 + \lambda_1 + \lambda_2 + \mu_1 + \mu_2}{3}.$$

The solution for $N = 2$ is given by

$$P_{00}(t) = \frac{e^{\alpha_1 t}[(\lambda_0 + \alpha_2)(\lambda_0 + \alpha_3) + \lambda_0\mu_1]}{(\alpha_1 - \alpha_2)(\alpha_1 - \alpha_3)}$$
$$+ \frac{e^{\alpha_2 t}[(\lambda_0 + \alpha_1)(\lambda_0 + \alpha_3) + \lambda_0\mu_1]}{(\alpha_2 - \alpha_1)(\alpha_2 - \alpha_3)}$$
$$+ \frac{e^{\alpha_3 t}[(\lambda_0 + \alpha_1)(\lambda_0 + \alpha_2) + \lambda_0\mu_1]}{(\alpha_3 - \alpha_1)(\alpha_3 - \alpha_2)},$$

$$P_{01}(t) = \frac{e^{\alpha_1 t}\lambda_0[-(\lambda_0 + \alpha_3) - (\lambda_1 + \mu_1 + \alpha_2)]}{(\alpha_1 - \alpha_2)(\alpha_1 - \alpha_3)}$$
$$- \frac{e^{\alpha_2 t}\lambda_0[(\lambda_0 + \alpha_3) + (\lambda_1 + \mu_1 + \alpha_1)]}{(\alpha_2 - \alpha_1)(\alpha_2 - \alpha_3)}$$
$$- \frac{e^{\alpha_3 t}\lambda_0[(\lambda_0 + \alpha_2) + (\lambda_1 + \mu_1 + \alpha_1)]}{(\alpha_3 - \alpha_1)(\alpha_3 - \alpha_2)},$$

$$P_{02}(t) = \frac{e^{\alpha_1 t}\lambda_0\lambda_1}{(\alpha_1 - \alpha_2)(\alpha_1 - \alpha_3)} + \frac{e^{\alpha_2 t}\lambda_0\lambda_1}{(\alpha_2 - \alpha_1)(\alpha_2 - \alpha_3)}$$
$$+ \frac{e^{\alpha_3 t}\lambda_0\lambda_1}{(\alpha_3 - \alpha_1)(\alpha_3 - \alpha_2)},$$ (5-47)

$$P_{10}(t) = - \frac{e^{\alpha_1 t}\mu_1[(\lambda_0 + \alpha_2) + (\lambda_1 + \mu_1 + \alpha_3)]}{(\alpha_1 - \alpha_2)(\alpha_1 - \alpha_3)}$$
$$- \frac{e^{\alpha_2 t}\mu_1[(\lambda_0 + \alpha_1) + (\lambda_1 + \mu_1 + \alpha_3)]}{(\alpha_2 - \alpha_1)(\alpha_2 - \alpha_3)}$$
$$- \frac{e^{\alpha_3 t}\mu_1[(\lambda_0 + \alpha_1) + (\lambda_1 + \mu_1 + \alpha_2)]}{(\alpha_3 - \alpha_1)(\alpha_3 - \alpha_2)},$$

$$P_{11}(t) = \frac{e^{\alpha_1 t}[\lambda_0\mu_1 + (\lambda_1 + \mu_1 + \alpha_2)(\lambda_1 + \mu_1 + \alpha_3) + \lambda_1\mu_2]}{(\alpha_1 - \alpha_2)(\alpha_1 - \alpha_3)}$$
$$+ \frac{e^{\alpha_2 t}[\lambda_0\mu_1 + (\lambda_1 + \mu_1 + \alpha_1)(\lambda_1 + \mu_1 + \alpha_3) + \lambda_1\mu_2]}{(\alpha_2 - \alpha_1)(\alpha_2 - \alpha_3)}$$
$$+ \frac{e^{\alpha_3 t}[\lambda_0\mu_1 + (\lambda_1 + \mu_1 + \alpha_1)(\lambda_1 + \mu_1 + \alpha_2) + \lambda_1\mu_2]}{(\alpha_3 - \alpha_1)(\alpha_3 - \alpha_2)},$$

$$P_{12}(t) = -\frac{e^{\alpha_1 t}\lambda_1(\lambda_1 + \mu_1 + \alpha_3 + \lambda_2 + \mu_2 + \alpha_2)}{(\alpha_1 - \alpha_2)(\alpha_1 - \alpha_3)}$$
$$- \frac{e^{\alpha_2 t}\lambda_1(\lambda_1 + \mu_1 + \alpha_1 + \lambda_2 + \mu_2 + \alpha_3)}{(\alpha_2 - \alpha_1)(\alpha_2 - \alpha_3)}$$
$$- \frac{e^{\alpha_3 t}\lambda_1(\lambda_1 + \mu_1 + \alpha_1 + \lambda_2 + \mu_2 + \alpha_2)}{(\alpha_3 - \alpha_1)(\alpha_3 - \alpha_2)},$$

$$P_{20}(t) = \frac{e^{\alpha_1 t}\mu_1\mu_2}{(\alpha_1 - \alpha_2)(\alpha_1 - \alpha_3)} + \frac{e^{\alpha_2 t}\mu_1\mu_2}{(\alpha_2 - \alpha_1)(\alpha_2 - \alpha_3)}$$
$$+ \frac{e^{\alpha_3 t}\mu_1\mu_2}{(\alpha_3 - \alpha_1)(\alpha_3 - \alpha_2)},$$

$$P_{21}(t) = -\frac{e^{\alpha_1 t}\mu_2(\lambda_1 + \mu_1 + \alpha_2 + \lambda_2 + \mu_2 + \alpha_3)}{(\alpha_1 - \alpha_2)(\alpha_1 - \alpha_3)}$$
$$- \frac{e^{\alpha_2 t}\mu_2(\lambda_1 + \mu_1 + \alpha_1 + \lambda_2 + \mu_2 + \alpha_3)}{(\alpha_2 - \alpha_1)(\alpha_2 - \alpha_3)}$$
$$- \frac{e^{\alpha_3 t}\mu_2(\lambda_1 + \mu_1 + \alpha_1 + \lambda_2 + \mu_2 + \alpha_2)}{(\alpha_3 - \alpha_1)(\alpha_3 - \alpha_2)},$$

(5-47)
(cont.)

$$P_{22}(t) = \frac{e^{\alpha_1 t}[\lambda_1\mu_2 + (\lambda_2 + \mu_2 + \alpha_2)(\lambda_2 + \mu_2 + \alpha_3)]}{(\alpha_1 - \alpha_2)(\alpha_1 - \alpha_3)}$$
$$+ \frac{e^{\alpha_2 t}[\lambda_1\mu_2 + (\lambda_2 + \mu_2 + \alpha_1)(\lambda_2 + \mu_2 + \alpha_3)]}{(\alpha_2 - \alpha_1)(\alpha_2 - \alpha_3)}$$
$$+ \frac{e^{\alpha_3 t}[\lambda_1\mu_2 + (\lambda_2 + \mu_2 + \alpha_1)(\lambda_2 + \mu_2 + \alpha_2)]}{(\alpha_3 - \alpha_1)(\alpha_3 - \alpha_2)}.$$

Similarly, one can obtain the solution for $N = 3$ on solving a quartic equation.

Exercise: Derive the foregoing solution.

5-5. Spectral-theory Solution of the Birth-Death Equations

The following analysis is a brief summary of a deep approach to the solution of the birth-death equations. The ideas are not unlike the case treated above but go beyond in an attempt to produce the general solution.

Ledermann and Reuter [426] have worked with the nth section of the matrix A of Chaps. 3 and 4 which we denote by $A^{(n)}$. The solution of the equation $dX(t)/dt = X(t)A^{(n)}$, $X(0) = I^{(n)}$, is written in the spectral form, and in the limit, i.e., as $n \to \infty$, it tends to a solution of the general birth-death-process problem.

By putting $\lambda_n = 0$ in the section $A^{(n)}$ we obtain the nth modified section of A which is denoted by $B^{(n)}$. $B^{(n)}$ is simply B of the previous sections with N replaced by n and $\lambda_n = 0$. Again a solution of the differential equation

$$\frac{dY(t)}{dt} = Y(t)B^{(n)}, \qquad Y(0) = I^{(n)},$$

is written in spectral form and has a subsequence of values of n which converges to a solution of the birth-death problem. This need not coincide with the previous solution.

The authors show that the elements of the first case are nonnegative for $t \geq 0$ and, being probabilities, they cannot exceed unity. To obtain the spectral resolution they first examine relations among determinants $\phi_n(\xi)$, that is,

$$\phi_n(\xi) - (\xi + \lambda_n + \mu_n)\phi_{n-1}(\xi) + \lambda_{n-1}\mu_n\phi_{n-2}(\xi) = 0, \qquad n \geq 1, \quad (5\text{-}48)$$

$\phi_{-1}(\xi) = 0$, $\phi_0(\xi) = \xi + \lambda_0$, of sections of order corresponding to the subscript of ϕ of the matrix obtained by taking the Laplace transform with parameter ξ of the birth-death problem. They show that the characteristic roots, or eigenvalues, of $A^{(n)}$ given by $\phi_n(\xi) = 0$ are distinct and are negative or zero, where the largest eigenvalue is zero if, and only if, $\lambda_0 = 0$, and the eigenvalues of $A^{(n-1)}$ separate those of $A^{(n)}$; that is, one has the largest eigenvalue of $A^{(n)}$ exceeding or equal to the largest eigenvalue of $A^{(n-1)}$ which exceeds an eigenvalue of $A^{(n)}$ which exceeds an eigenvalue of $A^{(n-1)}$ and so on, terminating in the smallest eigenvalue of $A^{(n)}$.

The eigenvalues of the modified section in their descending order are greater than or equal to the corresponding eigenvalues of the unmodified section. The elements of the solution have the same relation (according to their subscripts).

Write

$$l_{-1} = \lambda_0, \qquad l_0 = 1, \qquad l_\nu = (\lambda_1\lambda_2 \cdots \lambda_\nu)^{-1}, \qquad \nu \geq 1,$$
$$m_0 = 1, \qquad m_\nu = (\mu_1\mu_2 \cdots \mu_\nu)^{-1}, \qquad \nu \geq 1,$$

$\alpha_i^{(n)}$, $n = 0, 1, \ldots, n$, for the eigenvalues; $x_r^{(n)}$, $y_r^{(n)'}$, $r = 0, 1, \ldots, n$, for the eigenvectors; that is,

$$x_i^{(n)} A^{(n)} = x_i^{(n)}\alpha_i^{(n)}, \qquad A^{(n)}y_i^{(n)'} = \alpha_i^{(n)}y_i^{(n)'}, \qquad i = 0, 1, \ldots, n,$$
$$\theta_r^{(n)} = x_r^{(n)}y_r^{(n)'}, \qquad r = 0, 1, \ldots, n,$$
$$\theta_0^{(n)} = 1 \quad \text{if } \lambda_0 = 0.$$

Also, for the orthogonal idempotents write

$$D_r^{(n)} = \frac{y_r^{(n)'}x_r^{(n)}}{\theta_r^{(n)}} \equiv d_r^{(n)}(i,j).$$

Then it is shown that

$$d_r^{(n)}(i,j) = \frac{l_{i-1}m_j\phi_{j-1}[\alpha_r^{(n)}]\phi_{j-1}[\alpha_r^{(n)}]}{\theta_r^{(n)}}, \qquad r = 0, 1, \ldots, n.$$

If we write

$$y_0^{(n)'} = [\zeta_0^{(n)}, \zeta_1^{(n)}, \ldots, \zeta_n^{(n)}];$$

then

$$d_0^{(n)}(i,j) = \zeta_i^{(n)}\delta_{0j} \quad \text{if } \lambda_0 = 0.$$

The spectral resolution of an element of the nth-section solution is

$$f_{ij}^{(n)}(t) = \sum_{r=0}^{n} e^{t\alpha_r^{(n)}} d_r^{(n)}(i,j), \qquad (5\text{-}49)$$

etc., for the elements of the modified section $g_{ij}^{(n)}(t)$. By limiting processes these results are used to construct solutions $f_{ij}(t)$, $g_{ij}(t)$ of the infinite system of equations.

For the section $A^{(n)}$ and for $\lambda_0 > 0$ a spectral function $\rho^{(n)}(x)$ is defined for $-\infty < x < \infty$ as a nondecreasing step function, with discontinuities $1/\theta_r^{(n)}$ at $x = \alpha_r^{(n)}$ ($r = 1, \ldots, n$) and $\rho^{(n)}(0) = 0$.

Thus

$$\rho^{(n)}(x) = \begin{cases} 0, & x \geq 0, \\ -\left(\dfrac{1}{\theta_0^{(n)}} + \cdots + \dfrac{1}{\theta_r^{(n)}}\right), & \alpha_{r+1}^{(n)} \leq x < \alpha_r^{(n)}, \quad 0 \leq r \leq n-1, \\ -\left(\dfrac{1}{\theta_0^{(n)}} + \cdots + \dfrac{1}{\theta_n^{(n)}}\right), & x < \alpha_n^{(n)}. \end{cases}$$

$$(5\text{-}50)$$

Note that $\rho^{(n)}(x) = 0$ for $x \geq 0$ because $0 > \alpha_0^{(n)}$.

Using the foregoing results, we may write

$$f_{ij}^{(n)}(t) = l_{i-1} m_j \int_{-\infty}^{\infty} \phi_{i-1}(x) \phi_{j-1}(x) e^{tx} \, d\rho^{(n)}(x), \qquad 0 \leq i \leq n,$$
$$0 \leq j \leq n, \quad (5\text{-}51)$$

and, since $f_{ij}^{(n)}(0) = \delta_{ij}$, this result equals δ_{ij} on putting $t = 0$.

If $a_{ij}^{(n)}$ is an element of $A^{(n)}$, we also have

$$a_{ij}^{(n)} = l_{i-1} m_j \int_{-\infty}^{\infty} x \phi_{i-1}(x) \phi_{j-1}(x) \, d\rho^{(n)}(x), \qquad 0 \leq i \leq n,$$
$$0 \leq j \leq n. \quad (5\text{-}52)$$

It is shown that $\rho^{(n)}(x) \to \rho(x)$ as $n \to \infty$, and the resulting $f_{ij}(t)$ satisfies the equations of the birth and death process. A similar argument is given for $g_{ij}(t)$, and the problem is also studied for $\lambda_0 = 0$. Then, by letting $t \to \infty$, asymptotic values (steady state) are studied. The results are then applied to those processes for which

$$\frac{\lambda_{n+1}}{\lambda_n} = 1 + \frac{a}{n} + O\left(\frac{1}{n^2}\right),$$
$$\frac{\lambda_n}{\mu_n} = c\left[1 + \frac{b}{n} + O\left(\frac{1}{n^2}\right)\right], \qquad c > 0,$$

$$(5\text{-}53)$$

where a, b, and c are constants. Such a process is termed analytical with parameters a, b, c. It is shown that, for certain values of a, b, c, the two approximations, i.e., the solutions from the section and from the modified section, yield different values. Examples are given for the constant-coefficients case $\lambda_n = \lambda$, $\mu_n = \mu$, both for $\lambda_0 = 0$ and for $\lambda_0 > 0$, and an attempt to study the linear-coefficient case is also made.

PROBLEMS

1. Using the characteristic roots, show that if

$$A = \begin{bmatrix} 2 & 1 \\ 1 & 2 \end{bmatrix}$$

then

$$A^{100} = \begin{bmatrix} \dfrac{3^{100} + 1}{2} & \dfrac{3^{100} - 1}{2} \\ \dfrac{3^{100} - 1}{2} & \dfrac{3^{100} + 1}{2} \end{bmatrix}.$$

Compute e^A.

2. Given

$$\frac{dx}{dt} = x + y, \qquad \frac{dy}{dt} = x - y,$$

formulate the problem in matrix notation and give its solution by matrix methods.

3. Show that

$$\lim_{n \to \infty} T^n = \frac{1}{1 - p + q} \begin{bmatrix} q & 1 - p \\ q & 1 - p \end{bmatrix},$$

where

$$T = \begin{bmatrix} p & 1 - p \\ q & 1 - q \end{bmatrix}$$

with $0 < p, q < 1$.

4. J. Giltay has studied the following problem: Let $a_{ij}\, dt$ be the probability of a transition from state j ($j = 1, \ldots, m$) to state i ($i = 1, \ldots, m, i \neq j$) during a time interval dt. Let $P_i(t)$ be the probability that the system is in state j at time t. Then

$$P_i(t + dt) = P_i(t)\Big(1 - \sum_{k \neq i} a_{ki}\, dt\Big) + \sum_{j \neq i} a_{ij} P_j(t)\, dt.$$

We write $\displaystyle\sum_{k \neq i} a_{ki} = -a_{ii}$. Simplifying yields

$$P'_i(t) = \sum_{j=1}^{m} a_{ij} P_j(t), \qquad i = 1, \ldots, m.$$

We also have

$$\sum_{j=1}^{m} P_j(t) = 1.$$

If one considers the matrix A of the a_{ij}, one can show that the nonzero roots of the characteristic equation $|A - zI| = 0$ (where I is the identity matrix) all have a negative real part. Thus if

$$|a_{ii}| > \sum_{i \neq j} |a_{ij}|, \qquad i = 1, \ldots, m$$

(which is true in this case), then D, the determinant of A, differs from zero. [If it is zero, there is a linear relation between the rows; i.e., there exist numbers $b_1, \ldots, b_m$

not all zero such that

$$\sum_{i=1}^{m} b_i a_{ij} = 0, \qquad j = 1, \ldots, m.$$

Choose b_k such that $|b_k| = \max |b_i|$ $(i = 1, \ldots, m)$; then the equation with $j = k$ gives

$$a_{kk} = -\frac{\sum_{i \neq k} |a_{ik}| |b_i|}{|b_k|} \leq \sum_{i \neq k} |a_{ik}|,$$

a contradiction.]

Now if $z^* = x + iy$ is a nonzero complex characteristic root, then the principal diagonal elements of $[A - z^*I]$ have the moduli

$$[(a_{kk} - x)^2 + y^2]^{\frac{1}{2}}, \qquad k = 1, \ldots, m.$$

The determinant $|A - zI|$ (which must vanish at the characteristic value z^*) cannot be zero unless the following relation holds for at least one k:

$$[(a_{kk} - x)^2 + y^2]^{\frac{1}{2}} \leq \sum_{k \neq j} |a_{kj}| < |a_{kk}|,$$

where the first inequality must hold for $|A - z^*I|$ and the second always holds for D. This gives

$$(a_{kk} - x)^2 + y^2 < a_{kk}^2.$$

But a_{kk}, from its definition, is never positive; hence x must be negative for the inequality to be satisfied.

Apply this theorem to an example.

PART 3

NON-POISSON QUEUES

CHAPTERS 6 TO 10

In this part of the book we treat cases where either the input or the service distribution or both are non-Poisson; i.e., the inter-arrival or the service-time distributions, or both, are nonexponential. In Chap. 6 we study the Poisson-input constant-service-time multiple-channel problem. Then for a single channel we treat a Poisson input with Erlangian service and Erlangian input with exponential service. In Chap. 7 we introduce the concept of the imbedded Markoff chain as applied to bulk queues. In Chap. 8 we study the Poisson-input arbitrary-service-time problem. The study of this case is continued in Chap. 9, which concentrates on waiting-time studies and carries through to the general-input and arbitrary-service-time-distribution case. In Chap. 10 the general independent-input exponential-service-times problem is studied for a single channel.

CHAPTER 6

SPECIAL NON-POISSON CASES

6-1. Introduction

Having investigated various queues with Poisson input and exponential service times, which we shall simply call Poisson queues, the reader may have begun to wonder whether all queues are Poisson queues. Obviously this need not be the case. For example, from a practical standpoint, a queue might have constant service times instead of exponential service times. We shall return to Poisson queues in different parts of the book to study ramifications in the assumptions of cases already investigated.

Here our investigations will be mostly in the steady state since transient solutions are difficult to obtain. We study cases of Poisson input with constant and with Erlangian-type service. In the latter a customer or a unit may be considered to undergo several services, one after another, each with its service distribution, but here all are considered as identical exponential distributions. This may be the case, for example, where cars arriving at random at a service station demand several types of service from one attendant, e.g., gasoline, water, battery check, etc.

A third case with an Erlangian input and exponential service times is studied for an ordered single-channel queue.

The question is what should we investigate. One natural measure of effectiveness is to compute the probability of a given number in the system from which the average number in the system may be obtained. Another is the waiting time. We shall develop expressions for either one or for both in the sequel.

6-2. Constant Service Times, Multiple Channels

We now study with Crommelin [142, 143] the case of a Poisson input with parameter λ, c channels in parallel, and identical constant service times for each channel. We take the service time as the unit of time. Note that for equilibrium we have $c > \lambda$.

In the steady state let p_n be the probability that there are precisely n units in line and in service at a given time and let a_n be the probability

153

that there are no more than n units in the system at that time. Thus

$$a_n = \sum_{i=0}^{n} p_i. \qquad (6\text{-}1)$$

Since we have taken the service time as the unit of time, we can express the probability of the number present at the end of a time interval in terms of the probability of the number present at the beginning of the interval, multiplied by the probability of a given number arriving during the interval. As this can occur in different combinations, we add the probabilities. Note that during such a service-time interval all items in service would finish their service and leave the system.

Then in a unit interval of time (which is the constant service time) we have

$$p_0 = a_c e^{-\lambda},$$
$$p_1 = a_c \lambda e^{-\lambda} + p_{c+1} e^{-\lambda},$$
$$\cdots \cdots \cdots \cdots \cdots \qquad (6\text{-}2)$$
$$p_n = a_c \frac{\lambda^n}{n!} e^{-\lambda} + p_{c+1} \frac{\lambda^{n-1}}{(n-1)!} e^{-\lambda} + \cdots + p_{n+c} e^{-\lambda}.$$

The first equation describes the case when the queue is vacant if no more than c are in service at the beginning of the interval and no arrivals occur during the interval. The second equation describes the case where one unit is in the queue if no more than c are in service at the beginning and one arrives in the meantime or there are $c + 1$ in the system at the beginning and nothing arrives in the meantime, etc. The factor $(\lambda^n/n!)e^{-\lambda}$ gives the probability of n units originating in the interval.

1. We begin by calculating p_n. Define

$$P(z) = \sum_{n=0}^{\infty} p_n z^n.$$

Then
$$P(1) = 1.$$

If we also define

$$P_c(z) = \sum_{n=0}^{c} p_n z^n, \qquad (6\text{-}3)$$

then, by multiplying the equations by appropriate powers of z and summing, we have

$$P(z) = \frac{P_c(z) - a_c z^c}{1 - z^c e^{\lambda(1-z)}}. \qquad (6\text{-}4)$$

Since $0 \le p_n \le 1$, $P(z)$ is regular and bounded within the unit circle $|z| < 1$. Thus the numerator has zeros that coincide within and on the unit circle with all zeros of the denominator within and on the unit circle.

By Rouché's theorem, there are c such zeros, which we denote by z_1, . . . , z_c. Obviously, $z_c = 1$ is a zero.

Since the numerator is a polynomial of degree c we may replace it by

$$K(z - 1)(z - z_1) \cdots (z - z_{c-1}), \tag{6-5}$$

where we determine the constant K from

$$\lim_{z \to 1} P(z) = 1.$$

We then have

$$P(z) = -\frac{c - \lambda}{(1 - z_1) \cdots (1 - z_{c-1})} \frac{(z - 1)(z - z_1) \cdots (z - z_{c-1})}{1 - z^c e^{\lambda(1-z)}}. \tag{6-6}$$

p_n may now be obtained as the coefficient of z^n in the power series expansion of $P(z)$. For example, for $c = 1$, by differentiating

$$P(z) = \frac{(1 - \lambda)(1 - z)}{1 - ze^{\lambda(1-z)}} \tag{6-7}$$

n times with respect to z, dividing by $n!$, and putting $z = 0$, we have, on writing $\rho = \lambda/\mu$ (note that we have been using the service rate $\mu = 1$ but the analysis holds for any μ),

$$p_0 = (1 - \rho), \qquad p_1 = (1 - \rho)(e^\rho - 1),$$

$$p_n = (1 - \rho) \sum_{k=1}^{n} (-1)^{n-k} e^{k\rho} \left[\frac{(k\rho)^{n-k}}{(n - k)!} + \frac{(k\rho)^{n-k-1}}{(n - k - 1)!} \right],$$

$$n \geq 2, \tag{6-8}$$

ignoring the second factor for $k = n$.

2. We now compute $P(t = 0)$, the probability of zero delay. It obviously equals the sum of the probabilities that $i < c$ channels are occupied. Thus

$$P(t = 0) = a_{c-1} = \sum_{i=0}^{c-1} p_i. \tag{6-9}$$

Let

$$Q(z) = \sum_{k=0}^{\infty} a_k z^k; \tag{6-10}$$

then, since $a_k - a_{k-1} = p_k$, we have

$$Q(z)(1 - z) = P(z), \qquad \text{or} \qquad Q(z) = \frac{P(z)}{1 - z}. \tag{6-11}$$

Since $P(z)$ is known, it is easy to compute a_{c-1} as the coefficient of z^{c-1} in $Q(z)$. It is also not difficult to see that z^{c-1} appears in the numerator with coefficient K; hence

$$P(t = 0) = \frac{c - \lambda}{(1 - z_1) \cdots (1 - z_{c-1})},$$

and $$\log P(t = 0) = \log (c - \lambda) - \sum_{k=1}^{c-1} \log (1 - z_k). \tag{6-12}$$

We now evaluate this expression explicitly.

From Whittaker and Watson [759, p. 119] we have

$$\frac{1}{2\pi i} \int_C f(z) \frac{g'(z)}{g(z)} dz = \Sigma r_i f(a_i) - \Sigma s_j f(b_j), \tag{6-13}$$

where C is a contour in the z plane and $f(z)$ and $g(z)$ are analytic inside and on C, except that $g(z)$ has a finite number of poles b_j with corresponding multiplicities s_j inside C. Also $g(z)$ has the zeros a_i with corresponding multiplicities r_i inside C.

If we write

$$g(z) = 1 - \frac{e^{\lambda(z-1)}}{z^c} = -\frac{e^{\lambda(z-1)}}{z^c} [1 - z^c e^{\lambda(1-z)}], \tag{6-14}$$

we note that $z_1, \ldots, z_{c-1}$ are zeros each of multiplicity 1, while $z = 0$ is a pole of multiplicity c. Thus

$$\frac{1}{2\pi i} \int_\Gamma \log (z - 1) \frac{g'}{g} dz = \sum_{k=1}^{c-1} \log (z_k - 1) - c \log (-1)$$

$$= (c - 1)\pi i + \sum_{k=1}^{c-1} \log (1 - z_k) - c\pi i = -\pi i + \sum_{k=1}^{c-1} \log (1 - z_k)$$

$$= -\pi i + \log (c - \lambda) - \log P(t = 0). \tag{6-15}$$

Thus, to obtain an explicit expression for $P(t = 0)$, we evaluate the integral on the left.

We split Γ [the contour which excludes the singularity at $z = 1$ and the external zeros of $g(z)$] into two parts: (1) along C and (2) along the remainder, as shown in Fig. 6-1. We first evaluate the integral along C. Integration by parts gives

$$\frac{1}{2\pi i} \int_C \log (z - 1) \frac{g'}{g} dz = \frac{1}{2\pi i} [\log (z - 1) \log g(z)]_C - \frac{1}{2\pi i} \int_C \frac{\log g(z)}{z - 1} dz,$$

where the first quantity on the right is to be evaluated at the extremes of C, that is, that at $\theta = 0$ and that at $\theta = 2\pi$, and the value at $\theta = 0$

subtracted from that at $\theta = 2\pi$. To see this, let

$$z = 1 + Re^{i\theta};$$

then
$$\log (z - 1) = \log R + i\theta,$$

using the principal part; and

$$\log g(z) = \log \left[1 - \frac{e^{\lambda Re^{i\theta}}}{(1 + Re^{i\theta})^c} \right]$$

$$= \log \left[1 - \frac{e^{\lambda R(\cos \theta + i \sin \theta)}}{(1 + R \cos \theta + iR \sin \theta)^c} \right]. \quad (6\text{-}16)$$

One can show that as θ changes from zero to 2π the argument of $\log g(z)$

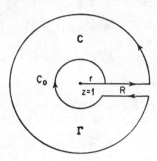

Fig. 6-1

returns to its initial value zero. Thus

$$[\log (z - 1) \log g(z)]_0^{2\pi} = (\log R + 2\pi i) \log \left[1 - \frac{e^{\lambda R}}{(1 + R)^c} \right]$$

$$- \log R \log \left[1 - \frac{e^{\lambda R}}{(1 + R)^c} \right] = 2\pi i \log g(1 + R). \quad (6\text{-}17)$$

Along the straight parts we have

$$\int \log (z - 1) \frac{g'(z)}{g(z)} \, dz = \int_{1+r}^{1+R} \log (z - 1) \frac{g'(z)}{g(z)} \, dz$$

$$+ \int_{1+R}^{1+r} \log (z - 1) \frac{g'(z)}{g(z)} \, dz, \quad (6\text{-}18)$$

where $\theta = 0$ for the first expression on the right and $\theta = 2\pi$ for the second. When $\log (z - 1)$ is expanded into real and imaginary parts, it is clear that, after cancellation, the above equals

$$\int_{1+r}^{1+R} - 2\pi i \frac{g'(z)}{g(z)} \, dz = -2\pi i [\log g]_{1+r}^{1+R}$$

$$= -2\pi i [\log g(1 + R) - \log g(1 + r)]. \quad (6\text{-}19)$$

On the small circle we have $z = 1 + re^{i\theta}$, $dz = ire^{i\theta}\, d\theta$, and one has as $r \to 0$, that is, as $z \to 1$, from the series expansion about $z = 1$ that $g(z)$ behaves like the first term since the zeroth term vanishes at $z = 1$, that is,

$$\lim_{z \to 1} g(z) \sim (z - 1)g'(1) \sim (c - \lambda)(z - 1) \qquad (6\text{-}20)$$

and

$$\lim_{z \to 1} \frac{g'(z)}{g(z)} \sim \frac{1}{z - 1} = \frac{e^{-i\theta}}{r}. \qquad (6\text{-}21)$$

Thus

$$\int_{C_0} \log (z - 1) \frac{g'}{g}\, dz = \int_{2\pi}^{0} (\log r + i\theta)i\, d\theta[1 + o(r)]$$

$$= 2\pi^2 - 2\pi i \log r, \qquad \text{using } \lim_{r \to 0} r \log r = 0,$$

$$= 2\pi^2 - 2\pi i[\log g(1 + r) - \log (c - \lambda)]. \qquad (6\text{-}22)$$

This is because

$$g(z) \sim (c - \lambda)(z - 1) = (c - \lambda)re^{i\theta}, \qquad (6\text{-}23)$$

since we have, for $z = 1 + r$, $\theta = 0$,

$$\log g(1 + r) \sim \log (c - \lambda)r = \log (c - \lambda) + \log r,$$

or

$$\log r \sim \log (1 + r) - \log (c - \lambda). \qquad (6\text{-}24)$$

When the foregoing results are combined we have

$$\frac{1}{2\pi i} \int_{\Gamma} \log (z - 1) \frac{g'}{g}\, dz = -\frac{1}{2\pi i} \int_{C} \frac{\log g}{z - 1}\, dz - \pi i + \log (c - \lambda). \qquad (6\text{-}25)$$

Therefore

$$\log P(t = 0) = \frac{1}{2\pi i} \int_{C} \frac{\log g}{z - 1}\, dz. \qquad (6\text{-}26)$$

But

$$\log g(z) = \log \left[1 - \frac{e^{\lambda(z-1)}}{z^c} \right] = -\sum_{n=1}^{\infty} \frac{e^{n\lambda(z-1)}}{nz^{nc}}, \qquad (6\text{-}27)$$

valid since

$$\left| \frac{e^{\lambda(z-1)}}{z^c} \right| < 1 \text{ on } C.$$

Therefore

$$\log P(t = 0) = -\frac{1}{2\pi i} \int_{C} \frac{dz}{z - 1} \sum_{n=1}^{\infty} \frac{e^{n\lambda(z-1)}}{nz^{nc}}$$

$$= -\sum_{n=1}^{\infty} \frac{1}{n} \frac{1}{2\pi i} \int_{C} \frac{e^{n\lambda(z-1)}}{z^{nc}} \frac{dz}{z - 1}. \qquad (6\text{-}28)$$

Here we have a pole of order nc at $z = 0$ and another of the first order at $z = 1$. We integrate term by term. By the theory of residues, we have that, if $z = z_0$ is a pole of order m of $h(z)$, the residue is the product of $1/(m - 1)!$ and the $(m - 1)$st derivative of $(z - z_0)^m h(z)$ evaluated at

z_0. All the residues are then added. We finally have the logarithm of the expression we seek:

$$\log P(t = 0) = - \sum_{n=1}^{\infty} \frac{1}{n} \left[1 - \sum_{i=0}^{nc-1} \frac{(n\lambda)^i}{i!} e^{-n\lambda} \right]$$

$$= - \sum_{n=1}^{\infty} \frac{1}{n} \sum_{i=nc}^{\infty} \frac{(n\lambda)^i}{i!} e^{-n\lambda}, \tag{6-29}$$

since

$$\lim_{z \to 1} (z - 1) \frac{e^{n\lambda(z-1)}}{z^{nc}(z - 1)} = 1, \tag{6-30}$$

and by the Leibnitz theorem

$$\lim_{z \to 0} \frac{1}{(nc - 1)!} \frac{d^{nc-1}}{dz^{nc-1}} z^{nc} \frac{e^{n\lambda(z-1)}}{z^{nc} (z - 1)} = - \sum_{j=0}^{nc-1} \frac{(n\lambda)^j}{j!} e^{-n\lambda}. \tag{6-31}$$

3. Pollaczek has given the following approximation for large c:

$$P(t > 0) = 1 - P(t = 0) = \frac{1}{1 - \rho} \frac{\rho^c e^{(1-\rho)c}}{\rho \sqrt{2\pi c}}, \tag{6-32}$$

where $\rho = (\lambda/c) < 1$. This is another important measure giving the probability of any waiting. He also gives for large c

$$W_q = \frac{1}{c \sqrt{2\pi c}} \frac{\rho^c e^{(1-\rho)c}}{(1 - \rho)^2} \left[1 + O \left(\frac{1}{c} \right) \right].$$

4. We now compute the probability of waiting less than time t, that is, $P(<t) = 1 - P(>t)$. Let $t = T + s$, where T is an integral multiple of the time unit and s is the excess fraction of a unit. Let b_n be the probability that at a given time not more than n units which are in service will be in progress at time s. b_{Tc+c-1} is the probability that not more than $Tc + c - 1$ units will be in progress at time s later and therefore not more than $c - 1$ units at time $s + T = t$ later (since c units complete service in one interval and hence Tc units in T intervals). Hence

$$P(<t) = b_{Tc+c-1}. \tag{6-33}$$

Note that

$$a_n = e^{-\lambda s} \sum_{k=0}^{n} b_k \frac{(\lambda s)^{n-k}}{(n - k)!}. \tag{6-34}$$

Let

$$R(z) = \sum_{k=0}^{\infty} b_k z^k, \quad \text{for } |z| < 1, \tag{6-35}$$

and
$$Q(z) = \sum_{n=0}^{\infty} a_n z^n = e^{-\lambda s} \sum_{k=0}^{\infty} b_k z^k e^{\lambda z s}. \tag{6-36}$$

Thus

$$R(z) = Q(z)e^{\lambda s(1-z)} = a_{c-1}(z - z_1) \cdots (z - z_{c-1}) \sum_{j=0}^{\infty} z^{jc} e^{j\lambda} e^{-\lambda z(j+s)}.$$

$$\tag{6-37}$$

The coefficient of z^{Tc+c-1} gives

$$b_{Tc+c-1} = \sum_{m=0}^{T} \sum_{k=0}^{c-1} a_k \frac{[-\lambda(m + s)]^{T-mc+c-1-k}}{(T - mc + c - 1 - k)!} e^{\lambda(m+s)}. \tag{6-38}$$

A more convenient way to obtain b_{Tc+c-1} is to use

$$R(z) = \frac{a_{c-1}(z - z_1) \cdots (z - z_{c-1})e^{\lambda s(1-z)}}{1 - z^c e^{\lambda(1-z)}} \tag{6-39}$$

which has a pole at $z = 1$ [although $Q(z)$ did not, since its numerator had a zero at $z = 1$]. We expand $R(z)$ in a Laurent series up to the next pole which is real. To show that the next pole is real consider the expression $z^c e^{\lambda(1-z)}$. Then its derivative is positive as x increases from $x = 1$. But the expression vanishes at ∞; hence there is a value r_0 at which it assumes the value unity and thus the denominator of $R(z)$ vanishes. That r_0 is the first pole follows from

$$z^c e^{\lambda(1-z)} = r^c e^{\lambda(1-r \cos \theta)} e^{i(c\theta - \lambda r \sin \theta)}, \qquad z = re^{i\theta},$$
$$|z^c e^{\lambda(1-z)}| > r^c e^{\lambda(1-r)} > 1 \qquad \text{for } 1 \le r \le r_0.$$

Since $z = 1$ is a singularity, we now write

$$R(z) = -\frac{1}{z - 1} + H(z),$$

where $H(z)$ is analytic in $|z| < r_0$. Let $H(z) = \sum_{n=0}^{\infty} h_n z^n$. For $1 < |z| < r_0$, $R(z)$ may be expanded in a Laurent series by writing

$$-e^{\lambda(1-z)} z^c \left[1 - \frac{1}{z^c e^{\lambda(1-z)}} \right]$$

for its denominator. Thus

$$R(z) = -a_{c-1}(z - z_1) \cdots (z - z_{c-1}) \sum_{n=0}^{\infty} z^{-(n+1)c} e^{-(n+1-s)\lambda} e^{z\lambda(n+1-s)}.$$

$$\tag{6-40}$$

Then

$$H(z) = R(z) + \frac{1}{z - 1} = R(z) + \sum_{n=1}^{\infty} \frac{1}{z^n} \quad \text{for } 1 < |z| < r_0.$$

Note that for $|z| > 1$

$$\frac{1}{z - 1} = \frac{1}{z(1 - 1/z)} = \frac{1}{z}\left(1 + \frac{1}{z} + \frac{1}{z^2} + \cdots\right) = \sum_{n=1}^{\infty} \frac{1}{z^n}.$$

This helps determine $H(z)$. But $H(z)$ is also analytic for $|z| < 1$ and $(1 - z)^{-1} = \sum_{n=0}^{\infty} z^n$ is analytic for $|z| < 1$; hence we may obtain the expansion of $R(z)$ for $|z| < 1$, from which the desired coefficient is obtained.

5. On using $\sum_{n=1}^{\infty} np_n$ by a similar analysis, the average delay may be. shown to be

$$W_q = \frac{1}{\lambda} \sum_{k=1}^{c-1} \frac{1}{1 - z_k} + \frac{\lambda^2 - c^2 + c}{2\lambda(c - \lambda)} = \sum_{i=1}^{\infty} e^{-i\lambda} \left[\sum_{j=ic}^{\infty} \frac{(i\lambda)^j}{j!} \right.$$

$$\left. - \frac{c}{\lambda} \sum_{j=ic+1}^{\infty} \frac{(i\lambda)^j}{j!} \right]. \quad (6\text{-}41)$$

One may replace λ by ρc if it is desired to use $\rho = \lambda/\mu c$, where $1/\mu$ is the constant service time which has been taken as the time unit. The reader may justify this by applying dimensional analysis to W_q.

6. For the case $c = 1$ we have (see Chap. 3)

$$P(<t) = (1 - \rho) \sum_{i=0}^{k} e^{\rho(\mu t - i)} \frac{[-\rho(\mu t - i)]^i}{i!}, \quad (6\text{-}42)$$

where k is the largest integer in μt, and

$$W_q = \int_0^{\infty} t \, dP(<t) = \frac{\rho}{2\mu(1 - \rho)}. \quad (6\text{-}43)$$

We note the interesting fact that, at a service channel with Poisson arrivals, the waiting time of constant arrivals is one-half that of exponential service, keeping the arrival rate the same. Thus there is an advantage in having a constant service. The expected waiting time of those delayed is

$$\frac{W_q}{P(t > 0)} = \frac{1}{2\mu(1 - \rho)}. \quad (6\text{-}44)$$

7. For the general case, Molina has given the approximation

$$W_q = P(t > 0) \frac{1}{\mu c(1 - \rho)} \frac{c}{c + 1} \frac{1 - \rho^{c+1}}{1 - \rho^c},$$ (6-45)

where $\quad P(t > 0) = \dfrac{\dfrac{(\lambda^c e^{-\lambda}/c!)c}{c - \lambda}}{1 - e^{-\lambda} \displaystyle\sum_{n=c}^{\infty} (\lambda^n/n!) + \dfrac{(\lambda^c e^{-\lambda}/c!)c}{c - \lambda}}.$ (6-46)

In the last expression it has been assumed that $\mu = 1$. To obtain a result involving μ, one replaces λ by λ/μ.

8. As an illustration, we compute the average waiting time W_q for $c = 1$ by an alternative method due to Fry [236]. Now $W_q = L_q/\lambda$, where L_q is the expected number in line. But $L_q = L - \rho$, where ρ is the utilization factor (if we take $\mu = 1$, then $\rho = \lambda$) and $L = \displaystyle\sum_{n=0}^{\infty} n p_n$. The value of this expression may be obtained by evaluating $\displaystyle\sum_{n=0}^{\infty} n^2 p_n$, which enables one to solve for L.

Now

$$S = \sum_{n=0}^{\infty} n^2 p_n = \sum_{n=0}^{\infty} n^2 \left(\sum_{k=0}^{n} p_{n+1-k} \frac{\lambda^k e^{-\lambda}}{k!} + p_0 \frac{\lambda^n e^{-\lambda}}{n!} \right).$$ (6-47)

Note that

$$\sum_{n=0}^{\infty} \sum_{k=0}^{n} = \sum_{k=0}^{\infty} \sum_{n=k}^{\infty}.$$ (6-48)

Now $D \equiv \displaystyle\sum_{n=0}^{\infty} n^2(\lambda^n e^{-\lambda}/n!)$ is evaluated as follows:

$$\sum_{n=0}^{\infty} \frac{\lambda^n}{n!} = e^{\lambda},$$

$$\frac{d}{d\lambda} \sum_{n=0}^{\infty} \frac{\lambda^n}{n!} = \sum_{n=1}^{\infty} n \frac{\lambda^{n-1}}{n!} = e^{\lambda},$$

$$\frac{d}{d\lambda} \sum_{n=1}^{\infty} n \frac{\lambda^{n-1}}{n!} = \sum_{n=2}^{\infty} n(n - 1) \frac{\lambda^{n-2}}{n!} = e^{\lambda}.$$ (6-49)

Thus $D = \lambda(1 + \lambda)$. Hence

$$p_0 \sum_{n=0}^{\infty} n^2 \frac{\lambda^n e^{-\lambda}}{n!} = p_0 \lambda(1 + \lambda).$$ (6-50)

In the double sum we put $m = n - k + 1$; then

$$n^2 = m^2 + 2m(k - 1) + (k - 1)^2$$

and

$$\sum_{k=0}^{\infty} \sum_{n=k}^{\infty} n^2 \, p_{n+1-k} \frac{\lambda^k e^{-\lambda}}{k!} = \sum_{k=0}^{\infty} \frac{\lambda^k e^{-\lambda}}{k!} \sum_{m=1}^{\infty} m^2 p_m$$

$$+ 2 \sum_{k=0}^{\infty} (k - 1) \frac{\lambda^k e^{-\lambda}}{k!} \sum_{m=1}^{\infty} m p_m + \sum_{k=0}^{\infty} (k - 1)^2 \frac{\lambda^k e^{-\lambda}}{k!} \sum_{m=1}^{\infty} p_m.$$ (6-51)

Thus finally

$$S = S + 2\lambda L - 2L + (1 - p_0)\lambda(1 + \lambda) - 2(1 - p_0)\lambda$$
$$+ (1 - p_0) + p_0 \lambda(1 + \lambda).$$ (6-52)

But $p_0 = 1 - \lambda.$ (6-53)

Therefore $L = \dfrac{\lambda^2 - 2\lambda}{2(\lambda - 1)}$ (6-54)

and $W_q = \dfrac{(\lambda^2 - 2\lambda)/2(\lambda - 1) - \lambda}{\lambda} = \dfrac{\lambda}{2(1 - \lambda)}.$ (6-55)

If we desire a result for a general value of the service time $1/\mu$ we replace λ by $\lambda/\mu \equiv \rho$ in (6-54) and we have $W_q = \rho/2\mu(1 - \rho)$.

9. J. L. Everett [195] has given a method of computing p_n for large values of n, using the fact that, for large n, $p_{n+1} \sim p_n K$ or, more generally, $p_{n+c} \sim K^c p_n$, where K is a constant depending on the number of channels c and on λ (again the service time is taken as the time unit). On substituting in the system of equations in p_n and simplifying, one can sum the resulting series, which, for large n, may be replaced by an infinite series. This gives the following relation which enables determining K, $K = e^{(\lambda/c)(1-1/K)}$. This constant K is shown to be analogous to ρ, the utilization factor in the exponential-service-time multiple-channel case. By replacing ρ by K in the solution of the latter problem, an estimate is obtained for the constant-service-time case which is then substituted in the system of equations to obtain p_n, which is now a new estimate. In turn, this is used in the equations, etc., until stability is reached. This iterative procedure has been verified successfully by Everett.

6-3. Poisson Input, Erlangian Service Times

In this and in the next section we study $M/E_k/1$ and $E_k/M/1$.

By using E_k, the Erlangian distribution, as the service distribution, an item entering service may be considered to generate a set of k phases of service. The phases have identical exponential distributions with parameter μk. As we have seen in Chap. 3, the density function of the sum of k random variables which are independently and identically distributed according to the exponential distribution with parameter μk is E_k.

We also use E_k for arrivals in the following section, except that we require that its mean be λ/k, which essentially says that only $(1/k)$th of the arrival rate of a Poisson distribution is used.

In either case, the Erlangian distribution indicates a possibility of variation on randomness assumptions, as already explained in Chap. 3.

In Chap. 4 we studied the transient solution of a phase-type service problem. For historical reasons and with some modification of the service distribution, we shall treat a related case for a single channel in the steady state with ordered service and with Poisson input and parameter λ. For the total service time of a unit we have the Erlangian distribution

$$\frac{(\mu k)^k}{\Gamma(k)} e^{-\mu kt} t^{k-1}, \qquad 0 \le t < \infty.$$

We recall that this is a scale-modified chi-square distribution with mean $1/\mu$ (which is k times the mean of the distribution of any phase) and $2k$ degrees of freedom.

Let $P_n(t)$ be the probability that there are n phases waiting and in service at time t. Each arrival increases the number of phases in the system by k, and each time a phase is completed, the number of phases in the system is decreased by unity. Thus

$$P_n(t + \Delta t) = P_n(t)[1 - (\lambda + k\mu)\,\Delta t] + P_{n+1}(t)k\mu\,\Delta t + P_{n-k}(t)\lambda\,\Delta t,$$
$$n \ge 1, \quad (6\text{-}56)$$

etc., for $n = 0$, and where $P_j(t) \equiv 0$ for $j < 0$.

In the usual manner, this system gives rise to the following steady-state system of equations:

$$\lambda p_0 = k\mu p_1,$$
$$(\lambda + k\mu)p_n = k\mu p_{n+1} + \lambda p_{n-k}, \qquad n \ge 1. \qquad (6\text{-}57)$$

We write $\rho = \lambda/k\mu$ for the traffic intensity of the phases. Note that $p_j \equiv 0$ for $j < 0$. If we divide through by μ and introduce the generating function

$$P(z) = \sum_{n=0}^{\infty} p_n z^n$$

in the usual manner, we have, according to A. Craw,

$$P(z) = \frac{p_0}{1 - \rho z[(1 - z^k)/(1 - z)]} = p_0 \sum_{n=0}^{\infty} (\rho z)^n \left(\frac{1 - z^k}{1 - z}\right)^n$$

$$= p_0 \sum_{n=0}^{\infty} \rho^n (z + z^2 + \cdots + z^k)^n. \quad (6\text{-}58)$$

Since $P(1) = 1$, we have $p_0 = 1 - k\rho$.

Note that the problem of calculating p_n is a familiar one whose solution is readily obtained if $k = 1$. We easily have

$$(1 - z^k)^n = \sum_{i=0}^{n} (-1)^i \binom{n}{i} z^{ik},$$

$$(1 - z)^{-n} = \sum_{j=0}^{\infty} (-1)^j \binom{-n}{j} z^j = \sum_{j=0}^{\infty} (-1)^{2j} \binom{n+j-1}{j} z^j,$$

since $\quad \binom{-n}{j} = (-1)^j \binom{n+j-1}{j}.$

Hence

$$z^n(1 + z + \cdots + z^{k-1})^n = \sum_{j=0}^{\infty} \sum_{i=0}^{n} (-1)^i \binom{n}{i} \binom{n+j-1}{j} z^{j+ik+n},$$

and $\quad P(z) = (1 - k\rho) \sum_{m=0}^{\infty} \rho^m \sum_{j=0}^{\infty} \sum_{i=0}^{m} (-1)^i \binom{m}{i} \binom{m+j-1}{j} z^{j+ik+m}.$

$$(6\text{-}59)$$

The coefficient of z^n, where $n = j + ik + m$, is

$$p_n = (1 - k\rho) \sum \rho^m (-1)^i \binom{m}{i} \binom{m+j-1}{j}, \quad (6\text{-}60)$$

where the summation is taken over the partitions of n in the indicated manner. For example, for $k = 2$, $n = 7$,

$$p_7 = (1 - 2\rho)(4\rho^4 + 10\rho^5 + 6\rho^6 + \rho^7).$$

The mean number of phases, that is, L_{phases}, may be obtained directly from the intermediate quantity equations by computing $\sum_{n=1}^{\infty} n^2 p_n$, as we have done in the constant-service-time case.   Thus we have

$$(1 + \rho) \sum_{n=1}^{\infty} n^2 p_n = \sum_{n=1}^{\infty} n^2 p_{n+1} + \rho \sum_{n=k}^{\infty} n^2 p_{n-k},$$

which may be written as

$$(1 + \rho) \sum_{n=1}^{\infty} n^2 p_n = \sum_{n=1}^{\infty} (n - 1)^2 p_n + \rho \sum_{n=0}^{\infty} (n + k)^2 p_n.$$

On expanding and simplifying, we have

$$2(1 - k\rho)L_{\text{phases}} = (1 - p_0) + \rho k^2. \tag{6-61}$$

The average number of phases in the system is

$$L_{\text{phases}} = \frac{k\rho(k + 1)}{2(1 - k\rho)}, \tag{6-62}$$

and the average number of phases in the queue is L phases in queue $=$ L phases $- E$ (No. of phases in service)

$$= \frac{k\rho(k + 1)}{2(1 - k\rho)} - \frac{\lambda(k + 1)}{2\mu} = \frac{(k + 1)(k\rho)^2}{2(1 - k\rho)} \tag{6-63}$$

from which the average number in queue can be found by dividing by k. This gives

$$L = \frac{(k + 1)k\rho^2}{2(1 - k\rho)} \tag{6-64}$$

The expected wait in queue is easily computed as

$$W = \frac{L}{\lambda} = \frac{\rho(k + 1)}{2\mu(1 - k\rho)} \tag{6-65}$$

6-4. Erlangian Arrivals, Exponential Service, Single Channel

Suppose that arrivals occur by the Erlangian distribution

$$dA(t) = \frac{\lambda^k}{\Gamma(k)} e^{-\lambda t} t^{k-1} dt \tag{6-66}$$

with mean k/λ.   Arrivals by this distribution are equivalent to arrivals by a Poisson distribution with parameter λ from which every kth unit is admitted into the waiting line.

Suppose that the service-time intervals are exponentially distributed with parameter μ.   R. R. P. Jackson and D. G. Nickols [319] start out by simplifying the approach.   Let the states of the system be represented by (n_1, n_2), where n_1 gives the number of arrivals from the original Poisson input since the last acceptance into service and n_2 is the number of units actually in the system. Thus $0 \le n_1 < k$, and the transition $(k - 1, \; n_2) \to (0, \; n_2 + 1)$ characterizes the input scheme. If one writes $n = kn_2 + n_1$, then a unique pair (n_1, n_2) is defined by the greatest integers in the quantities indicated in brackets:

$$n_2 = \left[\frac{n}{k} \right]$$

$$n_1 = n - \left[\frac{n}{k} \right] k$$

We use $P_n(t)$ to indicate the probability that the system is in state n at time t. The steady-state equations are then given by

$$\begin{aligned}
\lambda p_0 - \mu p_k &= 0, & n &= 0, \\
\lambda p_n - \mu p_{n+k} - \lambda p_{n-1} &= 0, & 1 &\le n \le k - 1, \\
(\mu + \lambda) p_n - \mu p_{n+k} - \lambda p_{n-1} &= 0, & n &\ge k.
\end{aligned} \quad (6\text{-}67)$$

Let

$$P(z) = \sum_{n=0}^{\infty} p_n z^n \quad \text{and} \quad \rho = \frac{\lambda}{\mu}.$$

To obtain $P(z)$, we multiply the above equations by z^n, sum over the indicated values of n, and add. This gives

$$(\lambda p_0 - \mu p_k) + \sum_{n=1}^{k-1} (\lambda p_n - \mu p_{n+k} - \lambda p_{n-1}) z^n$$

$$+ \sum_{n=k}^{\infty} (\mu p_n + \lambda p_n - \mu p_{n+k} - \lambda p_{n-1}) z^n = 0. \quad (6\text{-}68)$$

On simplifying, this becomes

$$\lambda \sum_{n=0}^{\infty} p_n z^n - \mu \sum_{n=0}^{\infty} p_{n+k} z^n - \lambda \sum_{n=1}^{\infty} p_{n-1} z^n + \sum_{n=k}^{\infty} \mu p_n z^n = 0,$$

or $\quad \lambda P(z) - \dfrac{\mu}{z^k} \displaystyle\sum_{n=0}^{\infty} p_{n+k} z^{n+k} - \lambda z P(z) + \mu \sum_{n=k}^{\infty} p_n z^n = 0. \quad (6\text{-}69)$

Further simplifying yields

$$\lambda P(z) - \lambda z P(z) - \left(\frac{\mu}{z^k} - \mu \right) \sum_{n=k}^{\infty} p_n z^n = 0,$$

or $\quad \lambda P(z) - \lambda z P(z) - \left(\dfrac{\mu}{z^k} - \mu \right) \left[- \displaystyle\sum_{n=0}^{k-1} p_n z^n + P(z) \right] = 0. \quad (6\text{-}70)$

from which one has

$$P(z) = \frac{(\mu/z^k - \mu) \sum\limits_{n=0}^{k-1} p_n z^n}{\lambda z - \lambda + \mu/z^k - \mu}.$$

Finally,
$$P(z) = \frac{\mu(1 - z^k) \sum\limits_{n=0}^{k-1} p_n z^n}{\lambda z^{k+1} - \lambda z^k + \mu - \mu z^k}$$

$$= \frac{(1 - z^k) \sum\limits_{n=0}^{k-1} p_n z^n}{\rho z^{k+1} - (\rho + 1)z^k + 1}. \tag{6-71}$$

One must eliminate the unknown probabilities appearing on the right. The function $P(z)$ must converge at least inside the unit circle. Now the denominator has $k + 1$ zeros. Of these zeros, k lie on or interior to the unit circle by Rouché's theorem as applied to the denominator.

Thus, if $f(z) = -z^k(\rho + 1)$ and $g(z) = \rho z^{k+1} + 1$, the absolute values on the circle $|z| = 1 + \delta$ satisfy Rouché's theorem, and since $f(z)$ has k zeros inside this circle, then so does the sum of f and g, which is the denominator. One zero of the denominator is $z = 1$, and $k - 1$ zeros must coincide with those of $\sum\limits_{n=0}^{k-1} p_n z^n$. Otherwise $P(z)$ would not converge for values of z inside the unit circle, owing to the vanishing of the denominator at a zero which is not canceled by one in the numerator. This leaves one zero of the denominator lying outside this circle; we denote it by z_0. Then we may write (equating common factors of numerator and denominator)

$$(z - 1)(z - z_0) \sum\limits_{n=0}^{k-1} p_n z^n = A[\rho z^{k+1} - (1 + \rho)z^k + 1], \tag{6-72}$$

where A is a constant. This gives

$$P(z) = \frac{A \sum\limits_{n=0}^{k-1} z^n}{z_0 - z}.$$

A is determined by $P(1) = 1$ to give

$$P(z) = \frac{(z_0 - 1) \sum\limits_{n=0}^{k-1} z^n}{(z_0 - z)k}. \tag{6-73}$$

From this generating function it can be shown by differentiation that

$$p_n = \frac{1 - z_0^{-n-1}}{k}, \quad n \leq k - 1,$$

and

$$p_n = \frac{z_0^{-n-1}(z_0^k - 1)}{k}, \quad n \geq k.$$

(6-74)

The probability that there are n units in the system is given by

$$P_n = \sum_{j=nk}^{nk+k-1} p_j,$$

(6-75)

which gives

$$P_0 = 1 - \rho, \quad P_n = \rho(1 - z_0^{-k})z_0^{-(n-1)k}.$$

(6-76)

The probability Q_n that an arriving unit finds n units in the system is equal to P_n for the random-arrival case, i.e., for $k = 1$; otherwise

$$Q_n = kp_{kn+k-1} = (1 - z_0^{-k})z_0^{-nk},$$
$$Q_0 = 1 - z_0^{-k}.$$

(6-77)

The waiting-time distribution is given by

$$w(t) = \sum_{n=1}^{\infty} Q_n \frac{\mu^n t^{n-1}}{(n-1)!} e^{-\mu t} = z_0^{-k}\lambda(z_0 - 1) \exp[-\lambda(z_0 - 1)t]. \quad (6-78)$$

The waiting-time distribution for the units that must wait is $z_0^k w(t)$, which is an exponential distribution, as asserted by Smith (see Chap. 9).

We now obtain results for a regular input distribution

$$A(x) = \begin{cases} 0, & x < 1, \\ 1, & x \geq 1, \end{cases}$$

by allowing k and $\lambda \to 0$ so that $k/\lambda = 1$. Let $z_0 = 1 + y/k$, and let $\rho y = 1 - e^{-y} + O(k^{-1})$, from which $z_0 = 1 + y_0/k + O(k^{-2})$, where y_0 is the real positive root of $\rho = (1 - e^{-y})/y$. On substituting in the expression for P_n, we have

$$P_0 = 1 - \rho,$$
$$P_n = \rho(1 - e^{-y_0})e^{-(n-1)y_0}, \quad n \geq 1,$$
$$Q_n = \rho y_0(1 - \rho y_0)^n = (1 - e^{-y_0})e^{-ny_0},$$
$$w(t) = e^{-y_0}\mu(1 - e^{-y_0}) \exp[-\mu(1 - e^{y_0})t].$$

(6-79)

6-5. Erlangian Input, Constant Service Time

The procedure given in the example of Chap. 3 to compute the waiting-time distribution for Poisson input, constant service time of unit length, may be generalized to the case where

$$dA(x) = \lambda \frac{(\lambda x)^{c-1}}{(c-1)!} e^{-\lambda x} dx,$$

of which $c = 1$ is a special case. Here one defines an operator

$$\Omega = \frac{(-1)^{c-1}}{(c-1)!} \lambda^c \frac{\partial^{c-1}}{\partial \lambda^{c-1}}$$

which gives the desired expression after making a transformation of variables:

$$P(w) = \begin{cases} \int_0^\infty P(u)\Omega e^{-\lambda(u+1-w)} du, & w < 1, \\ \int_{w-1}^\infty P(u)\Omega e^{-\lambda(u+1-w)} du, & w \geq 1. \end{cases}$$

The Laplace-Stieltjes transform is given by

$$\gamma(s) = \frac{\lambda^c e^{-s}}{(\lambda - s)^c} \gamma(s) - s \int_0^\infty P(u)\Omega e^{-\lambda(u+1)} du$$

$$= \frac{s f_{c-1}(s)}{\lambda^c e^{-s} - (\lambda - s)^c},$$

where $f_{c-1}(s)$ is a polynomial of degree $(c-1)$ in s. It can be shown that for $(\lambda/c) < 1$ the denominator has exactly c zeros [in the half plane $R(s) \geq 0$], which we denote by

$$s_0 = 0, \qquad s_1, \ldots, s_{c-1},$$

with $\qquad R(s_i) > 0, \qquad i = 1, \ldots, c-1.$

This leads to the vanishing of the numerator at these zeros, etc., finally giving

$$\gamma(s) = \frac{\lambda^{c-1}(c - \lambda)}{\lambda^c e^{-s} - (\lambda - s)^c} s \prod_{k=1}^{c-1} \frac{s_k - s}{s_k}.$$

If we let $s \to \infty$, we obtain (see definition)

$$P(=0) = \frac{\lambda^{c-1}(c - \lambda)}{\prod\limits_{k=1}^{c-1} s_k}.$$

By expanding $\gamma(s)$ in series in powers of s we obtain the moments. Thus the expected waiting time is given by

$$W_q = \sum_{k=1}^{c-1} \frac{1}{s_k} + \frac{1}{2} \frac{\lambda^2 - c^2 + c}{\lambda(c - \lambda)}.$$

CHAPTER 7

BULK QUEUES

7-1. Introduction

So far we have looked at queues with the idea that arrivals and service occur singly. But there are many queueing activities in which arrivals and service can occur in a group, i.e., in bulk or in batches. Several people may go to a restaurant together and obtain service as a batch. A number of long-distance telephone calls may present themselves simultaneously before an operator. An elevator serves a group of people at the same time. Even though aircraft may arrive singly at an airport, if one considers the queueing operations which arriving passengers must undergo, it is clear that for some purposes one must apply the bulk-arrival assumption, as, for instance, supplying customs officers to handle the arriving groups. Note that the passengers themselves do not appear all at once for service but tend to move in small groups.

It is clear that we are dealing with a more general type of queueing problem which can be specialized to yield the single-arrival and single-service cases. For many practical applications, bulk arrivals and service are the only realistic assumptions.

We develop formulas for the expected number waiting and the expected waiting time in the steady state, for a single-channel service queue with Poisson input and arbitrary holding times in which items are served in batches not exceeding a certain number s. Thus, whenever the service facility is idle, s items are served if the length of the queue is greater than s. Otherwise, the entire queue is served in a batch. We follow the works of Bailey and Downton [20, 162, 163]. In order to do this, we require the simple idea of an imbedded Markoff chain due to D. G. Kendall [362].

Kendall has utilized the fact that one may study a single-channel queue with Poisson input at epochs of departure which are points of regeneration (a concept due to Palm). According to him, an epoch is a point of regeneration if a knowledge of the state of the process at a particular epoch has the characteristic Markovian consequence: that a statement of the past history of the process has no predictive value. Thus a

Markoff process is precisely a process for which every epoch is a point of regeneration. The advantage of this is the fact that the problem is reduced to one with a Markoff chain in discrete time, despite the fact that the congestion process itself is not Markovian.

The concept of regeneration points may be applied to epochs of arrivals if the service times are exponentially distributed.

For arbitrary input and service times the only possible regeneration points are epochs at which an arrival and a departure occur simultaneously (which are extremely rare) and epochs at which a new customer arrives and finds the counter free (which occur occasionally), but here there is no simple way of relating regeneration points.

Note (as we have seen in the derivation of the Pollaczek-Khintchine formula) the need for a Poisson input, which implies the independence of the queue size at the moment of departure of a customer and of the number of items arriving during the next service period. For more general inputs, dependence between these two quantities might be expected.

7-2. Poisson Arrivals, Service in Bulk

We shall be concerned with the case of a Poisson-input, arbitrary-service-time distribution $B(t)$.

First, we introduce the transition probabilities r_{ij}, which give the conditional probability that the next state is E_j (i.e., there are j items in the line), given that the previous state was E_i. The queue length is measured at those instants (regeneration points) just before service of a batch takes place or, equivalently, after a batch has just completed service. If one denotes by π_j ($j = 0, 1, \ldots$) the probability of being in state E_j at times before service begins, obviously π_j is obtained from π_i by multiplying π_i by r_{ij} and then summing over i to account for all possible ways of transiting to state E_j from E_i. Thus when the system is in statistical equilibrium one has

$$\pi_j = \sum_{i=0}^{\infty} \pi_i r_{ij}. \tag{7-1}$$

See Sec. 7-6 for the relation to p_j. Note that in the transient case we have π_j^{n+1} on the left and π_i^n on the right.

Since the purpose is to find π_j for all j, we must first find a method of determining r_{ij}. Note that we shall work with π_j which are not necessarily the actual steady-state probabilities at any time.

Let the Poisson process $(\lambda t)^n e^{-\lambda t}/n!$ ($n = 0, 1, 2, \ldots$) denote the arrival distribution of n items during any given service interval t. Let $B(t)$ be the cumulative-service-time distribution $[dB(t)/dt = b(t)$, when it exists]. We assume that the service times, which are also the times between successive epochs of departures, are independently distributed. Then the probability of n arrivals during a randomly chosen service time is

$$k_n = \frac{1}{n!} \int_0^\infty e^{-\lambda t} (\lambda t)^n \, dB(t), \qquad n = 0, 1, 2, \ldots , \qquad (7\text{-}2)$$

where we have taken the integral over all intervals of length t, whose distribution is $B(t)$.

Define

$$K(z) = \sum_{n=0}^\infty k_n z^n = \int_0^\infty e^{-(1-z)\lambda t} \, dB(t) \equiv \beta[\lambda(1 - z)]. \qquad (7\text{-}3)$$

Thus $K(z)$ is simply related to the Laplace-Stieltjes transform $\beta(z)$ of the service-time distribution.

We now multiply Eq. (7-1) through by z^j and sum over j. This gives

$$P(z) \equiv \sum_{j=0}^\infty \pi_j z^j = \sum_{j=0}^\infty z^j \sum_{i=0}^\infty \pi_i r_{ij}. \qquad (7\text{-}4)$$

Since units are served in batches of s or less, a change from i to j units in the system is obtained either by having the i units all served in one batch (if $i \leq s - 1$) and j units arrive or (if $i \geq s$) by having s units served and $i - s$ remaining with $j - (i - s)$ new arrivals to bring the total to j. Finally, if $i > j + s$, then $r_{ij} = 0$, since even if s units are served it is impossible to reduce the number to j. Thus

$$r_{ij} = \begin{cases} k_j & \text{for } 0 \leq i \leq s - 1, \\ k_{j-(i-s)} & \text{for } j + s \geq i \geq s, \\ 0 & \text{for } i > j + s, \end{cases} \qquad (7\text{-}5)$$

and $k_j = 0$ for $j < 0$.

We now have

$$P(z) = \sum_{j=0}^\infty z^j \left(\sum_{i=0}^{s-1} \pi_i k_j + \sum_{i=s}^{j+s} \pi_i k_{j-i+s} \right)$$

$$= K(z) \sum_{i=0}^{s-1} \pi_i + \sum_{j=0}^\infty z^j (\pi_s k_j + \pi_{s+1} k_{j-1} + \cdots + \pi_{s+j} k_0). \qquad (7\text{-}6)$$

Note that $\pi_s k_j + \cdots + \pi_{s+j} k_0$ is the convolution of two sequences. We use the fact that the generating function of the convolution is the product of the generating functions of the separate sequences. The generating function of the k_j is $K(z)$ and that of the π_{s+j} is

$$\sum_{j=0}^\infty \pi_{j+s} z^j = \sum_{i=s}^\infty \pi_i z^{i-s}.$$

Also

$$z^{-s} \sum_{i=s}^{\infty} \pi_i z^i = \left[P(z) - \sum_{i=0}^{s-1} \pi_i z^i \right] z^{-s}. \qquad (7\text{-}7)$$

And we finally have

$$P(z) = K(z) z^{-s} \left[P(z) + \sum_{i=0}^{s-1} \pi_i (z^s - z^i) \right], \qquad (7\text{-}8)$$

which on solving for $P(z)$ gives

$$P(z) = \frac{\sum_{i=0}^{s-1} \pi_i (z^s - z^i)}{z^s / K(z) - 1}. \qquad (7\text{-}9)$$

Here the unknown probabilities $\pi_0, \ldots, \pi_{s-1}$ are determined by considering the fact that, in order for $P(z)$ to converge on and inside the unit circle (which follows from its definition), the zeros of the numerator must coincide with those of the denominator within the circle $|z| = 1 + \delta$ for $\delta > 0$ and small. This condition is satisfied for the examples considered below.

To illustrate, we shall assume in the rest of this section that the service intervals are distributed according to chi square with $2k$ degrees of freedom, i.e., have the Erlangian distribution

$$b(t) = \frac{\mu^k}{\Gamma(k)} t^{k-1} e^{-\mu t}, \qquad 0 \le t < \infty, \qquad (7\text{-}10)$$

where the expected service time k/μ defines the traffic intensity ρ by the relation

$$\rho \equiv \frac{\lambda k}{\mu s} = \frac{\lambda}{s} \left(\sum_{j=0}^{s} j \pi_j + s \sum_{j=s+1}^{\infty} \pi_j \right)^{-1}.$$

That is, the average number arriving in a service interval equals the average batch size actually served. Note that such a queue can attain a steady state if $\rho < 1$, which we assume to hold.

Now, using (7-2) and (7-3), one has

$$K(z) = \left[1 + \frac{\rho s (1 - z)}{k} \right]^{-k}. \qquad (7\text{-}11)$$

If one considers the denominator in (7-9), one can show that it has $s - 1$ simple zeros z in $|z| \le 1$, other than $z = 1$.

Remark: To determine the number of zeros, one applies Rouché's theorem to z^s and to $z^s - K(z)$ inside and on a circle $|z| = 1 + \delta$. Each of the zeros, when substituted for z in the numerator, makes it vanish. On and inside the unit circle there can be no multiple zeros; otherwise both the denominator and its derivative would vanish and on dividing the two expressions one obtains the condition that $\rho > 1$, which we excluded.

Clearly, as $z \to 1$, $P(z) \to 1$. Thus, applying L'Hospital's rule to (7-9), one has as $z \to 1$

$$\sum_{i=0}^{s-1} (s - i)\pi_i = s - \rho s. \tag{7-12}$$

This, together with the $s - 1$ numerator equations obtained by substituting, respectively, the zeros of the denominator and equating to zero, forms a consistent set (i.e., the coefficient determinant does not vanish) from which the π_j can be determined and normalized. If the coefficient set of each π_j is written in a column, a new determinant is obtained by subtracting from each column the column to its right, and finally the common factors from each row are taken out as multipliers, there results the determinant that does not vanish, because all the z_i are different and none is equal to unity:

$$\begin{vmatrix} 1 & 1 & \cdots & 1 \\ 1 & z_1 & \cdots & z_1^{s-1} \\ \cdots & \cdots & \cdots & \cdots \\ 1 & z_{s-1} & \cdots & z_{s-1}^{s-1} \end{vmatrix} \prod_{i=1}^{s-1} (z_i - 1). \tag{7-13}$$

By observing (7-11), one concludes that the denominator in (7-9) has k zeros z_j outside the unit circle since there are $s + k$ zeros in all; consequently, after canceling the s common factors of the numerator and the denominator, that is, $z - 1$ and $z - z_i$ $(i = 1, \ldots, s - 1)$ [otherwise $P(z)$ would not exist in the unit circle for these values], one has

$$P(z) = \frac{A}{\displaystyle\prod_{j=s}^{s+k-1} (z_j - z)} \tag{7-14}$$

where A is a proportionality constant. On putting $z = 1$ and using $P(1) = 1$ one can also write

$$P(z) = \prod_{j=s}^{s+k-1} \frac{z_j - 1}{z_j - z}, \tag{7-15}$$

from which the π_j may be obtained by partial-fraction expansion.

Remark: For a brief discussion of computing the zeros, see Downton [163].

The moment-generating function $M(\theta)$ is obtained by putting $z = e^\theta$ in $P(z)$, and the cumulant generating function $\psi(\theta)$ is obtained by taking the logarithm of the moment-generating function. As usual, the expected number in the queue is obtained by taking the first derivative of the cumulant generating function with respect to θ and setting $\theta = 0$. This gives

$$\sum_{j=s}^{s+k-1} (z_j - 1)^{-1}. \tag{7-16}$$

The second derivative gives the variance

$$\sum_{j=s}^{s+k-1} z_j(z_j - 1)^{-2}. \tag{7-17}$$

Exercise: Verify the last two expressions.

For exponential-service-times distribution one puts $k = 1$ in (7-15), which gives

$$P(z) = \frac{z_s - 1}{z_s - z},$$

where z_s is the single zero outside the unit circle. The probabilities form a simple geometric series.

For constant-service time, one lets $k \to \infty$ and $\mu \to \infty$, keeping their ratio equal to unity, and sets $\rho s = \lambda$. This leads to infinite products. Direct analysis gives, on letting $k \to \infty$ in $K(z)$,

$$P(z) = \frac{\sum_{j=0}^{s-1} \pi_j(z^s - z^j)}{z^s e^{\rho s(1-z)} - 1}. \tag{7-18}$$

Because the denominator in (7-18) has s zeros z_i inside and on the unit circle, the numerator can be written as follows:

$$\sum_{j=0}^{s-1} \pi_j(z^s - z^j) = \left(\sum_{j=0}^{s-1} \pi_j \right) (z - 1) \prod_{i=1}^{s-1} (z - z_i). \tag{7-19}$$

This is used (together with the s equations previously used to determine π_j) to obtain an expression for $\sum_{j=0}^{s-1} \pi_j$. Here again the coefficient determinant is used with $\sum_{j=0}^{s-1} \pi_j$ appearing as a constant.

Since there are $s + 1$ equations to determine s unknowns, the determinant of the augmented matrix must vanish. Subtracting from each column the column to its left; leaving the second-to-the-last column, i.e., the sth column, as it is; factoring out the common factors from each row and setting the determinant without the outside factor equal to zero (since the product of the factors taken out is not zero); then expanding according to the coefficients of the first row; and canceling common factors give

$$\sum_{j=0}^{s-1} \pi_j = (s - \rho s) \prod_{i=1}^{s-1} (1 - z_i)^{-1}. \tag{7-20}$$

Thus Eqs. (7-18) to (7-20) give

$$P(z) = \frac{(s - \rho s)(z - 1) \prod_{i=1}^{s-1} (z - z_i)/(1 - z_i)}{z^s e^{\rho s(1-z)} - 1}. \tag{7-21}$$

In a similar way, one obtains for the expected number in the queue

$$L_q = \frac{s - (s - \rho s)^2}{2(s - \rho s)} + \sum_{i=1}^{s-1} (1 - z_i)^{-1}, \tag{7-22}$$

and for the variance

$$\frac{s(s + 2\rho s) + 6\rho s(s - \rho s)^2 - (s - \rho s)^4}{12(s - \rho s)^2} - \sum_{i=1}^{s-1} z_i(1 - z_i)^2. \tag{7-23}$$

7-3. The Laplace Transform of the Waiting Times

Denote by $C_r(w)$ the probability that, if a unit is rth in a batch at the beginning of its service, its waiting time is $\leq w$. The probability of a unit which is rth in a batch arriving during $(t, t + dt)$ is $w_r \, dt$, where w_r is the expected number of rth units arriving in time $1/\lambda$ (also the expected number of rth units served in time $1/\lambda$, since the queue is in statistical equilibrium).

Now if $E(t)$ is the expected service time, we have

$$w_r = \sum_{j=r}^{\infty} \pi_j/\lambda E(t), \tag{7-24}$$

since the computation is made for r units and the probability that there are at least r units in the queue (also the expected number of rth units being served at each service epoch) is $\sum_{j=r}^{\infty} \pi_j$ and in statistical equilibrium the expected number of service epochs per time $1/\lambda$ is $1/\lambda E(t)$. Note that $\sum_{r=1}^{s} w_r = 1$.

We can now write the waiting-time distribution when non-waiting customers are excluded as

$$C(w) = \sum_{r=1}^{s} w_r C_r(w). \tag{7-25}$$

The probability that j units have arrived during time w is $(\lambda w)^j e^{-\lambda w}/j!$, and consequently

$$\frac{1}{j!} \int_0^\infty (\lambda w)^j e^{-\lambda w}\, dC_r(w) = \frac{\pi_{j+r}}{\displaystyle\sum_{i=r}^\infty \pi_i} \tag{7-26}$$

gives the conditional probability that there are $j + r$ units in the queue at time w, given that there are at least r. Hence

$$P(z) = \sum_{j=0}^{r-1} \pi_j z^j + \left(\sum_{i=r}^\infty \pi_i\right) z^r \int_0^\infty e^{-\lambda w} \sum_{j=0}^\infty \frac{(\lambda w z)^j}{j!}\, dC_r(w)$$

$$= \sum_{j=0}^{r-1} \pi_j z^j + \left(\sum_{i=r}^\infty \pi_i\right) z^r M_r(z). \tag{7-27}$$

Here
$$M_r(z) = \int_0^\infty e^{-(1-z)\lambda w}\, dC_r(w). \tag{7-28}$$

Multiplying Eq. (7-28) through by w_r and summing over $1 \le r \le s$, we have

$$M(z) \equiv \int_0^\infty e^{-(1-z)\lambda w}\, dC(w) \equiv \gamma[\lambda(1-z)], \tag{7-29}$$

which is related to the Laplace-Stieltjes transform of the waiting-time distribution. Note that

$$M(z) = \sum_{r=1}^s w_r M_r(z). \tag{7-30}$$

Again multiplying (7-27) by z^{s-r} and summing over $1 \le r \le s$ after using (7-24) and (7-9), we have

$$M(z) = \frac{[1/K(z)] - 1}{\lambda E(t)(1-z)} P(z). \tag{7-31}$$

From (7-29) and (7-31) it is possible to write the Laplace-Stieltjes transform with parameter p of the waiting-time distribution as

$$\gamma(p) = \frac{P(1 - p/\lambda)}{pE(t)} \left[\frac{1}{\beta(p)} - 1\right], \tag{7-32}$$

where $\beta(p)$ is the Laplace-Stieltjes service-distribution transform.

From (7-11) and (7-15) and using the fact that $M(1) = 1$, one has

$$M(z) = \frac{\mu}{\lambda k} \frac{[1 + \lambda(1-z)/\mu]^k - 1}{1-z} \prod_{i=s}^{s+k-1} \frac{z_i - 1}{z_i - z}, \tag{7-33}$$

where z_i are the zeros that lie outside the unit circle. This gives for the desired Laplace-Stieltjes transform

$$\gamma(p) = \frac{\mu}{kp}\left[\left(1 + \frac{p}{\mu}\right)^k - 1\right] \prod_{i=s}^{s+k-1} \frac{\lambda(z_i - 1)}{\lambda(z_i - 1) + p}. \tag{7-34}$$

Let $\lambda = 1$; then the mean waiting time is obtained from (7-34). It is

$$E(w) = \sum_{i=s}^{s+k-1} \frac{1}{z_i - 1} - \frac{k - 1}{2\mu} \tag{7-35}$$

and

$$\text{var } (w) = \sum_{i=s}^{s+k-1} \frac{1}{(z_i - 1)^2} + \frac{(k - 1)(k - 5)}{12\mu^2}. \tag{7-36}$$

Thus for regular service we have, as $\mu \to \infty$, $k \to \infty$, so that $k/\mu = b$ is constant,

$$M(z) = \frac{s - b}{-b} \frac{e^{b(1-z)} - 1}{z^s e^{b(1-z)} - 1} \prod_{i=1}^{s-1} \frac{z - z_i}{1 - z_i}. \tag{7-37}$$

The Laplace-Stieltjes transform is

$$\frac{s - b}{-b} \frac{e^{bp} - 1}{(1 - p)^s e^{bp} - 1} \prod_{i=1}^{s-1} \frac{1 - p - z_i}{1 - z_i} \tag{7-38}$$

and

$$E(w) = \sum_{i=1}^{s-1} \frac{1}{1 - z_i} + \frac{s(b + 1 - s)}{2(s - b)}, \tag{7-39}$$

$$\text{var } (w) = \frac{s(6 - s + 2b)}{12} - \frac{s(5s - 8b)}{12(s - b)^2} - \sum_{i=1}^{s-1} \frac{1}{(1 - z_i)^2}. \tag{7-40}$$

Downton [163] gives for q the queue-size variable, using chi-square service distribution with $2k$ degrees of freedom and $\rho = 1/\mu s$ with $\lambda = 1$:

$$\lim_{s \to \infty} \frac{E(q)}{s} = \sum_{j=1}^{k} \frac{1}{s(z_j - 1)},$$

$$\lim_{s \to \infty} \frac{\text{var } (q)}{s^2} = \sum_{j=1}^{k} \frac{1}{[s(z_j - 1)]^2},$$

$$\lim_{s \to \infty} \frac{E(w)}{s} = \sum_{j=1}^{k} \frac{1}{s(z_j - 1)} - \frac{(k - 1)\rho}{2k},$$

$$\lim_{s \to \infty} \frac{\text{var } (w)}{s^2} = \sum_{j=1}^{k} \frac{1}{[s(z_j - 1)]^2} + \frac{(k - 1)(k - 5)\rho^2}{12k^2}.$$

The first two quantities can be obtained from the generating function of the moments of the queue-size distribution, that is, $E(q)/s$, $E(q^2)/s^2$, and so on. In the limit it becomes the Laplace transform of the distribution of $\lim_{s \to \infty} (q/s)$. It is given by

$$P_s(z) = \sum_{j=0}^{\infty} \pi_j \left(1 - \frac{z}{s} \right)^j = \prod_{j=1}^{k} \frac{s(z_j - 1)}{s(z_j - 1) + z}. \qquad (7\text{-}41)$$

The second two quantities above are obtained from the Laplace-Stieltjes transform of the distribution of the ratio of waiting time and batch size s. It is obtained from (7-34) by replacing z by z/s.

For the constant-service-time case, by a similar but somewhat elaborate analysis, computing residues mostly at simple poles, he gives the Laplace-Stieltjes transform of the waiting time, as $s \to \infty$, as

$$\frac{1 - e^{-\rho z}}{\rho z}, \qquad (7\text{-}42)$$

and consequently

$$\lim_{s \to \infty} \frac{E(w)}{s} = \frac{\rho}{2},$$

$$\lim_{s \to \infty} \frac{\text{var }(w)}{s^2} = \frac{\rho^2}{12},$$

$$\lim_{s \to \infty} \frac{E(q)}{s} = \rho = \lim_{s \to \infty} \frac{\text{var }(q)}{s}.$$

Of course, $\lim_{s \to \infty} \dfrac{\text{var }(q)}{s^2} = 0.$

The limiting properties of the queue-size distribution apply to the queue-size distribution of the multiserver queue with random arrivals and regular service, as shown by Bailey. The method can be used to obtain similar results for the waiting-time distribution in this case.

Remark 1: Bulk Service and Bulk Arrivals.   Recently R. G. Miller, Jr. [487], in a novel treatment of the subject, has considered the case of bulk inputs with bulk services, single channel, first come, first served. A late arrival may either join a batch already in service, as long as its size is less than the quota number, or he may not and hence wait for the next batch service. The entire quota need not be met for service to begin. No easy result is given in the second case.

For the first case where late arrivals can join a batch in service, Miller obtains

$$K(z) = \sum_{j=0}^{\infty} z^j \int_0^{\infty} a_j(t) \, dB(t) = \int_0^{\infty} e^{-\lambda t[1 - Q(z)]} \, dB(t), \qquad (7\text{-}43)$$

where $a_j(t)$ is the probability that j items arrive in $[0,t]$ and $Q(z) = \sum_{j=1}^{\infty} q_j z^j$, where $q_j = \text{Prob } (N = j) = 0$, for $j > n_0$, for some n_0, and N is the number of group arrivals in $[0,t]$. He obtains the generating function of the steady-state probabilities π_j [where the input has the geometric distribution $(1 - p)p^{v-1}$, $0 < p < 1$, and exponential service for one item at a time, where $B(t) = 1 - e^{-\mu t}$] from

$$K(z) = \frac{1}{1 + (\lambda/\mu)[(1 - z)/(1 - pz)]}. \tag{7-44}$$

The desired generating function is

$$\sum_{j=0}^{\infty} \pi_j z^j = \frac{1 - p - \lambda/\mu}{1 - (p + \lambda/\mu)z}. \tag{7-45}$$

Hence the stationary distribution is also geometric with parameter $p + \lambda/\mu$.

7-4. Binomial Input, Arbitrary Service

This section is included here because of the use it makes of the idea of an imbedded Markoff chain. It serves as a useful interlude which should aid the reader to achieve a better understanding of the ideas.

As a preliminary discussion to this section, we give a simplified version of an important problem to which no analytical answers are available. We then give the analytical development of a simple queueing problem which might be useful in solving the problem.

Given: one aircraft carrier with N intercept aircraft aboard and the requirement that $m < N$ aircraft be on station in the air at all times for an interval of time of length T. Aircraft on station are relieved. There are c maintenance shops on the aircraft carrier. The duration of maintenance times for each aircraft follows an exponential probability distribution which is identical for all aircraft and all shops. Aircraft require maintenance independently with probability p; thus the number requiring maintenance follows a binomial probability law. An aircraft landing at the end of its flying period T requires maintenance. How many aircraft are required on board the carrier to maintain m aircraft on station with given probability π at all times for a total period of operation of length t?

The study of binomial input might be applicable to the foregoing example. For general study of binomial input the reader is referred to most works of Pollaczek. In the following we give a simplified discussion.

T. Meisling [480] considers a first-come–first-served, single-channel queue with single customers arriving at discrete "time marks." The probability of a customer arriving during a time mark is a constant p independent of arrivals at previous time marks. The probability of no arrival at a time mark is $q = 1 - p$.   The service times are independently and identically distributed.

After obtaining the expected number in the queue and the expected waiting time, he passes to the limit to obtain continuous time results for Poisson input.   If a time period t_n is considered with n time marks, each of length Δt, the probability of $0 \leq m \leq n$ customers arriving during t_n is

$$A_{m,n} = \binom{n}{m} p^m q^{n-m} \tag{7-46}$$

and zero for $m > n$.   It is easily shown that $E[m] = np$ and

$$E[m(m - 1)] = n(n - 1)p^2.$$

The mean number of arrivals λ is given by

$$\lambda = \frac{E(m)}{n \, \Delta t} = \frac{p}{\Delta t}. \tag{7-47}$$

The service time t has a distribution which assumes values that are multiples of Δt and is given by the (discrete) probability b_n $(n = 0, 1, \ldots)$ that the service time equals $n \, \Delta t$ (an interval which is open on the left and closed on the right).   Note that

$$\sum_{n=0}^{\infty} b_n = 1, \qquad E(t) = \Delta t \sum_{n=0}^{\infty} n b_n, \qquad E[t(t - \Delta t)] = \overline{\Delta t^2} \sum_{n=0}^{\infty} n(n - 1) b_n,$$

where $E(t)$ and $E[t(t - \Delta t)]$ are the mean and variance of the service time and the traffic intensity, which is the ratio of the mean arrival and the mean service rates, is $\rho = \lambda E(t)$, here assumed <1.

Using the imbedded-Markoff-chain theory outlined above, the probability k_m that m items arrive during a service time is

$$k_m = \sum_{n=0}^{\infty} A_{m,n} b_n, \tag{7-48}$$

and r_{ij} are as previously given, except that the number in a batch to be served is unity.   Consequently, from our study of bulk queues we have

$$P(z) = \frac{\pi_0 K(z)(z - 1)}{z - K(z)}, \tag{7-49}$$

where
$$K(z) = \sum_{m=0}^{\infty} k_m z^m = \sum_{m=0}^{\infty} z^m \sum_{n=m}^{\infty} \binom{n}{m} p^m q^{n-m} b_n$$

$$= \sum_{n=0}^{\infty} b_n \sum_{m=0}^{n} \binom{n}{m} (zp)^m q^{n-m} = \sum_{n=0}^{\infty} b_n (zp + q)^n. \tag{7-50}$$

Now, since $P(1) = \sum_{j=0}^{\infty} \pi_j = 1$ and $K(1) = 1$, $K'(1) = \rho$, on applying L'Hospital's rule to the above expression in $P(z)$ (while $z \to 1$), one has $\pi_0 = 1 - \rho$. The expected number in the line is given by $P'(1)$, which requires applying L'Hospital's rule a second time, using

$$K''(1) = \sum_{n=0}^{\infty} b_n p^2 n(n-1)(p+q)^{n-2} = \lambda^2 E[t(t - \Delta t)],$$

having used the two expressions above defining λ and $E[t(t - \Delta t)]$. This gives

$$L = P'(1) = \rho + \frac{\lambda^2 E[t(t - \Delta t)]}{2(1 - \rho)}. \tag{7-51}$$

Now as $\Delta t \to 0$ the binomial-input distribution becomes a Poisson distribution, the service distribution becomes continuous, and one has

$$L = \rho + \frac{\lambda^2 E(t^2)}{2(1 - \rho)}. \tag{7-52}$$

The mean number of customers arriving during the mean waiting time plus the mean service time equals the mean queue length. This is used here to obtain the mean waiting time:

$$W_q = \frac{\lambda E[t(t - \Delta t)]}{2(1 - \rho)}. \tag{7-53}$$

If the service-time distribution is a geometric distribution

$$b_k = (1 - d)d^k,$$

then $E(t) = \Delta t \dfrac{d}{1 - d};$ $E[t(t - \Delta t)] = \overline{\Delta t^2} \dfrac{2d^2}{(1 - d)^2}$

$$= 2[E(t)]^2;$$

and consequently, as $\Delta t \to 0$,

$$L = \frac{\rho}{1 - \rho},$$

$$W_q = E(t) \frac{\rho}{1 - \rho}, \tag{7-54}$$

which are the results obtained for Poisson input and exponential service time.

For constant service time T_0,

$$E[t(t - \Delta t)] = E(t^2) - \Delta t E(t) = T_0^2 - \Delta t T_0,$$

$$L = \rho + \frac{\rho^2(1 - \Delta t/T_0)}{2(1 - \rho)}, \tag{7-55}$$

$$W_q = T_0 \frac{\rho(1 - \Delta t/T_0)}{2(1 - \rho)}. \tag{7-56}$$

7-5. Bulk Arrivals and Arbitrary Service Times (Transient Case)

Gaver [248] has studied the problem of arrival in batches to a single channel with arbitrary service distribution $B(t)$ by the imbedded-Markoff-chain method in the transient case. Arrivals are assigned numbers to be served in an ordered manner. The reader might profit by reading a few subsequent chapters before studying this section. Here we give a summary of his results.

Let $\{\Delta_k\}$ be a sequence of positive independent random variables where

$$\text{Prob } (\Delta_k \leq x) = 1 - e^{-\lambda x}, \quad \lambda > 0. \tag{7-57}$$

Δ_k is the period between the arrival times of two successive batches. The time of arrival of the kth batch α_k is given by

$$\alpha_k = \sum_{i=1}^{k} \Delta_i. \tag{7-58}$$

The number of customers in a batch is C_k, where

$$\text{Prob } (C_k = j) = c_j. \tag{7-59}$$

The number of customers which arrive in the interval $(0,t)$ is

$$A(0,t) = \sum_{0 < \alpha_i < t} C_i. \tag{7-60}$$

We have

$$\text{Prob } [A(0,t) = k] = a_k(t) = \sum_{n=0}^{\infty} e^{-\lambda t} \frac{(\lambda t)^{n}}{n!} c_k^{n*}, \tag{7-61}$$

where c_k^{n*} is the n-fold convolution of $\{c_k\}$.

One may also introduce the generating function

$$a(z,t) = \sum_{k=0}^{\infty} a_k(t)z^k = \exp \{-\lambda t[1 - c(z)]\}, \tag{7-62}$$

where
$$c(z) = \sum_{j=0}^{\infty} c_j z^j. \tag{7-63}$$

The state of the system is given by a pair, the first of which is the number in the system at the nth departure time after an initial time t_0 and the second is the time of the nth departure. The sequence of such states forms an imbedded Markoff chain. One first examines the busy period. If $P_{ij}^{(n)}(t)$ is the (transition) probability that the number of customers in the system passes from $i > 0$ at t_0 to $j > 0$ immediately after the nth departure (which occurs before $t + t_0$) without having passed zero in the meantime, then, by direct enumeration, $P_{ij}^{(n)}(t)$ satisfy the forward Chapman-Kolmogorov equation:

$$P_{ij}^{(n+1)}(t) = \sum_{h=-1}^{j-1} \int_0^t P_{i,j-h}^{(n)}(t - x)a_{h+1}(x) \, dB(x). \tag{7-64}$$

A formal solution of this equation is obtained using the Laplace-Stieltjes transform and generating function. The joint distribution of the number present in the system at a time t from the beginning of a busy period and the number served by that time are then obtained.

The mean and variance of a busy period are, respectively,

$$\frac{\rho}{\lambda\delta_1(1 - \rho)}, \tag{7-65}$$

$$\frac{\sigma^2 + \rho\mu_2(\delta_2/\delta_1)}{1 - \rho}, \tag{7-66}$$

where $\rho = \lambda\delta_1\mu_1$ is the traffic intensity, μ_1 is the mean service time, μ_2 is its second moment, $\sigma^2 = \mu_2 - \mu_1^2$, and δ_i $(i = 1, 2)$ are the first and second moments, respectively, of the $\{c_k\}$.

The mean of the distribution of the number discharged during a busy period and its variance are, respectively,

$$\frac{1}{1 - \rho} \tag{7-67}$$

and
$$\frac{\rho(\delta_2/\delta_1) + \rho^2(\sigma^2/\mu_1^2)}{(1 - \rho)^3}. \tag{7-68}$$

Finally, the general process which includes the busy and idle periods is studied, based on the results for a busy period, and its ergodic properties investigated, using renewal theory. The generating function of the steady-state probabilities p_j (with $\rho < 1$) corresponding to $P_{ij}(t)$, the probability that there are j in the system at time t, given that initially there were i in the system at time zero, is given by

$$P(z) = \sum_{j=0}^{\infty} p_j z^j = \frac{(1 - \rho)(1 - z)\beta[\lambda\{1 - c(z)\}]}{\beta[\lambda\{1 - c(z)\}] - z}, \tag{7-69}$$

where β is the Laplace-Stieltjes transform of $B(t)$.

The mean number in the system is

$$\rho + \frac{\rho}{2(1 - \rho)}\left[\left(\frac{\delta_2}{\delta_1} - 1\right) + \rho\left(1 + \frac{\sigma^2}{\mu_1^2}\right)\right]. \tag{7-70}$$

The waiting-time distribution has the Laplace-Stieltjes transform

$$(1 - \rho)\left(1 + \lambda \frac{1 - c\{\beta(z)\}}{z - \lambda[1 - c\{\beta(z)\}]}\right). \tag{7-71}$$

If all departures from pure Poisson arrivals are due to batches arriving together, this becomes

$$\frac{1 - \rho}{1 - \lambda([1 - c\{\beta(z)\}]/z)}. \tag{7-72}$$

The mean waiting time in this last case is

$$\frac{\rho}{2(1 - \rho)}\left(\frac{\delta_2}{\delta_1} + \frac{\sigma^2}{\mu_1^2}\right)\mu_1, \tag{7-73}$$

and its variance is

$$\frac{\rho}{(1 - \rho)^2}[\rho C_1^2 + (1 - \rho)C_2], \tag{7-74}$$

where C_1 and C_2 are the first and second moments about the origin of the distribution whose transform is

$$\frac{1}{\mu_1 \delta_1}\frac{1 - c\{\beta(z)\}}{z}. \tag{7-75}$$

7-6. Khintchine's Argument

Consider a single-channel queue with Poisson input and arbitrary-service-time distribution. Let π_n $(n = 0, 1, 2, \ldots)$ be the probability that, after the moment of termination of service of a customer, there are n customers waiting in line. The probability that at the instant of its arrival a unit finds n other units in the system is also denoted by π_n, because it is equal to the previous π_n. This is true for a stationary process with only single step transitions (note that service in batches does not belong to this category) in which the zero state is a recurrent nonnull state.

To see this, note that in long periods of time the number of arrivals tends to the same value as the number of departures and thus transitions from state E_n to E_{n+1} are followed some time later by transitions from state E_{n+1} to state E_n. Thus the law of large numbers yields the identity of the two distributions.

The following proof is a modified version of a proof given by Khintchine which shows the equality of π_n and the steady-state probability p_n. Note that π_n is taken at epochs of departure, whereas p_n is taken at an arbitrary moment. One first obtains a relation among the π_n.

Let π_n be the probability that n customers are in the system after the termination of service of a customer. At this moment a customer is about to go into service, thus reducing the number waiting from n to $n - 1$. π_0 is the probability that a randomly chosen customer will begin service without waiting. A necessary and sufficient condition for this to happen is that the preceding customer began his service with no one waiting (the probability of which is $\pi_0 + \pi_1$) and nothing arrived during his service.

The probability of n arrivals during a service time is given by

$$k_n = \frac{1}{n!} \int_0^\infty (\lambda t)^n e^{-\lambda t}\, dB(t), \qquad n \geq 0. \qquad (7\text{-}76)$$

Hence

$$\pi_0 = (\pi_0 + \pi_1)k_0. \qquad (7\text{-}77)$$

Generally,

$$\pi_n = \pi_{n+1}k_0 + \pi_n k_1 + \cdots + \pi_2 k_{n-1} + (\pi_1 + \pi_0)k_n, \qquad n \geq 0;$$
$$\sum_{n=0}^\infty \pi_n = 1. \qquad (7\text{-}78)$$

Thus the probability that the departing customer leaves n behind is equal to the probability that the previous customer left $n + 1$ behind and nothing arrived during the service time of the present customer, plus the probability that the previous customer left n behind and one arrived during the service time of the present customer, and so on.

Let p_n be the steady-state probability that there are n units in the system at any time. Khintchine gives the following proof, which shows that $p_n = \pi_n$.

One can show (see also Chap. 9) that the probability that the server continues to be occupied by the same customer after time y is

$$\frac{1}{1/\mu} \int_y^\infty B_c(x)\, dx, \qquad (7\text{-}79)$$

where $B_c(x) = 1 - B(x)$ and where $1/\mu$ is the mean service time. Thus for a duration of occupancy between y and $y + dy$ this probability is $\mu B_c(y)\, dy$. Now $\rho = \lambda/\mu$ is the probability that the channel is occupied. Thus if h_n is the probability of n arrivals during the occupancy of the channel, then

$$h_n = \frac{\rho \mu}{n!} \int_0^\infty (\lambda y)^n e^{-\lambda y} B_c(y)\, dy. \qquad (7\text{-}80)$$

Since p_n is considered at an arbitrary moment, whereas π_n is considered at instants of departure, we must include arrivals that can occur during all possible durations of occupancy of the channel.

Thus

$$p_n = \pi_n h_0 + \pi_{n-1} h_1 + \cdots + (\pi_1 + \pi_0) h_{n-1}$$

or simply $$p_n = \lambda \int_0^\infty B_c(x) u_{n-1}(x)\, dx, \qquad (7\text{-}81)$$

where $u_{n-1}(x) = \pi_n e^{-\lambda x} + \pi_{n-1}(\lambda x) e^{-\lambda x} + \cdots$

$$+ (\pi_1 + \pi_0) \frac{(\lambda x)^{n-1} e^{-\lambda x}}{(n-1)!}, \quad (7\text{-}82)$$

which may be shown to satisfy the differential-difference equation

$$\frac{du_n(x)}{dx} = \lambda[u_{n-1}(x) - u_n(x)]. \qquad (7\text{-}83)$$

Now from the relation (7-78) we have

$$\pi_n = \int_0^\infty u_n(x)\, dB(x), \qquad (7\text{-}84)$$

and integration by parts gives

$$\pi_n = u_n(0) + \int_0^\infty u_n'(x)[1 - B(x)]\, dx$$

$$= \pi_{n+1} + \lambda \int_0^\infty [u_{n-1}(x) - u_n(x)][1 - B(x)]\, dx$$

$$= \pi_{n+1} + p_n - p_{n+1}. \qquad (7\text{-}85)$$

From this we have

$$\pi_n - p_n = \pi_{n+1} - p_{n+1}. \qquad (7\text{-}86)$$

This relation shows that the difference $\pi_n - p_n$ must be the same constant for all n. But π_n and p_n are members of convergent series; therefore, if we sum $\pi_n - p_n = \text{const}$, over n, we obtain a finite quantity on the left (in fact, zero, since π_n and p_n are probabilities) and hence also on the right if, and only if, the constant is equal to zero. We therefore conclude that $\pi_n = p_n$.

For many practical purposes

$$p_n = \int_0^\infty u_n(x)\, dB(x) \qquad (7\text{-}87)$$

is a useful representation.

Remark 2: The Semi-Markoff-process Approach to Bulk Queues. By means of the theory of semi-Markoff processes Fabens [197] has studied the system $E_s/G/1$ and the system of Poisson arrivals and service in batches of exactly size s at a single channel. For $E_s/G/1$, where arrivals occur by a gamma (or Erlangian) distribution with integral order s and parameter λ, let $B(t)$ be the cumulative service distribution where items are served singly. Then, using the imbedded-Markoff-chain method, it turns out that π_n, the probability that at the termination of his service a customer leaves n customers behind him in the system, is the same as π_n

in the bulk-service queue. Thus, even though the two problems are not the same, their imbedded Markoff chains are identical. The difference between them does not show at the special instants examined in the imbedded chain.

From π_n it is possible to determine p_n, the probability of n items in the system at any time, for the problem. It is here that the concept of a semi-Markoff process is used. [Let $X(t)$ be a stochastic process—in continuous or discrete time—which assumes at any instant one of a countable number of states. Let the sequence of states, or values of $X(t)$, be a Markoff chain X_n with transition probabilities

$$\text{Prob } \{X_{n+1} = j | X_n = i\} = q_{ij}. \tag{7-88}$$

Let the length of time T spent by $X(t)$ in state i before the next transition, given that the next transition is into state j, be a random variable with distribution $W_{ij}(t)$. It is always assumed that a transition occurred at $t = 0$. Let $X(t)$ be defined as continuous on the right, except for states occupied a zero length of time. Such a process is a semi-Markoff process.]

The arguments yield

$$p_n = \frac{c}{\lambda} \sum_{j=0}^{n} \pi_j \tag{7-89}$$

for $n < s$, and for $n \geq s$

$$p_n = c \sum_{j=s}^{n} \int_0^\infty \frac{(\lambda t)^{n-j}}{(n-j)!} e^{-\lambda t} [1 - B(t)] \, dt$$
$$+ c\pi \int_0^\infty \frac{(\lambda t)^{n-s} e^{-\lambda t}}{(n-s)!} [1 - B(t)] \, dt, \tag{7-90}$$

where
$$\pi = \sum_{j=0}^{s-1} \pi_j, \qquad c = ab, \qquad a = \frac{1}{1+\pi} \tag{7-91}$$

$$b = \frac{1}{\Sigma \mu_i r_i}, \qquad r_i = \begin{cases} a\pi_i, & i < s, \\ a(\pi + \pi_s), & i = s, \\ a\pi_i, & i > s, \end{cases} \tag{7-92}$$

and $\mu_i = \sum_j q_{ij}\mu_{ij}$, where μ_{ij} is the mean of $W_{ij}(T)$. The probability of waiting no longer than time t for $E_s/G/1$ is

$$P(\leq t) = s \sum_{j=0}^{\infty} p_{sj+s-1} \tilde{B}^j(t), \tag{7-93}$$

where
$$\tilde{B}^1(t) = \int_0^t \frac{1 - B(x)}{\mu_2} \, dx$$

and $\tilde{B}^{n+1}(t)$ is the convolution of B and $\tilde{B}^n$ for $n \geq 1$.

For the batch-service queue with Poisson arrivals we have

$$P(\leq t) = \sum_{i=0}^{\infty} \sum_{j=1}^{s} V(t - w | W, i, j) \, d\tilde{B}^i(w) p_{is+j-1}, \tag{7-94}$$

where
$$V(u | w, i, j) = \sum_{m=j}^{s-1} \Gamma_{s-m}(u) \frac{(\lambda w)^{m-j} e^{-\lambda w}}{(m-j)!} + H(u) \sum_{n=s-j}^{\infty} \frac{(\lambda w)^n e^{-\lambda w}}{n!}, \tag{7-95}$$

where
$$\Gamma_n(u) = \int_0^u \frac{\lambda^n x^{n-1} e^{-\lambda x}}{(n-1)!} \tag{7-96}$$

and $H(u)$ is the unit step function at the origin.

For batch service with gamma input, the waiting time can be determined by a combination of the two methods used to derive the above quantities. The expected length of a busy period is the same for $E_s/G/1$ and the batch-service queue. It is given by

$$\frac{\sum_{i=0}^{s-1} \pi_i(s-i)}{\lambda \sum_{j=0}^{s-1} \pi_j} \left(\frac{1}{\sum_{i=0}^{s-1} p_i} - 1 \right). \tag{7-97}$$

If the number of states is a finite number N and if the input is Poisson with batch service, the transition matrix of the imbedded chain is truncated; each transition probability Q_i in the last column (i.e., the Nth column) is the sum of the omitted probabilities in its row in the infinite matrix, starting with the element that occupies the same position as Q_i in the infinite matrix. Thus, using an asterisk to identify the probabilities of the finite case, we have

$$\pi_n^* = \pi_n, \tag{7-98}$$

$$\pi_N^* = \frac{\sum_{j=0}^{N-s-1} Q_{N-j} \pi_{s+j}^* + Q_N \pi^*}{1 - Q_s} \tag{7-99}$$

and
$$p_N^* = c \sum_{j=s}^{N} \pi_j^* \int_0^{\infty} \sum_{m=N}^{\infty} \frac{(\lambda t)^{m-j} e^{-\lambda t}}{(m-j)!} [1 - B(t)] \, dt$$

$$+ c\pi^* \int_0^{\infty} \sum_{m=N}^{\infty} \frac{(\lambda t)^{m-s} e^{-\lambda t}}{(m-s)!} [1 - B(t)] \, dt. \tag{7-100}$$

PROBLEMS

1. If for service in batches of size s with Poisson arrivals we have the constant-service distribution

$$B(t) = \begin{cases} 0, & 0 \leq t < s, \\ 1, & s \leq t < \infty, \end{cases}$$

show that $K(z) = e^{-\rho s(1-z)}$.

2. Jaiswal [326] defines p_{nr} as the steady-state probability that n customers are waiting in queue and service is in the rth phase. This is done in attempting to study bulk service in sizes of s or the entire queue if it is less than s and the service facility becomes vacant. He assumes service to consist of j imaginary phases. The group can choose to enter any of the phases. If the choice is on the rth phase with probability C_r, then, after finishing service, it must go to the $(r-1)$st phase and so on down to the first phase. Arrivals are Poisson with parameter λ. Service at each phase is exponential with parameter μ; first-come–first-served discipline for the batches is used at a single channel. Show that the equations are

$$-(\lambda + \mu)p_{nr} + \lambda p_{n-1,r} + \mu C_r p_{n+s,1} + \mu p_n, r+1 = 0, \qquad n > 0, \quad 1 \leq r < j,$$
$$-(\lambda + \mu)p_{nj} + \lambda p_{n-1,j} + \mu C_j p_{n+s,1} = 0, \qquad n > 0,$$

$$-(\lambda + \mu)p_{0r} + \mu p_{0,r+1} + \mu C_r \sum_{m=1}^{s} p_{m1} + \lambda C_r p_0 = 0, \qquad 1 \leq r < j,$$

$$-(\lambda + \mu)p_{0j} + \mu C_j \sum_{m=1}^{s} p_{m1} + \lambda C_j p_0 = 0,$$

$$-\lambda p_0 + \mu p_{01} = 0,$$

when p_0 is the probability that the facility is empty. Jaiswal obtains a generating-function equation.

More recently he has studied the time-dependent solution of this problem [843]. The reader is asked to write the transient equations.

Let the generating function of the Laplace transform with parameter α of $P_{nr}(t)$ be

$$Q_n^*(x,\alpha) = \sum_{r=1}^{j} x^r P_{nr}^*(\alpha)$$

and let

$$F(x,y,\alpha) = \sum_{n=0}^{\infty} y^n Q_n^*(x,\alpha),$$

$$x = \frac{\mu}{\mu + \alpha + \lambda - \lambda y}.$$

Then, on multiplying the equations by appropriate powers of x and y and summing over $1 \leq r \leq j$ and $0 \leq n < \infty$, one has, on simplifying, that

$$G(y,\alpha) \equiv \sum_{n=0}^{\infty} y^n P_{n1}^*(\alpha) = \frac{\mu \sum_{n=1}^{s-1} P_{n1}^*(\alpha)(y^s - y^n) + P_0^*(\alpha)(\lambda y^s - \alpha - \lambda) + 1}{\mu \left\{ \dfrac{y^s}{\displaystyle\sum_{r=1}^{j} C_r[\mu/(\mu + \alpha + \lambda - \lambda y)]^r} - 1 \right\}}.$$

Rouché's theorem is then applied to the denominator, showing that the denominator has s distinct zeros inside the unit circle.   One finally has

$$\sum_{n=0}^{\infty} y^n Q_n^*(1,\alpha) = \frac{\mu\left\{\dfrac{1}{\displaystyle\sum_{r=1}^{j} C_r[\mu/(\mu+\alpha+\lambda-\lambda y)]^r} - 1\right\}}{\alpha+\lambda-\lambda y} G(y,\alpha).$$

Note that $Q_n^*(1,\alpha)$ is the Laplace transform of the probability that there are n units waiting in the queue at time t.

Attempts are made by Jaiswal to specialize the solution for the Erlangian service distribution, but no explicit results are possible.   He successfully reproduces the known result due to Clarke, given in Chap. 4, for a single-channel queue with exponential service where $C_r = 1$ for $r = 1$ and $C_r = 0$ for $r \neq 1$, $s = 1$.

Derive this result from the foregoing expressions.

CHAPTER 8

POISSON INPUT, ARBITRARY SERVICE DISTRIBUTION, DISTRIBUTIONS OF THE NUMBER IN THE SYSTEM, THE BUSY PERIOD, AND THE NUMBER SERVED

8-1. Introduction

For a single channel with Poisson input and arbitrary service-time distribution we study the distribution of the number in the system, that of a busy period, and the probability of a given number being served in this period.

8-2. The Number in the System in the Steady State with Ordered Service

Suppose that, at an arbitrary moment in the steady state, p_j is the probability that the system is in state E_j ($j = 0, 1, 2, \ldots$); that is, there are j items waiting and being served.

The following derivation depends upon Khintchine's argument given in Chap. 7 which shows for the steady state that p_j is the same as π_j, the probability that the state is E_j at the end of a service time. Hence, from our discussion of bulk queues, we use for the generating function

$$P(z) = \frac{\sum_{i=0}^{s-1} \pi_i(z^s - z^i)}{z^s/K(z) - 1}. \tag{8-1}$$

If we put $s = 1$, $\pi_i = p_i$, $K(z) = \beta[\lambda(1 - z)]$, we have for the generating function of the p_j

$$P(z) = \frac{p_0(z - 1)\beta[\lambda(1 - z)]}{z - \beta[\lambda(1 - z)]}. \tag{8-2}$$

But as $z \to 1$, the reader can show that

$$\frac{1 - K(z)}{1 - z} \to \frac{\lambda}{\mu},$$

where $1/\mu$ is the mean service time.   Hence

$$1 = P(1) = \frac{p_0}{1 - \lambda/\mu}.$$

Thus finally $P(z) = \dfrac{(1 - \lambda/\mu)\beta[\lambda(1 - z)]}{1 - \{1 - \beta[\lambda(1 - z)]\}/(1 - z)},$ (8-3)

valid for $\lambda/\mu < 1$.

The p_j are all zero if $\lambda/\mu \geq 1$, since every state is a recurrent-null state. This can be seen from the fact that we have an irreducible chain; hence all its states belong to the same class.   (In particular, E_0 is null-recurrent.)

8-3. The Distribution of a Busy Period

We have already encountered the problem of determining the distribution of the periods in which a single channel is occupied.   These periods with random variables $X_1, X_2, \ldots$ are identically and independently distributed since nothing happens during the idle periods to cause a difference in the occupancy of the channel from one busy period to another.   A busy period begins when an item goes into service; it ends when the last item in the queue formed during this operation has completed service, with nothing arriving immediately after, and hence the channel becomes idle.

Let $G(x) = \text{Prob } (X_n \leq x)$ be the distribution of a busy period for a single-channel queue, regardless of the type of service selection, with Poisson input and parameter λ and arbitrary service-time distribution $B(t)$.

Note that the distribution of the sum of n busy periods is an n-fold convolution of $G(x)$.   From Chap. 3, the n-fold convolution of the distribution function $G(x)$ is obtained by taking the convolution of $G(x)$ with itself, and then again taking the convolution of the resulting function with $G(x)$, and so on.   Thus, if we define $G_1(x) = G(x)$, we have for the convolution of $G(x)$ with itself

$$G_2(x) = \int_0^x G_1(x - y) \, dG(y),   (8-4)$$

etc., finally yielding for the n-fold convolution

$$G_n(x) = \int_0^x G_{n-1}(x - y) \, dG(y),   n \geq 2.   (8-5)$$

We now compute $G(x)$.   The duration of a busy period may be divided into the duration of service of the first item and the duration of the busy period for subsequently arriving items.   Thus if y is the duration of service of the first item and n items arrive during this period y (the probability of this is given by the Poisson process with time y), then, as we indicated in the deterministic model of Chap. 2, these n items may be thought to stand aside. The first goes into service immediately after the termination of service of the initial item; if any arrivals occur during this service, they are all served first before another item of the remaining $n - 1$ is admitted. The same argument applies when this other item goes into service, etc.

Thus the busy period from time $x - y$ to time x is that caused by the n items together and the respective arrivals that occur during their service times. But the busy period for each of these n items for the period $x - y$ is $G(x - y)$, and an n-fold convolution of all n gives $G_n(x - y)$. This must be weighted by the Poisson process giving the probability of n arrivals during y and summed over n.

To obtain $G(x)$, the result must be multiplied by $dB(y)$ and integrated over the possible range of y. Hence finally

$$G(x) = \int_0^x \sum_{n=0}^{\infty} e^{-\lambda y} \frac{(\lambda y)^n}{n!} G_n(x - y) \, dB(y), \tag{8-6}$$

with
$$G_0(x) = \begin{cases} 1 & \text{if } x \geq 0, \\ 0 & \text{if } x < 0. \end{cases} \tag{8-7}$$

If we apply the Laplace-Stieltjes transform to the last equation, after defining

$$\Gamma(s) \equiv \int_0^{\infty} e^{-sx} \, dG(x), \tag{8-8}$$

we have with Takács [689]

$$\begin{aligned}
\Gamma(s) &= \sum_{n=0}^{\infty} \frac{[\Gamma(s)]^n}{n!} \int_0^{\infty} e^{-(s+\lambda)x} (\lambda x)^n \, dB(x) \\
&= \sum_{n=0}^{\infty} \frac{(-1)^n}{n!} \lambda^n [\Gamma(s)]^n \beta^{(n)}(s + \lambda) \\
&= \beta[s + \lambda - \lambda \Gamma(s)], \tag{8-9}
\end{aligned}$$

where β is the Laplace-Stieltjes transform of the service-time distribution. This gives a functional equation which the Laplace-Stieltjes transform of $G(x)$ must satisfy, for $\text{Re}(s) \geq 0$. Takács gives a proof that this equation determines $\Gamma(s)$ uniquely [with $\Gamma(\infty) = 0$] and also $G(x)$ from the latter uniquely. He shows that $G(x)$ is a proper distribution if $\lambda/\mu \leq 1$; otherwise it is not, and the service period can be infinite with probability equal to $[1 - \lim_{x \to \infty} G(x)]$. The mean service time is $1/\mu$.

Now $\beta(s)$ is decreasing for $0 \leq s < \infty$ with $\beta(0) = 1$, $\beta(\infty) = 0$. It is differentiable for $0 \leq s < \infty$, and $\beta'(s)$ increases monotonically. We have $\beta'(0) = -1/\mu$ and $\beta'(\infty) = 0$. Since $\beta(s)$ is a monotone function, it has a unique inverse $\beta^{-1}(x)$ for $0 \leq x \leq 1$.

8-4. The Number Served in a Busy Period

Let f_j be the probability of j items receiving service during a busy period, and let the generating function be given by $F(u) = \sum\limits_{j=1}^{\infty} f_j u^j$. We multiply f_j by the duration of service of j items, which is the n-fold convolution of the service-time distribution, and sum from one to infinity. This also gives $G(x)$. If again we denote the Laplace-Stieltjes transform of $G(x)$ by $\Gamma(s)$, we have $\Gamma(s) = \sum\limits_{j=1}^{\infty} f_j[\beta(s)]^j$, since the transform of an n-fold convolution of a function is the nth power of the transform of the function.

But this quantity is $F(u)$, since we can write $\beta(s) = u$ or $s = \beta^{-1}(u)$, which exists for $0 \leq u \leq 1$, as we have just pointed out. Thus, from the relation on the distribution of a busy period, we have

$$F(u) = \Gamma[\beta^{-1}(u)] = \beta\{\beta^{-1}(u) + \lambda - \lambda\Gamma[\beta^{-1}(u)]\}. \qquad (8\text{-}10)$$

By the same argument used to derive (8-6) using (7-2), (7-3), and $f_1 = k_0$, $f_j = \sum\limits_{n=1}^{j-1} k_n \sum\limits_{j_1 + \cdots + j_n = j-1} f_{j_1} \cdots f_{j_n}$ we have

$$F(u) = u \sum\limits_{n=0}^{\infty} k_n[F(u)]^n = u\beta[\lambda - \lambda F(u)]. \qquad (8\text{-}11)$$

Thus from the definition of β we have

$$F(u) = u \int_0^{\infty} e^{-\lambda t[1 - F(u)]} \, dB(t). \qquad (8\text{-}12)$$

Takács gives a proof of the existence and uniqueness of an analytic solution $F(u)$ for $|u| \leq 1$ with $F(0) = 0$ and where $\lim\limits_{u \to 1} F(u)$ equals the smallest positive root of $\beta[\lambda(1 - x)] = x$.

PROBLEMS

1. Show that, if the service-time distribution is (a) $1 - e^{-\mu x}$, then

$$p_j = \left(1 - \frac{\lambda}{\mu}\right)\left(\frac{\lambda}{\mu}\right)^j,$$

and (b) if it is a constant α, then

$$p_j = (1 - \lambda\alpha) \sum\limits_{k=1}^{j} (-1)^{j-k} e^{k\lambda\alpha} \left[\frac{(k\lambda\alpha)^{j-k}}{(j-k)!} + \frac{(k\lambda\alpha)^{j-k-1}}{(j-k-1)!}\right],$$

ignoring the second factor in the brackets for $k = j$.

2. Obtain the Pollaczek-Khintchine formula by differentiating $P(z)$ with respect to z at $z = 1$.

3. For the distribution of a busy period, show that, if $B(x) = 1 - e^{-\mu x}$ $(x \geq 0)$, then $\beta(s) = 1/(1 + s/\mu)$ and

$$\Gamma(s) = \frac{\lambda + \mu + s - \sqrt{(\lambda + \mu + s)^2 - 4\lambda\mu}}{2\lambda},$$

using the negative sign before the square root because $\Gamma(\infty) = 0$. Also show that

$$G'(x) = \frac{1}{x} \sqrt{\frac{\mu}{\lambda}} \exp[-(\lambda + \mu)x] I_1(2\sqrt{\lambda\mu}\,x).$$

Also show that the average length of a busy period is $1/(\mu - \lambda)$ if $\lambda/\mu < 1$ and infinity otherwise.

4. Show that, if the service times are constant with value α, the generating function $F(u)$ of f_j, the probability of j items obtaining service in a busy period is

$$F(u) = \sum_{j=1}^{\infty} e^{-\lambda\alpha j}(\lambda\alpha j)^{j-1} \frac{u^j}{j!}$$

by putting $y = \lambda\alpha F(u)$, $x = ue^{-\lambda\alpha}\lambda\alpha$, obtaining $ye^{-y} = x$, and writing

$$y(x) = \sum_{j=1}^{\infty} \frac{j^{j-1}}{j!} x^j.$$

(See the discussion on applications of queues to car traffic.) Show that

$$G(x) = \sum_{j=1}^{[x/\alpha]} \frac{e^{-\lambda\alpha j}}{j!} (\lambda\alpha j)^{j-1}$$

and that

$$f_j = \frac{e^{-\lambda\alpha j}}{j!} (\lambda\alpha j)^{j-1}.$$

If the service time is exponentially distributed with mean α, show that

$$F(u) = \frac{(1 + \lambda\alpha) - \sqrt{(1 + \lambda\alpha)^2 - 4\lambda\alpha u}}{2\lambda\alpha}$$

and hence

$$f_j = \frac{1}{2}\binom{2j}{j} \frac{(\lambda\alpha)^{j-1}}{(1 + \lambda\alpha)^{2j-1}} \frac{1}{2j - 1}.$$

CHAPTER 9

POISSON AND ARBITRARY INPUT, ARBITRARY SERVICE; THE WAITING TIME FOR SINGLE AND MULTIPLE CHANNELS

9-1. Introduction

In this chapter we shall study two types of equations which give the waiting-time distribution in the line (not including service) for a single and for multiple channels with ordered queueing. The first type, an integrodifferential equation originally due to Takács [689] and independently derived by Descamps [149], gives the waiting-time distribution for Poisson input and arbitrary-service-time distribution in the transient state. Descamps's derivation is extended to c channels in the steady state.

The second type is a Wiener-Hopf type of integral equation due to Lindley. It gives the waiting-time distribution for a general independent (i.e., a general distribution identical for all interarrival intervals which are independently distributed) input distribution and arbitrary holding time. A result for many channels and arbitrary inputs and service times has been developed by Kiefer and Wolfowitz [380]. Of course these results also apply to Poisson queues.

The first type of result is derived in the time-dependent form; then one obtains the steady state. The general result mentioned above reduces to the single-channel case but is difficult to compute for other cases.

9-2. The Integrodifferential Equation of Takács

1. *Single Channel*

We assume Poisson input with parameter λ, arbitrary holding time (the same for all customers), at a single channel with ordered queueing. We shall follow Descamps's intuitive argument to derive an equation for this process. Before doing this, let us make the interesting observation that the function $W(t)$, which denotes the waiting time in the queue at time t, jumps upward discontinuously at the arrival of a new customer (with

198

service time greater than zero) by an amount equal to the service time of the new arrival; otherwise $W(t)$ approaches zero with a slope equal to minus unity.   We have a Markoff process with a continuous parameter t. We remark that $W(t)$ is not the probability of waiting any specified moment, but is the time that a customer arriving at time t has to wait before entering service in an ordered queue.   Similar remarks apply to the case of several channels.

Now we develop the equations for a single channel.   Let $P(w,t)$ be the probability of an arrival waiting in queue an interval of time $W(t) \leq w$ given that the arrival occurs at time t.

The following argument pictures a large number N of single-channel queues with identical input and service distributions operating simultaneously.   We wish to derive an expression for $P(w, t + \Delta t)$ in terms of $P(w,t)$ and $P(w + \Delta t, t)$.   At time t one may divide these waiting lines into two sets:

1. Those for which the waiting time is $W(t) \leq w$.   This number is $NP(w,t)$ [i.e., one multiplies N by the fraction which waits for a time $W(t) \leq w$].

2. Those for which the waiting time is $W(t) > w$.   Their number is $N - NP(w,t)$.

At time $t + \Delta t$ the number in the first set becomes $NP(w + \Delta t, t)$ [there is no loss in generality by not writing $P(w + \Delta t, t + \Delta t)$] minus the number of queues whose waiting time has changed from $W(t) \leq w$ to $W(t) > w$, because of arrivals during Δt.   We now obtain the latter quantity.

We first take the number of queues whose waiting time lies between x and $x + dx$ at time t.   Since $P(x,t)$ is a cumulative distribution, we obtain the density function for $x > 0$ by differentiation.   This number of queues is given by (1) $N[\partial P(x,t)/\partial x]\, dx$, if $x > 0$, or (2) $NP(0,t)$, if $x = 0$.

Now the waiting time is made to exceed w during $(t, t + \Delta t)$ if there is an arrival during Δt and if the service time y of this arrival, when added to x, exceeds w, that is, $x + y > w$, from which $y > w - x$.   We must therefore multiply this number of queues whose waiting time is x by the probability of an arrival during Δt, that is, $\lambda \Delta t$, and by the probability that the service time of the arrival exceeds $w - x$.   If we assume that the density function of y is $b(y)$, the probability of the last event is $\int_{w-x}^{\infty} b(y)\, dy$.

Therefore, for a fixed value $w > 0$ of waiting time, the number of waiting lines in the first case which change to the second set is given by

$$\left[N\lambda\, \Delta t\, \frac{\partial P(x,t)}{\partial x}\, dx \right]\left[\int_{w-x}^{\infty} b(y)\, dy \right] = N\lambda\, \Delta t\, \frac{\partial P(x,t)}{\partial x}\, dx\, B_c(w - x), \quad (9\text{-}1)$$

which must be summed over all $0 < x \leq w$.   Here

$$B_c(y) \equiv 1 - \int_0^y b(u) \, du \equiv 1 - B(y)$$

In case $x = 0$, the number of waiting lines which change to the second set [i.e., wait $W(t) > w$] is given by

$$N\lambda \, \Delta t P(0,t) \int_w^\infty b(y) \, dy = N\lambda \, \Delta t P(0,t) B_c(w). \tag{9-2}$$

Hence

$$NP(w, t + \Delta t) = NP(w + \Delta t, t) - N\lambda \, \Delta t \int_{0+}^w \frac{\partial P(x,t)}{\partial x} B_c(w - x) \, dx \\ - N\lambda \, \Delta t P(0,t) B_c(w). \tag{9-3}$$

The fact that we have considered a large number of queues N has served its purpose to yield an equation. We now divide through by N. Subtracting $P(w,t)$ from both sides, dividing by Δt, and passing to the limit yield

$$\frac{\partial P(w,t)}{\partial t} = \frac{\partial P(w,t)}{\partial w} - \lambda \int_{0+}^w \frac{\partial P(x,t)}{\partial x} B_c(w - x) \, dx - \lambda P(0,t) B_c(w). \tag{9-4}$$

We have used the expansion for small $\Delta t > 0$:

$$P(w + \Delta t, t) = P(w,t) + \frac{\partial P(w,t)}{\partial w} \Delta t + o(\Delta t).$$

In the integrodifferential equation, the following conditions must be satisfied: $P(w,0) = 1$ for all w, $P(\infty,t) = 1$ for all t.

Integrating by parts gives

$$\frac{\partial P(w,t)}{\partial t} = \frac{\partial P(w,t)}{\partial w} - \lambda P(w,t) + \lambda \int_0^w P(w - v, t) \, dB_c(v). \tag{9-5}$$

Note that, if in (9-4) we write $B(t) = 1 - B_c(t)$ and take the lower limit from zero instead of zero plus (thus including the value of the integral at the origin), we have

$$\frac{\partial P(w,t)}{\partial t} = \frac{\partial P(w,t)}{\partial w} - \lambda P(w,t) + \lambda \int_0^w B(w - x) \, d_x P(x,t). \tag{9-6}$$

This integrodifferential equation due to Takács is also valid if λ is replaced by $\lambda(t)$. This equation has also been investigated by Beneš.

On taking the Laplace transform of (9-4) with respect to w (when it exists)—noting the presence of a convolution—we have

$$\frac{\partial P^*(s,t)}{\partial t} = sP^*(s,t) - P(0,t) - \lambda[sP^*(s,t) - P(0,t)]B_c^*(s) - \lambda P(0,t) B_c^*(s)$$

$$= sP^*(s,t)[1 - \lambda B_c^*(s)] - P(0,t),$$

where

$$P^*(s,t) = \int_0^\infty e^{-sw} P(w,t)\, dw;$$

$$\int_0^\infty e^{-sw} \frac{\partial P(w,t)}{\partial w}\, dw = -P(0,t) + sP^*(s,t).$$

The steady-state result is obtained by setting the left side equal to zero, suppressing t, and solving for $P^*(s)$. This yields

$$sP^*(s) = \frac{P(0)}{1 - \lambda B_c^*(s)} = \frac{P(0)}{1 - \lambda[1/s - b^*(s)/s]}. \tag{9-7}$$

But $\lim_{w \to \infty} P(w) = 1$, and from Laplace-transform theory

$$\lim_{w \to \infty} P(w) = \lim_{s \to 0} sP^*(s).$$

Remark: Note that $P(w)$ plays the same role here as $P(<t)$, a symbol of frequent occurrence indicating the probability of a wait less than t.

Since $\lim_{s \to 0} sP^*(s) = 1$, we now have

$$P(0) = \lim_{s \to 0} \frac{s - \lambda[1 - b^*(s)]}{s}.$$

This is an indeterminate form to which we must apply L'Hospital's rule. Recall that

$$\lim_{s \to 0} b^*(s) = \lim_{s \to 0} \int_0^\infty e^{-sy} b(y)\, dy = \int_0^\infty b(y)\, dy = 1.$$

Thus, differentiating numerator and denominator separately with respect to s gives

$$P(0) = 1 - \frac{\lambda}{\mu} \equiv 1 - \rho,$$

where $\quad \dfrac{db^*(s)}{ds} = \displaystyle\int_0^\infty \frac{d}{ds} e^{-sy} b(y)\, dy = -\int_0^\infty y e^{-sy} b(y)\, dy,$

and, as $s \to 0$, this gives the mean of $b(y)$, which we denote by $1/\mu$. We finally have the solution of our problem given by

$$sP^*(s) = \frac{1 - \rho}{1 - \lambda\{[1 - b^*(s)]/s\}}. \tag{9-8}$$

On taking the inverse transform, this yields the expression that we seek for $P(w)$, giving the probability of not waiting in queue longer than w in the steady state.

Example: Let $b(y) = \mu e^{-\mu y}$. Then

$$B_c(y) = 1 - \int_0^y \mu e^{-\mu t}\, dt = e^{-\mu y}$$

and $\quad B_c^*(s) = \displaystyle\int_0^\infty e^{-sy} e^{-\mu y}\, dy = \frac{1}{s + \mu}.$

Therefore, from (9-7),

$$P^*(s) = P(0) \left[\frac{1}{s + \mu - \lambda} + \frac{\mu}{s(s + \mu - \lambda)} \right].$$

On observing that the second term is the Laplace transform of a convolution of an exponential and unity, we have

$$P(w) = P(0)[e^{-(\mu-\lambda)w} + \mu \int_0^w e^{-(\mu-\lambda)t}\, dt].$$

It is easy to verify that $P(0) = 1 - \lambda/\mu$ since $\lim_{w \to \infty} P(w) = 1$. Also,

$$W_q \equiv \int_0^\infty w \frac{dP(w)}{dw}\, dw = \frac{\lambda/\mu}{\mu(1 - \lambda/\mu)},$$

as can be seen by calculating dP/dw and integrating. This is a well-known result for the expected waiting time in the queue also derivable from the Pollaczek-Khintchine formula.

In case the density function $b(x)$ does not exist, we can then use the Laplace-Stieltjes transform, i.e.,

$$\gamma(s,t) = \int_0^\infty e^{-sw}\, dP(w,t),$$

$$\beta(s) = \int_0^\infty e^{-sx}\, dB(x),$$

where $B(x)$ is the cumulative-service-time distribution.

It is clear from the definitions of this transform and the Laplace transform that, as $t \to \infty$, $sP^*(s) = \gamma(s)$, and $b^*(s) = \beta(s)$ when $b(t)$ exists. Thus we must have

$$\gamma(s) = \frac{1 - \rho}{1 - \lambda\{[1 - \beta(s)]/s\}}, \qquad (9\text{-}9)$$

which is the result obtained by Takács by directly working on (9-6).

Takács obtains for the time-dependent case

$$\frac{\partial \gamma(s,t)}{\partial t} = \gamma(s,t)[s - \lambda + \lambda\beta(s)] - sP(0,t). \qquad (9\text{-}10)$$

He gives the following solution of this differential equation with $\lambda(t)$ instead of λ and

$$\Lambda(t) = \int_0^t \lambda(u)\, du:$$

$$\gamma(s,t) = \exp\{st - [1 - \beta(s)]\Lambda(t)\}$$
$$\left(1 - s \int_0^t \exp\{-sx + [1 - \beta(s)]\Lambda(x)\} P(0,x)\, dx\right). \qquad (9\text{-}11)$$

He shows that it is actually the Laplace-Stieltjes transform of the original differential equation.

He also shows that, if $1/\mu$ is the mean service time, if $\lim\limits_{t\to\infty} \lambda(t) = \lambda$, and if $\lambda/\mu < 1$, then $\lim\limits_{t\to\infty} P(w,t) = P(w)$ exists, is independent of the initial distribution $P(w,0)$, and is uniquely determined by its Laplace-Stieltjes transform (9-9), thus again providing a solution of the problem. If $\lambda/\mu \geq 1$, then $P(w)$ does not exist, and $\lim\limits_{t\to\infty} P(w,t) = 0$ for all w.

Instead of using transforms, a numerical method of calculation has been suggested by R. Descamps.   In the steady state, one may write

$$p(w) = \frac{dP(w)}{dw}.$$

For the steady state this leads to

$$p(w) - \lambda \int_0^w p(x)B_c(w - x)\, dx = \lambda\left(1 - \frac{\lambda}{\mu}\right)B_c(w), \qquad (9\text{-}12)$$

and, when $w \to 0$, to

$$p(0) = \lambda\left(1 - \frac{\lambda}{\mu}\right).$$

By writing $q(x) = p(x)B_c(w - x) = p(x)B_c(n\Delta - x)$ one attempts to evaluate the above Volterra-type integral equation by breaking the interval of integration into n small subintervals, each of length Δ, by applying the trapezoidal method of approximation, i.e.,

$$\int_0^{n\Delta} q(x)\, dx = \Delta\left\{\frac{q(0)}{2} + q(\Delta) + q(2\Delta) + \cdots + q[(n - 1)\Delta] + \frac{q[n\Delta]}{2}\right\}.$$

Since $B_c(w)$ is known, using the equation in $p(w)$, one can derive step by step $p(0)$, $p(\Delta)$, $p(2\Delta)$, etc.

In the transient case, one may assume that $P(w,t) = P_1(w)P_2(t)$.   Then one substitutes this in the equation, obtaining a differential equation the left side of which depends only on t, while the right side depends on w, which implies that each of these must equal a constant.   This gives rise to two equations each in one variable, etc., if it works out at all.

2. Several Channels, Steady-state Waiting Time

Here again we assume Poisson arrivals, ordered queueing, and arbitrary service distributions which are the same for all channels.   (The number of channels is c.)   The analysis is similar to the foregoing, except that here the increase in $W(t)$ is not the service time of a single arrival during Δt.   It is the time between the beginning of service of this arrival (the arrival occurs at time t while all other channels are occupied) and the time that a channel first becomes idle, since for a delay all channels must be occupied.   The analysis is approximately valid.

The probability that an occupied channel, other than the one occupied by the arrival during Δt, continues to be occupied after time y from the start of the arrival's service is

$$\frac{1}{1/\mu} \int_y^\infty B_c(x)\, dx, \tag{9-13}$$

where $1/\mu$ is the average time of occupation of the channel. The integral in the expression is the average time during which it would continue to be occupied after time y. The ratio is the desired probability. As a further elaboration, note that

$$
\begin{aligned}
\int_0^\infty B_c(y)\, dy &= \int_0^\infty \left[1 - \int_0^y b(x)\, dx \right] dy \\
&= \lim_{s \to 0} \int_0^\infty e^{-sy} \left[1 - \int_0^y b(x)\, dx \right] dy \\
&= \lim_{s \to 0} \left[\frac{1}{s} - \int_0^\infty b(x)\, dx \int_x^\infty e^{-sy}\, dy \right] \\
&= \lim_{s \to 0} \frac{1 - b^*(s)}{s} = \frac{1}{\mu},
\end{aligned}
$$

after applying L'Hospital's rule.

Thus the probability that all channels are occupied is the product of the probabilities that the other $(c - 1)$ channels are occupied for a time greater than y and the probability that the channel occupied by the arrival is occupied for a time greater than y [that is, $B_c(y)$]. This is given by

$$H(y) \equiv \left[\frac{1}{1/\mu} \int_y^\infty B_c(x)\, dx \right]^{c-1} B_c(y). \tag{9-14}$$

The function $H(y)$ will play the same role in this analysis as $B_c(y)$ did in the single-channel case.

As before, we study the change during Δt, but in equilibrium. Starting with N queues, the number of queues in which the waiting time is between w and $w + \Delta t$ is

$$N \frac{dP(w)}{dw} \Delta t.$$

This is to be reduced by the number of queues whose waiting time exceeds w, which now is

$$N\lambda\, \Delta t \int_{0+}^w \frac{dP(x)}{dx} H(w - x)\, dx, \tag{9-15}$$

when $x > 0$.

When $x = 0$, the number of lines which change to the second set during Δt and whose wait is greater than w is computed as follows:

The arriving item has waiting time zero in those queues where the number of items present is 0, 1, . . . , $c - 1$. The number of these queues is $NP(0)$. But among them only those with $c - 1$ items can, during Δt, give rise to the positive waiting-time category; i.e., the arrival would occupy one channel and the $c - 1$ others are in the remaining. Hence at that time all channels would be occupied, and there is a positive delay for an arrival. Among the last category, those that will give rise during $(t, t + \Delta t)$ to a wait greater than w are

$$N p_{c-1} \lambda \, \Delta t \, H(w), \tag{9-16}$$

where p_{c-1} is the steady-state probability that there are $c - 1$ items in the system.

Hence the steady-state equation in the c-channel case is given by

$$0 = \frac{dP(w)}{dw} - \lambda \int_{0+}^{w} \frac{dP(x)}{dx} H(w - x) \, dx - \lambda p_{c-1} H(w). \tag{9-17}$$

The Laplace transform of $P(w)$ from this equation is

$$P^*(s) = \frac{P(0)}{s} + p_{c-1} \frac{\lambda H^*(s)}{s[1 - \lambda H^*(s)]}. \tag{9-18}$$

Both $P(0)$ and p_{c-1} are unknown.

We use the following series expansion for small values of s:

$$H^*(s) = \int_0^\infty e^{-us} H(u) \, du = \sum_{n=0}^{\infty} \left[\int_0^\infty \frac{(-u)^n}{n!} H(u) \, du \right] s^n$$

$$\equiv a_0 + a_1 s + \cdots, \tag{9-19}$$

where

$$a_0 = \int_0^\infty H(u) \, du, \tag{9-20}$$

together with

$$\lim_{s \to 0} s P^*(s) = 1.$$

We then have

$$P(0) = 1 - \frac{\lambda a_0}{1 - \lambda a_0} p_{c-1}, \tag{9-21}$$

which leaves p_{c-1} to be determined. Of course, this approach does not help this determination. Experiment or other approaches may yield a value for p_{c-1}.

Remark: If we write $Q(w) = 1 - P(w)$, we have

$$Q^*(s) = \frac{p_{c-1}}{s} \left[\frac{\lambda a_0}{1 - \lambda a_0} - \frac{\lambda H^*(s)}{1 - \lambda H^*(s)} \right]. \tag{9-22}$$

Taking inverse transforms gives

$$Q(w) = p_{c-1} \left[\frac{\lambda a_0}{1 - \lambda a_0} - R(w) \right], \tag{9-23}$$

where $R(w)$ is the inverse transform of the product of $1/s$ and the second term in brackets. From the form of this product, the derivative $r(w) = dR(w)/dw$ satisfies

$$r(w) - \lambda \int_0^w r(x)H(w-x)\,dx = \lambda H(w), \qquad (9\text{-}24)$$

which can be solved numerically as previously outlined. Note that

$$\frac{dP(w)}{dw} = p_{c-1}r(w); \qquad (9\text{-}25)$$

i.e., we have the solution to within a constant of proportionality.

3. Gnedenko's Generalization of Waiting Times

B. Gnedenko [824a] has given a general formulation of the waiting-time problem. An arriving customer to a single-channel queue with ordered discipline may be lost, or he may leave after joining, or he may wait for service and then leave before its termination, or he may stay to finish service. Under this regime, let $D(w)$ be the distribution function for waiting in line an interval of time of length w, and let $G_x(w)$ be the conditional probability distribution that having waited in the line for a time x the customer will remain in service for a time of length w (i.e., without finishing service, and thus he waits an additional amount w), $G_x(+0) = 0$. Let $B(y)$ be the service-time distribution function.

Let $W(t)$ be the stochastic variable indicating the waiting time to the beginning of service of a customer arriving at time t. Let

$$P(w,t) = \text{Prob } [W(t) \leq w].$$

Suppose that arrivals occur by a Poisson distribution with parameter λ. We consider the conditions for $W(t + \Delta t) \leq w$ to hold. They are:

1. $W(t) < w + \Delta t$, and no arrival occurs during $(t, t + \Delta t)$.

2. $W(t) = x < w + \Delta t$; there is an arrival at time $t + \epsilon$, $0 < \epsilon < \Delta t$, who leaves before the beginning of service; i.e., he waits for a time less than $x - \epsilon$. (This is governed by D.)

3. The same as in (2) except that the arrival stays for partial service and leaves before time $w - x + \Delta t$. Note that before service he waited longer than $x - \epsilon$. (This is governed by D and G.)

4. The same as in (3) except that the arrival remains to complete service, and hence he stays in service for longer than time $w - x + \Delta t$. Clearly he waits before service longer than $x - \epsilon$. (This is governed by D, G, and B.)

Using probabilities we have

$$P(w, t + \Delta t) = (1 - \lambda \Delta t)P(w + \Delta t, t)$$
$$+ \int_0^{\Delta t} d\epsilon \int_0^{w + \Delta t} \{ D(x - \epsilon) + [1 - D(x - \epsilon)]G_{x-\epsilon}(w - x + \Delta t)$$
$$+ [1 - D(x - \epsilon)][1 - G_{x-\epsilon}(w - x + \Delta t)]B(w - x + \Delta t)\} d_x P(x, t)$$
$$+ o(\Delta t). \quad (9\text{-}26a)$$

In the usual way, by simplifying and dividing by Δt, which is then allowed to tend to zero, we have

$$\frac{\partial P(w,t)}{\partial t} = \frac{\partial P(w,t)}{\partial w} - \lambda P(w,t) + \lambda \int_0^w \{ D(x) + [1 - D(x)]G_x(w - x)$$
$$+ [1 - D(x)][1 - G_x(w - x)]B(w - x)\} d_x P(x, t) \quad (9\text{-}26b)$$

of which Takács' equation is a special case.

As $t \to \infty$ this equation yields the corresponding equation for the stationary distribution $P(w)$, which can be shown to exist. It is obtained by equating the left side to zero and suppressing t as the argument on the right. The resulting equation may be simplified to yield

$$\frac{dP(w)}{dw} = \lambda \int_0^w [1 - D(x)][1 - G_x(w - x)][1 - B(w - x)] \, dP(x) \quad (9\text{-}26c)$$

which has a unique solution with a discontinuity at $w = 0$ and is absolutely continuous for $w \neq 0$.

We now introduce some useful functions:

1. The distribution function of the effective waiting time is

$$P_1(w) = P(w) + [1 - P(w)]D(w), \quad (9\text{-}26d)$$

obtained as the joint occurrence of the two events of waiting in line and leaving before service.

2. The distribution function of the total time in the system is

$$P_2(w) = P(w) - \frac{P'(w)}{\lambda} + [1 - P(w)]D(w). \quad (9\text{-}26e)$$

3. The distribution function of the ratio of effective service time to the time necessary to complete service is

$$P_3(w) = 1 - \int_0^\infty \int_0^\infty [1 - G_x(wy)] \, dP(x) \, dB(y), \quad 0 < w < 1. \quad (9\text{-}26f)$$

4. The distribution function for the complete satisfaction of the demand for service (i.e., customer remains to complete service) is given by

$$P_4(w) = \int_0^\infty [1 - G_x(w)] \, dP(x). \quad (9\text{-}26g)$$

5. The pure loss probability is

$$\alpha_1 = \int_0^\infty D(x) \, dP(x). \quad (9\text{-}26h)$$

6. The probability that service will be partially fulfilled is

$$\alpha_2 = \int_0^\infty [1 - D(x)] \int_0^\infty G_x(y) \, dB(y) \, dP(x). \tag{9-26i}$$

Exercise 1: For lost calls the waiting density function is a Dirac delta function, and hence

$$D(w) = \begin{cases} 0, & w \le 0, \\ 1, & w > 0. \end{cases}$$

For a fixed wait in line of length τ, after which a customer reneges, i.e., leaves the line (obtained by translating the origin in the lost-call case to τ),

$$D(w) = \begin{cases} 0, & w \le \tau, \\ 1, & w > \tau, \end{cases} \qquad G_x(w) \equiv 0.$$

If τ is a random variable, then $G_x(w) = 0$. If the wait in the system is bounded by τ, then $G_x(w) = 1$ for $w + x > \tau$; and if τ is a stochastic variable, then

$$G_x(w) = \frac{D(w + x) - D(x)}{1 - D(x)}. \tag{9-26j}$$

Show for exponential service times with parameter μ for the case of reneging (note that if a customer enters service he remains to complete it) that

$$P(w) = \begin{cases} \dfrac{P(0)}{1 - \rho} (1 - \rho e^{-(\mu-\lambda)w}) & \text{for } w < \tau \\ 1 - \rho P(0) e^{\lambda \tau - \mu w} & \text{for } w > \tau \end{cases} \tag{9-26k}$$

where $\rho = \lambda/\mu$.

As we shall see in a later chapter, Barrer has shown that

$$P(0) = p_0 = \frac{1 - \rho}{1 - \rho^2 e^{-\tau(\mu-\lambda)}} \tag{9-26l}$$

Also show that

$$P_1(w) = \begin{cases} (1 - \rho e^{-(\mu-\lambda)w})/(1 - \rho^2 e^{-(\mu-\lambda)\tau}) & \text{for } w < \tau \\ 1 & \text{for } w > \tau \end{cases} \tag{9-26m}$$

$$\alpha_1 = \frac{\rho(1 - \rho)}{e^{(\mu-\lambda)\tau} - \rho^2} \tag{9-26n}$$

which is the saltus of $P(w)$ at τ. In the service system there are no partial losses (i.e., once a unit enters service it completes service); hence, $\alpha_2 = 0$.

$$P_2(w) = \begin{cases} (1 - e^{-(\mu-\lambda)w})/(1 - \rho^2 e^{-(\mu-\lambda)\tau}), & 0 < w < \tau, \\ 1 - \rho \left[1 - \dfrac{P(0)}{1 - \rho} (1 - \rho e^{-(\mu-\lambda)\tau}) \right] e^{-\mu(w-\tau)}, & w > \tau. \end{cases}$$

$$\tag{9-26o}$$

Exercise 2: Study the solvability of the stationary equation.

9-3. General Independent Input and Arbitrary Holding Times

1. *The Integral Equation of Lindley for a Single Channel*

We derive the waiting-time (in queue) distribution for a general independent input distribution and general independent service time for a single-channel, first-come–first-served queue.

Suppose that units arrive in order before a single channel. We denote by t_n the arrival interval between the nth and $(n + 1)$st unit and by s_n the service time of the nth unit. The t_n are assumed to be independent random variables with identical probability distributions and finite mean. So are the s_n. The sequences $\{t_n\}$, $\{s_n\}$, $n = 1, 2, \ldots$, need not be completely independent but are often assumed to be so for many practical applications.

Let w_n be the waiting time of the nth unit. Brief reflection will show the following to hold:

$$w_{n+1} = \begin{cases} w_n + u_n & \text{if } w_n + u_n > 0, \\ 0 & \text{if } w_n + u_n \leq 0, \end{cases} \qquad (9\text{-}27a)$$

where $u_n = s_n - t_n$. Since w_n is necessarily positive, we have from the above

$$
\begin{aligned}
P_{n+1}(w) &\equiv \text{Prob } (w_{n+1} \leq w) \\
&= \text{Prob } (w_{n+1} = 0) + \text{Prob } (0 < w_{n+1} \leq w) \\
&= \text{Prob } (w_n + u_n \leq 0) + \text{Prob } (0 < w_n + u_n \leq w) \\
&= \text{Prob } (w_n + u_n \leq w) = \int_{w_n + u_n \leq w} dP_n(w_n)\, dU(u_n) \\
&= \int_{u_n \leq w} P_n(w - u_n)\, dU(u_n), \qquad (9\text{-}27b)
\end{aligned}
$$

where $P_{n+1}(w)$ is the cumulative-waiting-time distribution of the $(n + 1)$st unit and $U(u_n)$ is the cumulative distribution of u_n which is the same for all values of n.

Now $P_1(w) = 1$ for $w \geq 0$, since the first item does not wait in line. We also have

$$P_2(w) = \int_{u_1 \leq w} dU(u_1) = \text{Prob } (u_1 \leq w),$$

$$P_3(w) = \int_{u_2 \leq w} \int_{u_1 \leq w - u_2} dU(u_1)\, dU(u_2) = \text{Prob } (u_1 + u_2 \leq w, u_2 \leq w),$$

and in general

$$
\begin{aligned}
P_{n+1}(w) &= \text{Prob } (u_1 + \cdots + u_n \leq w, u_2 + \cdots + u_n \leq w, \\
&\qquad\qquad \ldots, u_n \leq w) \\
&= \text{Prob } (u_1 + \cdots + u_n \leq w, u_1 + \cdots + u_{n-1} \leq w, \\
&\qquad\qquad \ldots, u_1 \leq w). \qquad (9\text{-}28)
\end{aligned}
$$

The second equality holds since the u_n are independently and identically distributed; hence the probability is unaffected if they are relabeled.

If we write

$$U_n = \sum_{k=1}^{n} u_n, \tag{9-29}$$

then

$$P_{n+1}(w) = \text{Prob } (U_k \leq w, \, k = 1, 2, \ldots, n) \tag{9-30}$$

Thus $P_{n+1}(w)$ is a monotonic decreasing function of n lying between zero and unity.   Hence it has a limit, i.e.,

$$\lim_{n \to \infty} P_{n+1}(w) = \lim_{n \to \infty} P_n(w) = P(w). \tag{9-31}$$

This limit can be shown to uniquely satisfy Lindley's integral equation, which depends on the distribution of the difference of input times and service times:

$$P(w) = \int_{u \leq w} P(w - u) \, dU(u) = \int_0^\infty P(y) \, dU(w - y). \tag{9-32}$$

The infinite upper limit implies that u goes to minus infinity and y to plus infinity under the transformation $y = w - u$.   However, $u = s - t$ permits this fact.   One can show that $P(w)$ is a distribution by proving that it is nonnegative, nondecreasing, and continuous on the right; that it is equal to zero for $w < 0$; and that it tends to unity as $w \to \infty$.

If $U(w - y)$ is absolutely continuous and hence has a derivative, we have the Wiener-Hopf-type integral equation:

$$P(w) = \int_0^\infty P(y) U'(w - y) \, dy \tag{9-33}$$

with $$U'(w) = \int_0^\infty b(y) a(y - w) \, dy = \int_0^\infty b(y + w) a(y) \, dy. \tag{9-34}$$

Note that $b(y)$ vanishes for negative argument.   We also have

$$U(w) = \int_0^\infty B(y + w) a(y) \, dy = \int_0^\infty B(y) a(y - w) \, dy$$

$$= 1 - \int_0^\infty b(y) A(y - w) \, dy$$

$$= 1 - \int_0^\infty b(y + w) A(y) \, dy. \tag{9-35}$$

Here $a(y)$ and $A(y)$ are the density and cumulative-interarrival-times distributions and $b(y)$ and $B(y)$ are the corresponding service-times density and distribution.

Example: Consider the case of arrivals at constant intervals of length T. The cumulative distribution of input times is then given by

$$A(\tau) = \begin{cases} 0 & \text{for } \tau < T, \\ 1 & \text{for } \tau \geq T. \end{cases}$$

Suppose that the service times are exponentially distributed with

$$b(\tau) = \mu e^{-\mu\tau}, \qquad \tau \geq 0,$$

and zero otherwise.   Then

$$U(\tau) = 1 - \int_T^\infty b(y + \tau) \, dy = B(T + \tau).$$

Thus
$$U'(\tau) = b(T + \tau).$$
This yields

$$P(\tau) = \int_0^{T+\tau} P(x)b(T + \tau - x) \, dx.$$

[Note that $b(T + \tau - x)$ vanishes for negative argument.]   This may be easily transformed into

$$P(\tau - T) = \int_0^\tau P(\tau - y)b(y) \, dy$$

by replacing τ by $\tau' - T$, then putting $y = \tau' - x$, and finally using τ for τ'.

Substituting for $b(y)$ and trying as solution

$$P(\tau) = 1 + c_1 e^{c_2\tau},$$

the following conditions, which must be satisfied, are obtained:

$$\frac{1}{\mu} + \frac{c_1}{\mu + c_2} = 0,$$

$$\frac{\mu}{\mu + c_2} = e^{-c_2 T}.$$

These give
$$c_1 = -[1 - P(0)] = -(1 - p_0),$$
$$c_2 = -\mu p_0.$$

Here p_0 is the nonzero root of $1 - p_0 = e^{-\mu p_0 T}$.   Thus the probability $P(\leq\tau) \equiv P(\tau)$ of waiting in queue less than time τ is given by

$$P(\leq\tau) = 1 - (1 - p_0)e^{-\mu p_0 \tau}. \tag{9-36}$$

Finally, the waiting time is given by

$$W_q = \int_0^\infty \tau \, dP(\leq\tau) = \frac{1 - p_0}{\mu p_0}. \tag{9-37}$$

For a single-channel queue with chi-square input with parameters λ and n and Erlang service with parameters μ and m, Syski finds in his book the solution of Lindley's equation as given in the following expression for the distribution of waiting time:

$$P(w) = 1 + \sum_{k=1}^{m} \alpha_k e^{-\beta_k w}.$$

This solution is unique, and β_k are the distinct roots of the characteristic equation

$$\left(\frac{\mu}{\mu - \beta}\right)^m \left(\frac{\lambda}{\lambda + \beta}\right)^n = 1,$$

and

$$\alpha_k = -\prod_{\substack{j=1 \\ j \neq k}}^{m} \frac{\beta_j}{\beta_j - \beta_k} \left(\frac{\mu - \beta_k}{\mu}\right)^m.$$

The mean waiting time is

$$W_q = -\sum_{k=1}^{m} \frac{\alpha_k}{\beta_k}.$$

W. Smith [661] has investigated the nature of the waiting-time distribution for a single channel in the steady state. We have already seen that for the Poisson- and constant-input case, the waiting-time distribution is exponential. It turns out that the exponential nature of this distribution occurs under less restrictive conditions.

If the generating function of a distribution is the reciprocal of a polynomial of the nth degree, the distribution is said to be K_n.

If $\tilde{U}(s)$ is the Laplace-Stieltjes transform of the cumulative distribution function $U(u)$, where u is the difference between the service and arrival times of a customer and where we write $s = \sigma + i\tau$, then we have the following theorem, whose proof is too long to give (due to Smith):

Theorem 9-1: If the service-time distribution is K_n, the waiting-time (in line and in service) distribution is K_n for all arrival distributions satisfying fairly general conditions.

The frequency function of the waiting time is given by

$$\sum_{i=1}^{n} A_i e^{a_i w},$$

where

$$A_i^{-1} = -\frac{1}{a_i} \prod_{j \neq i} \left(1 - \frac{a_i}{a_j}\right)$$

and a_j are the n zeros (supposed distinct) of $1 - \tilde{U}(s)$ in the open half plane $\sigma < 0$.

Another useful fact which is a corollary of the foregoing theorem and which we denote as a theorem due to its importance is the following:

Theorem 9-2: If the service time is distributed exponentially, so is the waiting time in the line for those who have to wait, whatever the arrival-time distribution.

Exercise: Prove Theorem 9-2, using Theorem 9-1.

2. *A General Contour Representation*

E. Ventura obtains the Laplace transform of the waiting-time distribution of the nth item by means of an ingenious device [908]. Using Eq. (9-27a) and defining $\gamma_n(s)$ as the transform of the waiting time of the nth item, then

$$\gamma_n(s) = E[e^{-sw_n}]. \tag{9-38}$$

Now one may represent e^{-sw_n} for positive real values of s by means of a contour integral as follows:

$$e^{-sw_n} = -\frac{1}{2\pi i} \oint_C \exp\left[-z(w_{n-1} + u_{n-1})\right]\frac{dz}{z-s} + \varepsilon_n, \tag{9-39}$$

where
$$\varepsilon_n = \begin{cases} 0 & \text{if } w_{n-1} + u_{n-1} > 0, \\ \tfrac{1}{2} & \text{if } w_{n-1} + u_{n-1} = 0, \\ 1 & \text{if } w_{n-1} + u_{n-1} < 0. \end{cases} \tag{9-40}$$

The contour C extends from $-i\infty$ to $i\infty$, bypassing the origin with a small semicircle to the right of the imaginary axis. (This is needed since, later, s will be allowed to tend to zero.)

We now justify this representation. For negative values of w_n the integral can be shown to be zero. This is done as follows: Consider the integral along the imaginary axis as described but going from $-ia$ to ia. If this contour is closed by a semicircle in the left half plane, the integral will be zero since the integrand is analytic inside and on this contour. Thus the integral along the imaginary axis and the small semicircle is the negative of the integral along the large semicircle in the left half plane.

Now one must show that the integral along the large semicircle tends to zero as $a \to \infty$. This is done in the process of showing that the integral actually converges in a sector with vertex at the origin but excluding the imaginary axis. This is delicately shown by using the polar representation of z, taking absolute values, and showing that the integral converges as the radius a tends to infinity. Consequently, as $a \to \infty$ the integral along C is zero for $w_{n-1} + u_{n-1} < 0$. Since the left side of the equation gives unity for this value [Eq. (9-27a)], we must have $\varepsilon_n = 1$.

For $w_{n-1} + u_{n-1} > 0$ a similar procedure is used except that the large semicircle is taken in the right half plane and consequently, for large values of a, the integral has a simple pole at $z = s$. From the theory of residues, the value of the integral is $-2\pi i e^{-sw_n}$; hence $\varepsilon_n = 0$. (Note that the closed contour has clockwise direction and hence the negative sign.) In this case one shows that the integral along the large semicircle. in the right half plane vanishes as $a \to \infty$, thus leaving the above value for the integral along C as defined along the imaginary axis.

A problem remains as to what happens for $w_{n-1} + u_n = 0$. In this case simple integration will not do. However, a closed contour is used, as in the positive case, yielding a residue of $-2\pi i$. In this case the integral does not vanish along the semicircle because of the absence of the exponential factor. Now the integral evaluated around a circle would be $-2\pi i$; hence around a semicircle its value is $-\pi i$. Thus $\varepsilon_n = \frac{1}{2}$. The last part requires careful justification by the reader. If we take expected values on both sides of the above expression (we can interchange the contour integral and the expected-value signs) and then note that the probability that $w_{n-1} + u_{n-1} = 0$ is generally zero, therefore, from the definition of ε_n, we have $E[\varepsilon_n] = E[\text{Prob } (w_n = 0)] = p_0$. The reader need not be troubled by this as we shall derive an expression for the quantity p_0.

We now have the useful expression which holds in general:

$$\gamma_n(s) = -\frac{1}{2\pi i} \oint_C E[\exp (-z w_{n-1}) \exp (-z u_{n-1})] \frac{dz}{z - s} + p_0. \quad (9\text{-}41)$$

If t_{n-1}, s_{n-1}, and w_{n-1} are assumed to be independent random variables (note that they can be dependently distributed and the representation still holds), then the expected value would distribute with respect to the exponentials in the integral, i.e.,

$$E[e^{-z w_{n-1}} e^{-z u_{n-1}}] = E[e^{-z w_{n-1}}] E[e^{-z s_{n-1}}] E[e^{z t_{n-1}}]. \quad (9\text{-}42)$$

Furthermore, if the distributions are identical for all items, the first expression on the right is $\gamma(z)$ and the middle expression is $\beta(z)$, the transform of the service-time distribution. If the interarrival distribution is exponential with parameter λ, then its transform is $[\lambda/(\lambda - z)]$, and $\gamma(s)$, which now is independent of n, is given by

$$\gamma(s) = \frac{\lambda}{2\pi i} \oint_C \frac{\gamma(z)\beta(z)}{(z - \lambda)(z - s)} \, dz + p_0. \quad (9\text{-}43)$$

Since both functions in the numerator of the integrand are regular in the right half plane and can be shown to yield convergence to zero on a semicircle in the right half plane with radius tending to infinity, C can be closed by such a semicircle enclosing two simple poles at λ and s.

The residue is simply obtained in the usual manner, yielding

$$\gamma(s) = -\frac{\lambda \gamma(\lambda)\beta(\lambda)}{\lambda - s} + \frac{\lambda \gamma(s)\beta(s)}{\lambda - s} + p_0. \quad (9\text{-}44)$$

On simplifying, this becomes

$$\gamma(s) = \frac{\lambda\gamma(\lambda)\beta(\lambda) + p_0(s - \lambda)}{s - \lambda + \lambda\beta(s)}. \tag{9-45}$$

Now, as $s \to 0$, both transforms tend to unity. If we multiply through by the denominator, the left side goes to zero with s. Thus the right side gives

$$p_0 = \gamma(\lambda)\beta(\lambda). \tag{9-46}$$

Now with this value of p_0, as $s \to 0$, the right side of the above equation as it stands is indeterminate. If we apply L'Hospital's rule we have

$$1 = \lim_{s \to 0} \frac{p_0}{1 + \lambda\beta'(s)}. \tag{9-47}$$

But $\lim_{s \to 0} \beta'(s) \equiv -1/\mu$, the mean waiting time; therefore

$$\gamma(s) = \frac{s(1 - \lambda/\mu)}{s - \lambda + \lambda\beta(s)}, \tag{9-48}$$

which is a familiar expression.

3. *A New Approach*

Again, if we let $W(t)$ denote the time that a customer arriving at time t would have to wait before entering service and if both arrival times and service times are described by a single function $K(t)$ for $t \geq 0$ [it is the work load submitted to the service channel during (u,t), and hence $W(t)$ can be thought of as the work remaining to be done at t], then Beneš [783] gives

$$W(t) = K(t) - t + \int_0^t U(-W(u))\,du, \qquad t \geq 0, \tag{9-49}$$

where U is the unit step function, i.e.,

$$U(x) = \begin{cases} 1, & x \geq 0, \\ 0, & x < 0. \end{cases}$$

Note that $W(0) = K(0)$.

This approach to queues makes no independence assumptions on arrival and service distributions and uses no special distributions. It also need not give rise to a Markoff process. Note that the solution of the foregoing equation involves the supremum functional whose distribution is very difficult to determine.

It can be shown that with probability 1 we have

$$\exp[-sW(t)] = \exp\{-s[K(t) - t]\}$$
$$\left(1 - s\int_0^t \exp\{s[K(u) - u]\}P(u,0)\,du\right), \tag{9-50}$$

where
$$P(u,0) = \begin{cases} 1 & \text{if } W(u) = 0. \\ 0 & \text{otherwise.} \end{cases}$$

We also have

$$\text{Prob } [W(t) \leq w] = \begin{cases} 0 & \text{for } w < 0, \\ \text{Prob } [K(t) - t \leq w] - \dfrac{\partial}{\partial w} \displaystyle\int_0^t \text{Prob } [K(t) \\ \qquad - K(u) - t + u \leq w \text{ and } W(u) \\ \qquad\qquad\qquad = 0]\, du \quad \text{for } w \geq 0. \end{cases} \quad (9\text{-}51)$$

The reader may have become accustomed to the idea that one must always specify arrival and service distributions. However, this alternative approach requires that one determine the work load function $K(t)$.

9-4. Multiple Channels, General Independent-input and Service Distribution

We shall give only the result. We refer the reader to the works of Kiefer and Wolfowitz on this subject. The authors point out the desirability of solving their equation for special values of the input and service distributions but also state that the task is likely to be difficult. Lindley's equation is a special case of their result. We have for the steady state

$$P(w) = \int P_\infty \{ \psi(w,b,c) \}\, dB(b)\, dA(c), \qquad (9\text{-}52)$$

where P_∞ is the measure according to $P(w)$ and where A and B are the input (interarrival) and service-time distributions, respectively, etc. It is shown that $P(w)$ is a distribution function when the traffic intensity ρ is less than unity. However, this is not generally the case, for $\rho \geq 1$. Here ρ is the ratio of the expected service time to the product of the number of channels and the expected arrival time.

PROBLEMS

1. The influx of papers to the editor of a journal is given by a Poisson distribution with parameter λ. The service distribution is truncated exponential, that is,

$$b(t) = \begin{cases} Ce^{-\mu t}, & t \leq T, \\ 0, & t > T, \end{cases}$$

where C is a normalizing constant which depends on T. Develop an expression for the expected waiting time, using Lindley's equation.

2. Apply the integral equation of Lindley for the waiting-time distribution of a single-channel queue to the following cases:

a. Poisson arrivals with parameter λ and exponential service times with parameter μ.

b. The arrival distribution $[\lambda^n/(n - 1)!]t^{n-1}e^{-\lambda t}$ and exponential service time with parameter μ. Obtain for the waiting-time distribution

$$w(t) = \mu \left(\frac{\lambda}{\lambda + \mu} \right)^n \left[\sum_{k=1}^{n} \frac{(\lambda + \mu)^{n-k}}{(n-k)!} \int_0^\infty w(t+y)e^{-\lambda y} y^{n-k} \, dy + \int_0^\infty w(t-y)e^{-\mu y} \, dy \right].$$

By assuming a solution of the form $w(t) = 1 - \alpha e^{-\beta t}$, show that

$$\alpha = \frac{\mu - \beta}{\mu}, \qquad \left(\frac{\lambda}{\lambda + \beta} \right)^n = \frac{\mu - \beta}{\mu},$$

c. For constant arrivals with unit separation and for the service-time distribution $\mu^{n+1} t^n e^{-\mu t}/n!$, assuming $w(t) = 1 + \sum_{k=1}^{n} c_k e^{z_k t}$. Determine c_k and x_k.

3. By reference to Smith's paper, obtain a proof of his main theorem.

4. By reference to Lindley's paper, show that the desired limiting function is a distribution function.

5. In a Poisson input with parameter λ, arbitrary-service-time density $b(x)$, single-channel queue, let $w_n(x,t)$ be the probability density that, at time t, n items are waiting in line (not including the one in service) and a time of length x has elapsed in servicing the item already in service. Let $w_n(x,0)$ be given. The probability that this service will terminate in $(x, x + \Delta)$ is $b(x)\Delta$, where [358]

$$b(x) = \eta(x) \exp \left[- \int_0^x \eta(y) \, dy \right],$$

and $\eta(x)\Delta$ is the probability that a system which survives to x will have completion in $(x, x + \Delta)$.

$$w_n(x + \Delta, t + \Delta) = w_n(x,t)(1 - \lambda\Delta)[1 - \eta(x)\Delta] + w_{n-1}(x,t)\lambda\Delta.$$

Expanding the left side in series about (x,t) up to the first power of Δ and simplifying gives

$$\frac{\partial w_n}{\partial t} + \frac{\partial w_n}{\partial x} + [\lambda + \eta(x)]w_n = \lambda w_{n-1}, \qquad n = 1, 2, \ldots,$$

$$\frac{\partial w_0}{\partial t} + \frac{\partial w_0}{\partial x} + [\lambda + \eta(x)]w_0 = 0.$$

If $E(t)$ is the probability that the system is empty, one may similarly derive

$$\frac{dE(t)}{dt} + \lambda E(t) = \int_0^\infty \eta(x)w_0(x,t) \, dt$$

with a given initial condition $E(0)$.

Justify the boundary conditions

$$w_n(0,t) = \int_0^\infty w_{n+1}(x,t)\eta(x) \, dx, \qquad n = 1, 2, \ldots,$$

$$w_0(0,t) = \int_0^\infty \dot{w}_1(x,t)\eta(x) \, dx + \lambda E(t).$$

We also have

$$w_n(x,0) = \delta_{nN}\delta(x - x_0), \qquad n = 0, 1, 2, \ldots,$$

where the first quantity on the right is the Kronecker delta and the second is the Dirac delta function.

The conservation of probability leads to

$$\frac{d}{dt} \left[E(t) + \sum_{n=0}^{\infty} \int_0^\infty w_n(x,t) \, dx \right] = 0.$$

One may introduce a generating function $G(z,x,t)$ for $w_n(x,t)$ and reduce the system of equations as usual. Then by writing

$$G(z,x,t) = H(z,x,t)e^{-N(x)},$$

where $N(x) = \int_0^x \eta(y) \, dy$, show that one has

$$\frac{\partial H}{\partial t} + \frac{\partial H}{\partial x} + \lambda(1 - z)H = 0$$

with the solution

$$H(z,x,t) = H_0(z, t - x)e^{-\lambda(1-z)x}$$

from which $G(z,x,t)$ may be obtained.

The condition on $w_n(x,0)$ may be used to yield

$$H_0(z, -x) = e^{N(x)}e^{\lambda(1-z)x}z^N\delta(x - x_0).$$

Show that the boundary conditions combine into

$$zG(z,0,t) = \int_0^\infty \eta(x)G(z,x,t) \, dx + \lambda zE(t) - \int_0^\infty \eta(x)G(0,x,t) \, dx.$$

One may also show the following condition to hold:

$$\frac{dE}{dt} + \lambda E + zH_0(z,t) = \int_0^\infty b(x)H_0(z, t - x)e^{-\lambda(1-z)x} \, dx + \lambda zE.$$

If one starts at $t = 0$ with an empty queue so that $E(0) = 1$, $G(z,x,0) = 0$, this equation may be written with the integral on the right taken from zero to t. Taking the Laplace transform with respect to x, with parameter s, yields, after simplifying,

$$H_0^*(z,s) = \frac{[s + \lambda(1 - z)]E^*(s) - 1}{b^*[s + \lambda(1 - z)] - z}.$$

By determining the relevant zero of the denominator for $0 \le z \le 1$ for which the numerator must vanish, the inverse transform of $E^*(s)$ may be computed.

There is a unique zero z_s of the denominator when $\lambda/\eta < 1$, with η representing the mean service time. This ensures the existence of an equilibrium solution with transform

$$E^*(s) = \frac{1}{s + \lambda(1 - z_s)}.$$

Using the Tauberian theorem,

$$\lim_{t \to \infty} E(t) = \lim_{s \to 0} sE^*(s) = 1 - \frac{\lambda}{\eta},$$

show that

$$\lim_{t \to \infty} H(z,t) = \frac{\lambda(1 - s)(1 - \lambda/\eta)}{b^*[\lambda(1 - z)] - z}.$$

Compute $E^*(s)$ for $b(x) = \eta e^{-\eta x}$.

Also show that for $E(0) = 1$ and $w_n(x,0) = 0$, $E(t)$ satisfies

$$\frac{dE}{dt} = -\lambda E + \text{the convolution of } \lambda E \text{ and } \frac{dG(t)}{dt}$$

where $G(t)$ is the busy-period distribution.

CHAPTER 10

GENERAL INDEPENDENT INPUT, EXPONENTIAL OR ERLANGIAN SERVICE

10-1. Introduction

In this chapter our main concern is the study of the number in the system and the distribution of the length of a busy period for systems with general independent input and exponential or Erlangian service distributions. We do this by following Conolly's several papers on the subject. The mathematics of this chapter is somewhat involved and requires patient study. One considers arrival instants and specifies sufficient variables to ensure that the process is Markovian.

In the case of the exponential service distribution, we introduce quantities $P_n(u,t)$ which are a means to an end; i.e., they are used to calculate $P_n(t)$, the number in the system at time t. However, we obtain the Laplace transform which may, by numerical application of the inversion integral, yield $P_n(t)$. The result has been applied to the Poisson-input case to verify correctness, which is reassuring. The distribution of the length of a busy period is also derived for this case and applied to a constant-input assumption.

The Erlangian-service-distribution case is described briefly for a joint distribution of the number served and the length of a busy period. The argument basically follows along lines similar to those used in the exponential-service case.

10-2. Queue-length Probabilities (Exponential Case)

Consider a single-channel ordered-service queue in which the interarrival-time intervals are independently and arbitrarily distributed according to the density function $a(t)$ and the service times are exponentially distributed with parameter μ. We define $P_n(t)$ as the probability of n units in the system (including the one in service) at time t. We define $P_n(u,t)$ as the conditional probability density of n units in the system at time t, given that the last arrival occurred during $(t - u - du, t - u)$, and $P_n(0,t)$ as the probability density that an arrival during $(t - dt, t)$ changed the state of the system from E_{n-1} to E_n. We have $P_0(0,t) = 0$. We also define $A(t) = \int_0^t a(x)\ dx$.

During an interval of length u in which no arrivals occur, the probability that n units complete service is $b_n(u) = e^{-\mu u}[(\mu u)^n/n!]$, provided that the state of the system was greater than n at the beginning of the interval. If there are N units in the system at the beginning of u, the probability that they all complete service before u expires is

$$\bar{b}_N(u) = 1 - \sum_{j=0}^{N-1} b_j(u). \tag{10-1}$$

It is assumed that the first unit enters the system at $t = 0$ and that there are no other units in the system. The study for an arbitrary initial state follows similar lines.

We have $P(0,t) = \sum_{n=1}^{\infty} P_n(0,t)$ as the probability that an arrival takes place in $(t - dt, t)$. Its Laplace transform is

$$P^*(0,s) = \sum_{n=1}^{\infty} P_n^*(0,s). \tag{10-2}$$

Note that $a(t - 0)$ is the probability density of a next arrival during $(t - dt, t)$ starting from $t = 0$. We show that the series converges at least for Re $(s) > 0$ by deriving an expression for $P(0,t)$. We then derive expressions for $P_n(0,t)$.

An arrival during $(t - dt, t)$ may be the first after the initial arrival at $t = 0$ with probability $a(t - 0)\ dt$, or the last arrival before this arrival may have occurred during $(t - u - du, t - u)$ with the probability $P(0, t - u)a(u - 0)\ du\ dt$. Thus we have

$$P(0,t) = a(t) + \int_0^t P(0, t - u)a(u)\ du. \tag{10-3}$$

The Laplace transform is

$$P^*(0,s) = \frac{a^*(s)}{1 - a^*(s)}, \tag{10-4}$$

which converges at least for Re $(s) > 0$, assuming that no difficulty occurs by writing $a(t - 0) = a(t)$.

We also have

$$P_1(0,t) = a(t)\bar{b}_1(t) + \int_0^t a(u) \sum_{m=1}^{\infty} P_m(0, t - u)\bar{b}_m(u) \, du,$$

$$P_2(0,t) = a(t)b_0(t) + \int_0^t a(u) \sum_{m=0}^{\infty} P_{m+1}(0, t - u)b_m(u) \, du, \qquad (10\text{-}5)$$

$$P_n(0,t) = \int_0^t a(u) \sum_{m=0}^{\infty} P_{n+m-1}(0, t - u)b_m(u) \, du, \qquad n \geq 3.$$

The first expression on the right of the first of these equations gives the probability density that the arrival during $(t - dt, t)$ is the first since time $t = 0$. The second expression on the right gives the probability density that an arrival occurred before this time, say during $(t - u - du, t - u)$, creating state E_m, all units of which were served before the end of u. The probability of this event must be summed over all m and all $0 \leq u \leq t$. This also applies for the other equations.

Taking the Laplace transform of the last equation yields the homogeneous difference equation with coefficients independent of n:

$$P_n^*(0,s) = \sum_{m=0}^{\infty} \lambda_m(s) P_{n+m-1}^*(0,s), \qquad n \geq 3. \qquad (10\text{-}6)$$

Here

$$\lambda_m(s) = \int_0^{\infty} e^{-su} a(u) b_m(u) \, du = \int_0^{\infty} e^{-u(s+\mu)} (\mu u)^m a(u) \frac{du}{m!}, \qquad (10\text{-}7)$$

since we have a convolution. It is also equal to the coefficient of x^m in the expansion in powers of x of

$$\int_0^{\infty} e^{-u[s+(1-x)\mu]} a(u) \, du = a^*[s + (1 - x)\mu]. \qquad (10\text{-}8)$$

The reader can verify this by expanding the right side and writing the integral of which the coefficient of x^m is the transform.

On substituting $x^n = P_n^*(0,s)$ in the difference equation (10-6) and simplifying [in order to obtain the general solution, which takes the form

$$P_n^*(0,s) = \sum_{j=1}^{\infty} k_j \xi_j^n$$

if the roots ξ_j of the following equation (10-9) obtained in the process are all simple], we have

$$a^*[s + (1 - x)\mu] = x. \qquad (10\text{-}9)$$

The coefficients k_j are independent of n and are determined by the requirement that the solution must hold for all values of n.   Now, even if the ξ_j are all simple, the solution would break down if none had modulus less than unity since $P^*(0,s)$ is finite.

Solving for x in (10-9), we find a single zero ξ with modulus less than unity which is the only zero that we use to write

$$P_n^*(0,s) = k\xi^n. \tag{10-10}$$

This solution in fact holds for $n = 2$.

On taking the Laplace transforms, the first of the integrodifference equations gives

$$P_1^*(0,s) = k\xi - 1. \tag{10-11}$$

Thus by substitution into $P^*(0,s)$ we have

$$\frac{a^*(s)}{1 - a^*(s)} = -1 + \frac{k\xi}{1 - \xi} \tag{10-12}$$

or

$$k = \frac{1 - \xi}{\xi[1 - a^*(s)]}. \tag{10-13}$$

The integrodifference equations for $P_n(t)$ are identical with those given above in (10-5) except that on the left-hand side $P_1(0,t)$ is replaced by $P_0(t)$, $P_2(0,t)$ by $P_1(t)$, $P_n(0,t)$ by $P_{n-1}(t)$ (holds for $n \geq 2$), and on the right $a(\cdot)$ by $1 - A(\cdot) \equiv A_c(\cdot)$.   Note that the integrands now are the conditional probabilities $P_0(u,t)$, $P_1(u,t)$, and $P_n(u,t)$, respectively.   Thus

$$P_n(u,t) = A_c(u) \sum_{m=0}^{\infty} P_{n+m}(0,\, t - u)b_m(u).$$

Taking Laplace transforms leads to

$$P_n^*(s) = \frac{(1 - \xi)\xi^{n-1}}{1 - a^*(s)} A_c^*[s + (1 - \xi)\mu], \qquad n \geq 2. \tag{10-14}$$

Note that

$$A_c^*(s) = \frac{1 - a^*(s)}{s}. \tag{10-15}$$

Thus

$$P_n^*(s) = \frac{(1 - \xi^2)\xi^{n-1}}{\mu[1 - a^*(s)](s/\mu + 1 - \xi)} \tag{10-16}$$

which holds for $n \geq 1$ and

$$P_0^*(s) = \frac{1}{s} - \frac{1 - \xi}{\mu[1 - a^*(s)](s/\mu + 1 - \xi)}. \tag{10-17}$$

As a check it may be noted that these results satisfy the relation $\sum_{n=0}^{\infty} P_n^*(s) = 1/s$.   They also yield the desired results for the case of

Poisson arrivals.   The zero ξ is the same as $\mu\alpha_2$ encountered in the analysis of that case in Chap. 4.   Some steady-state properties may be deduced, using the facts

$$\lim_{t \to \infty} P_n(t) = \lim_{s \to 0} sP_n^*(s), \qquad a \equiv \int_0^\infty ua(u)\, du. \qquad (10\text{-}18)$$

Near $s = 0$ we have

$$a^*(s) = 1 - as + O(s^2), \qquad\qquad\qquad (10\text{-}19)$$
$$\xi(s) = \xi_0 + \xi_1 s + O(s^2), \qquad \xi_0 \equiv \xi(0), \qquad \xi_1 \equiv \xi'(0). \quad (10\text{-}20)$$

When $a > 1/\mu$ (note that in the Poisson case $\lambda = 1/a$), we have $\xi_0 < 1$ and when $a \leq 1/\mu$, $\xi_0 = 1$.   In the first case we have the main result:

$$\lim_{s \to 0} sP_n^*(s) = \begin{cases} \dfrac{1}{\mu a}(1 - \xi_0)\xi_0^{n-1}, & n \geq 1, \\[2ex] 1 - \dfrac{1}{\mu a}, & n = 0. \end{cases} \qquad (10\text{-}21)$$

Hence the first case is the condition for statistical equilibrium.   In the case $a < 1/\mu$ the limits are zero.   Finally, if $a = 1/\mu$, $1 - \xi$ behaves like a fractional power of s, that is,

$$\xi(s) = 1 - cs^{1/k}, \qquad\qquad\qquad (10\text{-}22)$$

where $k > 1$ and c is a constant, and the limits are also zero.

10-3. The Busy Period (Exponential Case)

The first part of the discussion concentrates on the basic theory of a busy period and the second on computing the distribution.

Let $p_m(t)\, dt$ be the probability that the busy period will terminate in $(t,\ t + dt)$, m customers having been served.   Then the probability of serving m customers in a busy period is

$$p_m = \int_0^\infty p_m(t)\, dt, \qquad\qquad\qquad (10\text{-}23)$$

and
$$p(t)\, dt = \sum_{m=1}^\infty p_m(t)\, dt \qquad\qquad\qquad (10\text{-}24)$$

is the probability that a busy period terminates in $(t,\ t + dt)$, however many customers are served.   For the steady state we must have

$$\int_0^\infty p(t)\, dt = 1; \qquad\qquad\qquad (10\text{-}25)$$

otherwise the integral is <1.

If $c_n(t)$ is the joint probability and density function that an interarrival interval lasts for time t during which time exactly n units complete service and the next unit in line is in service at the end of the interval [conditional upon there being at least $(n + 1)$ units available at the beginning of the interval], then

$$c_n(t) = a(t)b_n(t). \tag{10-26}$$

From its definition, $b_n(t)$ gives the probability of n consecutive service periods in $(0,t)$, the $(n + 1)$st being in progress at t.

Suppose that a customer arrives during $(t, t + dt)$ after the start of the busy period and brings the total number waiting in the queue to $n \geq 1$. Let the mth customer to be served since the beginning of the period be in service. Let $f_{mn}(t)$ be the probability density of this event. We have for $n = 1$

$$f_{m1}(t) = \int_0^t \sum_{k=1}^{m-1} f_{m-k,k}(s)c_k(t - s)\, ds, \qquad m \geq 2, \tag{10-27a}$$

$$f_{11}(t) = c_0(t). \tag{10-27b}$$

The first equation shows that if the previous arrival occurred in $(s, s + ds)$ any customer from the first to the $(m - 1)$st could have been in service. If it was the $(m - k)$th, the previous arrival must have created a queue of size k, the last of which must be in service at t. The arrival at t is then the only one waiting. For $n \geq 2$, similar equations may be written, i.e.,

$$f_{mn}(t) = \int_0^t \sum_{k=0}^{m-1} f_{m-k,n-1+k}(s)c_k(t - s)\, ds. \tag{10-28}$$

Noting that one has a convolution, one applies the Laplace transform with parameter z. Then one introduces a generating function $f_n^*(y,z)$, which itself depends on n, for the Laplace transform of the $f_{mn}(t)$. This results in a difference equation in which, if x^n is substituted for the generating function $f_n^*(y,z)$, one has an equation of the form $x = c^*(xy,z)$, where

$$c^*(y,z) = \sum_{k=0}^{\infty} y^k c_k^*(z) = \int_0^{\infty} a(t) \exp\{-t[z + (1 - y)\mu]\}\, dt$$
$$= a^*[z + (1 - y)\mu]. \tag{10-29}$$

It can be shown that there is one root ξ with $|\xi| < 1$ of the equation in x in terms of which, for $n \geq 1$, $|y| \leq 1$, we have for Re $(z) > 0$, provided that ξ is simple,

$$f_n^*(y,z) = y\xi^n(y,z). \tag{10-30}$$

We now determine the desired distribution $p_m(t)$. Let the generating function of its Laplace transform be

$$P^*(y,z) = \sum_{m=1}^{\infty} y^m p_m^*(z). \tag{10-31}$$

Let
$$A_c(s) = \int_s^{\infty} a(t)\, dt, \tag{10-32}$$

$$d_n(t) = A_c(t)b_n(t), \tag{10-33}$$

$$D^*(y,z) = \sum_{n=1}^{\infty} y^n d_n^*(z) = \int_0^{\infty} A_c(s) \exp\{-[z + (1 - y)\mu]s\}\, ds$$

$$= \frac{1 - c^*(y,z)}{(z/\mu + 1 - y)\mu}, \tag{10-34}$$

where
$$A_c^*(z) = \frac{1 - a^*(z)}{z}. \tag{10-35}$$

We now compute the distribution of the length of a busy period. The probability that the busy period terminates in $t + dt$ after the completion of service of the first customer is the joint probability that no one arrives in $(0,t)$ and the service of the first customer terminates in $(t, t + dt)$. Thus

$$p_1(t) = A_c(t)\mu e^{-\mu t}. \tag{10-36}$$

If, at the arrival before t, the $(m - k)$th customer was in service, the arrival itself must have created a queue of k $(1 \leq k \leq m - 1)$ customers. Everyone including the one in service must be served in the remaining interval, and the last customer was served in $(t, t + dt)$, with no further arrivals. This gives

$$p_m(t) = \mu \int_0^t \sum_{k=1}^{m-1} f_{m-k,k}(s) d_k(t - s)\, ds. \tag{10-37}$$

As before, one may take Laplace transforms and introduce the generating function of the transformed equations. We have

$$P^*(y,z) = y d_0^*(z) + \mu \sum_{k=1}^{\infty} y^k d_k^*(z) f_k^*(y,z)$$

$$= \mu y D^*(\xi y,z) = \frac{y(1 - \xi)}{z/\mu + 1 - \xi y}. \tag{10-38}$$

Note that $P^*(y,0)$ is the generating function of p_m and $P^*(1,0)$ is the probability of termination in finite time.

The average number of customers served in a busy period is

$$N = \frac{\partial P^*(y,0)}{\partial y}\bigg|_{y=1}.$$

If $y = 1$,

$$P^*(1,z) = \frac{1 - \xi}{z/\mu + 1 - \xi}, \tag{10-39}$$

where ξ is the smallest root of

$$x = c^*(x,z) = a^*[z + (1 - x)\mu]. \tag{10-40}$$

If we can write

$$\xi(z) = \sum_{n=0}^{\infty} k_n z^n, \tag{10-41}$$

where the k_n are obtained from $\xi(z)$ by differentiation, then

$$P^*(1,z) = \frac{1 - k_0 - k_1 z - O(z)}{z/\mu + 1 - k_0 - k_1 z - O(z)}. \tag{10-42}$$

When $\rho = \lambda/\mu < 1$, where λ is the arrival rate, $k_0 < 1$ and $P(1,z) \to 1$ as $z \to 0$. When $\rho > 1$, $k_0 = 1$ and

$$P^*(1,z) \to \frac{k_1}{(k_1 - 1/\mu)} = \frac{1}{\rho} \qquad \text{as } z \to 0.$$

When $\rho = 1$, $\xi = 1$ is a double root for $z = 0$ and $P^*(1,z) \to 1$.

Using the definitions, one can similarly show that

$$N = \begin{cases} \dfrac{1}{1 - h_0}, & \rho < 1, \\[2ex] \dfrac{(h_2 - h_1)\rho}{h_1{}^2}, & \rho > 1, \end{cases} \tag{10-43}$$

where h_0, h_1, and h_2 are the coefficients in the expansion of ξ, the smallest root of $x = a^*[(1 - xy)\mu]$ in terms of $(1 - y)^n$.

For constant input with interarrival interval of length a, we have

$$a^*(z) = e^{-az}, \tag{10-44}$$

$$\xi = \exp\left(-\frac{1 - \xi y}{\rho}\right) = \frac{\rho}{y} \sum_{n=1}^{\infty} \frac{(ny/e^{1/\rho})^n}{n(n!)}, \tag{10-45}$$

and Conolly gives for all values of ρ

$$p_1 = 1 - e^{-1/\rho},$$
$$p_2 = e^{-1/\rho} - e^{-2/\rho}\left(1 + \frac{1}{\rho}\right), \tag{10-46}$$
$$\cdots \cdots \cdots \cdots \cdots \cdots$$
$$h_1 = -\frac{1}{\rho - 1}, \qquad h_2 = \frac{3\rho - 2}{2(\rho - 1)^3},$$

etc.; he also gives a table of values of the first five p_m and of N for different values of ρ both for constant and for exponential inputs.

10-4. The Busy Period (Erlangian Case)

This case has also been studied by Conolly in a paper whose principal interest is the computation of the joint probability and joint density that during a busy period exactly m groups each of size k (the Erlangian parameter) receive service and the whole process occupies time t. The marginal distributions of m (obtained by integrating t out of the joint density) and t (obtained by summing m out) are also discussed; the ideas are applied to known special cases already discussed in this book. The auxiliary functions that he uses for this case include added information on the member of a group and on the group that is undergoing service at the time considered.

If we let $p_m(t)$ be this joint probability, take its Laplace transform with parameter z with respect to t, and then form the generating function of the Laplace transforms $p_m^*(z)$, using a variable y raised to the power mk for the general term (note that m is the group number and multiplying it by k reduces it to single customers), we have for the generating function

$$\frac{y \sum_j (1 - \xi_j^k)}{(bz + 1 - y\xi_j)H'(\xi_j)}. \tag{10-47}$$

We give this only to arouse the reader's interest in the result. It is best to leave it to him to read the paper to obtain the undefined quantities. In any case, he has had experience with a treatment of a similar case.

PART 4

QUEUEING RAMIFICATIONS, APPLICATIONS, AND RENEWAL THEORY

CHAPTERS 11 TO 15

In this part we have five chapters which cover a wide variety of ideas. Chapter 11 studies queueing situations with different queue disciplines, particularly because the preceding parts concentrated on queues with first-come–first-served queue discipline. Chapter 12 attempts to cover queueing networks but considers elementary forms of connection. Chapter 13 examines various queueing situations described in Chap. 1. Chapter 14 contains a number of important applications of the theory to fields of operational endeavor. Many queueing processes may be considered as renewal processes, and the latter theory has received sufficient attention in the past several years to yield interesting results. Some of these are included in Chap. 15 because of increased application value.

CHAPTER 11

OTHER QUEUE DISCIPLINES

11-1. Introduction

We have so far in our study considered queues in which units obtained service on a first-come–first-served basis. Obviously, this is not the only possible type of queue discipline. For example, units may also be served on a priority basis. Thus, if the priority of a unit is high, regardless of its position in the line, it enters service before lower priorities. An illustration of this is the telegraph system in which urgent messages go first or an emergency room in a hospital where high priority is given to cases requiring immediate attention.

On the other hand, there are systems that are indifferent to the order of arrival of a unit and select units for service at random. An example of this case is a primitive-type telephone switchboard in which the operator selects a call at random from the many indicated on the board and gives it the desired connection.

A fourth type of queue discipline is one in which arriving units are allocated to specific channels. Yet another is a last-come–first-served queue discipline, as when arriving units (e.g., manufacturing items) are stacked and the last item on top is used first.

In each of the foregoing cases, our main desire is to compute waiting-time distributions and often the expected waiting time. The latter may be more readily obtained without first deriving the waiting-time distribution of which it is the expected value.

11-2. Priorities

An important subject in waiting lines is that of priorities. Units with different priorities arrive as input of the same or different arrival distributions, wait to be served on a first-come–first-served basis within each priority, and are served by one or more channels. A low-priority unit may (preemptive service) or may not (nonpreemptive service) be ejected back into the line when a higher-priority item arrives into the system.

The problem of priority gives rise to important cost questions which should be incorporated into the model but so far rarely have been.

231

In priority problems, cost considerations may be used to compare the loss due to nonpreempting the lower-priority unit already in service with the total or partial loss of service due to ejecting the lower priority out of service. Returning to service by this priority could mean recommencing the service operation from the beginning. One may wish to minimize the total cost of the waiting time, the cost of idle time for the service facility, and the cost of interrupted service.

1. Nonpreemptive Priorities

a. Single Channel [116]. Let rth-priority items arrive before a single channel (the smaller the number, the higher the priority) by different Poisson distributions. Let the means of these distributions be λ_k $(k = 1, \ldots, r)$ (and consequently the combined input is Poisson with parameter $\lambda = \sum_{k=1}^{r} \lambda_k$), and assume that the units wait on a first-come–first-served basis *within the priority*. They are then served by a combined cumulative-service-time distribution $F(t) = (1/\lambda) \sum_{k=1}^{r} \lambda_k F_k(t)$, where $F_k(t)$ is an arbitrary cumulative-service-time distribution with mean service time $1/\mu_k$ for units of priority k. The highest priority has index unity, etc., to the lowest priority, which has index r. A higher priority obtains service before a lower priority, no matter when it arrives. However, whatever the priority of the item in service, it completes its service before another item is admitted, i.e., nonpreemptive service.

Let

$$\rho_k = \frac{\lambda_k}{\mu_k}, \qquad\qquad 1 \leqq k \leqq r, \qquad\qquad (11\text{-}1)$$

and

$$\sigma_k = \sum_{i=1}^{k} \rho_i, \qquad \sigma_0 = 0, \qquad 1 \leqq k \leqq r. \qquad\qquad (11\text{-}2)$$

Suppose that an item of priority p enters the system at time t_0 and enters service at time t_1. Its waiting time then is $T = t_1 - t_0$. At t_0 let there be one or no item in service and let there be n_k $(k = 1, \ldots, p)$ items of kth-priority class in the line ahead of the arriving item. Let T_0 be the time required to finish the item already in service, and let T_k be the time required to serve n_k $(k = 1, \ldots, p)$. During the waiting time T, n_k' $(k = 1, \ldots, p - 1)$ items of type k will enter the line and go to service ahead of this item. Let T_k' be the service time of n_k' (i.e., all items of priority k which arrive during time T). Note that

$$T \geqq \sum_{k=1}^{p-1} T_k' + \sum_{k=1}^{p} T_k + T_0.$$

In fact, the equality must hold; otherwise the channel would be idle when the item is waiting in line.   Taking expected values and using the equality sign, we have

$$E(T) = \sum_{k=1}^{p-1} E(T'_k) + \sum_{k=1}^{p} E(T_k) + E(T_0). \tag{11-3}$$

By definition, $E(T) \equiv W_p$, the waiting time of the pth-type unit.

$E(T_k)$ is the expected value of the product of the service-time variable for priority k and n_k.   Since the two are independent, $E(T_k)$ is the product of their expected values, i.e.,

$$E(T_k) = \frac{1}{\mu_k} \lambda_k W_k. \tag{11-4}$$

Here $1/\mu_k$ is the kth-priority service-time expectation, and $\lambda_k W_k$ is the expected value of n_k. $E(n_k) = \lambda_k W_k = (1/\lambda_k)^{-1} W_k$ is the expected number of arrivals multiplied by the expected waiting time for priority k.

Similarly,

$$E(T'_k) = \frac{1}{\mu_k} \lambda_k W_p. \tag{11-5}$$

Here we use W_p, since the expected value of n'_k is the product of the average number of arrivals and the average time in which these arrivals occurred, i.e., the time T during which the item of priority p under consideration must wait.   We then have

$$W_p = \frac{\displaystyle\sum_{k=1}^{p} \rho_k W_k + E(T_0)}{1 - \sigma_{p-1}}. \tag{11-6}$$

By induction on p we have

$$W_p = \frac{E(T_0)}{(1 - \sigma_{p-1})(1 - \sigma_p)}. \tag{11-7}$$

For this to hold we require $\sigma_p < 1$.

Now from the Pollaczek-Khintchine formula, we have

$$W = \frac{L - \lambda/\mu}{\lambda} = \frac{\lambda \int_0^\infty t^2 \, dF(t)}{(2/\mu)(\mu - \lambda)}. \tag{11-8}$$

If we put $p = 1$, we have the classical case of a single-queue channel without priority but with first-come–first-served discipline.   This gives

$$W = W_1 = \frac{E(T_0)}{1 - \rho}, \tag{11-9}$$

which must coincide with the above, and consequently we have determined the constant $E(T_0)$. It is given by

$$E(T_0) = \frac{\lambda}{2} \int_0^\infty t^2 \, dF(t). \tag{11-10}$$

Thus we have obtained expressions for different-priority waiting times.

It is also possible to derive $E(T_0)$ by a feasibility argument and show that the single-channel expected-waiting-time result is obtained by putting $p = 1$. But the foregoing approach is more rigorous.

b. Multiple Channels. Here we assume the same type of inputs as above and that the service times' random variables at each of a total of c channels are identically exponentially distributed with parameter μ. The argument proceeds as before, except that when all channels are full, units are discharged at random at the average rate $c\mu$. We have

$$W_p = \frac{E(T_0)}{\left[1 - (1/c\mu) \sum_{k=1}^{p-1} \lambda_k\right]\left[1 - (1/c\mu) \sum_{k=1}^{p} \lambda_k\right]}. \tag{11-11}$$

Again $E(T_0)$ may be derived from the results of the single-priority multiple-channel steady-state average-waiting-time result. It is given by

$$\frac{(c\rho)^c/c\mu}{c!(1 - \rho) \sum_{j=0}^{c-1} [(c\rho)^j/j!] + \sum_{j=c}^{\infty} (c^c\rho^j/c!)}, \tag{11-12}$$

where $\rho = \lambda/c\mu$. We require that $(1/c\mu) \sum_{k=1}^{p} \lambda_k < 1$. Then W_p holds even if $\rho > 1$.

If a priority system operates under its capacity, the difference in the average waiting times for different priorities is small but becomes quite marked as the load on the system is increased. In the single-channel case, increase in arrival of the highest priority, for example, increases the delay for that priority and for all other priorities. It is therefore useful to control the assignment of high priority and to speed service times when possible.

c. Generalization of Single Channel. Kesten and Runnenburg [373] have developed the ideas further. Among their several results we give the following. Let

$$G_k(t) = \begin{cases} 0 & \text{for } t < 0, \\ 1 - e^{-\lambda_k t} & \text{for } t \geq 0 \text{ with } \lambda_k > 0, \end{cases} \tag{11-13}$$

be the cumulative distribution of arrival times for the kth priority in a set of r priorities, and let $F_k(t)$ be the cumulative-holding-time distribution for that priority in a single-channel queueing operation. Assume independence among all arrival intervals and among arrivals in each priority and among service times. Let

$$\lambda = \sum_{i=1}^{r} \lambda_i, \qquad \mu_k^{(l)} = \int_{0-}^{\infty} t^l \, dF_k(t), \qquad (11\text{-}14)$$

$$\varphi_k(\alpha) = \int_{0-}^{\infty} e^{-\alpha t} \, dF_k(t), \qquad \psi_k(\alpha) = \int_{0-}^{\infty} e^{-\alpha t} \, dH_k'(t), \qquad (11\text{-}15)$$

where $H_k(t)$ is the cumulative-waiting-time distribution function for the kth priority. Note that, if $\psi_k(\alpha)$ is determined, by taking inverse transforms, $H_k(t)$ will also be determined.

In the nonsaturated case

$$\sum_{i=1}^{r} \lambda_i \mu_i^{(1)} < 1,$$

$$\psi_k(\alpha) = \frac{\left[1 - \sum_{1}^{r} \lambda_i \mu_i^{(1)}\right]\left(-\sum_{1}^{k-1} \lambda_i + z_k^* - \alpha\right) - \sum_{k-1}^{r} \lambda_i \left[1 - \varphi_i\left(\sum_{1}^{k-1} \lambda_j - z_k^* + \alpha\right)\right]}{\lambda_k - \alpha - \lambda_k \varphi_k\left(\sum_{1}^{k-1} \lambda_j - z_k^* + \alpha\right)}$$

$$(11\text{-}16)$$

where $z_1^* = 0$ and $z_k^* = z_k^*(\alpha)$ satisfies

$$z_k^* = \sum_{1}^{k-1} \lambda_i \varphi_i\left(\sum_{1}^{k-1} \lambda_j - z_k^* + \alpha\right) = 0, \qquad k \geq 2. \qquad (11\text{-}17)$$

Thus, for example,

$$\psi_1(\alpha) = -\frac{\left[1 - \sum_{1}^{r} \lambda_i \mu_i^{(1)}\right]\alpha + \sum_{0}^{r} \lambda_i \left[1 - \varphi_i(\alpha)\right]}{\lambda_1 - \alpha - \lambda_1 \varphi_1(\alpha)}. \qquad (11\text{-}18)$$

The expected waiting time is given by

$$W_k \equiv E(w_k) = \frac{\sum_{1}^{r} \lambda_i \mu_i^{(2)}}{2\left[1 - \sum_{1}^{k-1} \lambda_i \mu_i^{(1)}\right]\left[1 - \sum_{1}^{k} \lambda_i \mu_i^{(1)}\right]}, \qquad (11\text{-}19)$$

and the second moment is given by

$$E(w_k^2) = \frac{\sum_1^r \lambda_i \mu_i^{(3)}}{3 \left[1 - \sum_1^{k-1} \lambda_i \mu_i^{(1)} \right]^2 \left[1 - \sum_1^k \lambda_i \mu_i^{(1)} \right]}$$

$$+ \frac{\sum_1^r \lambda_i \mu_i^{(2)} \sum_1^{k-1} \lambda_j \mu_j^{(2)}}{2 \left[1 - \sum_1^{k-1} \lambda_i \mu_i^{(1)} \right]^2 \left[1 - \sum_1^k \lambda_i \mu_i^{(1)} \right]^2}$$

$$+ \frac{\sum_1^r \lambda_j \mu_i^{(2)} \sum_1^{k-1} \lambda_j \mu_j^{(2)}}{2 \left[1 - \sum_1^{k-1} \lambda_i \mu_i^{(1)} \right]^3 \left[1 - \sum_1^k \lambda_i \mu_i^{(1)} \right]},$$

$$k = 1, \ldots, r. \quad (11\text{-}20)$$

In the saturation case, $\sum_1^r \lambda_i \mu_i^{(1)} \geq 1$, and one can find an integer s ($0 \leq s < r$) such that $\sum_{i=1}^s \lambda_i \mu_i^{(1)} < 1$, $\sum_1^{s+1} \lambda_i \mu_i^{(1)} \geq 1$,

$$W_k \equiv E(w_k) = \frac{\sum_{i=1}^s \lambda_i \mu_i^{(2)} + (\mu_{s+1}^{(2)}/\mu_{s+1}^{(2)}) \left[1 - \sum_1^s \lambda_i \mu_i^{(1)} \right]}{2 \left[1 - \sum_1^{k-1} \lambda_i \mu_i^{(1)} \right] \left[1 - \sum_1^k \lambda_i \mu_i^{(1)} \right]},$$

$$k = 1, 2, \ldots, s, \quad (11\text{-}21)$$

$$W_k \equiv E(w_k) = \infty, \quad k = s + 1, \ldots, r.$$

d. A Continuous Number of Priorities.

Phipps [556] has extended the results of a single-channel priority queueing system to machine break-downs. The number of machines available is assumed to be infinite. Priorities are assigned according to the length of time needed to service a machine—higher priorities to shorter jobs. Since the length of service time may correspond to any real number, so do the priorities, i.e., a continuum number of priorities.

Suppose that new repair jobs are generated by a Poisson law with an average of λ jobs per unit time. Also, suppose that the arrival rate of jobs of priority t (i.e., t is the amount of time required to service these units) is given by λ_t. Let $F(t)$ be the average cumulative-job-repair-time distribution for units of all priorities. Then

$$\lambda_t \, dt = \lambda \, dF(t). \quad (11\text{-}22)$$

The expected waiting time W_t of a job of duration t by analogy with the discrete priority case is

$$W_t = \frac{W_0}{\left[1 - \lambda \int_0^t s\, dF(s)\right]^2}, \qquad t \leqq \tau, \qquad (11\text{-}23)$$

$$W_t = \infty, \qquad\qquad\qquad t > \tau,$$

where

$$W_0 = \frac{\lambda}{2} \int_0^\tau t^2\, dF(t) \qquad\qquad (11\text{-}24)$$

and τ, the critical job duration, which is analogous to s in (11-21), is obtained from the equation

$$\lambda \int_0^\tau t\, dF(t) = 1. \qquad\qquad (11\text{-}25)$$

If a job has duration not less than τ, saturation is obtained and hence an infinite length queue.

The expected number of jobs in the waiting line L_q is given by

$$L_q = \lambda W_0 \int_0^\tau \frac{dF(t)}{\left[1 - \lambda \int_0^t s\, dF(s)\right]^2}. \qquad (11\text{-}26)$$

If, for example, $F(t) = 1 - e^{-\mu t}$, corresponding to exponential repair times, then for $\lambda < \mu$, τ is infinite and we have

$$W_0 = \frac{\lambda}{\mu^2}, \qquad \lambda < \mu \qquad \text{(unsaturated case)}. \qquad (11\text{-}27)$$

The expected waiting time is

$$W_t = \frac{W_0}{\{1 - \lambda/\mu[1 - e^{-\mu t}(1 + \mu t)]\}^2}, \qquad (11\text{-}28)$$

and the expected number of jobs waiting is

$$L_q = \frac{\lambda^2}{\mu} \int_0^\infty \frac{e^{-\mu t}\, dt}{\{1 - \lambda/\mu[1 - e^{-\mu t}(1 + \mu t)]\}^2}. \qquad (11\text{-}29)$$

L_q is infinite in the saturated case.

2. Preemptive Priorities

In preemptive-priority discipline the arrival of a higher priority while a lower priority is in service requires the return to the queue of the lower priority. There are two cases of preemptive priorities:

1. Repeat rule: The ejected item returns to service, having lost all service performed before ejection (this is the case we treat here).

2. Preemptive resume rule: The item resumes at the point of service where it was interrupted.

The resume-rule problem for two priorities with Poisson arrivals and general service distribution has been investigated by Jaiswal. He used the procedure of Prob. 5 in Chap. 9, representing the service-time distributions as indicated there and using their means in a set of differential-difference equations along with the arrival means to represent the probability of given numbers of each priority in line, given that the item at the head of the lower-priority queue was preempted after receiving service of length y and that the service time of the item in service is of length x. One must also consider expressions of these probabilities if no lower-priority item was preempted. Generating-function methods and Laplace transforms are used as in the procedure to be demonstrated below for the repeat-rule case.

We now study the case of preemptive priorities with the repeat rule; for each priority the arrivals are Poisson with ordered-service discipline within a priority, and the service at a single channel is exponential for each priority. When a higher priority arrives, a lower-priority unit in service is returned to the front of the line of its priority to await service.

We give below the differential-difference equations for three preemptive priorities. To obtain the result for two priorities it suffices to put λ_3 and μ_3 equal to zero and ignore the third subscript and redundant equations.

Let there be:

n units of highest priority with arrival rate λ_1 and service rate μ_1
m units of middle priority with arrival rate λ_2 and service rate μ_2
k units of lowest priority with arrival rate λ_3 and service rate μ_3

Let $P_{nmk}(t)$ be the probability that there are n units of the first priority, m of the second, and k of the third in the system at time t. We have

$$
\begin{aligned}
P'_{nmk}(t) &= -(\lambda_1 + \lambda_2 + \lambda_3 + \mu_1)P_{nmk}(t) \\
&\quad + \lambda_1 P_{n-1,mk}(t) + \lambda_2 P_{n,m-1,k}(t) + \lambda_3 P_{nm,k-1}(t) + \mu_1 P_{n+1,mk}(t), \\
P'_{000}(t) &= -(\lambda_1 + \lambda_2 + \lambda_3)P_{000}(t) + \mu_1 P_{100}(t) \\
&\qquad\qquad\qquad\qquad\quad + \mu_2 P_{010}(t) + \mu_3 P_{001}(t), \\
P'_{n00}(t) &= -(\lambda_1 + \lambda_2 + \lambda_3 + \mu_1)P_{n00}(t) + \lambda_1 P_{n-1,00}(t) \\
&\qquad\qquad\qquad\qquad\quad + \mu_1 P_{n+1,00}(t), \\
P'_{0m0}(t) &= -(\lambda_1 + \lambda_2 + \lambda_3 + \mu_2)P_{0m0}(t) + \mu_1 P_{1m0}(t) \\
&\qquad\qquad\qquad + \lambda_2 P_{0,m-1,0}(t) + \mu_2 P_{0,m+1,0}(t), \\
P'_{00k}(t) &= -(\lambda_1 + \lambda_2 + \lambda_3 + \mu_3)P_{00k}(t) + \mu_1 P_{10k}(t) \\
&\qquad + \mu_2 P_{01k}(t) + \lambda_3 P_{00,k-1}(t) + \mu_3 P_{00,k+1}(t), \\
P'_{nm0}(t) &= -(\lambda_1 + \lambda_2 + \lambda_3 + \mu_1)P_{nm0}(t) + \lambda_1 P_{n-1,m0}(t) \\
&\qquad\qquad\qquad + \lambda_2 P_{n,m-1,0}(t) + \mu_1 P_{n+1,m0}(t), \\
P'_{n0k}(t) &= -(\lambda_1 + \lambda_2 + \lambda_3 + \mu_1)P_{n0k}(t) + \lambda_1 P_{n-1,0k}(t) \\
&\qquad\qquad\qquad + \lambda_3 P_{n0,k-1}(t) + \mu_1 P_{n+1,0k}(t),
\end{aligned}
\tag{11-30}
$$

$$P'_{0mk}(t) = -(\lambda_1 + \lambda_2 + \lambda_3 + \mu_2)P_{0mk}(t) + \mu_1 P_{1mk}(t)$$
$$+ \lambda_2 P_{0,m-1,k}(t) + \mu_2 P_{0,m+1,k}(t) + \lambda_3 P_{0m,k-1}(t).$$

In general, one has 2^p equations describing p preemptive priorities.

Note that as long as there are high-priority units waiting, no lower priorities can go into service.

Exercise: Write the two priority-case equations in n and m. They are needed below.

Heathcote [293] has studied this problem for two priorities with an initial number $m_0 = h$ of the second priority at $t = 0$ and $n_0 = 0$ of the first priority. On taking the Laplace transform $g_{nm}(s)$ of the four equations describing the two priority problems involving the probabilities $P_{nm}(t)$, one introduces the generating function defined for $|z| < 1, |x| < 1$:

$$F(z,x,s) = \sum_{n=0}^{\infty} \sum_{m=0}^{\infty} g_{nm}(s) z^n x^m. \qquad (11\text{-}31)$$

The partial generating function

$$H_n(x,s) = \sum_{m=0}^{\infty} g_{nm}(s) x^m \qquad (11\text{-}32)$$

is also used. One multiplies by x^m and sums over m to obtain

$$\mu_1 H_1 - \left(s + \mu_2 + \lambda_1 + \lambda_2 - \lambda_2 x - \frac{\mu_2}{x}\right) H_0$$
$$= g_{00}\mu_2\left(\frac{1}{x} - 1\right) - x^h, \qquad (11\text{-}33)$$

$$\mu_1 H_{n+1} - (s + \mu_1 + \lambda_1 + \lambda_2 - \lambda_2 x) H_n + \lambda_1 H_{n-1} = 0.$$

Then multiplying by z^n and summing over n give

$$F(z,x,s)$$
$$= \frac{g_{00}\mu_2[(1/x) - 1] - x^h + \{\mu_1[(1/z) - 1] - \mu_2[(1/x) - 1]\}H_0(x,s)}{\lambda_1 z - (s + \mu_1 + \lambda_1 + \lambda_2 - \lambda_2 x) + \mu_1/z},$$
$$(11\text{-}34)$$

where $H_0(x,s)$ is obtained by solving the system of equations in H_n. This gives

$$H_0(x,s) = \frac{g_{00}\mu_2[(1/x) - 1] - x^h}{\mu_1 v_1 - (s + \mu_2 + \lambda_1 + \lambda_2 - \lambda_2 x - \mu_2/x)}. \qquad (11\text{-}35)$$

If $v_1(x,s)$ is the reciprocal of the zero of the denominator of $F(z,x,s)$ which is outside the unit circle $|z| = 1$, then we may write

$$F(z,x,s) = \frac{[\mu_2(1 - x)g_{00}(s) - x^{h+1}][1 - v_1(x,s)]}{\{\mu_2(1 - x)[1 - \rho_2 x - v_1(x,s)] - sx\}[1 - zv_1(x,s)]}, \qquad (11\text{-}36)$$

where $\rho_i = \lambda_i/\mu_i$, $i = 1, 2$. Note that $v_2(x,s)$ is the reciprocal of the other zero. Also $s + \mu_1 + \lambda_1 + \lambda_2 - \lambda_2 x = \mu_1(v_1 + v_2)$, $\rho_1 = v_1 v_2$. Now one can show that, in the limit $\rho_1 + \rho_2 < 1$,

$$\lim_{t \to \infty} P_{00}(t) = \lim_{s \to 0} s g_{00}(s) = 1 - \rho_1 - \rho_2.$$

We also have for the generating function of the steady-state probabilities

$$\lim_{s \to 0} s F(z,x,s) = \frac{(1 - \rho_1 - \rho_2)[1 - v_1(x,0)]}{[1 - \rho_2 x - v_1(x,0)][1 - z v_1(x,0)]}. \quad (11\text{-}37)$$

$g_{00}(s)$ is determined from $F(z,x,s)$, this time by finding the zero ζ that lies within the unit circle $|x| = 1$. In this case one requires that the first factor of the denominator vanish. The numerator must also vanish at ζ. This gives $g_{00}(s) = \zeta^{h+1}/\mu_2(1 - \zeta)$. Since we have determined the generating function, we can obtain the probabilities by differentiation. If $\mu_1 = \mu_2 \equiv \mu$, then still working within $|z| < 1$, $|x| < 1$, we have $\zeta = \alpha_2$ encountered in the case of a single-channel queue except that λ is replaced by $\lambda_1 + \lambda_2$.

$$F(z,x,s) = \frac{\zeta^{h+1}(1 - x)(1 - \zeta)^{-1} - x^{h+1}}{\mu[1 - z v_1(x,s)][1 - x v_2(x,s)]}. \quad (11\text{-}38)$$

On inverting $\zeta^{h+1}/\mu(1 - \zeta)$, we have

$$P_{00}(t) = \frac{e^{-t(\mu+\lambda_1+\lambda_2)}}{\mu t} \sum_{j=0}^{\infty} \left(\sqrt{\frac{\mu}{\lambda_1 + \lambda_2}} \right)^{j+h+1}$$
$$(j + h + 1) I_{j+h+1}[2t \sqrt{\mu(\lambda_1 + \lambda_2)}]. \quad (11\text{-}39)$$

The Laplace transform of the expected value $E(m|h,t)$ of the lower priority with initial number h is given by

$$\frac{\partial F(z,x,s)}{\partial x} \bigg|_{\substack{z=1 \\ x=1}}$$

from which the inverse transform gives

$$E(m|h,t) = h + (\lambda_2 - \mu)t + \mu \int_0^t P_{00}(u) \, du$$
$$+ \sqrt{\lambda_1 \mu} \int_0^t (t - u) \frac{e^{-u(\mu+\lambda_1)}}{u} I_1(2u \sqrt{\lambda_1 \mu}) \, du. \quad (11\text{-}40)$$

When $\lambda_1 = 0$, the results reduce to those of the single-service queue with parameters λ_2 and μ_2.

The distribution of the duration of a busy period for the nonpriority queue is computed from the original system of equations except that there is an absorbing barrier at the origin so that the process stops on first reaching this point.   We use $Q_{nm}(t)$ instead of $P_{nm}(t)$.   The space of the random walk associated with the problem is bounded by reflecting barriers along $n = 0$, $m = 1$, and the point $(0,0)$.   By taking the Laplace transform as above, one is able to determine the desired distribution. Now $Q_{00}^*(s) = \zeta^h/s$, where ζ is the zero of $F(z,x,s)$ when $\mu_1 \neq \mu_2$.

Hence the desired distribution dQ_{00}/dt is obtained by computing the inverse transform of ζ^h.   Its mean and variance are, respectively,

$$E(t|h) = \frac{h}{\mu_2(1 - \rho_1 - \rho_2)}, \tag{11-41}$$

$$V(t|h) = \frac{h[1 + \rho_2 - \rho_1(1 - 2\mu_2/\mu_1)]}{\mu_2^2(1 - \rho_1 - \rho_2)^3}. \tag{11-42}$$

When $\mu_1 = \mu_2 = \mu$,

$$\frac{dQ_{00}}{dt} = h \left(\sqrt{\frac{\mu}{\lambda_1 + \lambda_2}}\right)^h \frac{e^{-t(\mu+\lambda_1+\lambda_2)}}{t} I_h[2t \sqrt{\mu(\lambda_1 + \lambda_2)}]. \tag{11-43}$$

For several preemptive priorities whose total number is r, the kth priority arriving by a Poisson process with parameter λ_k and obtaining exponential service with parameter μ, the mean number of priority k waiting in the system in the steady state, that is, $\sum_{k=1}^{r} \lambda_k/\mu < 1$, is given by [807]

$$\frac{\rho_k}{\left(1 - \sum_{n=1}^{k-1} \rho_n\right) \left(1 - \sum_{n=1}^{k} \rho_n\right)}.$$

The variance is

$$\frac{2 \left(\sum_{n=1}^{k-1} \rho_n\right) \rho_k^3}{\left(1 - \sum_{n=1}^{k-1} \rho_n\right) \left(1 - \sum_{n=1}^{k} \rho_n\right)} + \frac{\rho_k^2 + \rho_k \left(1 - \sum_{n=1}^{k-1} \rho_n\right) \left(1 - \sum_{n=1}^{k} \rho_n\right)}{\left(1 - \sum_{n=1}^{k-1} \rho_n\right)^2 \left(1 - \sum_{n=1}^{k} \rho_n\right)},$$

where $\rho_k = \lambda_k/\mu$.   The general case has also been studied by Heathcote. H. White and L. Christie [758] and F. Stephan [674] have also studied this problem, mostly in equilibrium.   The latter gives some computations for the problem, having considered the mean waiting time and other moments for the lower priority.   He also considered the time in the system and the waiting time, including service interruptions.

3. *Dynamic Priorities*

J. R. Jackson has studied dynamic priorities for a single-channel queue in discrete time.   An entering unit is assigned an urgency-class number according to a given probability distribution. A newcomer takes precedence over a unit in the queue if, and only if, the difference between the former's class number and that of the latter is not less than the time the latter has spent waiting. This is an example in which management becomes concerned about a unit which has waited long. This reduces to the usual priority-study "state priority" as the above probability distribution is spread out, and to the first-come–first-served case if it shrinks to a single value.

Denote by p the probability of an arrival in a cycle in which the problem is studied, q the probability of completing service of a unit when the channel is occupied, π_n the probability that an entering unit receives an urgency number $\leq n$, $\rho = p/q$ the utilization factor, $\rho_n = \rho\pi_n$ the utilization factor by units with urgency numbers $\leq n$,

$$Z = \frac{1-q}{1-p}, \qquad A = \rho Z.$$

In a cycle let s_k be the steady-state probability of n units in the system just after a unit has been discharged (a low point) and just before a new item enters, if this occurs. In the usual manner, we have

$$p(1-q)s_0 = q(1-p)s_1,$$
$$[p(1-q) + q(1-p)]s_k = p(1-q)s_{k-1} + q(1-p)s_{k+1},$$
$$k = 1, 2, \ldots , \quad (11\text{-}44)$$

with solution
$$s_k = (1-A)A^k. \qquad (11\text{-}45)$$

If $x_i(t)$ is the equilibrium probability that i units remain in the system after receiving t units of service time, by units in the system at a given low point, then

$$x_0(t) = x_0(t-1) + qx_1(t-1),$$
$$x_i(t) = (1-q)x_i(t-1) + qx_{i+1}(t-1), \qquad i = 1, 2, \ldots , \quad (11\text{-}46)$$

and the solution is given by

$$x_i(t) = (1-A)A^i Z^t, \qquad t = 0, 1, \ldots , \qquad i = 1, 2, \ldots ,$$
$$x_0(t) = 1 - \sum_{i=1}^{\infty} x_i(t) = 1 - \rho Z^{t+1}. \qquad (11\text{-}47)$$

Let W be the over-all equilibrium mean waiting time for units of all urgency classes together. It equals the mean total service time for units in the system at the point of occurrence of arrivals in the cycle.

In the first-come–first-served case we have the usual expression for waiting no more than time t,

$$P(\leq t) = 1 - \rho Z^{t+1}, \qquad (11\text{-}48)$$

and in the static-priority case where W_j is the equilibrium mean waiting time for customers with urgency number j,

$$\frac{W_j}{W} = \frac{1 - \rho}{(1 - \rho_{j-1})(1 - \rho_j)},$$ (11-49)

established by analogy with Cobham's derivation but for discrete time.

Jackson gives the following bounds for W'_j, the equilibrium mean waiting time of units with urgency number j in the dynamic priority case:

$$\frac{W'_j}{W} \geq \frac{1 - \rho}{1 - \rho_j},$$

$$\frac{W'_j}{W} \leq \frac{1}{1 - \rho_{j-1}},$$

$$\frac{W'_j}{W} \leq 1 + (1 - Z) \sum_{i<j} \rho_i,$$ (11-50)

and

$$\frac{1 - \rho}{1 - \rho_j} \leq \frac{W_j}{W'_j} \leq \frac{1}{1 - \rho_{j-1}}.$$

The first inequality, for example, is proved by replacing the system by an identical one, except that all urgency classes with a number $<j$ are increased to j, and then replacing this system by a static-priority system with the same p, q, and urgency-number distribution.

Jackson presents a procedure for calculating successive waiting-time probabilities in the dynamic-priority case.

11-3. Random Selection for Service

1. *Exponential Service Times, Multiple Channels*

Mellor was the first to study the problem of random selection for service. However, it was left to Vaulot to give an exact formulation of the problem. Pollaczek provided asymptotic formulas based on Vaulot's works. Palm, Pollaczek, and Riordan each attempted solution of Vaulot's problem. In most cases the solution is given as an adequate approximation. Our object is to find the waiting-time distribution. We do this by introducing auxiliary concepts.

Here we assume Poisson input with parameter λ and the same exponential-holding-time distribution with parameter μ for multiple channels c. A customer is assumed to remain until he completes service. The waiting items are selected at random for service in one of the channels. Consequently the probability for a waiting item to obtain service depends on the number of items waiting at the same time and is independent of how long the item has already waited. It turns out that the longer one waits, the less will be the probability of being served within another interval t. This paradox shows that one might leave and reenter the system, improving the probability for obtaining service. Its explanation is that the longer the wait is, the more likely it is that one has arrived in a queue with a large number of items waiting. Regarding reentry into the system, Palm points out that the distribution already computed for the waiting time for calling at random during congestion would not apply to computing the additional waiting time t. In the case of an ordered queue (the waiting time is independent of the number of items which arrive after the given item) with the same assumptions as above, the waiting-time distribution is exponential, and the remaining waiting time does not increase with increased lapse in waiting time but remains the same. Thus when the conditions are the same, it is generally more convenient for a customer if an ordered queue is used; this idea is contested for cost and other reasons, such as relative improvement in waiting time.

We denote by $f_n(t)$ the probability that, among n items found waiting in the system at an arbitrary time, a given item will subsequently wait greater than t. We may therefore assume that the arbitrary time point is at the origin. Although the item considered waits for time t, starting this wait with $n - 1$ other items in the system, arrivals and departures can occur during this period. Consequently he may continue to wait with a greater or a smaller number of items, and the probability that he is waiting at time t will depend on the number of items present. Because of the Markovian property of the process, i.e., the semigroup property $P(t + s) = P(t)P(s)$, $f_n(t)$ may be determined by studying what happens during the interval $(0, \Delta t)$ followed by what happens during $(\Delta t, t - \Delta t)$.

1. During Δt an arrival may occur with probability $\lambda \, \Delta t$, and then the given item starts its wait for the period $t - \Delta t$ with n others; i.e., there are a total of $n + 1$ items waiting. This occurs with probability $f_{n+1}(t - \Delta t)$, and $\lambda \, \Delta t f_{n+1}(t - \Delta t)$ is the probability that both events will happen.

2. One of the other $n - 1$ items waiting obtains service during Δt. The probability that any item obtains service during Δt is $\mu c \, \Delta t$, and the probability that it is other than the given item is $(n - 1)/n$. Hence during Δt the probability of another item obtaining service is $[(n - 1)/n]\mu c \, \Delta t$. The given item begins its wait in the period $t - \Delta t$ with $n - 2$ others with probability $f_{n-1}(t - \Delta t)$, and the probability of both events then is $[(n - 1)/n]c\mu \, \Delta t \, f_{n-1}(t - \Delta t)$.

3. Nothing arrives and nothing is served during Δt with probability $(1 - \lambda \, \Delta t)(1 - \mu c \, \Delta t)$, and the given item begins its wait for the period $t - \Delta t$ with $n - 1$ other items with probability $f_n(t - \Delta t)$. Again the product of these two events is taken.

Note that in the first case we could have multiplied by the probability of nothing receiving service $(1 - \mu c\, \Delta t)$ and in the second case by the probability of nothing arriving $(1 - \lambda\, \Delta t)$; we could also have considered cases of more than a single occurrence, but, because of the Poisson nature of input and service distribution, these quantities would have been negligible.

By the law of total probabilities, we have for the cases described

$$f_n(t) = (1 - \lambda\, \Delta t)(1 - \mu c\, \Delta t)f_n(t - \Delta t) + \lambda f_{n+1}(t - \Delta t)\, \Delta t$$
$$+ \frac{n-1}{n}\, \mu c f_{n-1}(t - \Delta t)\, \Delta t.$$

This equation is valid for $n > 1$. Of course, if $n = 1$ and a channel becomes free, this item goes into service and we have

$$f_1(t) = (1 - \lambda\, \Delta t)(1 - \mu c\, \Delta t)f_1(t - \Delta t) + \lambda f_2(t - \Delta t)\, \Delta t.$$

On simplifying in the usual way, then dividing by Δt, and letting $\Delta t \to 0$, these equations may be reduced to the following:

$$f_n'(t) = -(\lambda + \mu c)f_n(t) + \frac{n-1}{n}\, \mu c f_{n-1}(t) + \lambda f_{n+1}(t), \qquad n > 1, \quad (11\text{-}51)$$
$$f_1'(t) = -(\lambda + \mu c)f_1(t) + \lambda f_2(t).$$

We discuss the solution of this system subject to $f_n(0) = 1\ (n = 1, 2, \ldots)$. Note that the condition $0 \le f_n(t) \le 1$ must hold for all n since $f_n(t)$ express probability.

These equations reduce to the system considered by Vaulot, Pollaczek, Riordan, and Wilkinson if the item under consideration is not counted in the definition (i.e., subscript) of the conditional probabilities $f_n(t)$.

In that case we use $F_{n-1}(t) = f_n(t)$ and, on substituting in Eqs. (11-51) and replacing n by $n + 1$, we have

$$F_n'(t) = -(\lambda + \mu c)F_n(t) + \frac{n}{n+1}\, \mu c F_{n-1}(t) + \lambda F_{n+1}(t), \qquad n \ge 1, \quad (11\text{-}52)$$
$$F_0'(t) = -(\lambda + \mu c)F_0(t) + \lambda F_1(t).$$

In (11-51) let $\rho = \lambda/\mu c$ after dividing through by λ and putting $\lambda = 1$, that is, using λ as the time unit. Then introduce the generating function

$$\varphi(z,t) = \sum_{n=1}^{\infty} f_n(t)z^n \qquad (11\text{-}53)$$

which has the boundary conditions

$$\varphi(0,t) = 0, \qquad \varphi(z,0) = \sum_{n=1}^{\infty} z^n = \frac{z}{1-z}. \qquad (11\text{-}54)$$

Multiplying the nth equation of (11-51) by nz^n and summing yield the hyperbolic differential equation

$$\frac{\partial^2 \varphi(z,t)}{\partial z\, \partial t} + \frac{(1-z)(z-\rho)}{\rho z} \frac{\partial \varphi(z,t)}{\partial z} + \frac{1}{z^2}\, \varphi(z,t) = 0. \qquad (11\text{-}55)$$

By taking the Laplace transform with respect to t and solving the resulting first-order linear partial differential equation, one obtains the Laplace transform of the generating function. We leave this as an exercise for the reader.

Palm [551], working with the system (11-51), writes in powers of t

$$\varphi(z,t) = \sum_{n=0}^{\infty} A_n(z)\, \frac{t^n}{n!}, \qquad (11\text{-}56)$$

where $A_n(z) = \varphi^{(n)}(z,0)$, according to Maclaurin's series expansion, is the nth derivative of φ at $t = 0$. Thus

$$A_n(z) = \sum_{m=1}^{\infty} f_m^{(n)}(0) z^m, \qquad (11\text{-}57)$$

and the coefficient $f_m^{(n)}(0)$ may be determined from the system (11-51) with $\lambda = 1$ by differentiating $n-1$ times. Since $f_m(0) = 1$, one has, for $n = 1$, $f_m^{(1)}(0) = -[1/(m-\rho)]$, $\rho = \lambda/\mu c$. This can be used to determine $f_m^{(2)}(0)$, etc. Thus, with $\varphi(z,t)$ determined, $f_m(t)$ are obtained as the coefficients in the series expansion of φ in terms of increasing powers of z. By introducing the series expansion in the partial differential equation in the generating function and equating to zero each power of t, one has

$$A_n(\rho) = -\lim_{z\to 0} \frac{A_{n-1}(z)}{z} - \frac{1}{\rho} \int_0^\rho A_{n-1}(z)\, dz \qquad (11\text{-}58)$$

which is complicated for computing all the A_n. The moments of the generating function are given by

$$\mu_n(z) = -\int_0^\infty t^n \frac{\partial \varphi(z,t)}{\partial t}\, dt = n \int_0^\infty t^{n-1} \varphi(z,t)\, dt. \qquad (11\text{-}59)$$

By integrating the differential equation in φ with respect to t and using the second of the boundary conditions, one has

$$\mu_1(z) = \frac{\rho}{2-\rho}\, z\, \frac{2-z}{(1-z)^2}. \qquad (11\text{-}60)$$

Now the probability of waiting longer than t is given by

$$P(>t) = (1-\rho)\rho^{-1} \sum_{n=1}^{\infty} \rho^n f_n(t) = \frac{1-\rho}{\rho}\, \varphi(\rho,t). \qquad (11\text{-}61)$$

The mean waiting time is obtained by integrating over t. We have

$$W_q = \frac{1-\rho}{\rho} \mu_1(\rho). \tag{11-62}$$

The second moment for the waiting-time distribution is given by

$$\frac{1-\rho}{\rho} \mu_1(\rho) = \frac{\rho}{(1-\rho)^2} \frac{4}{2-\rho}. \tag{11-63}$$

By expanding $\mu_1(z)$ and $\mu_2(z)$ in series, one obtains the first and second moments for the separate $f_n(t)$. They are the coefficients of z^n in each series, respectively.

Pollaczek [591] employs another method of attack, using the system (11-52), with $\rho = \lambda/c$ and $\mu = 1$ as the time unit.

We first note that

$$P(<t) = 1 - \sum_{j=0}^{\infty} p_{c+j} F_j(t) = 1 - p_c \sum_{j=0}^{\infty} \rho^j F_j(t)$$

$$= 1 - \left[\sum_{n=0}^{c-1} \frac{\lambda^n}{n!} + \frac{\lambda^c}{c!(1-\rho)} \right]^{-1} \frac{\lambda^c}{c!} \sum_{j=0}^{\infty} \rho^j F_j(t), \tag{11-64}$$

where p_{c+j} is the steady-state probability that $c + j$ units are in the system. This probability is identical with the result obtained for an ordered queue.

If we multiply (11-52) by $(1/\sqrt{\lambda c})\rho^{n/2}e^{(c+\lambda)t}$ and write $s = t\sqrt{\lambda c}$, $g_n(s) = \rho^{n/2}e^{(c+\lambda)t}F_n(t)$, Eqs. (11-52) become

$$\frac{dg_n(s)}{ds} = \frac{n}{n+1} g_{n-1}(s) + g_{n+1}(s), \qquad n = 0, 1, \ldots, \tag{11-65}$$

and $\qquad g_n(0) = \rho^{n/2}, \qquad\qquad\qquad n = 0, 1, \ldots. \tag{11-66}$

Since $0 \le F_n \le 1$, the transform

$$\varphi_n(z) = \int_0^{\infty} e^{-2zs}g_n(s)\, ds \tag{11-67}$$

converges for $\text{Re}\,(z) > (c+\lambda)/2\sqrt{\lambda c} = (1+\rho)/2\sqrt{\rho}$. When this is applied to the last system of equations, one has

$$2z\varphi_n(z) - \rho^{n/2} = \frac{n}{n+1} \varphi_{n-1}(z) + \varphi_{n+1}(z), \qquad n = 0, 1, \ldots. \tag{11-68}$$

Multiplying the nth equation by $(n + 1)x^n$ and summing, we have

$$2z \sum_{n=0}^{\infty} (n+1)x^n\varphi_n(z) - \frac{1}{(1 - x\sqrt{\rho})^2}$$

$$= \sum_{n=1}^{\infty} nx^n\varphi_{n-1}(z) + \sum_{n=0}^{\infty} (n+1)x^n\varphi_{n+1}(z). \tag{11-69}$$

By writing

$$\phi(x,z) = \sum_{n=0}^{\infty} x^n \varphi_n(z), \tag{11-70}$$

we have, on simplifying, for $|x| < 1/\sqrt{\rho}$

$$(x^2 - 2xz + 1)\frac{d\phi}{dx} + (x - 2z)\phi = -\frac{1}{(1 - x\sqrt{\rho})^2}. \tag{11-71}$$

We require solutions of this equation that are continuous in x for $|x| < 1/\sqrt{\rho}$ and real z sufficiently large. If we factor the coefficient of $d\phi/dx$ into $(x - \alpha)(x - \beta)$, where α has the positive sign before the radical, and then expand β in series for $|z| > 1$, we seek the unique solution for $\beta \sim 1/2z$ which lies in $|x| < 1\sqrt{\rho}$. We expand $\phi(x,z)$ and the right side in Taylor series in $(x - \beta)$. By fixing the constant which appears in the solution of first-order linear equations so that $\phi(x,z)$ remains finite at $x = \beta$, we have

$$\phi(x,z) = (\alpha - x)^m (\beta - x)^{-m-1} \int_x^\beta (\alpha - y)^{-m-1} (\beta - y)^m \frac{dy}{(1 - y\sqrt{\rho})^2}, \tag{11-72}$$

where $m = \beta/(\alpha - \beta) = (z/2\sqrt{z^2 - 1}) - \frac{1}{2}$ and the quantities $(\alpha - y)^m$ and $(\beta - y)^m$ are defined to coincide for $y = x$ with $(\alpha - x)^m$ and $(\beta - x)^m$.

We note that

$$\phi(x,z) = \int_0^\infty e^{-2zs} \sum_{n=0}^\infty x^n g_n(s)\, ds. \tag{11-73}$$

By inversion we have

$$\sum_{n=0}^\infty (x\sqrt{\rho})^n F_n(t) = \exp\left[-c(1 + \rho)t\right] \frac{1}{\pi i} \int_{-i\infty+a}^{i\infty+a} e^{2ct\sqrt{\rho}z} \phi(x,z)\, dz, \tag{11-74}$$

where $a > (1 + \rho)/2\sqrt{\rho}$, $t > 0$.

By argument in the complex plane, one obtains

$$\sum_{n=0}^\infty \rho^n F_n(t) = 2 \int_0^\pi \exp\left[-(1 + \rho - 2\sqrt{\rho}\cos w)ct\right]$$

$$(1 - \sqrt{\rho}e^{iw})^{i\cot w - 2}(1 - \sqrt{\rho}e^{-iw})^{-i\cot w - 2}e^{w\cot w}(e^{\pi\cot w} + 1)^{-1} \sin w\, dw. \tag{11-75}$$

As $c \to \infty$, we have

$$P(<t) = 1 - \left(1 + 2be^{b^2}\int_\infty^b e^{-t^2}\, dt\right)^{-1} 2\sqrt{2b\theta}K_1(2\sqrt{2b\theta})$$

$$\left[1 + O\left(\frac{1}{\sqrt{c}}\right)\right], \tag{11-76}$$

where $K_1(z) = \int_0^\infty e^{-z\cosh t} \cosh t \, dt$ is Schläfli's integral representation of the modified Bessel function of the third kind of order one, with

$$b = (1 - \rho) \sqrt{c/2} \quad \text{and} \quad \theta = t \sqrt{c/2}.$$

Note that the average waiting time must be the same as that for an ordered queue.

Riordan [631] observed and developed the solution for $P(>t)$, using the system (11-52) in exponential series. He gives for all t and for $\rho = \lambda/c$, $\mu = 1$, valid for $\rho < 0.7$,

$$P(>t) = \frac{1}{2}\left(1 - \sqrt{\frac{\rho}{2}}\right) e^{-t(1-\rho)\left(1 - \sqrt{\frac{\rho}{2}}\right)} + \frac{1}{2}\left(1 + \sqrt{\frac{\rho}{2}}\right) e^{-t(1-\rho)\left(1 + \sqrt{\frac{\rho}{2}}\right)}.$$

$$(11\text{-}77)$$

He assumed the general form

$$P(>t) = A_1 e^{-(1-\rho)t/x_1} + A_2 e^{-(1-\rho)t/x_2} + \cdots, \quad (11\text{-}78)$$

computed the moments M_k expanding in series of the form $A_1 x_1^k + A_2 x_2^k + \cdots$, and determined the A's by equating coefficients.

Useful computations were developed by Wilkinson [764]. Le Roy [431] has used matrix methods to study this problem.

2. Constant Service Times, Single Channel

Burke [90] has studied the problem of random selection for service for a single channel with Poisson input with parameter λ and constant service time in equilibrium. (This holds when the ratio of the arrival rate to the service rate is less than unity.) The service time is taken as the time unit. According to Khintchine's argument the steady-state probabilities are the same whether considered over the entire process or over a set of discrete instances comprising the times of departures of served units. This is seen from the fact that the generating function for the probabilities derived on the basis of the departure epochs by Kendall coincides with the generating function for those of the entire process developed by Crommelin. Of course what is required is the waiting-time distribution $P(\leq t)$.

At the next departure epoch after its arrival a delayed unit will be in the system with $n - 1$ others [with conditional probability Prob $(n|\lambda)$], if at the departure epoch immediately before its arrival there were k units in the system ($k = 0, \ldots, n$) and $n - k$ units arrived (including the one under consideration) during the service interval. We have

$$\text{Prob } (n|\lambda) = (p_0 + p_1)\frac{\lambda^{n-1}e^{-\lambda}}{(n - 1)!} + \cdots + p_n e^{-\lambda}, \quad (11\text{-}79)$$

which is Crommelin's equation for p_{n-1}, and hence

$$\text{Prob}\,(n|\lambda) = p_{n-1}. \tag{11-80}$$

If $q(t|n)$ denotes the conditional probability of a delay $\leq t$, given that the delayed unit is one of n units waiting for service at the first departure epoch following its arrival, then

$$P(\leq t) = \sum_{n=1}^{\infty} \text{Prob}\,(n|\lambda) q(t|n). \tag{11-81}$$

But the delay between an arrival and the first departure epoch is uniformly distributed (because of the Poisson random input during a service interval) and is independent of subsequent delay because of random selection for service.

If the delay T is expressed as the sum of its part T', which is an integral number of time units, and T'', which is the remaining fractional part, and if the time t is similarly divided into t' and t'', we have, using the foregoing remarks on independence,

$$q(t|n) = \text{Prob}\,(T \leq t|n) = \text{Prob}\,(T' < t'|n)$$
$$+ \text{Prob}\,(T' = t'|n)\,\text{Prob}\,(T'' \leq t'') \tag{11-82}$$

or, since T''' is uniformly distributed, we have

$$q(t'|n) = \sum_{i=0}^{t-1} \text{Prob}\,(T' = i|n) + t''\,\text{Prob}\,(T' = t'|n). \tag{11-83}$$

If we define $\text{Prob}\,(T' = i|n) \equiv Q_i(n)$, then

$$Q_0(n) = \frac{1}{n} \tag{11-84}$$

and $\quad Q_i(n) = [1 - Q_0(n)] \sum_{j=0}^{\infty} \frac{\lambda^j e^{-\lambda}}{j!} Q_{i-1}(n+j-1), \qquad i = 1, \ldots, t',$

whose solution provides $q(t|n)$ and hence $P(\leq t)$, after p_n has also been obtained. Burke gives graphical computations for the distribution of delay up to 130 holding times and compares the results with those of ordered queues for constant holding times and with random-selection queues for exponential holding times, all for a large number of occupancy levels. It is sufficient to follow the above argument.

11-4. Allocation to Channels

Volberg [742], following Pollaczek, has studied a several-parallel-channels (there are c channels) problem with Poisson input with parameter λ and arbitrary service times $B(t)$ the same for all channels and all customers. The customers are assigned to the channels in the order of their arrival: the first arrival to the first channel, the second arrival to the second channel, and so on, the cth arrival to the last channel, and then the $(c + 1)$st arrival to the first channel, running through the channels for a second time, and so on. Volberg computes, for a channel, the waiting-time distribution in the steady state. As we have pointed out in Chap. 3, the ideas correspond to those of a filtered Poisson input from which every cth arrival is admitted to a single channel; i.e., the input is Erlangian, and the system studied is $E_c/G/1$.

Let the waiting-time variable for the $(n + 1)$st customer arriving at a channel at time t be $w_n(t)$, and let his service-time variable be u_n. If we denote by $\tau_{n,c}$ the time interval between his arrival and that of the next person in the same queue [i.e., the $(n + c + 1)$st arriving customer], then we have

$$w_{n+c}(t) = w_n(t - \tau_{n,c}) + u_n - \tau_{n,c}. \tag{11-85}$$

Let

$$P_n(w,t) = \text{Prob } [w_n(t) \leq w] \tag{11-86}$$

and

$$F_{n,c}(\tau,t) = \text{Prob } (\tau_{n,c} \leq \tau). \tag{11-87}$$

Volberg then proceeds as follows:

$$\frac{\partial}{\partial \tau} F_{n,c}(\tau,t) = \frac{(t - \tau)^n}{n!} \frac{c - 1}{(c - 1)!} \frac{(n + c)!}{t^{n+c}}, \qquad t > 0. \tag{11-88}$$

He uses this with the first relation to obtain a recursion formula which enables computing $P_n(w,t)$ (if the probability that the time of initial occupancy does not exceed a given length for each channel is known). For any arbitrary waiting customer before a channel, the waiting-time distribution is then

$$P(w,t) = \sum_{n=0}^{\infty} \frac{(\lambda t)^n e^{-\lambda t}}{n!} P_n(w,t). \tag{11-89}$$

After some detail and by allowing $t \to \infty$, Volberg obtains for the Laplace-Stieltjes transform $\psi(s)$ of the waiting-time distribution in the steady state (i.e., when the traffic intensity $\rho = \lambda/\mu c < 1$, where μ is the service rate)

$$\psi(s) = \frac{sQ(s)}{\beta(s) - (1 - \lambda s)^c}. \tag{11-90}$$

Here $\beta(s)$ is the transform of the service-time distribution,

$$Q(s) = \frac{c\lambda(1 - \rho)(s_1 - s)(s_2 - s) \cdots (s_{c-1} - s)}{s_1 s_2 \cdots s_{c-1}}, \tag{11-91}$$

and s_i $(i = 1, \ldots, c - 1)$ are the zeros of $[\beta(s) - (1 - \lambda s)^c]$ that are in the right half plane. Volberg notes that when $c = 1$ this function coincides with the single-channel result given by Khintchine. If $\rho > 1$, we have

$$\lim P[\sqrt{\lambda t}\, w + (\rho - 1)t,\, t] = \frac{1}{2\pi} \int_{-\infty}^{w/\sqrt{(\mu_2 - \mu_1)/c + \mu_1{}^2/c^2}} e^{-(x^2/2)}\, dx, \quad (11\text{-}92)$$

where $\mu_1 = 1/\mu$ and μ_2 is the second moment of the service-time distribution.

11-5. Last-come–First-served Queue Discipline

Consider the case in which arriving units used in manufacturing, for example, are stacked on top of other units already present. Thus, when a unit is needed for use it is taken (i.e., goes into service) from the top. As a unit is being used, other units arrive and are stacked in the order of arrival. This is a case of last-come–first-served queue discipline. It is desired to compute the waiting time in the steady state.

Assume that a unit arrives at a single-channel queue by a Poisson process with parameter λ, is nth in the line, and obtains service by an arbitrary service-time distribution $B(t)$ with mean $1/\mu$. Its waiting time is computed by a method similar to the one used to calculate the distribution of a busy period in Chap. 8. When the unit arrives, it waits until the unit in service terminates its service. If y is the length of the remaining service time, m units might arrive during this service time and must all be served, together with anything arriving in the meantime, before the unit in question goes into service. Thus essentially, the waiting-time distribution for the last-come–first-served queue discipline is the distribution of a busy period in which the length of the queue, starting with n, returns to n. But the transitions under consideration are independent of n, and the busy period can be considered equal in length to the busy period of an ordered queue, i.e., the time required for a first return to zero having started with zero in the queue is the same in both cases. Thus the computations proceed exactly as in the distribution of a busy period of an ordered queue with one modification.

The steady-state distribution of y is no longer the service-time distribution, since the length of y is also determined by the time at which the unit under consideration arrives at the queue. In any case, at least formally, the Laplace-Stieltjes transform of the waiting time is given by

$$\gamma(s) = \psi\{s + [1 - \Gamma(s)]\lambda\}, \quad (11\text{-}93)$$

where $\psi(s)$ and $\Gamma(s)$ are, respectively, the Laplace-Stieltjes transforms of y and of the distribution of a busy period (see Chap. 8). By introducing differential-difference equations in probabilities $p_n(y,t)$, giving the number n in the system at time t and the time y required by the unit in service to complete service, then introducing their generating function and its Laplace-Stieltjes transforms with respect to y and t, and passing to the steady state, Wishart [910] obtains

$$\psi(s) = \frac{1 - \rho + \lambda[1 - \beta(s)]}{s}. \qquad (11\text{-}94)$$

Here $\beta(s)$ is the Laplace-Stieltjes transform of $B(t)$. This finally gives

$$\gamma(s) = \frac{1 + \Gamma(s)}{[s/\lambda + 1 - \Gamma(s)]} + 1 - \rho \qquad (11\text{-}95)$$

Exercise: Show that the expected waiting time is the same as that for the first-come–first-served Poisson case but that the variance is greater for the case under consideration. (Recall that the moments are obtained by differentiating the Laplace-Stieltjes transform and evaluating at $s = 0$.)

If one has a single channel with a general independent input and with exponential service, the unexpended service time is exponentially distributed and thus the waiting time is the same as a busy period generated by a service time which is the same as a busy period. From Conolly's work in Chap. 10, we have for the Laplace-Stieltjes transform of the waiting-time distribution

$$\frac{1 - \lambda + \lambda(1 - \xi)}{s/\mu + 1 - \xi} \qquad (11\text{-}96)$$

which also has greater variance than the first-come–first-served case.

PROBLEMS

1. By a numerical example, study the nonpreemptive-priority problem.

2. In a c-channel queue with Poisson arrivals and parameter λ and identical exponential service times with parameter μ, suppose that arrivals are served in any manner. Let $f_n(t)$ be the probability that, during an interval of length t after a state of n waiting items has occurred, the congestion has not ceased and thus the state $c - 1$ has not occurred. Thus $f_0(t)$, the desired distribution of the duration of a congestion, gives the probability that a congestion lasts at least for the time t. Show that the equations are

$$\frac{1}{\lambda} f_n'(t) = -\frac{1 + \rho}{\rho} f_n(t) + f_{n+1}(t) + \frac{1}{\rho} f_{n-1}(t), \qquad n > 0,$$

$$\frac{1}{\lambda} f_0'(t) = -\frac{1 + \rho}{\rho} f_0(t) + f_1(t), \qquad n = 0,$$

$$f_n(0) = 1, \qquad \rho = \frac{\lambda}{c\mu}.$$

Also show that, if

$$F(x,t) = \sum_{n=0}^{\infty} f_n(t) x^n,$$

which converges for $|x| < 1$ since $0 \le f_n(t) \le 1$, one has

$$\frac{1}{\lambda} \frac{\partial F}{\partial t} + \frac{(1-x)(x-\rho)}{\rho x} F + \frac{f_0(t)}{x} = 0,$$

with

$$F(0,t) = f_0(t), \qquad F(x,0) = \frac{1}{1-x}.$$

Using Palm's series-expansion procedure for the random-selection case, show how one may obtain a solution to the above system. Also show that $(1 - \rho)F(\rho,t)$ gives the probability that, during time t, after a moment selected at random, the congestion does not cease. Taking the Laplace transform of the partial differential equation, show, by arguing on the zeros of the denominator of the transform, that $f_n(t)$ is given by the same formula as the distribution of a busy period in an ordered single-channel queue with μ replaced by $c\mu$ [684].

3. In the multiple-channel random-selection-for-service problem justify the expression given for $P(>t)$.

4. As an interesting idea but not specially related to this chapter, the second-order equation

$$\frac{\partial P(z,t)}{\partial t} = - \frac{\partial b(z)\, P(z,t)}{\partial z} + \frac{1}{2} \frac{\partial^2 a(z)\, P(z,t)}{\partial z^2}$$

yields the Fokker-Planck equation if $a(z)$ and $b(z)$ are taken as constants. The Fokker-Planck equation, which is a general heat or diffusion equation, occurs frequently in the theory of stochastic processes. The equation is valid only if the stochastic process is Markovian and has a Gaussian distribution. Here $P(z,t)$ is the probability that a random variable has the value z at time t.

Another well-known version of this equation by the same name is obtained by putting $b(z) = -\beta z$ and $a(z) = 2D$, where β and D are constants. If $D = 0$, we have a linear equation which we have encountered in the text frequently. Solve the heat equation which results from putting $b(z) = 0$ and $a(z) = a$, obtaining

$$P(z,t) = \frac{1}{(2\pi a t)^{1/2}} \exp\left(- \frac{x^2}{2at} \right).$$

Also write the solution for arbitrary constants $a > 0$ and b. The solution in this case is the same as above except that x^2 is replaced by $(x - bt)^2$.

CHAPTER 12

QUEUES IN TANDEM OR SERIES

12-1. Introduction

Most of the analytical work in queues with a number of facilities in series has been restricted to Poisson arrivals and exponential service times with ordered-queue discipline; once in the system, a unit either stays on for service through to the last phase (channel in series) or is able to go to the proper phase for service. Note that phase-type service corresponds to channels in series without waiting before every channel except before the first one. A new item is admitted into service after the previous item completes all phases. In this chapter we shall generally assume that waiting is allowed before the different phases. Studies have almost entirely concentrated on deriving steady-state solutions.

An important question to determine in studying tandem queues is the distribution of output from one channel which then comprises the input into the subsequent channel. Burke [89], Reich [616], and Cohen [123] have studied this problem. Note that the output of a steady-state queue is independent of the type of service discipline used. Burke showed for the steady state the independence of time intervals between departures for the system $M/M/c$. For this system, the departures satisfy a Poisson process with the same parameter λ as that of the input distribut on—a fact previously known on intuitive grounds based on steady-state arguments. He also showed the independence of the inter-departure-interval random variable and the state of the system at the end of the interval.

Reich, using different methods, has shown that, for interarrival and service periods having normalized chi-square distribution each with four degrees of freedom (a slight change from Burke's assumptions) and for a single-channel queue, the departure epochs do not constitute a normalized chi-square distribution with four degrees of freedom. Hence it is not generally reasonable to expect the outputs to match the inputs even in the steady state. He also gave a partial converse to Burke's result: that, for a single-channel queue, Poisson arrivals and departures imply an exponential-service-time distribution or a step function at zero.

255

Finch [214] has shown that it is only when an infinite-size queue is tolerated that the Poisson result for the output is correct. He gave $N = \infty$ and exponential service as a necessary and sufficient condition for the independence of the interdeparture intervals and the independence of the queue length left by a departing unit from the interval since the previous departure.

O'Brien [527] studied the case of two channels in series in the steady state with Poisson inputs and exponential service times and gave expected numbers and the expected waiting times.

R. R. P. Jackson [318, 318a] has methodically studied the problem of tandem queues in the steady state with Poisson assumptions (different exponential distribution for each phase), giving the probability of various numbers of items at the different phases, together with the average number and the waiting-time distribution in each phase.

J. R. Jackson [315] permits Poisson arrivals to each phase, both from the system and from outside. A unit arrives at different phases with different probabilities. The service distributions are exponential, and each phase consists of several parallel channels. He derives an expression for the probability of a given number of customers being in some phase, for Poisson input, exponential service times, no leaving of the system until all channels are entered, beginning at the first channel.

Assuming Poisson input, exponential service times, no leaving of the system until all channels are entered, beginning at the first channel, Hunt [311] studied utilization (ratio of mean arrival rate to mean service rate) for the cases of (1) an infinite queue allowed before each channel; (2) a finite nonzero queue, except that the first channel may have an infinite queue; (3) a zero queue except at the first channel, where it is infinite (with blocking at a channel until a unit which has finished service can move to be served immediately in the following channel); and finally (4) a zero queue (except at the first channel, where it may be infinite) and no vacant facility (the entire line moves as one unit). In the last case the probability of completion of service in time t after the line moves is also computed. When no queues are allowed before subsequent channels, a service channel's effective service rate is obtained as the product of the given service and the fraction of time during which the channel is unblocked.

By assuming a different exponential distribution for waiting longer than a prescribed time at each channel in a collection of channels in series, Nelson [516] has computed the probability of waiting longer than a prescribed time at all channels. DeBaun and Katz [148], using chi-square approximation to the sum of exponentials, simplify Nelson's computations. Sacks [646] has studied the ergodicity of tandem queues.

12-2. The Output of a Queue

It is advantageous in this study to know if the output of a queue with Poisson input is itself Poisson. Then we can study a number of queues in series, knowing that their inputs and outputs are Poisson. We now give a proof of this important fact which is due to Burke.

Theorem 12-1: The steady-state output of a queue with c channels in parallel, with Poisson input and parameter λ and the same exponential-service-time distribution with parameter μ for each channel, is itself Poisson-distributed.

Proof: The proof proceeds in several steps. The first part of the argument is to compute the number in the system at the time of departure of a unit. This is then used to give the initial conditions for a set of joint probabilities. These probabilities give the number in the system at a time t after a departure, and that t has a length less than the interdeparture-interval chance variable. By summing over the number in the system, one obtains a marginal distribution related to the interdeparture-interval distribution. It is then found that this distribution is exponential, which is the same thing as having the departures occurring by a Poisson process. It is also shown that the number in the system given at the time of a departure and the departure interval are independently distributed and thus their joint probability is the product of the separate probabilities. Note that the steady-state probabilities p_n of n units in the system for this case have been computed in a previous chapter. When a unit departs, the system moves from state E_{n+1} to E_n, while the converse holds for an arriving unit which finds the system in state E_n. Since the number of transitions $E_n \rightarrow E_{n+1}$ cannot differ by more than one from the number of transitions $E_{n+1} \rightarrow E_n$, the proportion of units leaving the system in state E_n approaches the same limit (the steady-state probability) as that of the proportion of arriving units which find it in state E_n.

Denote by τ the length of an arbitrary interval between departures and by $k(t)$ the state of the system at time t after the previous departure. Define

$$F_n(t) \equiv \text{Prob } [k(t) = n \text{ and } t < \tau], \tag{12-1}$$

$$F(t) \equiv \sum_{n=0}^{\infty} F_n(t), \qquad t < \tau. \tag{12-2}$$

Thus $F(t)$ is a marginal distribution since we have summed over all values of the other joint variable. Note that $1 - F(t)$ is the cumulative distribution of the length of time between departures.

We have the initial condition

$$F_n(0) \equiv p_n. \tag{12-3}$$

We now write

$$F_0(t + \Delta t) = (1 - \lambda \, \Delta t)F_0(t),$$
$$F_n(t + \Delta t) = (1 - \lambda \, \Delta t)(1 - n\mu \, \Delta t)F_n(t) + \lambda F_{n-1}(t) \, \Delta t, \quad n < c, \quad (12\text{-}4)$$
$$F_n(t + \Delta t) = (1 - \lambda \, \Delta t)(1 - c\mu \, \Delta t)F_n(t) + \lambda F_{n-1}(t) \, \Delta t, \quad n \geq c.$$

Simplifying in the usual manner yields a set of differential-difference equations with the solution

$$F_n(t) = p_n e^{-\lambda t}, \quad t < \tau. \quad (12\text{-}5)$$

Exercise: Verify that (12-5) satisfies the simplified form of (12-4).

Summing over n from zero to infinity yields the fact that the marginal distribution of the intervals between departures is the same as the distribution between arrivals.

From this we also have the independence of τ and $k(\tau)$, since

$$\text{Prob } [k(\tau + 0) = n \text{ and } t < \tau < t + dt] \quad (12\text{-}6)$$

is given by

$$\begin{matrix} F_{n+1}(t)(n + 1)\mu \, dt & \text{for } n + 1 \leq c, \\ F_{n+1}(t)c\mu \, dt & \text{for } n + 1 > c. \end{matrix} \quad (12\text{-}7)$$

When replaced by the expression for $F_n(t)$, this factors into the probability distribution functions of $k(\tau)$ and τ, proving the independence of the last two quantities. It is also true that τ is independent of the set of lengths of all (subsequent) intervals between departures following τ. This is proved using the last result above and the Markovian property of the system. The output distribution at any time t depends only on the state of the system at t and not on previous states.

A partial converse of the above theorem due to Reich follows:

Theorem 12-2: If the arrivals and departures of a single-channel queue are Poisson-distributed, the service-time distribution is exponential or a step function at zero.

One may also argue that if the input has a fixed arrival rate λ and if the output is Poisson uniformly for all values of the service parameter μ, the input must be Poisson. This fact follows from the uniformity assumption and by allowing $\mu \to \infty$. For then the service time is zero and the output distribution coincides with the input distribution.

If the interarrival and the service-time intervals for a multiple-channel queue are constant, the output (when there is a queue) is also constant, with intervals equal to the service time. The exponential case of arrivals and services and the case just mentioned correspond to a chi-square distribution with two and infinite degrees of freedom, respectively. Reich gives the following for an intermediate case:

Theorem 12-3: The interdeparture-time distribution from a single-channel queue whose arrival-times variable and service-times variable are each the sum of two identically exponentially distributed variables, with parameters λ and μ, respectively, is not the sum of two variables which are exponentially distributed.

The proof proceeds by contradiction.   If it were true that the distribution of the interdeparture interval τ is the sum of two exponentially distributed variables, it would have the same marginal distribution as the input variable.   This leads to $1 - \lambda/\mu$ as the probability that a departing customer leaves zero persons behind.   This is found not to be the case by referring to Volberg's work examined in the last chapter. There, arrivals from a Poisson input are assigned to channels for service.

Finch [214] has shown that, for a Poisson-input single-channel queue where an unlimited number of customers can wait in the waiting room, the following are true (in the steady state) if, and only if, the service time is exponential:

1. The queue size left by a departing customer is independent (in the limit) of the duration of the interval since the previous departure.

2. Two successive departure intervals are independent (in the limit).

12-3. Two Queues in Tandem

For an infinite input with Poisson distribution with parameter λ to the first of two channels in series, with exponential service times having parameter μ_1 and μ_2, respectively, at each channel (an item must go through both service channels), we have the following transient equations for $P(n_1,n_2,t)$, the probability that there are n_1 units in the first phase of the system (including service) and n_2 in the second (derived as usual) at time t [318]:

$$P'(0,0,t) = -\lambda P(0,0,t) + \mu_2 P(0,1,t), \qquad n_1, n_2 = 0, \qquad (12\text{-}8a)$$

$$
\begin{aligned}
P'(0,n_2,t) = &-(\lambda + \mu_2)P(0,n_2,t) \\
&+ \mu_1 P(1, n_2 - 1, t) \\
&+ \mu_2 P(0, n_2 + 1, t),
\end{aligned}
\qquad n_1 = 0, \qquad n_2 > 0, \quad (12\text{-}8b)
$$

$$
\begin{aligned}
P'(n_1,0,t) = &-(\lambda + \mu_1)P(n_1,0,t) \\
&+ \mu_2 P(n_1,1,t) \\
&+ \lambda P(n_1 - 1, 0, t),
\end{aligned}
\qquad n_1 > 0, \qquad n_2 = 0, \quad (12\text{-}8c)
$$

$$
\begin{aligned}
P'(n_1,n_2,t) = &-(\lambda + \mu_1 + \mu_2)P(n_1,n_2,t) \\
&+ \mu_1 P(n_1 + 1, n_2 - 1, t) \\
&+ \mu_2 P(n_1, n_2 + 1, t) \\
&+ \lambda P(n_1 - 1, n_2, t),
\end{aligned}
\qquad n_1, n_2 > 0. \qquad (12\text{-}8d)
$$

The steady-state equations have the solution

$$p(n_1,n_2) = \rho_1^{n_1}\rho_2^{n_2}p(0,0), \qquad \rho_1 = \frac{\lambda}{\mu_1}, \qquad \rho_2 = \frac{\lambda}{\mu_2}. \qquad (12\text{-}9)$$

Since
$$\sum_{n_1,n_2=0}^{\infty} p(n_1,n_2) = 1, \qquad (12\text{-}10)$$

we find
$$p(0,0) = (1 - \rho_1)(1 - \rho_2). \qquad (12\text{-}11)$$

The expected number in the system may be computed by multiplying $p(n_1,n_2)$ by $n_1 + n_2$ and summing over n_1 and then over n_2. This gives

$$L = \frac{\rho_1}{1 - \rho_1} + \frac{\rho_2}{1 - \rho_2}. \qquad (12\text{-}12)$$

Exercise: Verify that (12-9) satisfies the steady-state equations.
Exercise: Obtain (12-11) and (12-12).

The probability of n_1 units in the first phase is obtained by summing over n_2. This yields $\rho_1^{n_1}(1 - \rho_1)$. Similarly, $\rho_2^{n_2}(1 - \rho_2)$ is the probability of n_2 units in the second phase (verify this).

If there are a finite number of input sources N or, equivalently, if there are N identical input sources and hence each unit has a probability $\lambda\,\Delta t$ of joining the queue during Δt and if everything else remains the same, the equations are as follows:

1. The same as (12-8a).
2. The same as (12-8b) for $n_1 = 0$; $n_2 = 1, \ldots, N - 1$.
3. $P'(0,N,t) = -\mu_2 P(0,N,t) + \mu_1 P(1, N - 1, t)$; $n_1 = 0$; $n_2 = N$.
4. The same as (12-8c) for $n_1 = 1, \ldots, N - 1$; $n_2 = 0$.
5. $P'(N,0,t) = -\mu_1 P(N,0,t) + \lambda P(N - 1, 0, t)$; $n_1 = N$; $n_2 = 0$.
6. The same as (12-8d) for $n_1, n_2 > 0$; $n_1 + n_2 < N$.
7. $P'(n_1,n_2,t) = -(\mu_1 + \mu_2)P(n_1,n_2,t) + \mu_1 P(n_1 + 1, n_2 - 1, t) + \lambda P(n_1 - 1, n_2, t)$; $n_1, n_2 > 0$; $n_1 + n_2 = N$.

Again for the steady-state solution we have

$$p(n_1,n_2) = \rho_1^{n_1}\rho_2^{n_2}p(0,0). \qquad (12\text{-}13)$$

Summing over n_1 and n_2 from zero to N $(n_1 + n_2 \leq N)$ and equating to unity give

$$p(0,0) = \frac{(\rho_1 - \rho_2)(1 - \rho_1)(1 - \rho_2)}{(\rho_1 - \rho_2) - (\rho_1^{N+2} - \rho_2^{N+2}) + \rho_1\rho_2(\rho_1^{N+1} - \rho_2^{N+1})}. \qquad (12\text{-}14)$$

The average number in the system with $n_1 + n_2 \leq N$ is

$$L = \sum_{n_1=0}^{N} \sum_{n_2=0}^{N} (n_1 + n_2)p(n_1,n_2)$$

$$= \frac{1}{\rho_1 - \rho_2} \left\{ \frac{\rho_1^2[1 - (N + 1)\rho_1^N + N\rho_1^{N+1}]}{(1 - \rho_1)^2} \right.$$

$$\left. - \frac{\rho_2^2[1 - (N + 1)\rho_2^N + N\rho_2^{N+1}]}{(1 - \rho_2)^2} \right\} p(0,0). \qquad (12\text{-}15)$$

The average number being served is given by

$$
\sum_{n_1=1}^{N} p(n_1,0) + \sum_{n_2=1}^{N} p(0,n_2) + 2 \sum_{n_1=1}^{N} \sum_{n_2=1}^{N} p(n_1,n_2)
$$
$$
= \frac{(\rho_1 + \rho_2)[(\rho_1 - \rho_2) - (\rho_1^{N+1} - \rho_2^{N+1}) + \rho_1\rho_2(\rho_1^N - \rho_2^N)]p(0,0)}{(1 - \rho_1)(1 - \rho_2)(\rho_1 - \rho_2)}. \quad (12\text{-}16)
$$

The average number waiting is

$$
\sum_{n_1=2}^{N} (n_1 - 1)p(n_1,0) + \sum_{n_2=2}^{N} (n_2 - 1)p(0,n_2) + \sum_{n_1=2}^{N} (n_1 - 1)p(n_1,1)
$$
$$
+ \sum_{n_2=2}^{N-1} (n_2 - 1)p(1,n_2) + \sum_{\substack{n_1=2 \\ n_1+n_2 \leq N}}^{N-2} \sum_{n_2=2}^{N-2} (n_1 + n_2 - 2)p(n_1,n_2)
$$
$$
= \left\{ \frac{\rho_2^2(1 + \rho_1)[1 - N\rho_1^{N-1} + (N - 1)\rho_1^N]}{(1 - \rho_1)^2} \right.
$$
$$
+ \frac{\rho_1^2(1 + \rho_2)[1 - N\rho_2^{N-1} + (N - 1)\rho_2^N]}{(1 - \rho_2)^2}
$$
$$
+ \frac{\rho_1\rho_2}{\rho_1 - \rho_2} \left[\frac{1 - (N - 1)\rho_1^{N-2} + (N - 2)\rho_1^{N-1}}{(1 - \rho_1)^2} \right.
$$
$$
\left. \left. - \frac{1 - (N - 1)\rho_2^{N-2} + (N - 2)\rho_2^{N-1}}{(1 - \rho_2)^2} \right] \right\} p(0,0). \quad (12\text{-}17)
$$

The ideas, with unlimited input source, may now be applied to three phases.

Exercise: Derive (12-13) and then verify all the remaining expressions.

12-4. Multiple Channels in Series and in Parallel

The above ideas may be extended to k phases in series with queueing before each phase and an unlimited Poisson input to the first phase. The ith phase then consists of r_i parallel channels all with exponential service rate μ_i [thus the probability that a unit among n_i in the ith phase finishes service in Δt is $n_i\mu_i \Delta t + o(\Delta t)$ for $n_i < r_i$ and $r_i\mu_i \Delta t + o(\Delta t)$ for $n_i \geq r_i$], and for $\rho_i = \lambda/r_i\mu_i < 1$ $(i = 1, \ldots, k)$ the steady-state equations in R. R. P. Jackson's notation [318a] are

$$
\left[\lambda + \sum_{j=1}^{k} \delta(n_j)a(n_j)\mu_j \right] p(n_1, \ldots, n_k)
$$
$$
= \sum_{j=1}^{k} \delta(n_j + 1)a(n_j + 1)\mu_j p(n_1, \ldots, n_j + 1, n_{j+1} - 1, n_{j+2}, \ldots, n_k)
$$
$$
+ \lambda p(n_1 - 1, n_2, \ldots, n_k), \quad (12\text{-}18)
$$

where if any of the arguments are negative the probability is zero and the last term in the last sum involves no minus-one term and one plus-one term. Also,

$$a(n_j) = \begin{cases} n_j, & n_j < r_j, \\ r_j, & n_j \geq r_j, \end{cases} \tag{12-19}$$

$$\delta(n_j) = \begin{cases} 1, & n_j \neq 0, \\ 0, & n_j = 0, \end{cases} \quad j = 1, 2, \ldots, k, \tag{12-20}$$

$$\delta(n_{k+1}) = 1.$$

The solution is given by

$$p(n_1, \ldots, n_k) = p(0, \ldots, 0) \prod_{j=1}^{k} b(n_j), \tag{12-21}$$

$$b(n_j) = \begin{cases} \dfrac{1}{n_j!} (r_j \rho_j)^{n_j}, & n_j < r_j, \\ \dfrac{1}{r_j!} (r_j \rho_j)^{r_j} (\rho_j)^{n_j - r_j}, & n_j \geq r_j. \end{cases} \tag{12-22}$$

Since the sum over all n_j must equal unity, we have

$$\sum_{n_1=0}^{\infty} \cdots \sum_{n_k=0}^{\infty} \left[\prod_{j=1}^{k} b(n_j) \right] = \prod_{j=1}^{k} \sum_{j=0}^{\infty} b(n_j) \equiv \prod_{j=1}^{k} A_j, \quad j = 1, \ldots, k, \tag{12-23}$$

and

$$p(0, \ldots, 0) = \prod_{j=1}^{k} A_j^{-1}. \tag{12-24}$$

A proof of the uniqueness of the above solution is also given by Jackson. The distribution for any phase is obtained by summing out over the numbers in all the other phases. This gives for the marginal probability that there are n_j customers in the jth phase

$$p(n_j) = \frac{b(n_j)}{A_j}.$$

(Put $n_j = n$ to obtain the probability that there are n in the jth phase.) This can also be obtained as the solution of c channels in parallel with $c = r_j$.

Writing $r_j = 1$ $(j = 1, \ldots, k)$ gives for single queues in series

$$p(n_1, \ldots, n_k) = p(0, \ldots, 0) \prod_{j=1}^{k} \rho_j^{n_j}, \tag{12-25}$$

$$p(0, \ldots, 0) = \prod_{j=1}^{k} (1 - \rho_j), \tag{12-26}$$

and because of independence among the phases we have for the probability of n items in the jth phase

$$\rho_j^n(1 - \rho_j). \tag{12-27}$$

For the average number in that phase we have

$$\sum_{n=0}^{\infty} n\rho_j^n(1 - \rho_j) = \frac{\rho_j}{1 - \rho_j}. \tag{12-28}$$

The average number being served in the jth phase is

$$\sum_{n=1}^{\infty} \rho_j^n(1 - \rho_j) = \rho_j. \tag{12-29}$$

The average number waiting in the jth phase is

$$\rho_j^2(1 - \rho_j)^{-1}. \tag{12-30}$$

The average for the system is the sum of the averages of the phases.

The waiting-time distribution of a customer going from the $(j - 1)$st phase to the jth phase as illustrated in the single-channel case is

$$f_j(\xi) \, d\xi = \sum_{n=1}^{\infty} (1 - \rho_j)\rho_j^n \mu_j^n \xi^{n-1} \frac{e^{-\mu_j \xi}}{(n - 1)!} \, d\xi$$
$$= \lambda(1 - \rho_j)e^{-(\mu_j - \lambda)\xi} \, d\xi. \tag{12-31}$$

The probability of not waiting for the jth phase is $1 - \rho_j$, and the probability of waiting when the system is occupied is

$$(\mu_j - \lambda)e^{-(\mu_j - \lambda)\xi} \, d\xi. \tag{12-32}$$

For $\mu_j = \mu$ let $\Sigma n_j = n$ and $p(n)$ be the probability that n units are waiting in the system.   Then

$$p(n) = \binom{n + k - 1}{k - 1} \rho^n(1 - \rho)^k, \tag{12-33}$$

where $\rho = \lambda/\mu$.

Exercise: Justify the steady-state equations and verify all the remaining results of this section.

12-5. Nonsystematic Tandem-queueing Procedure

If Poisson arrivals to the jth phase (with r_j channels in parallel) are also allowed from outside the system with parameter λ_j and, finishing there, leave for the mth phase with probability q_{jm} and are then served in order of arrival, the average arrival rate of units at the jth phase is defined by

$$\bar{\lambda}_j = \lambda_j + \sum_m q_{mj}\bar{\lambda}_m \qquad (12\text{-}34)$$

and plays the same role in the jth phase as λ did in the previous study. We have

$$p(n_1, \ldots, n_k) = p_{n_1}^1 p_{n_2}^2 \cdots p_{n_k}^k, \qquad (12\text{-}35)$$

where

$$\bar{\lambda}_j < \mu_j, \qquad j = 1, \ldots, k,$$

and the probability that there are n_j in the jth phase is

$$p_{n_j}^j = \begin{cases} \dfrac{p_0^j (\bar{\lambda}_j/\mu_j)^{n_j}}{n_j!}, & n_j = 0, 1, \ldots, r_j, \\[2ex] \dfrac{p_0^j (\bar{\lambda}_j/\mu_j)^{n_j}}{r_j!(r_j)^{n_j-r_j}}, & n_j = r_j, \ldots. \end{cases} \qquad (12\text{-}36)$$

The constant p_0^j is obtained from $\sum\limits_{n_j} p_{n_j}^j = 1$. This result is based on the fact that the distribution of the number in each phase is independent of the distribution of the number in any other phase. By writing the transient equations and passing to the steady state, J. R. Jackson [315] obtains the above results.

12-6. No Queues Allowed before Subsequent Phases

Here there are two phases with Poisson input and parameter λ to the first phase and the same exponential service distribution in both with parameter μ and with first-come–first-served discipline. The output of the first phase comprises the input into the second, and no queue is allowed to form before the second phase, whereas an infinite queue is allowed before the first. This results in blocking of service (no units allowed into service) at the first channel even though the service of a unit has just been completed and there is a queue. The channel opens for service when the unit can go to service in the second channel when the latter becomes free. The utilization factor $\rho = \lambda/\mu < 1$ is now reduced in value since the effective service rate must be multiplied by the fraction of time during which the first-phase service is unblocked.

Let $P_n(.,.,t)$ be the probability that there are n units waiting and being served in the first phase at time t.

1. Let the first dot be replaced by zero if the unit is still in service in the first phase or by unity if it has completed service but is blocked because the second-phase service is occupied and hence the second dot is also replaced by unity.

2. Otherwise, let the second dot be replaced by zero.

Note that a subscript of zero indicates that there is nothing waiting or in service in the first phase; thus it is understood in this case that, when the first dot is replaced by zero, the first-phase service channel is vacant. We have for $n = 0$

$$P_0'(0,0,t) = -\lambda P_0(0,0,t) + \mu P_0(0,1,t),$$
$$P_0'(0,1,t) = -(\lambda + \mu)P_0(0,1,t) + \mu P_0(1,1,t) + \mu P_1(0,0,t), \quad (12\text{-}37)$$
$$P_0'(1,1,t) = -(\lambda + \mu)P_0(1,1,t) + \mu P_1(0,1,t).$$

For $n \geq 1$,

$$P_n'(0,0,t) = -(\lambda + \mu)P_n(0,0,t) + \lambda P_{n-1}(0,0,t) + \mu P_n(0,1,t),$$
$$P_n'(0,1,t) = -(\lambda + 2\mu)P_n(0,1,t) + \lambda P_{n-1}(0,1,t) + \mu P_n(1,1,t)$$
$$+ \mu P_{n+1}(0,0,t), \quad (12\text{-}38)$$
$$P_n'(1,1,t) = -(\lambda + \mu)P_n(1,1,t) + \lambda P_{n-1}(1,1,t) + \mu P_{n+1}(0,1,t).$$

Here μ and λ are the mean service and arrival rates, respectively.

One may obtain the steady-state equations from these by equating the left-hand sides to zero. In (12-38) we introduce the shifting operation $Ep_n = p_{n+1}$ after passing to the steady state. Solving the system leads to a vanishing determinant [311] and subsequently an equation whose roots are

$$r_1 = 1, \quad r_2 = \frac{\rho}{\rho + 1}, \quad r_3, r_4 = \frac{\rho}{4}[(\rho + 3) \pm \sqrt{\rho^2 + 6\rho + 1}], \quad (12\text{-}39)$$

where $\rho = \lambda/\mu$ is the utilization factor. All roots which exceed unity and all values of ρ which lead to values of roots exceeding unity must be ignored, for reasons of convergence, since the steady-state probability $p_n(.,.)$ of n units in the system is a linear combination of the nth powers of these roots.

The critical value of ρ for which it is possible to write the probabilities is the maximum utilization. Thus we discard r_1 and find that for $\rho \geq \frac{2}{3}$ we must also discard r_3. Therefore $\rho = \frac{2}{3}$ is the maximum utilization, and the general solution for each steady-state probability for $\rho < \frac{2}{3}$ and for $n \geq 4$ is given by

$$p_n(0,0) = \sum_{j=2}^{4} c_{1j}r_j^n, \quad p_n(0,1) = \sum_{j=2}^{4} c_{2j}r_j^n, \quad p_n(1,1) = \sum_{j=2}^{4} c_{3j}r_j^n. \quad (12\text{-}40)$$

For $n < 4$ they are directly obtained from the steady-state equations. $p_0(0,0)$ is computed from

$$\sum_{n=0}^{\infty} [p_n(0,0) + p_n(0,1) + p_n(1,1)] = 1. \quad (12\text{-}41)$$

The expected number of units in the system is given by

$$L = \sum_{n=0}^{\infty} [np_n(0,0) + (n + 1)p_n(0,1) + (n + 2)p_n(1,1)]. \quad (12\text{-}42)$$

A simple procedure for computing the maximum utilization for the two-phase system is to define

$$Q(0,0,t) = \sum_{n=0}^{\infty} P_n(0,0,t), \qquad Q(1,1,t) = \sum_{n=0}^{\infty} P_n(1,1,t),$$

$$Q(0,1,t) = \sum_{n=0}^{\infty} P_n(0,1,t).$$

On assuming service rates μ_1 and μ_2 in the first and second phases, respectively, one has

$$
\begin{aligned}
Q'(0,0,t) &= -\mu_1 Q(0,0,t) + \mu_2 Q(0,1,t) + \mu_1 P_0(0,0,t), \\
Q'(0,1,t) &= -(\mu_1 + \mu_2)Q(0,1,t) + \mu_1 Q(0,0,t) + \mu_2 Q(1,1,t) \\
&\qquad - \mu_1 P_0(0,0,t) + \mu_2 P_0(0,1,t), \\
Q'(1,1,t) &= -\mu_2 Q(1,1,t) + \mu_1 Q(0,1,t) - \mu_1 P_0(0,1,t).
\end{aligned}
\quad (12\text{-}43)
$$

Let $\mu_1 = \mu_2 = \mu$; then passing to the steady state we note that, with increased utilization, $P_0(0,0,t)$ and $P_0(0,1,t)$ are negligible. Thus

$$\lim_{t \to \infty} Q(.,.,t) \equiv q(.,.) \quad (12\text{-}44)$$

are all equal.

The fraction of time during which the first-phase service is free is

$$\frac{q(0,0) + q(0,1)}{q(0,0) + q(0,1) + q(1,1)} = \frac{2}{3}$$

because of the last assertion. This is the same value of maximum utilization obtained above by arguing on the roots.

In the case of k phases, more blocking occurs in the first phase than in any other phase, and the maximum utilization for the first phase denoted by ρ_{max} comprises the maximum utilization for the system.

For the above case with $\mu_1 \neq \mu_2$, we have

$$\rho_{max} = \frac{\mu_2(\mu_1 + \mu_2)}{\mu_1^2 + \mu_1\mu_2 + \mu_2^2}. \quad (12\text{-}45)$$

The limiting value of the arrival rate that can be tolerated in the steady state is $\lambda_{max} = \mu_1\rho_{max}$. This last procedure has been applied to three stages with eight state probabilities involved, giving ρ_{max}, which, when all service rates are equal, is

$$\rho_{max} = 22\!\!/\!\!39 \sim 0.5641.$$

For four stages there are 21 state probabilities and similarly

$$\rho_{max} = 0.5115.$$

When a finite queue of length $q - 1$ is allowed in the second phase, there are $q + 2$ state probabilities and

$$\rho_{max} = \frac{\mu_2(\mu_1^{q+1} - \mu_2^{q+1})}{\mu_1^{q+2} - \mu_2^{q+2}}, \tag{12-46}$$

which for $\mu_1 = \mu_2$ becomes $\rho_{max} = (q + 1)/(q + 2)$.   Fifteen state probabilities are obtained if a queue of size one is allowed in the second and third phases.   For equal service rate

$$\rho_{max} = \frac{8529}{12,721} \sim 0.6705.$$

For the case of an infinite queue allowed in the first phase and no queues in the remaining $k - 1$ phases and where a vacant service facility is not allowed (i.e., a unit waits in the service channel until the subsequent channel becomes available), we have

$$\rho_{max} = \frac{1}{\mu \tau_k}.$$

Here τ_k is the mean of the probability that service in all k phases has been completed in time t after the start, if the same service rate is assumed for all phases.   It is given by

$$\tau_k = k \int_0^\infty t\mu e^{-\mu t}(1 - e^{-\mu t})^{k-1} \, dt = -k \left[\frac{dB(p/\mu, N)}{dp} \right]\Bigg|_{p=\mu}. \tag{12-47}$$

The last expression on the right is obtained using the Laplace transform; B is the beta function.

The distribution found in the integrand is a generalization of the expression obtained for two phases where one argues that the probability of completing service in all phases is the sum of two identical expressions. Each expression is the product of two quantities, of which the first is the probability of completing service at a phase during $(t, t + dt)$ and the second is the probability of completing service in time $\leq t$ at the other phase.

For three phases $\rho_{max} = \frac{6}{11}$, and for four phases it is $\frac{12}{25}$.   The method is applicable to the case where the service rates are different.

12-7. Some Waiting-time Results

We assume

$$\text{Prob } (x_j > \tau) = K_j e^{c_j \tau} \tag{12-48}$$

for the probability of waiting longer than time τ at the jth phase, which consists of r_j channels in parallel.   In this expression, let [see (12-31)]

$$c_j = -r_j \mu_j(1 - \rho_j), \tag{12-49}$$

$$K_j = \text{Prob } (x_j > 0). \tag{12-50}$$

The cumulative probability distribution for the total waiting time through all k phases can then be shown to be

$$\text{Prob}\left(\sum_{j=1}^{k} x_j \leq \tau\right) = 1 - \sum_{j=1}^{k} A_{jk}e^{c_j\tau}, \qquad (12\text{-}51)$$

where

$$A_{jk} = K_j\left(\prod_{\substack{i=1 \\ i \neq j}}^{k} \frac{1 - K_i c_j}{c_j - c_i}\right), \qquad j = 1, \ldots, k, \qquad (12\text{-}52)$$

and

$$c_i \neq c_j \qquad \text{for } i \neq j.$$

The proof is inductive.   It assumes the result for $y = \displaystyle\sum_{j=1}^{k-1} x_j$ and proves it for k.   Thus

$$\text{Prob } (y + x_k \leq \tau) = \text{Prob } (y = 0, x_k = 0) + \text{Prob } (y = 0, 0 < x_k \leq \tau)$$
$$+ \text{Prob } (0 < y \leq \tau, x_k = 0) + \text{Prob } (0 < y \leq \tau, 0 < x_k \leq \tau). \quad (12\text{-}53)$$

When each quantity is replaced by its equivalent, the result follows.

Nelson [516] computes each c_j and each K_j, using the number of channels in parallel in that phase; employs the result for the probability of waiting longer than time zero from the several channels in parallel (independence assumed, based on the works of Burke and Reich); and then computes A_{jk} from the above formula.   A computational procedure is outlined for the case $c_i = c_j$ for some i in which each of the two quantities is perturbed by a small quantity Δ which is added to one and subtracted from the other.

DeBaun and Katz [148] use the chi-square distribution to approximate the sum of the exponentials by equating means and variances to those of the density function obtained from the cumulative distribution in (12-51). This gives a useful result in applications, with appropriate precautions for applicability (see, for example, Ref. 643).

Remark 1: J. Sacks [646] has studied the problem of ergodicity of queues in series.   In the case of two queues in series, if the waiting time in the first queue of the nth unit is w_n and w_n^* is its waiting time in the second (it must go through both queues) and if the joint distribution (w_n, w_n^*) converges to a probability distribution as $n \to \infty$, then the queueing system is said to be ergodic.

Remark 2: J. W. Cohen [123] has studied the situation of a Poisson input to multiple parallel channels whose output feeds into another set of multiple parallel channels, etc.   A number of such systems are taken in parallel.   The output of all these sets at a certain point forms the input to a single-channel queue.

CHAPTER 13

INTERESTING QUEUEING PHENOMENA

13-1. Introduction

In this chapter we shall consider several cases which constitute interesting aspects of the structure of a queue. We have already mentioned in the first chapter various phenomena which can occur in a queue. Here, however, we shall analyze some aspects of these phenomena sometimes briefly and sometimes in detail. We attempt to maintain some sort of order by starting with events that influence the input, the queue, the service channel, and then the output. For example, a customer may balk or renege. Also, a customer may require some orientation before starting service. Customers may arrive late even though they are scheduled. When there are several queues, customers may jockey back and forth among them. The number of channels may vary with the length of the queue. The channels may cooperate to serve the needs of a customer. They may provide special service. Also they may have different service rates. There may be a single server who moves with a moving service belt. Finally, customers may cycle back to the waiting line after having received service and wait for more service.

Our treatment of these subjects is by no means exhaustive. It is aimed at stimulating interest in the different subjects. To begin with, we point to recent work on calls for initial and extra service. O. Swensson [680] considers an interesting class of waiting-line problems. A finite number of customers make calls for service on a system with several servers. If no server is free, the customer waits until one is available. Each customer may call for additional service either while he is being served or while he is waiting for service. The arrivals of original calls, as well as those of calls for extra service, have Poisson distributions. The author allows for different rates for initial calls and the two types of extra service calls but, for the most part, considers special cases in which at least two of the three rates are equal. Service rates for initial and extra service he again considers mainly in the special, equal-rate case. Two types of service procedure are examined—one in which calls are processed in the order of their arrival and the other in which service is given in the order of the arrival of the initial calls of the customers.   Approximate solutions are obtained for both types.

13-2. Balking

In the case of balking, immediately on arrival a unit decides not to join the queue, perhaps because of the length of the queue or because of other information on the length of service.   A unit reneges (i.e., becomes impatient and leaves without having been served) after joining the queue if it is decided that the wait will be longer than can be tolerated.   The latter phenomenon is likely to happen if the queue discipline is one of random selection for service.   It is examined in the next section.

1. *First Case of Balking*

Here we have the usual Poisson assumptions with parameter λ for arrivals served in order at a single channel with exponential service times and parameter μ.

Let $P_n(t)$ denote the probability that the queue length is $\leq n$ at time t. The reader should note that here we are giving a new meaning to $P_n(t)$. Let K be the greatest length of the queue at which an arriving unit would not balk.   It has the same distribution for all arriving units.   Let $F(n) = \text{Prob } (K \leq n)$ be the balking distribution, i.e., the probability that the unit refuses to join the queue when n are in the queue.   Thus a unit joins the queue if the queue length is less than the balking value. This happens with probability $1 - F(n - 1)$.   Define $G(n) = 1 - F(n)$, $f(n) = F(n) - F(n - 1)$,   and $p_n(t) = P_n(t) - P_{n-1}(t)$.   In   general $p_0(t) = P_0(t) \neq 0$ and

$$f(0) = F(0) \neq 0,$$

which are, respectively, the probability of zero queue at time t and the probability that a unit is queue-resistant with certainty.   The equations are

$$p_n(t + \Delta t) = [1 - (\lambda + \mu) \Delta t]p_n(t) + \mu p_{n+1}(t) \Delta t \\ + \lambda p_{n-1}(t)G(n - 2) \Delta t + \lambda p_n(t)F(n - 1) \Delta t + O(\Delta t)^2. \quad (13\text{-}1)$$

The last two expressions before $O(\Delta t)^2$ describe an arrival during Δt which joins (the first) or balks (the second).

This, as usual, may be reduced to a differential-difference equation which must be adjusted for $n = 0$, using $F(-1) = 0 = F(-2)$.

If the equations are summed over n, one has

$$\frac{dP_n(t)}{dt} = \mu p_{n+1}(t) - \lambda p_n(t)G(n - 1), \quad\quad (13\text{-}2)$$

or, if one puts $\lambda G(n) \equiv \lambda_n$, this equation may be written as follows:

$$\frac{dP_n(t)}{dt} = \lambda_{n-1}P_{n-1}(t) - (\mu + \lambda_{n-1})P_n(t) + \mu P_{n+1}(t). \quad (13\text{-}3)$$

If $F(n) = 0$ for all n, the last equation becomes a special case of the birth-death-process equation.

The steady-state solution of (13-2) is given by

$$p_1 = \rho p_0, \qquad p_{n+1} = \rho G(n-1)p_n,$$
$$p_n = \rho^n c_n p_0. \qquad\qquad (13\text{-}4)$$

Here $\quad \rho = \dfrac{\lambda}{\mu}, \qquad c_0 = c_1 = 1, \qquad c_n = \prod_{i=0}^{n-2} G(i), \qquad i = 2, 3, \ldots,$

$$(13\text{-}5)$$

and p_0 is determined from $\displaystyle\sum_{n=0}^{\infty} p_n = 1$, which is a series the convergence of which [279] can be shown to be necessary and sufficient for equilibrium.

Note from the second relation before the solution that, if $p_n \neq 0$ and $p_{n+1} = 0$, then $G(n-1) = 0$; hence $p_m = 0$ for $m > n$, that is, complete balking for queue lengths $\geq n$. Also if $p_n, p_{n+1} \neq 0$, then

$$f(n) = G(n-1) - G(n) = \frac{p_0}{p_1}\left(\frac{p_{n+1}}{p_n} - \frac{p_{n+2}}{p_{n+1}}\right), \quad (13\text{-}6)$$

and $\qquad\qquad\qquad p_{n+1}^2 \geq p_n p_{n+2}. \qquad\qquad\qquad (13\text{-}7)$

Any infinite distribution of queue lengths satisfying the last inequality corresponds to a balking distribution.

We now introduce generating functions in (13-2), pass to the steady state, and relate the results with the solution previously obtained in order to derive L and the variance.

$$P(z,t) = \sum_{n=0}^{\infty} p_n(t)z^n = (1-z)\sum_{n=0}^{\infty} P_n(t)z^n \quad (13\text{-}8)$$

and $\qquad Q(z,t) = \displaystyle\sum_{n=0}^{\infty} p_{n+1}(t)F(n)z^n. \qquad\qquad (13\text{-}9)$

On multiplying (13-2) by z^n and summing, we have

$$\frac{\partial}{\partial t}\frac{P(z,t)}{1-z} = -\lambda p_0(t) + \left(\frac{\mu}{z} - \lambda\right)[P(z,t) - p_0(t)] + \lambda z Q(z,t) \quad (13\text{-}10)$$

which in the steady state (i.e., as $t \to \infty$) becomes

$$P(z) - \rho z P(z) - p_0 + \rho z^2 Q(z) = 0. \quad (13\text{-}11)$$

Hence, when $z = 1$,

$$P(0) \equiv p_0 = 1 - \rho + \rho Q(1). \quad (13\text{-}12)$$

Since $0 < p_0 \leq 1$, this implies that $(\rho - 1)/\rho < Q(1) \leq 1$. The effective value of the traffic intensity for those who join the queue is

$$\rho' = \rho - \rho Q(1), \tag{13-13}$$

since ρ was computed for both balking and joining units.

Now from the steady-state solution and the definition of $P(z)$ we have

$$P(z) = p_0 \sum_{n=0}^{\infty} \rho^n c_n z^n, \tag{13-14}$$

and $Q(z)$ may now be explicitly determined, using (13-11). Thus once p_0 is known, the queue length and balking distributions are uniquely determined.

Now

$$\rho \frac{\partial}{\partial \rho} \log p_n = \rho \frac{\partial}{\partial \rho} \left(\log c_n + n \log \rho - \log \sum_{n=0}^{\infty} \rho^n c_n \right) = n - L, \tag{13-15}$$

where L is the mean queue length.

The queue-length variance is easily seen to be

$$\sum_{n=0}^{\infty} (n - L)^2 p_n = \sum_{n=0}^{\infty} (n - L)\rho \frac{\partial p_n}{\partial \rho} = \rho \frac{\partial}{\partial \rho} L. \tag{13-16}$$

The two expressions for $P(z)$ and $Q(z)$ give

$$\left(\rho \frac{\partial}{\partial \rho} - z \frac{\partial}{\partial z} \right) \begin{cases} \log P(z) \\ \log Q(z) \end{cases} = \begin{cases} -L, \\ 1 - L. \end{cases} \tag{13-17}$$

Differentiation of (13-11) with respect to z once and setting $z = 0$ and then another time and setting $z = 0$ give for the mean and variance, respectively,

$$L = \frac{\rho}{1 - \rho} [1 - Q'(1) - 2Q(1)], \tag{13-18}$$

$$\sigma^2 = \frac{\rho}{(1 - \rho)^2} \{1 - \rho[Q'^2(1) + 4Q^2(1) + 4Q(1)Q'(1)]$$
$$- (1 - \rho)[Q''(1) + 5Q'(1) + 4Q(1)]\}. \tag{13-19}$$

Haight applies the concepts to distributions satisfying (13-7).

In the following example we briefly illustrate some of the foregoing ideas.

Example: The binomial case gives

$$p_k = \binom{n}{k} p^k (1 - p)^{n-k}, \qquad k = 0, \ldots, n,$$

where $0 < p < 1$. Then $\rho = np/(1 - p)$, $L = np/(n + \rho) \to n$ as

$\rho \to \infty$. Note that ρ is determined using (13-4) and the fact that $c_1 = 1$ after substituting for p_0. The balking distribution is

$$G(k) = \begin{cases} \dfrac{n - k - 1}{n(k + 2)}, & 0 \le k \le n - 1, \\ 0, & n \le k, \end{cases}$$

$$f(n) = \begin{cases} \dfrac{n + 1}{n} \dfrac{1}{(k + 1)(k + 2)}, & 0 \le k \le n - 1, \\ 0, & n \le k, \end{cases}$$

which is independent of p. On solving for p in terms of ρ, we have

$$p_0 = \left(\frac{n}{n + \rho}\right)^n, \qquad P(z) = \left(\frac{n + \rho z}{n + \rho}\right)^n, \qquad \sigma^2 = \left(\frac{n}{n + \rho}\right)^2.$$

Exercises: The reader may now verify the same ideas for the following distributions:

1. Negative binomial, $p_k = \dbinom{N + k - 1}{N - 1} x^k (1 + x)^{-N-k}$, $k = 0, 1,$ . . . , where $x > 0$, $N > 1$

2. Poisson, $p_k = \dfrac{\rho^k e^{-\rho}}{k!}$, $k = 0, 1, \ldots$

3. Type-III ordinate, $p_k = A(k + a)^v e^{-\lambda k}$, $k = 0, 1, \ldots,$ where $a, v, \lambda > 0$ and A is a function of these quantities

4. Normal ordinates, $p_k = A e^{-(k-m)2/2v}$, where m and v are nearly the mean and variance if v and m^2/v are large (i.e., both >9)

5. Deterministic balking in which all units have the same degree of queue resistance, that is, $f(n) = 1$ for $n = K$ and zero otherwise. In this case,

$$G(n) = \begin{cases} 1, & 0 \le n \le K - 1, \\ 0, & K \le n, \end{cases} \qquad c_n = \begin{cases} 1, & 0 \le n \le K + 1, \\ 0, & K + 2 \le n. \end{cases}$$

2. *Another Case of Balking*

For the single-channel case with Poisson arrivals with parameter λ and exponential service times with parameter μ, let $b_0 = 1$,

$$0 \le b_{i+1} \le b_i \le 1 \ (i = 0, 1, 2, \ldots),$$

give the probability that an arriving unit joins the queue when it finds i units in the system. The equation describing this problem for $P_n(t)$, the probability of n items in the system at time t [the customary $P_n(t)$], is given by T. Homma [308]:

$$P'_n(t) = \lambda[b_{n-1}P_{n-1}(t) - b_n P_n(t)] - \mu[P_n(t) - P_{n+1}(t)], \quad n \ge 1, \tag{13-20}$$
$$P'_0(t) = \mu P_1(t) - \lambda P_0(t).$$

Let $W(t)$ be the waiting time of a customer joining the queue at time t. Then define $P(w,t)$ and calculate its value:

$$P(w,t) \equiv \text{Prob } [W(t) \leq w] = \int_0^w \sum_{n=0}^\infty P_n(t) e^{-\mu y} \frac{(\mu y)^{n-1}}{(n-1)!} \mu \, dy. \quad (13\text{-}21)$$

The Laplace-Stieltjes transform of $P(w,t)$ with respect to w is given by

$$\gamma(s,t) = \int_0^\infty e^{-sw} \, dP(w,t) = \sum_{n=0}^\infty P_n(t) \left(\frac{\mu}{\mu + s} \right)^n. \quad (13\text{-}22)$$

On taking the derivative with respect to time, substituting for $P_n'(t)$ from the transient equations, and then putting

$$\frac{\partial \gamma(s,t)}{\partial t} = 0$$

to obtain the equilibrium transform of the waiting-time distribution, there results

$$\gamma(s) = \frac{\lambda}{\mu + s} \sum_{n=0}^\infty b_n p_n \left(\frac{\mu}{\mu + s} \right)^n + p_0. \quad (13\text{-}23)$$

By putting $s = 0$, this gives

$$p_0 = 1 - \rho \sum_{n=0}^\infty b_n p_n. \quad (13\text{-}24)$$

The mean $E(W)$ and variance $\sigma^2(W)$ of the waiting time W are obtained by using the coefficients of s and s^2 in the expression of $\gamma(s)$. That is,

$$E(W) = \frac{1 - p_0}{\mu} + \frac{\rho}{\mu} \sum_{n=1}^\infty n b_n p_n, \quad (13\text{-}25)$$

$$\sigma^2(W) = \frac{\rho}{\mu^2} \sum_{n=0}^\infty (n + 1)(n + 2) b_n p_n - [E(W)]^2. \quad (13\text{-}26)$$

Note that by putting $b_n = 1$ $(n \geq 0)$ the familiar result for the mean waiting time in the steady state is obtained, i.e.,

$$E(W) = \frac{\rho}{\mu(1 - \rho)},$$

since

$$\lim_{t \to \infty} P_n(t) = p_n = (1 - \rho)\rho^n, \qquad n \geq 0,$$

and

$$\sum_{n=1}^\infty n p_n = \frac{\rho}{(1 - \rho)}.$$

Homma has also studied this type of problem for a general input distribution and an exponential-service-time distribution in the multiple-channel case, giving conditions for equilibrium and using the imbedded-Markoff-chain concept.

Exercise 6: Justify the equations describing the above system and verify the expression for the mean and variance.

3. *Balking in a System with Arbitrary Input*

In the case of a single channel with arbitrary input and exponential service distributions, let $\tau_1, \ldots, \tau_n, \ldots$ denote the arrival times [with $A(x)$ denoting the distribution of $\tau_n - \tau_{n-1}$ assumed independently and identically distributed]. Let the random variable $\eta(t)$ denote the number in queue at time t. Define $\eta_n = \eta(\tau_n - 0)$.

Let $\{\gamma_n\}$ be a sequence of random variables which assume only integer values and are independently and identically distributed. The maximum queue length allowed is $N + 1$, including the unit in service. When this queue length is attained, an arriving unit will balk. Generally a unit arriving at τ_n will join the queue if, and only if, $\eta_n \leq \min (N, \gamma_n)$. A non-increasing sequence (the balking sequence) of nonnegative real numbers b_m with $b_0 = 1$ defined by

$$b_m = \text{Prob} (\gamma_n \geq m), \qquad m = 0, 1, \ldots ,$$

gives the probability that an arriving unit which finds m units ahead of it joins the queue. It is desired to determine the queue-length (including the unit in service) distribution

$$p_k = \lim_{n \to \infty} \text{Prob} (\eta_n = k), \qquad k = 0, 1, \ldots , N + 1.$$

Finch [213] proved the existence of p_k and its independence of the initial state.

For the arbitrary-input case we have

$$p_k = q_{N+1-k} \Big/ \sum_{j=0}^{N+1} q_j, \qquad k = 0, \ldots , N + 1, \qquad (13\text{-}27)$$

where q_i are defined by

$$q(z) = \frac{(1 - z)\, \tilde{A}[\mu(1 - z)]}{1 - \tilde{A}[\mu(1 - z)]} \, \overline{q \cdot \beta(z)}, \qquad q_0 = 1, \qquad (13\text{-}28)$$

and where $\quad q(z) = \sum_{i=0}^{\infty} q_i z^i, \qquad \overline{q \cdot \beta(z)} = \sum_{j=1}^{\infty} q_j \beta_j z^{j-1},$

$$\tilde{A}(s) = \int_0^{\infty} e^{-sx}\, dA(x),$$

and β_i $(i = 0, 1, \ldots)$, $0 < \beta_i \leq 1$, are arbitrary but yield convergence $(\beta_k = 1, k > N + 1)$. The q_i can be obtained by differentiating (13-28) or from the recursive relation

$$q_k = r_0\beta_{k+1}q_{k+1} + \sum_{i=0}^{k-1} (r_{i+1} - r_i)\beta_{k-i}q_{k-i} + \sum_{i=0}^{k} r_iq_{k-i}, \qquad k = 0, 1, \ldots ,$$
$$\tag{13-29}$$

where

$$r_i = \int_0^\infty e^{-\mu x} \frac{(\mu x)^i}{i!} \, dA(x). \tag{13-30}$$

This result for p_k agrees with a similar result due to Takács' for $b_m = 1$ ($m = 0, \ldots , N$) and with Haight for a Poisson input. Its verification is left as an exercise.

If the input is an Erlang E_2 process, i.e.,

$$dA(x) = \lambda^2 x e^{-\lambda x}, \qquad x \geq 0, \tag{13-31}$$

then $\tilde{A}(s) = \lambda^2/(\lambda + s)^2$. This gives, for $\rho = \lambda/\mu$,

$$p_k = \left(\frac{\rho^2}{2\rho + 1}\right)^k \frac{b_{k-1}b_{k-2} \cdots b_0}{w_{k-1}w_{k-2} \cdots w_0} p_0, \qquad k = 1, \ldots , N+1, \tag{13-32}$$

where p_0 is determined as usual and where w_k is given by the continued fraction

$$w_k = 1 - \frac{db_{k+1}}{1-} \frac{db_{k+2}}{1-} \cdots \frac{db_N}{1-}, \qquad k = 0, 1, \ldots , N - 1,$$
$$w_N = 1,$$

and $d = [\rho/(2\rho + 1)]^2$.

Theorem 13-1: If $A(x)$ is not a lattice distribution, i.e., if the distribution is not just defined for integer values of x, we can assert that the steady-state distribution of the number in the queue,

$$\lim_{t \to \infty} \text{Prob } [\eta(t) = k] = p_k, \qquad k = 0, \ldots , N + 1,$$

exists and is independent of the initial state. It is given by

$$p_k = \frac{\lambda}{\mu} b_{k-1}p_{k-1}, \qquad k = 1, 2, \ldots , N + 1, \tag{13-33}$$

$$p_0 = 1 - \frac{\lambda}{\mu} \sum_{n=0}^{N} b_n p_n.$$

Proof: The proof of existence depends on the following transition probabilities of the finite Markoff chain η_n ($n = 1, 2, \ldots$):

$$\text{Prob } (\eta_{n+1} = k | \eta_n = j) = b_j r_{j,k} + (1 - b_j) r_{j-1,k} \equiv \pi_{j,k},$$
$$j, k = 0, \ldots , N, \tag{13-34}$$
$$\pi_{N+1,k} = r_{N,k},$$

where

$$r_{N,k} = \begin{cases} 0, & j+1-k < 0, \\ \int_0^{\infty} e^{-\mu x} \dfrac{(\mu x)^{j+1-k}}{(j+1-k)!}\, dA(x) = r_{j+1-k}, & j+1-k \geq 0. \end{cases}$$

(13-35)

The Markoff chain is ergodic; hence p_k exist and are independent of the distribution of η_1, the first η_n.

The p_k are uniquely determined by

$$p_k = \sum_{j=k-1}^{N+1} \pi_{j,k} p_j, \qquad k = 0, \ldots, N,$$

(13-36)

$$\sum_{k=0}^{N+1} p_k = 1.$$

Substituting for $\pi_{j,k}$, writing $Q_k = p_{N+1-k}$ $(k = 0, 1, \ldots, N+1)$, $\beta_k = b_{N+1-k}$ $(k = 1, 2, \ldots, N+1)$, and solving for Q_k in terms of Q_0, we have $Q_k = p_0 q_k$, which gives the desired result.

13-3. Reneging

This is a case of units leaving the queue after having joined it, usually because of a long waiting time.

1. *Reneging and Balking*

Here we use the same notation as in the previously considered first case of balking for a single channel with Poisson arrivals and exponential service times. In addition, the probability of a unit reneging during Δt when there are n units in the queue is $r(n)\, \Delta t$.

We develop a set of differential-difference equations based on the possibilities of no transitions, an arrival balking, an arrival joining, a departure due to service, and a departure due to reneging. Neglecting higher-order transitions, we have

$$\begin{aligned} p_n(t + \Delta t) = {}& (1 - \lambda\, \Delta t)(1 - \mu\, \Delta t)[1 - r(n)\, \Delta t]p_n(t) \\ & + \lambda\, \Delta t(1 - \mu\, \Delta t)[1 - r(n)\, \Delta t]p_n(t)F(n-1) \\ & + \lambda\, \Delta t(1 - \mu\, \Delta t)[1 - r(n-1)\, \Delta t]p_{n-1}(t)G(n-2) \\ & + (1 - \lambda\, \Delta t)\mu\, \Delta t[1 - r(n+1)\, \Delta t]p_{n+1}(t) \\ & + (1 - \lambda\, \Delta t)(1 - \mu\, \Delta t)r(n+1)\, \Delta t p_{n+1}(t). \end{aligned}$$

(13-37)

When $n = 0$, μ is omitted in the first term and $r(0) = r(1) = 0$. This may be simplified as usual to yield

$$\frac{d}{dt} p_n(t) = -[\lambda + \mu + r(n)]p_n(t) + \lambda p_n(t)F(n-1)$$
$$+ \lambda p_{n-1}(t)G(n-2) + \mu p_{n+1}(t) + r(n+1)p_{n+1}(t), \quad (13\text{-}38)$$

which, on summing from zero to n and simplifying, yields

$$\frac{d}{dt} P_n(t) = \mu p_{n+1}(t) - \lambda G(n-1)p_n(t) + r(n+1)p_{n+1}(t). \quad (13\text{-}39)$$

Assuming that, as $t \to \infty$, $p_n(t)$ and $P_n(t)$ tend to a limit, we have

$$\mu p_{n+1} + r(n+1)p_{n+1} = \lambda G(n-1)p_n \quad (13\text{-}40)$$

which has the solution

$$p_n = c_n \left(\frac{\lambda}{\mu}\right)^n p_0, \quad (13\text{-}41)$$

where p_0 is determined from $\sum\limits_{n=0}^{\infty} p_n = 1$ and where, if we write

$$R(n) = \frac{\lambda}{\mu + r(n)}, \quad (13\text{-}42)$$

then

$$c_n = \prod_{j=0}^{n-2} G(j)R(j+2). \quad (13\text{-}43)$$

By appropriate specialization of the functions one has the solution only for reneging or only for balking [280].

Exercise 7: Verify the last statement for the case of balking previously obtained.

2. *Impatient Customers*

D. Y. Barrer [26, 27] has studied the problem of a unit leaving a queue after having waited longer than an acceptable time. He introduces the ideas of a unit being acquired for service, i.e., accepted to be served. A unit may wait in line yet not be acquired for service. A unit will wait if it is acquired for service before its prescribed waiting limit has been reached; otherwise it reneges. Another case is treated where the unit reneges once its prescribed waiting limit has been reached, whether acquired for service or not. The first case is treated with ordered (multiple-channels) selection for service and the second with random (single-channel) selection. The random-selection case is applicable to $c > 1$ channels (see the work of Gnedenko in Chap. 9).

In the case of impatient customers and random selection for service in a Poisson-input, single-channel, exponential-service-times queue, if $P_n(t)$ denotes the probability that there are n units in the queue at time t and if $C_n(\tau_0,t) \, \Delta t + o(\Delta t)$ is the conditional probability that a unit reneges in $(t, t + \Delta t)$ if the queue length is n at time t and its fixed waiting limit is τ_0, we have, in the usual manner,

$$P_0'(t) = -\lambda P_0(t) + [\mu + C_1(\tau_0,t)]P_1(t),$$
$$P_n'(t) = \lambda P_{n-1}(t) - [\lambda + \mu + C_n(\tau_0,t)]P_n(t)$$
$$+ [\mu + C_{n+1}(\tau_0,t)]P_{n+1}(t), \qquad n \geq 1. \quad (13\text{-}44)$$

Barrer points out that, unlike the usual case of remaining for service no matter what, specifying the number of units in the queue at time t does not generally completely determine the future behavior of the queue. This is, of course, true because units leave the queue by becoming impatient. Except for special cases, what is needed is to specify the time or the probability distribution of the time which a unit has waited. This is needed to determine $C_n(\tau_0,t)$ and subsequently leads to initial conditions required for solving the above system. To determine $C_n(\tau_0,t)$, let $C_n(\tau,t,\Delta\tau) \, \Delta\tau$ be the probability that, if the queue length was n at time t, the queue has a unit that has waited between $\tau - \Delta\tau$ and τ. Since the probability of obtaining service during $\Delta\tau$ by being chosen at random is

$$\frac{\mu \, \Delta\tau}{n} + o(\Delta\tau),$$

we have

$$C_n(\tau,t,\Delta\tau) \, \Delta\tau \left(1 - \frac{\mu \, \Delta\tau}{n}\right) = C_n(\tau + \Delta\tau, t + \Delta\tau, \Delta\tau) \, \Delta\tau. \quad (13\text{-}45)$$

On simplifying, this leads to

$$\frac{\partial}{\partial\tau} C_n(\tau,t,0) + \frac{\partial}{\partial t} C_n(\tau,t,0) = -\frac{\mu}{n} C_n(\tau,t,0) \quad (13\text{-}46)$$

which has the solution

$$C_n(\tau,t,0) = f_n(t - \tau)e^{-\mu\tau/n}, \quad (13\text{-}47)$$

where the function $f_n(t - \tau)$ is determined from appropriate initial conditions. If the left side is assumed continuous in $\Delta\tau$ at $\Delta\tau = 0$, one may write

$$C_n(\tau,t,\Delta\tau) \, \Delta\tau = f_n(t - \tau)e^{-\mu\tau/n} \, \Delta\tau + o(\Delta\tau).$$

The desired $C_n(\tau_0,t) \, \Delta\tau$ is the left side of the last relation with $\tau = \tau_0$; hence

$$C_n(\tau,t) = f_n(t - \tau)e^{-\mu\tau/n}. \quad (13\text{-}48)$$

The solution of the given system of equations in equilibrium may be obtained from our study of the equilibrium solution of the birth-death process, having written $C_n = \lim_{t \to \infty} C_n(\tau_0,t) = A_n e^{-\mu\tau_0/n}$, where A_n is determined from

$$\int_0^{\tau_0} C_n(\tau,t)\, d\tau = n = A_n \frac{n}{\mu} (1 - e^{-\mu\tau_0/n}).$$ (13-49)

A unit which reneges is said to survive, and the average rate for this is

$$C = \sum_{n=1}^{\infty} p_n C_n = \lambda - \mu(1 - p_0).$$ (13-50)

Thus the probability of a unit being lost is

$$\frac{1 - (1 - \rho) \sum_{n=1}^{\infty} \rho^{n-1} \prod_{k=1}^{n} (1 - e^{-\mu\tau_0/k})}{1 + \sum_{n=1}^{\infty} \rho^n \prod_{k=1}^{n} (1 - e^{-\mu\tau_0/k})}.$$ (13-51)

The case of ordered-queue discipline and c channels in parallel with Poisson input and identical exponential service is described by

$$
\begin{aligned}
P_0'(t) &= -\lambda P_0(t) + \mu P_1(t), \\
P_n'(t) &= \lambda P_{n-1}(t) - (\lambda + n\mu)P_n(t) + (n+1)\mu P_{n+1}(t), && n < c, \\
P_n'(t) &= \lambda P_{n-1}(t) - [\lambda + c\mu + C_n(t)]P_n(t) \\
&\quad + [c\mu + C_{n+1}(t)]P_{n+1}(t), && n \geq c.
\end{aligned}
$$ (13-52)

Here $C_n(t)\, \Delta t + o(\Delta t)$ gives the probability that the unit is lost in $(t,\, t + \Delta t)$ when the queue length is n at time t.

Palm has solved the reneging problem for the Poisson case with c channels in the steady state [551].   See also Cohen's work in Prob. 10, Chap. 14.

13-4. Oriented and Retarded Arrivals

R. Leroy and A. E. Vaulot [429] have studied the following problem: Suppose that, in a queueing system with c channels in parallel (telephone lines), a call requires a period of orientation (before obtaining service) which is exponentially distributed with average duration b.   The duration distribution of the service itself (of a conversation) is also exponential and has a mean which we take as the unit of time.   Let λ be the average number of arrivals per unit time from a Poisson input.   Let p_i be the probability that $0 \leq i \leq c$ channels are actually occupied by conversations and let q_j ($0 \leq j \leq c$) be the probability of occupation of c channels of which $j - 1$ are occupied by conversations and $c - j + 1$ are occupied by calls in orientation.   If all channels are occupied, a call is lost.   Then the steady-state equations are

$$cp_c = \frac{q_c}{b},$$

$$(\lambda + i)p_i = (i + 1)p_{i+1} + \frac{q_i}{b}, \qquad 0 \le i < c,$$

$$\left(c - 1 + \frac{1}{b}\right)q_c = \lambda p_{c-1},$$

$$\left(j - 1 + \frac{1}{b}\right)q_j = \lambda p_{j-1} + jq_{j+1}, \qquad 0 < j < c,$$

(13-53)

with
$$\sum_{i=0}^{c} p_i + \sum_{j=1}^{c} q_j = 1.$$

Summing the above system, using the last equation, yields

$$b\lambda \sum_{i=0}^{c-1} p_i = \sum_{j=1}^{c} q_j = 1 - \sum_{i=0}^{c} p_i. \tag{13-54}$$

The proportion of lost calls is

$$p_c + \sum_{j=1}^{c} q_j = \frac{b\lambda + p_c}{b\lambda + 1}. \tag{13-55}$$

The steady-state equation permits one to express p_i and q_j in terms of p_c. The q_j satisfy

$$\frac{q_j}{b} = j\left(\frac{1}{b\lambda} + \frac{\lambda + j}{\lambda}\right)q_{j+1} - j\frac{j+1}{\lambda}q_{j+2}, \qquad 0 < j < c. \tag{13-56}$$

Using the notation $(N,n) = N(N + 1) \cdots (N + n - 1)$, we have

$$\frac{q_{c-j}}{(c + 1, j + 1)bp_c} = \left(c + \lambda + \frac{1}{b}, 1\right)\left(\frac{b}{\lambda}\right)^j$$
$$- \frac{(j - 1, 1)}{(1,1)}\left(c + \lambda + \frac{1}{b}, 2\right)\left(\frac{b}{\lambda}\right)^{j-1}$$
$$+ \frac{(j - 3, 2)}{(1,2)}\left(c + \lambda + \frac{1}{b}, 3\right)\left(\frac{b}{\lambda}\right)^{j-2}$$
$$- \frac{(j - 5, 3)}{(1,3)}\left(c + \lambda + \frac{1}{b}, 4\right)\left(\frac{b}{\lambda}\right)^{j-3} + \cdots \tag{13-57}$$

which gives

$$\frac{\sum_{j=1}^{c} q_j}{bp_c} = (c,1) + (c - 1, 2)\left(\lambda + \frac{1}{b} + 1, 1\right)\left(\frac{b}{\lambda}\right)$$
$$+ (c - 2, 3)\left(\lambda + \frac{1}{b} + 1, 2\right)\left(\frac{b}{\lambda}\right)^2 + \cdots$$
$$+ (0,c)\left(\lambda + \frac{1}{b} + 1, c - 1\right)\left(\frac{b}{\lambda}\right)^c. \tag{13-58}$$

On denoting the right side by Q, we have for the probability that all channels are busy

$$p_c = \frac{\lambda}{\lambda + (1 + b\lambda)Q},\tag{13-59}$$

and the proportion of lost calls becomes

$$\frac{b\lambda}{b\lambda + 1} + \frac{\lambda}{b\lambda + 1}\frac{1}{\lambda + (1 + b\lambda)Q} = \frac{\lambda(1 + bQ)}{\lambda + (1 + b\lambda)Q}.$$

For fixed b and λ as $c \to \infty$, E tends to $b\lambda/(b\lambda + 1)$. As $\lambda \to \infty$, the proportion of lost calls tends to one for any c. As $b \to 0$, one has the well-known Erlang loss formula

$$\frac{\lambda^c/c!}{1 + \lambda + (\lambda^2/2!) + \cdots (\lambda^c/c!)}.\tag{13-60}$$

This expression tends to zero if $\lambda \to \infty$ and $c \to \infty$ in a fixed ratio.

For a more elegant formulation of this problem in a different context, see the common-control loss system in Chap. 14, where the ideas are also extended to waiting systems.

13-5. Scheduled but Inexact Arrivals

Winsten [768] has studied a single-channel queueing problem with constant input and exponential service time. He obtains the equilibrium distribution of the number in the queue by studying the different trajectories that the queue can follow between two points which constitute arrival times. He obtains for π_n the probability that there are n units in the system an instant before an arrival occurs.

$$\pi_n = \pi^n\pi_0,\tag{13-61}$$

where π is obtained from the relation

$$\pi e^\lambda = e^{\lambda\pi}\tag{13-62}$$

and as usual $\pi_0 = 1 - \pi$.

For the case of units arriving late even though the schedule of arrival times requires constant arrivals, if $P(\leq t)$ is the probability that the delay is not more than t and if we require that $0 \leq t < 1$ so that even though a unit is late it arrives before the arrival time of the unit immediately following, one can give expressions for the distribution of queue length (1) at the time in which arrivals are scheduled; (2) at the time they actually occur, just before the scheduled time; and (3) in general. The first is given by

$$\pi_n = \pi_1\pi^{n-1}, \qquad n \geq 1.\tag{13-63}$$

Thus
$$\sum_{n=0}^{\infty} \pi_n = 1$$

gives
$$\pi_1 = (1 - \pi)(1 - \pi_0), \qquad (13\text{-}64)$$

$$\pi_0 = 1 - \pi \frac{\int_0^1 e^{\lambda t}\, dP(\leq t)}{\int_0^1 e^{\lambda \pi t}\, dP(\leq t)}, \qquad (13\text{-}65)$$

where π is determined as before.

The length of the queue at the actual arrivals requires that we compute $\pi_n(t)$ in terms of π_n and average over t. This gives

$$\pi_n(t) = \sum_{i=0}^{\infty} \pi_{n+i} e^{-\lambda t} \frac{(\lambda t)^i}{i!}, \qquad n \geq 1. \qquad (13\text{-}66)$$

Thus
$$\pi_n(t) = \pi_1 \pi^{n-1} e^{-\lambda t}(1 - \pi),$$

$$\pi_0(t) = 1 - \frac{\pi_1}{1 - \pi} e^{-\lambda t}(1 - \pi). \qquad (13\text{-}67)$$

For the distribution of queue length just before the scheduled arrivals we have

$$\bar{\pi}_n = \bar{\pi}_1 \pi^{n-1}, \qquad n \geq 1, \qquad (13\text{-}68a)$$

$$\bar{\pi}_1 = \pi_1 \int_0^1 e^{-\lambda t(1-\pi)}\, dP(\leq t)$$

$$= \pi(1 - \pi) \frac{\int_0^1 e^{\lambda t}\, dP(\leq t) \int_0^1 e^{-\lambda t + \lambda \pi t}\, dP(\leq t)}{\int_0^1 e^{\lambda \pi t}\, dP(\leq t)} \qquad (13\text{-}68b)$$

$$\bar{\pi}_0 = \int_0^1 \pi_0(t)\, dP(\leq t). \qquad (13\text{-}68c)$$

Note that if $P(\leq t)$ has a single jump, i.e., all units are late by an equal amount, the problem reduces to the case of scheduled arrivals. Now they may be considered scheduled for a longer period which includes the delay.

Finally the general distribution p_n of queue length in the case of delayed arrivals under study is a weighted average:

$$p_n = \left(1 - \frac{\bar{\pi}_1}{1 - \pi}\right) \text{Prob (queue length} = 0)$$

$$+ \frac{\bar{\pi}_1}{1 - \pi} \text{Prob (queue length} \geq 1). \qquad (13\text{-}69)$$

But Prob (queue length ≥ 1) $= \pi^{n-1}(1 - \pi)$; hence Prob (queue length $= 0$) may be determined from $\sum\limits_{n=0}^{\infty} p_n = 1$.   The waiting time **may** be computed similarly.   The author indicates generalizations without giving explicit results.

13-6. Jockeying

The problem of jockeying back and forth among the several queues, each before one of several channels operating in parallel, is an interesting one, although it is analytically difficult to pursue very far.   We shall illustrate the ideas on a two-channel facility with Poisson input with parameter λ and exponential service times with parameters μ_1 and μ_2, respectively.   We assume that the two queues always maintain the same length unless they differ by one unit.   As soon as they differ by more than one unit, the last unit in the longer queue immediately moves to the shorter queue.   An arriving unit always joins the shorter queue unless they are equal, in which case it either joins the queue before the facility with the shorter service time, if such prior knowledge is available, or joins either queue with equal probability in the absence of prior knowledge.   We use the second possibility.

Let $P(n_1,\ n - n_1,\ t)$ be the probability at time t that there are n_1 units in the first channel system (including service) and $n - n_1$ in the second channel system, where n is the total number in both systems. We have

$$P(0, 0, t + \Delta t) = (1 - \lambda\,\Delta t)P(0,0,t) + \mu_1\,\Delta t\, P(1,0,t) + \mu_2\,\Delta t\, P(0,1,t),$$

$$P(1, 0, t + \Delta t) = [1 - (\lambda + \mu_1)\,\Delta t]P(1,0,\Delta t) + \frac{\lambda}{2}\,\Delta t\, P(0,0,t)$$
$$+ \mu_2\,\Delta t\, P(1,1,t),$$

$$P(0, 1, t + \Delta t) = [1 - (\lambda + \mu_2)\,\Delta t]P(0,1,\Delta t) + \frac{\lambda}{2}\,\Delta t\, P(0,0,t)$$
$$+ \mu_1\,\Delta t\, P(1,1,t).$$

For $n_1 > 2$ if $n = 2n_1$ we have

$$P(n_1,\ n - n_1,\ t + \Delta t) = [1 - (\lambda + \mu_1 + \mu_2)\,\Delta t]P(n_1,\ n - n_1,\ t)$$
$$+ \lambda\Delta t\,[P(n_1 - 1,\ n - n_1,\ t) + P(n_1,\ n - n_1 - 1,\ t)]$$
$$+ \mu_1\,\Delta t\, P(n_1 + 1,\ n - n_1,\ t) + \mu_2\,\Delta t\, P(n_1,\ n - n_1 + 1,\ t),$$

and if $n = 2n_1 + 1$, that is, if the second facility has one more unit than the first, we have

$$P(n_1,\ n - n_1,\ t + \Delta t) = [1 - (\lambda + \mu_1 + \mu_2)\,\Delta t]P(n_1,\ n - n_1,\ t)$$
$$+ \frac{\lambda}{2}\,\Delta t\, P(n_1,\ n - n_1 - 1,\ t) + \mu_1\,\Delta t\, P(n_1 + 1,\ n - n_1,\ t)$$

On the other hand, if the first facility has the extra unit, that is,

$$n = 2n_1 - 1,$$

we have

$$P(n_1, n - n_1, t + \Delta t) = [1 - (\lambda + \mu_1 + \mu_2) \Delta t]P(n_1, n - n, t)$$
$$+ \frac{\lambda}{2} \Delta t \, P(n_1 - 1, n - n_1, t) + \mu_2 \Delta t \, P(n_1, n - n_1 + 1, t).$$

13-7. Variable Number of Parallel Channels

In many practical situations the servers in a parallel-channel queueing process have other work to do whenever their channels are empty. An example is a tool crib where the clerks may sharpen tools and do other deferrable work when there is no one or only a few persons waiting to be served at the tool-crib window. When the number of persons waiting increases to some undesirable level, the second or third clerk (channel) will begin to assist in serving the queue. In such situations, which also exist in banks, stores, maintenance work, and other operations, the number of channels in operation is a random variable.

Romani [634] has derived steady-state probabilities for the case of Poisson input to a single queue with parameter λ and identical exponential service with parameter μ for a variable number of channels whose minimum number is one and which increase with the length of queue. The number of channels increases only after the maximum allowed length N of the queue has been reached; i.e., with each newly arriving unit after N a new channel is made available to serve the unit at the head of the queue. Channels are canceled at the termination of service if there are no units waiting, with the exception of one channel which remains open at all times. Let $P_n(m,t)$ be the probability that there are n units in the queue with m units in service (which equals the number of channels in operation) at time t. Then $p_n(m)$ will be the corresponding probability in the steady state. We have

$$P_n(m, t + \Delta t) = (1 - \lambda \, \Delta t)(1 - m\mu \, \Delta t)P_n(m,t)$$
$$+ \lambda \, \Delta t \, P_{n-1}(m,t) + m\mu \, \Delta t \, P_{n+1}(m,t),$$

$$P_N(m, t + \Delta t) = (1 - \lambda \, \Delta t)(1 - m\mu \, \Delta t)P_N(m,t)$$
$$+ \lambda \, \Delta t \, P_N(m - 1, t) + \lambda \, \Delta t \, P_{N-1}(m,t),$$

$$P_0(m, t + \Delta t) = (1 - \lambda \, \Delta t)(1 - m\mu \, \Delta t)P_0(m,t) \qquad (13\text{-}70)$$
$$+ m\mu \, \Delta t \, P_1(m,t) + (m + 1)\mu \, \Delta t \, P_0(m + 1, t),$$

$$P_0(1, t + \Delta t) = (1 - \lambda \, \Delta t)(1 - \mu \, \Delta t)P_0(1,t) + \lambda \, \Delta t \, P_0(0,t)$$
$$+ \mu \, \Delta t \, P_1(1,t) + 2\mu \, \Delta t \, P_0(2,t),$$

$$P_0(0, t + \Delta t) = (1 - \lambda \, \Delta t)P_0(0,t) + \mu \, \Delta t \, P_1(1,t).$$

When m channels operate, $m\mu\,\Delta t$ is the probability of termination of a single service during Δt. We have for initial conditions

$$P_n(m,0) = \delta_{ni}^{mj} \equiv \begin{cases} 1, & n = i, \quad m = j, \\ 0 & \text{otherwise.} \end{cases} \qquad (13\text{-}71)$$

The steady-state equations are obtained in terms of $p_n(m)$ in the usual manner, and then these equations are solved. The results are as follows, where $\rho = \lambda/\mu$ (rather than $\rho = \lambda/m\mu$, which is the usual definition in fixed multiple-channel models):

$$p_n(1) = \frac{(\rho/1!)\displaystyle\sum_{k=0}^{N-n}\rho^{-k}}{(N+1)e^\rho + \displaystyle\sum_{k=1}^{N}(N-k+1)\rho^{-k}}, \qquad (13\text{-}72)$$

$$p_n(m) = \frac{\rho^m/m!}{(N+1)e^\rho + \displaystyle\sum_{k=1}^{N}(N-k+1)\rho^{-k}}. \qquad (13\text{-}73)$$

The reader can verify the result for an infinite-channel queue which is obtained by putting $N = 0$ (ignore the second term of the denominator). This gives

$$p_n(m) = \frac{\rho^m}{m!}\,e^{-\rho}. \qquad (13\text{-}74)$$

If we let N approach infinity, the result is a single-channel queue, and it may be shown that

$$\lim_{N\to\infty} p_{n-1}(1) = \rho^n(1-\rho). \qquad (13\text{-}75)$$

One can also verify that the expected number of units in service is given by

$$\sum_{m=0}^{\infty} m p_n(m) = \rho. \qquad (13\text{-}76)$$

A variation on Romani's model has been studied by Phillips [555]. In this case the number of channels is limited to a maximum M, and when M channels are operating the queue is allowed to grow without limit. The process is otherwise the same as Romani's except for a difference in notation—in this model the "shift point" in the queue length is considered N rather than $N+1$. The process requires eight steady-state equations, which were studied by means of recursive solutions for several particular cases.   The general solution was found to require the following expressions:

$$p_n(1) = \frac{\rho^{n+1}(1 - \rho^{N-n})}{1 - \rho^N} \, p_0(0), \qquad \begin{array}{l} 0 \leq n \leq N + 1, \\ m = 1, \end{array} \quad (13\text{-}77a)$$

$$p_n(m) = \frac{\rho^{N+m-1}(1 - \rho)}{m!(1 - \rho^N)} \, p_0(0), \qquad \begin{array}{l} 0 \leq n \leq N - 1, \\ 2 \leq m \leq M - 1, \end{array} \quad (13\text{-}77b)$$

$$p_n(M) = \frac{\rho^{N+M-1}(1 - \rho)(M^{n+1} - \rho^{n+1})}{M^n M!(1 - \rho^N)(M - \rho)} \, p_0(0),$$

$$\begin{array}{l} 0 \leq n \leq N - 2, \\ m = M, \end{array} \quad (13\text{-}77c)$$

$$p_n(M) = \frac{\rho^{n+M}(1 - \rho)(M^N - \rho^N)}{M^n M!(1 - \rho^N)(M - \rho)} \, p_0(0), \qquad \begin{array}{l} N - 1 \leq n \leq \infty, \\ m = M, \end{array} \quad (13\text{-}77d)$$

where

$$p_0(0) = \left\{ 1 + \frac{\rho}{1 - \rho} - \frac{N\rho^{N-1}}{1 - \rho^N} + \frac{N(1 - \rho)\rho^{N-1}}{1 - \rho^N} \right.$$

$$\left. \times \left[\sum_{m=0}^{M-1} \frac{\rho^m}{m!} + \frac{\rho^M}{(M - 1)!(M - \rho)} \right] \right\}^{-1}. \quad (13\text{-}78)$$

It can be shown that these equations give the same result as fixed multiple-channel models if $N = 1$.

It is interesting to compare the above equation for $p_n(m)$ with the corresponding equation in Romani's solution.   In both cases the expressions are independent of n, implying that whenever a given number of channels (except one or M channels) are operating, all lengths of queue less than the shift point are equally likely.

A new measure of effectiveness appropriate to variable channel models, the mean interruption rate, was also developed by Phillips.   This is a measure of the frequency of channel starts, or the frequency of interruptions for the servers who are doing deferrable work.   This measure is derived from the fraction of time during which the system is vulnerable to an interruption.   In the limited channel case, this fraction is

$$\sum_{m=1}^{M-1} p_{N-1}(m).$$

This is then multiplied by λ and divided by the number of variable servers $M - 1$, to give the desired measure:

$$\frac{\lambda}{M - 1} \sum_{m=1}^{M-1} p_{N-1}(m) = \frac{\lambda\rho^{N-1}(1 - \rho)p_0(0)}{(M - 1)(1 - \rho^N)} \sum_{m=1}^{M-1} \frac{\rho^m}{m!} \quad (13\text{-}79)$$

which is in terms of interruptions per variable server per unit time.

Exercise 8: Under what conditions would the two variable-channel models described in this section give approximately the same results? (*Ans.* When M and N are large and ρ is relatively small.)

Exercise 9: In variable-channel processes, what is the meaning of idle time? [*Ans.* Since variable-channel models presuppose that deferrable work is always available, only the fixed channel can ever be idle. Thus $p_0(0)$ is the measure of idle time.]

13-8. Cooperating Parallel Channels

Fagen and Riordan [198] have studied, in the steady state, a Poisson-input queue with parameter λ, ordered service, c channels, where all channels cooperate to serve a unit's many needs.

The service distributions considered are:

1. The Erlangian distribution whose cumulative function is

$$B(t) = \int_0^t k \frac{(kx)^{k-1}}{(k-1)!} e^{-kx} dx, \qquad t \geq 0, \qquad k = 1, 2, \ldots,$$

which gives the cumulative exponential distribution for $k = 1$, and for $k \to \infty$ the constant-service-time distribution $B(t) = \begin{cases} 0, & t < 1, \\ 1, & t \geq 1. \end{cases}$

2. The uniform distribution

$$B(t) = \begin{cases} 0, & t < 1 - a, \\ \dfrac{t - 1 + a}{2a}, & 1 - a \leq t \leq 1 + a, \qquad 0 < a < 1, \\ 1, & t > 1 + a, \end{cases}$$

which also tends to the constant-service-time distribution as $a \to 0$.

Once the fastest server (channel) has finished, the next unit is served by all channels, and its service time then is the minimum of the service times of the c channels. The service times are independently and identically distributed according to the probability function $B(t)$. The probability that any channel has not completed its service on the unit by time t is $1 - B(t)$ and, because of the independence of this probability for all channels, we have $[1 - B(t)]^c$ as the probability that all channels have not finished service by time t. Therefore $1 - [1 - B(t)]^c = H(t)$ is the cumulative-service-time distribution of a unit. We may now treat the problem as a single-channel queue with the above service time.

Denote by μ the average service rate and let $\rho = \lambda/\mu$. Let $P(w)$ be the probability of waiting in line less than time t and define the Laplace-Stieltjes transforms

$$\gamma(s) = \int_0^\infty e^{-sw}\, dP(w),$$
$$\tilde{H}(s) = \int_0^\infty e^{-st}\, dH(t).$$

Then Eq. (9-9) gives

$$\gamma(s) = (1 - \rho)\left\{1 - \frac{\mu\rho}{s}[(1 - \tilde{H}(s))]\right\}^{-1}. \qquad (13\text{-}80)$$

Note that the probability of waiting in the system less than time w is the distribution of the sum of two random variables, the first for the time spent in queue and the second for the time in service. The Laplace-Stieltjes transform of the distribution of this sum (which is a convolution) is the product of the separate transforms, that is, $\gamma(s)\tilde{H}(s)$.

For the case $k = 1$, $H(t) = 1 - e^{-ct}$, and without difficulty we have

$$\gamma(s)\tilde{H}(s) = (c - \lambda)(c - \lambda + s)^{-1}; \qquad (13\text{-}81)$$

hence the probability of waiting in the system less than time w is given by

$$1 - e^{-(c-\lambda)w}. \qquad (13\text{-}82)$$

From this the jth moment is given by $\mu_j = j!(c - \lambda)^{-j}$. Of particular interest is the case $j = 1$, which gives the average waiting time $1/(c - \lambda)$. Waiting-time results are also computed for the case $k = 2$ of the Erlangian distribution.

The mean service time in the Poisson-input exponential-service-time multiple-channel operation where the channels serve different items simultaneously is, for unit mean service time, $1 + [C(c,\lambda)/(c - \lambda)]$, where

$$C(c,\lambda) = \frac{\lambda^n}{(c - \lambda)(c - 1)!}$$
$$\left[1 + \lambda + \cdots + \frac{\lambda^{c-1}}{(c - 1)!} + \frac{\lambda^c}{(c - \lambda)(c - 1)!}\right]^{-1}. \qquad (13\text{-}83)$$

Comparison of average waiting times by taking the ratio of the above waiting time to $(c - \lambda)^{-1}$ shows that the mean waiting time for single operation, i.e., several channels working independently, is greater than that for multiple operation by several channels on a single item. For $k = 2$ the comparison is difficult when λ exceeds a critical value, but in practice the single operation is preferred for such values. For values of λ smaller than the critical value, multiple operation is advantageous. The argument can be generalized to other cases of the Erlangian and to the uniform distribution. Constant service time is least favorable to multiple operation, whereas the most advantage is obtained from exponential or uniform service times with $a = 1$. The greater the chance of a small service time, the greater the advantage in using multiple operation.

13-9. Queueing with Special Service

E. Koenigsberg [388] has studied the problem in which c parallel channels have a subset c_1, which can provide a special type of service. Arriving units require either special service at the c_1 channels or general service at any of the c channels. The discipline is first-come–first-served. The first non-special-service unit in line may go to service at one of the remaining $c - c_1$ channels if the latter becomes vacant, even though a special-service unit may be waiting to enter service at one of the c_1 channels. Of course, a non-special-service unit may enter a special-service channel if it is first in line and the latter becomes idle. Therefore we note that no priorities are imposed. The ideas may be generalized to more than one special service, and some of the channels may specialize in one or more types of special service. At this writing, the ideas have not been sufficiently developed to give a full account. However, the reader will have no difficulty writing the equations for the Poisson arrivals with $r\lambda = \lambda_1$ for the special-service-unit arrival rate and $(1 - r)\lambda = \lambda_1$ for the non-special-service-unit arrival rate and exponential service times. A study of the equations of preemptive-priority queues would be a helpful guide. Koenigsberg also attempts a formulation for constant service times.

13-10. Different Service Rates at Parallel Channels

In the ordered-queue multiple-parallel-channel case with Poisson arrivals and parameter λ, let us assume that the exponential service differs at each of the channels. For simplicity, suppose that there are two channels with parameters μ_1 and μ_2 for the first and second, respectively. Let $P_n(t)$ be the probability that there are n units in the system at time t for $n \geq 2$. When $n = 1$ the unit may be in the first service channel or in the second. We use $P_1(1,0,t)$ to indicate that there is one unit in the system; it is in the first channel, and there is nothing in the second channel at time t. Similarly, $P_1(0,1,t)$ indicates the presence of the unit in the second channel when the first channel is empty. When both channels are empty, an arriving unit joins the first channel with probability $\mu_1/(\mu_1 + \mu_2)$ and the second with probability $\mu_2/(\mu_1 + \mu_2)$. If only one channel is available, it immediately enters service there. (Other assumptions may be that the unit joins each channel with probability $\frac{1}{2}$ when both are available for service. It may also always join the first channel.) The equations describing the system are

$$P_0'(t) = -\lambda P_0(t) + \mu_1 P_1(1,0;t) + \mu_2 P_1(0,1;t),$$

$$P_1'(1,0;t) = -(\lambda + \mu_1)P_1(1,0;t) + \frac{\mu_1}{\mu_1 + \mu_2}\lambda P_0(t) + \mu_2 P_2(t),$$

$$P_1'(0,1;t) = -(\lambda + \mu_2)P_1(0,1;t) + \frac{\mu_2}{\mu_1 + \mu_2}\lambda P_0(t) + \mu_1 P_2(t),\qquad (13\text{-}84)$$

$$P_2'(t) = -(\lambda + \mu_1 + \mu_2)P_2(t) + \lambda[P_1(1,0;t) + P_1(0,1;t)] + (\mu_1 + \mu_2)P_3(t),$$

$$P_n'(t) = -(\lambda + \mu_1 + \mu_2)P_n(t) + \lambda P_{n-1}(t) + (\mu_1 + \mu_2)P_{n+1}(t),\qquad n > 2.$$

On putting $\mu_1 = \mu_2 = \mu$, one obtains the classical equations for two parallel channels on using

$$P_1(t) = P_1(1,0;t) + P_1(0,1,t).$$

The steady-state equations hold under the assumption

$$\rho = \frac{\lambda}{\mu_1 + \mu_2} < 1$$

and are obtained by putting the time derivatives equal to zero. We have at the end of a time interval T, which tends to infinity, that the mean number of those served equals the mean number λT of those which arrived during T. The mean duration of activity of the first channel is

$$T[1 - p_0 - p_1(0,1)],\qquad (13\text{-}85)$$

and the average number of units served during T is

$$\mu_1 T[1 - p_0 - p_1(0,1)].\qquad (13\text{-}86)$$

For the second channel this number is

$$\mu_2 T[1 - p_0 - p_1(1,0)].\qquad (13\text{-}87)$$

By the foregoing statement we have

$$\lambda = (\mu_1 + \mu_2)(1 - p_0) - \mu_1 p_1(0,1) - \mu_2 p_1(1,0).\qquad (13\text{-}88)$$

This equation, together with the first (steady-state) equation, and defining $p_1 = p_1(1,0) + p_1(0,1)$, yields $p_1 = (1 - \rho)(1 - p_0)$. The other steady-state equations of the system give $p_n = \rho p_{n-1}$, which, on solving, yields $p_n = (1 - \rho)\rho^{n-1}(1 - p_0)$. The mean number in the system is $L = (1 - p_0)/(1 - \rho)$, where p_0 is determined as usual.

13-11. A Moving Single Server

B. McMillan and J. Riordan [476] have studied the following problem: A single server moves with an assembly line which itself moves at uniform speed with units spaced along it for service. Once the server finishes serving a unit, he moves back to the next unit without loss of time. The line has a barrier in which the server is absorbed; i.e., service is discontinued if he moves into the barrier. It is therefore desirable for him to move back immediately to the starting point. We assume that the service times are exponentially distributed with parameter μ, and we denote by $p(k,T)$ the probability that service is completed on k units, having started service on the first unit when it is T time units away from the barrier.

Karlin, Miller, and Prabhu [352] point out that this problem is analogous to a single queue with a single server with Poisson input and service-time distribution $F_s(t) = B(t)$. Note that $B(t)$ corresponds to the distribution of spacing between items. The time until absorption in the moving-single-server problem is equivalent to the length of a busy period for the simple queue in which the service distribution of the first item in line is $F_T(t) = \begin{cases} 1, & t > T, \\ 0, & t \leq T. \end{cases}$ The number served before absorption is one less than the number served during a busy period in which the first item is served according to $F_T(t)$. We define

$$P(x,T) = \sum_{k=0}^{\infty} p(k,T)x^k. \tag{13-89}$$

Let f_j be the probability that j items are served during a busy period in which the first item has the service distribution F_s, and let

$$F(x) = \sum_{j=1}^{\infty} f_j x^j. \tag{13-90}$$

If n items arrive during T, the service period of the first item, the probability distribution of the number of items served during the remainder of the busy period is the n-fold convolution of $\{f_j\}$. Thus

$$P(x,T) = \sum_{n=0}^{\infty} \frac{e^{-\mu T}(\mu T)^n}{n!} [F(x)]^n = e^{-\mu T[1-F(x)]}. \tag{13-91}$$

But we know from our study of the distribution of a busy period in Chap. 8 that, for $|x| \leq 1$, $F(x)$ is the unique analytic solution of the integral equation

$$F(x) = x \int_0^{\infty} e^{-\mu t[1-F(x)]} \, dB(t), \tag{13-92}$$

with $F(0) = 0$; therefore $P(x,t)$ is the unique analytic solution of

$$P(x,T) = \exp\left\{ -\mu T \left[1 - x \int_0^{\infty} P(x,t) \, dB(t) \right] \right\} \tag{13-93}$$

with $|P(x,t)| \le 1$, $|x| \le 1$, and for all t. For arrivals with constant spacing ε time units, that is, $B(t) = \begin{cases} 1, & t > \varepsilon, \\ 0, & t \le \varepsilon, \end{cases}$ we have, on putting $T = \varepsilon$ in the last equation,

$$P(x,\varepsilon) = e^{-\mu\varepsilon[1-xP(x,\varepsilon)]}. \tag{13-94}$$

From this, McMillan and Riordan obtain a Lagrange expansion

$$P(x,T) = e^{-\mu T} + \sum_{k=1}^{\infty} \frac{T(T + k\varepsilon)^{k-1}}{k!} e^{-\mu T}(\mu e^{-\mu\varepsilon})^k x^k; \tag{13-95}$$

hence we have the $p(k,T)$.

The mean is given by

$$L(T) = \Sigma k p(k,T) = \mu T(1 - \mu\varepsilon)^{-1}, \qquad \mu\varepsilon < 1, \tag{13-96}$$

and the variance is

$$\text{var }(T) = \mu T(1 - \mu\varepsilon)^{-3}. \tag{13-97}$$

If $B(t) = 1 - e^{-\lambda t}$, we determine $P(x,T)$ by integrating both sides of the equation with respect to $dB(T)$ and solving for $\int_0^{\infty} P(x,t)\, dB(t)$, which equals

$$\frac{\mu + \lambda - \sqrt{(\mu + \lambda)^2 - 4\mu\lambda x}}{2\mu x}. \tag{13-98}$$

Thus $P(x,T) = \exp[-(T/2)(\mu - \lambda) + \sqrt{(\mu + \lambda)^2 - 4\mu\lambda x}]. \tag{13-99}$

The mean in this case is

$$L(T) = \frac{\mu T}{1 - \mu/\lambda}, \tag{13-100}$$

and the variance is

$$\text{var }(T) = \frac{[1 - (\mu/\lambda)^2]\mu T}{(1 - \mu/\lambda)^3}. \tag{13-101}$$

If $G(u,T)$ is the probability that the server is absorbed prior to time u, then Karlin et al. give as a solution of its Laplace-Stieltjes transform

$$\tilde{G}(s,T) = \exp\left\{-sT - \mu T\left[1 - \int_0^{\infty} \tilde{G}(s,t)\, dB(t)\right]\right\} \tag{13-102}$$

which equals, for the first example above,

$$e^{-(s+\mu)T} + \sum_{k=1}^{\infty} \frac{T(T + k\varepsilon)^{k-1}}{k!} e^{-\mu T}(\mu e^{-\mu\varepsilon})^k e^{-s(T+k\varepsilon)} \tag{13-103}$$

and, for the second example,

$$\exp\left\{-\frac{T}{2}[s + \mu - \lambda + \sqrt{(s + \mu + \lambda)^2 - 4\mu\lambda}]\right\}. \tag{13-104}$$

13-12. Cyclic Queues

E. Koenigsberg [389] has studied a coal-cutting problem whereby a face in a coal mine is worked on by a cutting machine, a drilling machine, a blasting crew, etc., each of which, on finishing, and in the same order, proceeds to work on another face and, when finishing with all N faces, returns to the first face. The problem is regarded as a set of queues in tandem serving N units (the faces), using ordered service, where a unit leaving the last phase of service waits for service in the queue of the first phase and so on, going through the queue and service of the phases in a sequence; hence the name cyclic queues.

Finch [212] studied the same system except that he allows a unit a certain probability p_j to return to the jth phase on finishing the last phase. Another case is studied where there is a probability $p_j < 1$ of joining the jth-phase queue, having just finished service at the jth phase. However, from the definition of p_j, a unit has a chance of moving to the next phase.

1. *Cyclic Queues without Feedback*

If n_j is the number of units waiting and in service at the jth phase with a total of k phases, we have

$$\sum_{j=1}^{k} n_j = N.$$

The service distribution for the jth phase, which consists of one service channel, is exponential with rate μ_j; thus $\mu_j \, \Delta t$ is the probability of completing service in $(t, t + \Delta t)$ at the jth phase. The equations describing the system for the probability of n_j units in the jth phase $(j = 1, \ldots, k)$ at time t are given by

$$P(n_1, \ldots, n_k, t + \Delta t) = [1 - (\mu_1 + \cdots + \mu_k) \, \Delta t]P(n_1, \ldots, n_k, t)$$
$$+ \mu_1 \, \Delta t \, P(n_1 + 1, n_2 - 1, n_3, \ldots, n_k, t)$$
$$+ \mu_2 \, \Delta t \, P(n_1, n_2 + 1, n_3 - 1, n_4, \ldots, n_k, t) + \cdots$$
$$+ \mu_k \, \Delta t \, P(n_1 - 1, n_2, \ldots, n_k + 1, t), \quad (13\text{-}105)$$

which simplifies to

$$P'(n_1, \ldots, n_k, t) = - \sum_{j=1}^{k} \mu_j P(n_1, \ldots, n_k, t)$$

$$+ \sum_{j=1}^{k} \mu_j P(n_1, \ldots, n_j + 1, n_{j+1} - 1, \ldots, n_k, t). \quad (13\text{-}106)$$

If $n_m = 0$ for some m, the mth term does not appear, i.e., it is zero, and if $n_{m+1} - 1 < 0$, the mth term is again zero. Note that the kth phase is linked to the first phase; i.e., transitions are allowed from $(n_1 - 1, \ldots, n_k + 1)$ to $(n_1, \ldots, n_k)$. There is no question of arrivals since the entire population is distributed among the phases. The steady-state equations are obtained from the above in the usual way. We have for the solution

$$p(n_1, \ldots, n_k) = \frac{\mu_1^{N-n}}{\mu_2^{n_2} \cdots \mu_k^{n_k}} p(N, 0, \ldots, 0), \qquad (13\text{-}107)$$

where the last quantity on the right is obtained by equating to unity the sum over all partitions of N—into $(n_1, \ldots, n_k)$—of the expression on the left. The process of summing over all partitions of N will be used everywhere below unless indicated otherwise. (Note that one starts with the entire population waiting for the first-phase service.) Note also that D_j, the fraction of time during which the jth-phase service is idle, is obtained by putting n_j equal to zero in $p(n_1, \ldots, n_k)$ and summing over all partitions of N in which $n_j = 0$. When subtracted from unity, it gives the probability that the jth-phase service is occupied. Thus $1 - D_j$ is the utilization factor of the jth-phase service.

If we write $x_j = \mu_1/\mu_j$ and thus $x_1 = 1$, we have

$$p(N, 0, \ldots, 0) = \frac{1}{\Sigma x_1^{n_1} x_2^{n_2} \cdots x_k^{n_k}} = \left[\sum_{\substack{j=1}}^{k} \frac{x_j^{N+k-1}}{\prod_{\substack{m=1 \\ m \neq j}}^{k} (x_j - x_m)} \right]^{-1}. \qquad (13\text{-}108)$$

Koenigsberg has computed the reciprocal of the last quantity on the right for $k = 1, \ldots, 5$. The average number of units in the jth phase (including the one in service) is given by

$$\bar{n}_j = \Sigma n_j p(n_1, \ldots, n_j, \ldots, n_k)$$

$$= \frac{x_j}{\Sigma x_1^{n_1} \cdots x_k^{n_k}} \frac{d}{dx_j} \Sigma x_1^{n_1} \cdots x_k^{n_k}. \qquad (13\text{-}109)$$

The average number of units awaiting service at the jth phase is

$$\Sigma(n_j - 1) p(n_1, \ldots, n_k) = (\bar{n}_j - 1 + D_j), \qquad (13\text{-}110)$$

where the sum is taken over all partitions of N in which $n_j \geq 1$. We have

$$D_j = \Sigma p(n_1, n_2, \ldots, n_{j-1}, 0, n_{j+1}, \ldots, n_k). \qquad (13\text{-}111)$$

If $\mu_1 = \mu_2 = \cdots = \mu_k = \mu$, then $x_j = 1$ and

$$p(N, 0, \ldots, 0) = \frac{(N + k - 1)!}{(k-1)! N!},$$

$$D_j = \frac{k-1}{N+k-1} \qquad \text{(independent of } j \text{ and hence is the same for all } j),$$

$$\bar{n}_j = \frac{N}{k} \qquad \text{(independent of } j \text{ and hence is the same for all } j),$$

and the average number of units waiting in the jth phase is $N(N-1)/k(N+k-1)$.

The average delay at the jth phase is obtained by dividing the last quantity by μ. The time to go through all stages and thus complete a cycle is

$$\frac{1}{\mu}\left[k + \frac{N(N-1)}{N+k-1} \right]. \tag{13-112}$$

The fraction of time during which the jth-phase service is busy (i.e., the probability that it is occupied) is

$$(1 - D_j)\mu_j = \frac{N}{N+k-1} \mu. \tag{13-113}$$

If μ_j is the output rate per unit time and if the total time worked H is known, the total output may be obtained by multiplying the last expression by H.

To illustrate a special case for $m = j$, in the expression for $p(N,0, \ldots ,0)$ suppose that $x_1 = x_2 = x_3$ in the case $k = 4$; then

$$p(N,0,0,0) = \left[\sum_{n_4=0}^{N} x_4^{n_4} \frac{(N+k-n_4-2)!}{(k-2)!(N-n_4)!} \right]^{-1}, \tag{13-114}$$

having used the above discussion with equal service rate.

If the jth phase, for example, includes two channels in parallel with the same μ_j, the latter is replaced in the solution by $2\mu_j$ when $n_j \geq 2$; otherwise the equations remain the same. Thus

$$p(n_1, \ldots ,n_k) = x_1^{n_1} \cdots \frac{x_j^{n_j}}{2!2^{n_j-2}} \cdots x_k^{n_k} p(N,0, \ldots ,0). \tag{13-115}$$

In the evaluation of $p(N,0, \ldots ,0)$, x_j is replaced in the previous determination of it by $x_j/2$.

2. Cyclic Queues with Feedback

a. Terminal Feedback. The ideas [212] proceed along lines similar to those followed above. Let $p_j < 1$ $(j = 1, \ldots , k)$ be the probability that a unit returns to the jth phase on completing the last phase k, and let $p = \sum_{j=1}^{k} p_j$ and $q = 1 - p$ be the probability of a unit not returning to the system.   This procedure is called terminal feedback.   Let

$$\varepsilon_j = \begin{cases} 1, & n_j > 0, \\ 0, & n_j = 0, \end{cases} \qquad a = \begin{cases} 1, & \sum_j n_j < N, \\ 0, & \sum_j n_j = N. \end{cases} \qquad (13\text{-}116)$$

The equilibrium equations can be written as follows:

$$0 = -\left(\lambda a + \sum_{j=1}^{k} \mu_j \varepsilon_j\right) p(n_1, \ldots, n_k) + \lambda \varepsilon_1 p(n_1 - 1, n_2, \ldots, n_k)$$

$$+ \sum_{j=1}^{k-1} \mu_k p_j \varepsilon_j p(n_1, \ldots, n_{j-1}, n_j - 1, n_{j+1}, \ldots, n_{k-1}, n_k + 1)$$

$$+ \mu_k p_k \varepsilon_k p(n_1, \ldots, n_k)$$

$$+ \sum_{j=1}^{k-1} \mu_j \varepsilon_j p(n_1, \ldots, n_{j-1}, n_j + 1, n_{j+1} - 1, \ldots, n_{k-1}, n_k)$$

$$+ (1 - p)\mu_k a p(n_1, \ldots, n_{k-1}, n_k + 1) \qquad (13\text{-}117)$$

for $n_j = 0, \ldots, N$ and $\sum_{j=1}^{k} n_j \leq N$.   The system is uniquely satisfied by

$$p(n_1, \ldots, n_k) = x_1^{n_1} \cdots x_k^{n_k} p(0, \ldots, 0) \qquad (13\text{-}118)$$

with $x_j = \dfrac{\lambda(1 - p + p_1 + \cdots + p_j)}{\mu_j(1 - p)}, \qquad j = 1, \ldots, k. \qquad (13\text{-}119)$

Again $p(0, \ldots, 0)$ is determined by making the sum of the probabilities over $n_j = 1, \ldots, N$ and $\Sigma n_j \leq N$ equal to unity.

If $p_j = 0$ for all j, $\lambda/\mu_j < 1$, and $N \to \infty$, then the solution becomes that given by R. R. P. Jackson and developed in the case of queues in tandem. If $x_j < 1$ $(j = 1, \ldots, k)$, then $p(0, \ldots, 0)$ tends to a finite nonzero limit as $N \to \infty$; otherwise it tends to zero and hence $p(n_1, \ldots, n_k)$ also tends to zero.   In the first case we have

$$p(0, \ldots, 0) = \prod_{j=1}^{k} (1 - x_j)$$

and

$$p(n_1, \ldots, n_k) = \prod_{j=1}^{k} x_j^{n_j}(1 - x_j). \qquad (13\text{-}120)$$

b. Single-service Feedback.   Here $p_j < 1$ denotes the probability that, having finished service at the jth phase, the unit returns to the queue of the jth phase [we also define $q_j = 1 - p_j$, the probability of going to the $(j + 1)$st phase or of departing if $j = 1$]; then the equilibrium equations are

$$0 = -(\lambda a + \sum_{j=1}^{k} \mu_j q_j \varepsilon_j) p(n_1, \ldots, n_k) + \lambda \varepsilon_1 p(n_1 - 1, n_2, \ldots, n_k)$$

$$+ \sum_{j=1}^{k-1} \mu_j q_j \varepsilon_{j+1} p(n_1, \ldots, n_j + 1, n_{j+1} - 1, \ldots, n_k)$$

$$+ \mu_k q_k a p(n_1, \ldots, n_k + 1). \quad (13\text{-}121)$$

We have the unique solution

$$p(n_1, \ldots, n_k) = x_1^{n_1} \cdots x_k^{n_k} p(0, \ldots 0), \quad (13\text{-}122)$$

where
$$x_j = \frac{\lambda}{\mu_j q_j}. \quad (13\text{-}123)$$

The argument on $x_j < 1$ and as $N \to \infty$ again applies as before, and the previous equations hold, using the last expression just derived for x_j.

In the case of a single channel, both cases treated above coincide.

13-13. Multiqueues Served Cyclically

Consider the case of N queues which are numbered $(1, 2, \ldots, N)$ and a single server who starts serving all items in the first queue which were there at the moment of his arrival and then moves on to the second queue and so on to the Nth queue and then back to the first queue. All items which arrive at a queue after the server has started serving that queue must wait to be served not in that service cycle but in the following one. Thus, only those units present in the queue at the moment of arrival of the server are served during that cycle. It is assumed that there is a time lag between finishing the service of one queue and the start of service of another. This lag between two consecutive queues is described by N random variables which are identically distributed according to the density function $\omega(t)$ indicating this lag time or walking time. Also assume that if $N = 1$ then there is a lag in serving the units already present and those which arrive during service.

Let inputs to all queues be Poisson with identical parameters λ and let the service times of all units and all queues be identically distributed according to the density function $b(t)$. M. A. Leibowitz [864a] has computed the steady-state probability π_n that there are n units in a queue at the moment of arrival of the server at that queue.

The ideas are applicable to messages arriving at the terminals (queues) of a computer and are stored (wait) and then served (transmit its messages to the computer for processing) as described above. The ideas are also applicable to letters sorted into cells in a post office. Arrivals in each cell are Poisson-distributed. A server then takes the letters present in that cell and, after reading an address, drops the letter in the proper transportation sack. Having finished with that cell, with some delay, the server goes on to the next cell overlooking the fact that new letters might have arrived into the cell which he has just served.   These letters will be served in the next cycle.

Let $N = 1$, and let r_{in} be the probability that the server on his arrival finds n units in the queue, given that on the previous arrival he found i units.   Then

$$\pi_n = \sum_{i=0}^{\infty} r_{in}\pi_i. \qquad (13\text{-}124)$$

Note that the service-time density function of i units is an n-fold convolution of $b(t)$ which we denote by $b_i(t)$.   The density function of the sum of the time of serving i units and the time lag before service on the next set of items is the convolution of $b_i(t)$ and $\omega(t)$, i.e., conv $[b_i(t),\omega(t)]$. As in bulk queues we have

$$r_{in} = \int_0^{\infty} \frac{e^{-\lambda t}(\lambda t)^n}{n!} \text{ conv } [b_i(t),\omega(t)] \, dt. \qquad (13\text{-}125)$$

If we let

$$P(z) = \sum_{n=0}^{\infty} \pi_n z^n \qquad (13\text{-}126)$$

and use $b^*(s)$ and $\omega^*(s)$ for the Laplace transforms of these functions, then by applying (13-126) to (13-124) and using (13-125) and after interchanging summation and simplifying we have

$$P(z) = \omega^*[\lambda(1 - z)]P\{b^*[\lambda(1 - z)]\}. \qquad (13\text{-}127)$$

We may obtain the mean and variance of π_n by differentiating $P(z)$ and putting $z = 1$.   We have, for example, for the mean, $\lambda\bar{\omega}/(1 - \lambda/\mu)$, where $\bar{\omega}$ is the mean walking time and $1/\mu$ is the mean service time.   Note that taking logarithms in (13-127) simplifies the computations and that $\lambda/\mu < 1$ is necessary for the existence of a steady state.

Exercise 10: Compute the variance of π_n.

When $N > 1$, to find π_{n_1} at a specific queue requires knowledge of the probability that there are n_2 at the second queue, n_3 at the third, and so forth.   Thus π_n are no longer the stationary probabilities of a Markoff process.   Treatment of this case by means of a transition matrix leads to complicated results.   Thus, as an approximation, it is assumed that the same probability π_n applies at each queue at the moment of arrival of the server.   This assumption leads to the correct mean of the exact problem and to correct series expansion for π_n in terms of λ/μ up to terms in $(\lambda/\mu)^3$.

In this case the right side of (13-127) is raised to the Nth power, where, of course, $P(z)$ is defined for the new π_n.   We have

$$P(z) = (\omega^*[\lambda(1 - z)]P\{b^*[\lambda(1 - z)]\})^N. \qquad (13\text{-}128)$$

The reason for this is that during a cycle the number n which enters the queue under consideration is the sum

$$n = \delta_1 + \cdots + \delta_N,$$

where δ_k is the number arriving at the queue under consideration during the service of the units of the kth queue. Since the right side of (13-127) gives the generating function of the probability distribution of the random variable δ_k, therefore it is raised to the Nth power to yield (13-128). In this case the mean is given by $N\lambda\bar{\omega}/[1 - N(\lambda/\mu)]$ and for stationarity we must have $N\lambda/\mu < 1$.

Exercise 11: Compute the variance.

For $\mu = 1$, we have up to terms in $O(\lambda^3)$ the series expansion at $z = 1$:

$$\log P(z) = \frac{-N\lambda}{1 - N\lambda}(1 - z) + \left\{ \frac{N^2\lambda^3\omega(t)\bar{b}^2}{1 - N\lambda} \right.$$
$$+ N\lambda^2 \text{ variance } [\omega(t)] \left. \right\} \frac{(1 - z)^2}{2}$$
$$- N\lambda^3(\bar{\omega}^3 - 3\bar{\omega}\bar{\omega}^2 + 2\bar{\omega}^3)\frac{(1 - z)^3}{6} \quad (13\text{-}129)$$

where $\bar{\omega}^j$ is the jth moment about the origin of $\omega(t)$ and $\bar{b}^2$ is the second moment of $b(t)$ about the origin.

One takes the exponential of (13-129) and differentiates with respect to z as usual to obtain the π_n. One has $\pi_n = 0$ for $n > 3$.

Note that for large N, λ/μ must be small to preserve stationarity and hence again for $\mu = 1$, $\alpha = N\lambda$, we have

$$\log P(z) = \frac{\alpha\bar{\omega}}{1 - \alpha}(1 - z) + \frac{1}{N}\left\{ \frac{\alpha^2\bar{b}^2\bar{\omega}}{1 - \alpha} \right.$$
$$+ \alpha^2 \text{ variance } [\omega(t)] \left. \right\} \frac{(1 - z)^2}{2} + \cdots. \quad (13\text{-}130)$$

Hence expanding $\exp[\log P(z)]$ in power series in z, selecting π_n as the coefficient of z^n [to terms in $O(1/N)$], and observing that in (13-130) we let f indicate the coefficient of $(1 - z)$ and g the coefficient in braces, we have

$$\pi_0 = e^{-f}\left(1 + \frac{g}{2N}\right)$$
$$\pi_1 = e^{-f}\left(f + \frac{gf}{2N} - \frac{g}{N}\right)$$
$$\pi_n = e^{-f}\frac{f^n}{n!} + \frac{g}{N}e^{-f}\frac{f^{n-2}}{(n-2)!}\left[\frac{f^2}{2} - nf + \frac{n(n-1)}{2}\right], \quad n \geq 2. \quad (13\text{-}131)$$

As $N \to \infty$, π_n has a Poisson distribution with parameter f.

The exact solution for $N = 2$ is obtained as follows: we define π_{nm} as the joint probability that there are n units in the first queue and m in the second at the moment the server arrives at the first queue. We also let $r(n,m|i,j)$ be the corresponding probabilities given that there were i, j units in the first and second queues respectively at the server's last previous arrival at the first queues. Then

$$\pi_{nm} = \sum_{n=0}^{\infty} \sum_{m=0}^{\infty} r(n,m|i,j)\pi_{ij}. \tag{13-132}$$

We have

$$r(n,m|i,j) = \sum_{k=j}^{\infty} \int_0^{\infty} \int_0^{\infty} \left\{ \text{conv } [b_i(t_1),\omega(t_1)]e^{-\lambda t_1} \frac{(\lambda t_1)^{k-j}}{(k-j)!} \text{ conv } [b_k(t_2),\omega(t_2)] \right\}$$
$$e^{-\lambda t_1 t_2} \frac{[\lambda(t_1 + t_2)]^n}{n!} e^{-\lambda t_2} \frac{(\lambda t_2)^m}{m!} dt_1 \, dt_2, \tag{13-133}$$

where the quantity in braces is the conditional probability density that, given that there are i, j units in the first and second queues respectively when the server begins service at the first queue, a time t_1 is required to complete service and move on to the second queue in which he will have k units to serve, and it takes a time t_2 to serve the second queue and return to the first queue.

To solve (13-132) one first introduces the generating function

$$Q(y,z) = \sum_{n=0}^{\infty} \sum_{m=0}^{\infty} \pi_{nm} x^n y^m. \tag{13-134}$$

The analysis leads to the equation

$$P(x,y) = \omega^*(u)\omega^*(v)P[b^*(u),b^*(v)] \tag{13-135}$$

where $u = \lambda(2 - x - y)$ and $v = \lambda(1 - x) + \lambda - \lambda b^*(u)$. Taking logarithms, differentiating, and evaluating at $x = 1$, $y = 1$, yield for the mean $2\bar{\omega}\lambda/(1 - 2\lambda)$.

Exercise 12: Verify (13-135) and the mean and obtain the variance.

CHAPTER 14

APPLICATIONS

14-1. Introduction

In this chapter we first study queueing applications to telephone systems, which makes it convenient for us to examine loss systems, particularly since the main emphasis of the book has been on delay systems. This is followed by applications to car and air traffic. Sometimes the application is of a general nature without close involvement of queueing theory, as in Sec. 14-4. The application to machine interference permits investigation of the case where a finite population constitutes the possible input to a queue. Provision of spare machines has also been investigated, using queueing theory which is illustrated here. In another section we give an example of an inventory application.

Dams and storage systems are yet another field of interest to engineers, which we examine because of the interesting ideas related to queues that occur in this field. General remarks and a description of hospital and medical-care-facility congestions are also given. A similar attempt is made for cafeteria design. An application to coal mining is then made, where it is desired to minimize lost working time. A queueing model for semiconductor noise is given. Finally, there is a discussion of general applications, from actual and from possible case histories.

14-2. Applications to Telephone Congestion

We shall briefly examine a few of the useful ideas occurring in telephone congestion theory, to which the reader has been introduced in Chap. 1. We note that, in the delay and loss systems usually encountered, there is a full availability of trunks to all callers. However, when grading is introduced and various linkages are used, full availability is put aside and the actual structure of the system acquires considerable importance.

We shall first examine full availability, mostly with loss, since we have studied delay in general queueing theory. We give Erlang's well-known loss formula, follow it by Vaulot's relative-traffic-intensity work, and then briefly summarize some of Cohen's work. The case of repeated calls as studied by Cohen is discussed in a problem at the end.

1. *Full Availability*

a. Erlang's Loss Formula. Let $P_n(t)$ be the probability that exactly n among c channels are busy if all channels are free at $t = 0$. Let arrivals be from an unlimited input by a Poisson distribution with parameter λ, and let the service distributions at all channels be identical exponential with parameter μ. We then have, as usual,

$$P'_n(t) = -(\lambda + n\mu)P_n(t) + \lambda P_{n-1}(t) + (n + 1)\mu P_{n+1}(t),$$
$$1 \leq n \leq c - 1,$$

$$P'_0(t) = -\lambda P_0(t) + \mu P_1(t),$$
$$P'_c(t) = -c\mu P_c(t) + \lambda P_{c-1}(t).$$

One can show that a steady-state solution exists by examining the characteristic roots of the matrix of coefficients. One finds that it has one zero root. Since the solution is a linear combination of exponentials, one term will be a constant, and as $t \to \infty$ this constant is the only term which survives and is the equilibrium probability. The solution of the steady-state equations obtained from the above is

$$p_n = \frac{\rho^n/n!}{1 + \rho/1! + \rho^2/2! + \cdots + \rho^c/c!}, \qquad \rho = \frac{\lambda}{\mu}, \qquad (14\text{-}1)$$

which is Erlang's loss distribution. For $n = c$ this is the Erlang loss formula which is the time congestion. Note that the probability that a call is lost is equal to the probability that all channels are busy; i.e., call congestion equals time congestion, which is true only in the Poisson-input case. When $c \to \infty$ we have the Poisson distribution

$$p_n = \frac{\rho^n}{n!} e^{-\rho}.$$

Exercise: The reader is urged to write the steady-state equations and obtain Erlang's loss distribution.

Vaulot, Pollaczek, Palm, Kosten, Fortet, and Sevastyanov have given different proofs of the significant fact that *the form of Erlang's formula is independent of the service distribution*, provided that the input is Poisson.

In his paper on the generalized Engset formulas (see below), Cohen derives results similar to the Erlang loss formula except for a finite population and for an arbitrary-service-time distribution which, in fact, may be different for each source.

When the input population is finite, each unit of the population may be considered as a possible input source [usually with Poisson assumptions applied to the source, i.e., with probability $\lambda \, \Delta t$ of initiating a call during $(t, t + \Delta t)$ and a negligible probability of two sources arriving simultaneously].

b. The Engset-O'Dell Distributions. Suppose that c channels receive traffic by a Poisson input and that the traffic intensity, i.e., the ratio of the input rate to the service rate, depends on the number of free channels. Thus the traffic is y_0 if there are zero channels occupied, y_1 if there is one occupied channel, similarly to y_c, where $y_i = \lambda_i/\mu$.

Let p_n $(n = 0, \ldots ,c)$ be the probability that n channels are busy. Then for an identical exponential service time with parameter μ we have the following equations:

$$y_0 p_0 = p_1,$$
$$(y_i + i)p_i = (i + 1)p_{i+1} + y_{i-1}p_{i-1}, \qquad i = 1, \ldots , c - 1, \quad (14\text{-}2)$$
$$c p_c = y_{c-1}p_{c-1}.$$

The solution of this system is

$$p_i = p_0 \frac{y_0 \cdots y_{i-1}}{i!}, \qquad (14\text{-}3)$$

where p_0 is obtained from $\displaystyle\sum_{i=0}^{c} p_i = 1$.

The number of lost calls is given by $y_c p_c$. The total number of calls (arrivals) per unit time is $\displaystyle\sum_{i=0}^{c} y_i p_i = \sum_{i=1}^{c} i p_i + y_c p_c$. The proportion of lost calls is the quotient of lost calls to the total number of calls per unit time:

$$\frac{(y_1 \cdots y_c)/c!}{1 + y_1 + y_1 y_2/2! + \cdots + (y_1 \cdots y_c)/c!}, \qquad (14\text{-}4)$$

which is independent of y_0. One obtains Erlang's formula by putting $y_i = y$ $(i = 1, \ldots , c)$.

An interesting application is to assume a finite set of sources N and put $y_i = (N - i)\lambda$. This yields the Engset distribution or the O'Dell distribution, according to whether λ depends on loss or is a pure constant. The denominator of the resulting expression may be simplified in terms of the incomplete beta function for computation purposes. It becomes

$$(1 + \lambda)^{N-1} \frac{\displaystyle\int_{\lambda/(\lambda+1)}^{1} w^c (1 - w)^{N-c-2}\, dw}{\displaystyle\int_{0}^{1} w^c (1 - w)^{N-c-2}\, dw}.$$

If $N \to \infty$ so that $\lambda N = \text{const}$, then the O'Dell formula becomes the Erlang formula. For the Engset formula the same result holds.

c. Loss-Delay Systems with Priorities. We now assume with Cohen [121] full availability of c channels to two traffic sources with Poisson inputs and parameters λ_1, λ_2, respectively. The channels have identical exponential service times with parameter μ. The first-source calls may wait but the second-source calls are lost if all channels are busy.

Let p_n be the probability that n channels are busy when $n \leq c$ and the probability that $n - c$ units are waiting from the first source when $n > c$. Then we have in the usual manner for the steady state with traffic intensities $\rho_1 \equiv \lambda_1/\mu < c$, $\rho_2 \equiv \lambda_2/\mu < c$, respectively,

$$(n + 1)p_{n+1} + (\rho_1 + \rho_2)p_{n-1} - (\rho_1 + \rho_2 + n)p_n = 0,$$
$$0 \leq n < c,$$
$$cp_{c+1} + (\rho_1 + \rho_2)p_{c-1} - (\rho_1 + c)p_c = 0, \qquad \qquad (14\text{-}5)$$
$$cp_{n+1} + \rho_1 p_{n-1} - (\rho_1 + c)p_n = 0, \qquad \qquad c < n.$$

Also
$$\sum_{n=0}^{\infty} p_n = 1.$$

The solution is given by

$$p_n = \frac{c - \rho_1}{c - \rho_1 + \rho_1 E_c(\rho_1 + \rho_2)} \frac{(\rho_1 + \rho_2)^n/n!}{N_c(\rho_1 + \rho_2)}, \qquad 0 \leq n \leq c,$$
$$p_n = \frac{(c - \rho_1)E_c(\rho_1 + \rho_2)}{c - \rho_1 + \rho_1 E_c(\rho_1 + \rho_2)} \left(\frac{\rho_1}{c}\right)^{n-c}, \qquad c \leq n,$$
(14-6)

where
$$N_c(\rho) = \sum_{i=0}^{c} \frac{\rho^i}{i!},$$

$$E_c(\rho) = \frac{\rho^c/c!}{N_c(\rho)}.$$

The probability of a call being lost is found by summing p_n, $n \geq c$:

$$P(>0) = \frac{cE_c(\rho_1 + \rho_2)}{c - \rho_1 + \rho_1 E_c(\rho_1 + \rho_2)}. \qquad (14\text{-}7)$$

The number of lost calls from the second source is $\rho_2 P(>0)$. The probability that a call from the first source is delayed longer than t is

$$P(>t) = \int_t^{\infty} \sum_{n=c}^{\infty} p_n e^{-\mu c t} \frac{(\mu c t)^{n-c}}{(n - c)!} c\mu \, dt = P(>0)e^{-(c-\rho_1)\mu t}. \quad (14\text{-}8)$$

The average delay for all calls from the first source is

$$\frac{P(>0)}{\mu(c - \rho_1)}, \qquad (14\text{-}9)$$

and the average delay for delayed calls is

$$\frac{1}{\mu(c - \rho_1)}, \qquad (14\text{-}10)$$

which is independent of ρ_2.

Cohen has also studied the problem when the first-source calls have precedence over those of the second source and both-source calls can wait. The average delays for all calls from the first and from the second source are, respectively,

$$\frac{P(>0)}{\mu(c - \rho_1)} \qquad (14\text{-}11)$$

and

$$\frac{cP(>0)}{\mu(c - \rho_1)(c - \rho_1 - \rho_2)}. \qquad (14\text{-}12)$$

The average delay for delayed calls is

$$\frac{1}{\mu(c - \rho_1)} \qquad (14\text{-}13)$$

for the first priority and

$$\frac{c}{\mu(c - \rho_1)(c - \rho_1 - \rho_2)} \qquad (14\text{-}14)$$

for the second priority.

The case of a single channel with different exponential service times for the two sources with parameters μ_1 and μ_2, respectively, and where the first source has waiting facility but not the second source, is also investigated. Here $p(n,1)$ and $p(n,0)$ are, respectively, the probabilities for the occurrence of the state $(n,1)$, where the channel is occupied by a unit from the second source while n units from the first source are waiting, and the state $(n,0)$, where the channel is engaged by a unit from the first source and $n - 1$ units from that source are waiting.

Note that

$$\sum_{n=0}^{\infty} [p(n,1) + p(n,0)] = 1.$$

If

$$\alpha = \frac{\mu_2}{\mu_1}, \qquad \rho_1 = \frac{\lambda_1}{\mu_1} < 1, \qquad \rho_2 = \frac{\lambda_2}{\mu_2},$$

then the steady-state equations are

$$p(1,0) + \alpha p(0,1) - (\rho_1 + \alpha\rho_2)p(0,0) = 0, \qquad (14\text{-}15a)$$

$$\rho_2 p(0,0) - \left(1 + \frac{\rho_1}{\alpha}\right) p(0,1) = 0, \qquad (14\text{-}15b)$$

$$p(n + 1, 0) + \rho_1 p(n - 1, 0) + \alpha p(n,1) - (\rho_1 + 1)p(n,0) = 0,$$
$$n > 0, \qquad (14\text{-}15c)$$

$$\frac{\rho_1}{\alpha} p(n - 1, 1) - \left(\frac{\rho_1}{\alpha} + 1\right) p(n,1) = 0, \qquad n > 0. \qquad (14\text{-}15d)$$

Also

$$\lim_{n \to \infty} p(n,1) = 0, \qquad \lim_{n \to \infty} p(n,0) = 0.$$

The last equation in the system is used to simplify the equation before it and the latter is summed over n, bearing the first two equations in mind.

This gives

$$p(n,0) = \rho_1^n \left\{ 1 + \frac{\alpha\rho_2}{\rho_1 + \alpha - 1} \left[1 - \left(\frac{1}{\rho_1 + \alpha} \right)^n \right] \right\} p(0,0). \quad (14\text{-}16)$$

For $\rho_1 < 1$ we have from the fact that the sum of the probabilities must equal unity that

$$p(0,0) = \frac{1 - \rho_1}{1 + \rho_2} \quad (14\text{-}17)$$

and

$$p(n,1) = \frac{\alpha\rho_2}{\rho_1 + \alpha} \frac{1 - \rho_1}{1 + \rho_2} \left(\frac{\rho_1}{\rho_1 + \alpha} \right)^n. \quad (14\text{-}18)$$

Also

$$P(>0) = 1 - p(0,0) \quad (14\text{-}19)$$

gives the probability that a unit arriving from the first source must wait or a unit arriving from the second source is lost.

The average number of waiting units is

$$L = \sum_{n=1}^{\infty} (n - 1)p(n,0) + \sum_{n=0}^{\infty} np(n,1) = \frac{\rho_1\rho_2}{\alpha(1 + \rho_2)} + \frac{\rho_1^2}{1 - \rho_1}. \quad (14\text{-}20)$$

The average delay for units from the first source is given by L/λ_1.

If both sources have waiting facilities and the first-source calls have precedence over those of the second source one may use $p_1(n_1,n_2)$, $n_1 > 0$, $n_2 \geq 0$, for the probability that the channel is engaged by a unit from the first source and $n_1 - 1$ first-source and n_2 second-source units are waiting. Similarly, $p_2(n_1,n_2)$, $n_1 \geq 0$, $n_2 > 0$, may be used for the probability that a second-source unit is in the channel and that n_1 first-source units and $n_2 - 1$ second-source units are waiting, and $p(0,0)$ for the probability that the channel is free. The steady-state probabilities may be written.

The expected number of waiting calls from the first source is given by

$$\sum_{n_2=0}^{\infty} \sum_{n_1=1}^{\infty} (n_1 - 1)p_1(n_1,n_2) + \sum_{n_2=1}^{\infty} \sum_{n_1=0}^{\infty} n_2 p_2(n_1,n_2)$$
$$= \frac{\rho_1(\rho_1 + \rho_2/\alpha)}{1 - \rho_1}. \quad (14\text{-}21)$$

For the second source we similarly have

$$\sum_{n_2=0}^{\infty} \sum_{n_1=1}^{\infty} n_1 p_1(n_1,n_2) + \sum_{n_2=1}^{\infty} \sum_{n_1=0}^{\infty} (n_2 - 1)p_2(n_1,n_2)$$
$$= \frac{\rho_2(\alpha\rho_1 + \rho_2)}{(1 - \rho_1)(1 - \rho_1 - \rho_2)}. \quad (14\text{-}22)$$

The total average number of delayed units is the sum of the last two expressions. The average waiting times are obtained by dividing the first expression by λ_1 and the second by λ_2. To obtain the respective average waiting time for delayed calls from each source, the average total delay for the corresponding source is divided by $\rho_1 + \rho_2$.

Exercise: Write down the equations and derive (14-21) and (14-22).

If no precedence is assumed, the Pollaczek-Khintchine formula may be used to give the average number of waiting calls. Thus

$$L = \frac{\rho^2}{2(1 - \rho)} \int_0^\infty \frac{t^2}{h^2} b(t) \, dt = \frac{(\rho_1 + \alpha\rho_2)(\rho_1 + \rho_2/\alpha)}{1 - \rho_1 - \rho_2}, \quad (14\text{-}23)$$

where

$$\rho = \rho_1 + \rho_2, \qquad h = \frac{\rho_1 + \rho_2}{\lambda_1 + \lambda_2},$$

and $b(t)$ is the service-time density, i.e.,

$$b(s) = \mu_1 \frac{\lambda_1}{\lambda_1 + \lambda_2} e^{-\mu_1 t} + \mu_2 \frac{\lambda_2}{\lambda_1 + \lambda_2} e^{-\mu_2 t}.$$

In that case the average number of waiting calls from the first source is

$$\frac{\rho_1(\rho_1 + \rho_2/\alpha)}{1 - \rho_1 - \rho_2} \quad (14\text{-}24)$$

and that for the second source is

$$\frac{\rho_2(\rho_2 + \alpha\rho_1)}{1 - \rho_1 - \rho_2}. \quad (14\text{-}25)$$

The average waiting times are obtained by dividing the first quantity by λ_1 and the second by λ_2.

Comparison of L (total number waiting) in this case with that for the immediately preceding case shows that for $\alpha < 1$, if the precedence is given to the shorter-holding-time source calls, the result is a smaller number waiting than for the ordered-queueing case.

2. Link Systems

a. A Combinatorial Approach. In a link system, as indicated in Chap. 1, a set of sources may have limited access to a set of destinations. Link systems have been examined through the equations-of-state approach (very complicated) and by Jacobaeus through combinatorial methods, giving good approximations. For example, suppose that p links are blocked and $m - p$ are free (m is the total number of links). If $G(p)$ indicates the probability that the p links are blocked and $H(m - p)$ the probability that the remaining ones are blocked, the product $H(m - p) G(p)$ gives the probability that a telephone call is blocked, given that p links are busy. The total probability of a blocked call is obtained by summing over p. This gives

$$\sum_{p=0}^{m} H(m - p)G(p) \equiv E. \tag{14-26}$$

To determine $H(m - p)$, suppose that j calls arrive at random and are distributed also at random among the m links. Suppose that i of them go to the $m - p$ free links and $j - i$ to the p blocked ones. The probability of such an occurrence is the hypergeometric distribution

$$\frac{\binom{p}{j-i}\binom{m-p}{i}}{\binom{m}{j}}. \tag{14-27}$$

If $i = m - p$ is the only possibility, we have

$$H(m - p) = \sum_{j=m-p}^{m} \frac{\binom{p}{j-m+p}}{\binom{m}{j}} F(j), \tag{14-28}$$

where $F(j)$ is the input distribution. If, for example,

$$F(j) = \binom{m}{j} b^j(1 - b)^{m-j},$$

then

$$H(m - p) = b^{m-p}. \tag{14-29}$$

If $G(p) = \binom{m}{p} c^p(1 - c)^{m-p}$, then

$$E = (b + c - bc)^m. \tag{14-30}$$

This formula can be justified on intuitive grounds. Thus, in a set of m pairs of links, if b is the probability of the first item of a pair being blocked and c for the second, then $(b + c - bc)^m$ is the probability that all m are blocked.

b. Common-control Loss System. In this type of link system there is a common-control circuit. An arriving call is lost if all c channels are busy or if the common control is occupied. Every call first goes to the common-control operator (as if asking for permission to make a call) and then to a channel. Let P_j be the probability that j lines are busy and the common control is free and Q_j be the probability that j lines are busy and the common control is busy. Then, if arrivals occur by a Poisson process with parameter λ and service is exponential with parameter μ at the channels and exponential with parameter η at the common control, we have

$$(\lambda + j\mu)P_j = \eta Q_{j-1} + (j + 1)\mu P_{j+1},$$
$$(\eta + j\mu)Q_j = \lambda P_j + (j + 1)\mu Q_{j+1}. \tag{14-31}$$

Note, for example, that transitions from and to the state $(j,1)$, where lines are engaged and the common control is free, are (1) by having been in $(j,1)$ and staying there; (2) by going there from $(j + 1, 1)$, and from $(j - 1, 0)$, that is, when the common control is occupied and $j - 1$ channels are occupied; (3) by leaving $(j,1)$ to $(j,0)$ if an arrival occurs. In the equations one can eliminate either P_j or Q_j and obtain a fourth-order equation. The probability of blocking is the sum of the probabilities that all servers are engaged and the common control is engaged. It is given by

$$E = \frac{(\lambda/\mu)(r + K)}{(\lambda/\mu)r + (\lambda/\mu + r)K}, \qquad r = \frac{\eta}{\mu}. \qquad (14\text{-}32)$$

If c is the number of channels, then

$$K = \sum_{i=0}^{c-1} \frac{c!}{(c - i - 1)!} \frac{\Gamma(\lambda/\mu + r + 1 + i)}{\Gamma(\mu + r + 1)} \frac{1}{[(\lambda/\mu)r]^i}. \qquad (14\text{-}33)$$

This solution has been obtained by Vaulot and Leroy and has already been discussed in Chap. 13 under oriented arrivals.

c. Common-control Waiting System. If waiting is allowed for the above case, let $P(i,1)$ be the probability that there are i waiting calls in the queue and the common control is free. Let $Q(i,0)$ be the probability that there are i calls waiting and the common control is busy. Let there be a single channel to provide service. $P(0,0)$ indicates nothing in the system. We have [684]

$$
\begin{array}{ll}
(\lambda + \mu)P(i,1) = \lambda P(i - 1, 1) + \eta Q(i,0), & i \geq 1, \\
\lambda P(0,0) = \mu P(0,1), & i = 0, \\
(\lambda + \eta)Q(i,0) = \lambda Q(i - 1, 0) + \mu P(i + 1, 1), & i \geq 1, \qquad (14\text{-}34) \\
(\lambda + \eta)Q(0,0) = \lambda P(0,0) + \mu P(1,1), & i = 0, \\
\Sigma P(i,1) + Q(i,0) = 1. &
\end{array}
$$

This system has the solution

$$P(i,1) = \frac{\lambda}{\mu} \frac{(1 - \rho)}{\alpha_2 - \alpha_1} (\alpha_2^{i+1} - \alpha_1^{i+1}), \qquad i \geq 0,$$

$$P(0,0) = 1 - \rho, \qquad (14\text{-}35)$$

$$Q(i,0) = \frac{\lambda}{\mu\eta} \frac{1 - \rho}{\alpha_2 - \alpha_1} [(\lambda + \mu)(\alpha_2^{i+1} - \alpha_1^{i+1}) - \lambda(\alpha_2^i - \alpha_1^i)], \qquad i \geq 0.$$

Here α_1 and α_2 are the roots of

$$\mu\eta\alpha^2 - \lambda(\lambda + \mu + \eta)\alpha + \lambda^2 = 0,$$

$$\rho = \lambda \left(\frac{1}{\eta} + \frac{1}{\mu} \right).$$

As $\eta \to \infty$, one obtains the geometric distribution which is the solution of the single-channel queue. (Verify this.)

When the number of channels is infinite we have

$$P(j,0) = \left(1 - \frac{\lambda}{\eta}\right) e^{-\lambda/\mu} \frac{1}{j!} \left(\frac{\lambda}{\mu}\right)^j, \qquad\qquad j = 0, 1, \ldots ,$$

$$Q(j,i) = \left(1 - \frac{\lambda}{\eta}\right) \left(\frac{\lambda}{\eta}\right)^{i+1} e^{-\lambda/\mu} \frac{1}{j!} \left(\frac{\lambda}{\mu}\right)^j, \qquad i, j = 0, 1, \ldots ,$$

(14-36)

where $P(0,j)$ gives the probability that j lines are occupied and the common control is free and hence no one is waiting, and $Q(i,j)$ indicates j busy lines with i waiting since the common control is occupied. The infinite case has three equations. The c-channel case has nine equations, and the one-channel case that we considered above has four equations. The discussion of this case is continued in the problems.

3. *Limited Availability* (See Chap. 1)

The problems of grading have been treated by Berkeley, who formulated the so-called overflow principle. According to this principle, the traffic overflowing from a part of a grading is regarded as being Poisson-distributed with appropriate mean, and the Erlangian theory is then used for the evaluation of loss. The method is approximate but gives good results in practice.

Recently the method has been extended to a two-parameter case (mean and variance) by R. Wilkinson in his equivalent random theory. Equations of state have been used by Elldin and recently by Ekberg, but they are highly complicated and require the use of computers.

14-3. Applications to Car Traffic

We shall first consider a problem due to E. Borel [70]. Following the formal solution of the problem, an application to automobile traffic will be made.

On a half line OX an infinite sequence of points $\{A_i\}$ $(i = 1, 2, \ldots)$ is Poisson-distributed with mean $\lambda = 1$; that is, the unit of the scale is equal to the mean density. Thus the probability that there are n points on an interval of length L is $L^n e^{-L}/n!$.

Let ρ be a fixed length. The distribution of A_i may be transformed in terms of ρ as follows: Let A_1 be the first point to the right of the origin. Choose a point B_1 to the right of A_1 such that the length of the interval A_1B_1 equals ρ. If this interval contains no new A_i, the interval is said to form a series of one interval. If, on the other hand, the interval contains α_1 points A_i, we measure off, starting with and to the right of B_1, a new interval of length $\alpha_1\rho$ terminating in B_2. If B_1B_2 contains no new A_i, the interval A_1B_2 is said to form a series of $1 + \alpha_1$ intervals. Otherwise, if it contains α_2 of the A_i, then we measure off to the right of B_2 an interval of length $\alpha_2\rho$, and so on. The process is continued.

The series may terminate if we reach a last interval $B_{n-1}B_n$ that contains no new point of the A_i. For the beginning of the new series we take the first point of the sequence $\{A_i\}$ to the right of B_n. We now compute p_n, the probability that a series consists of exactly n intervals of length ρ, and we compute p_∞, the probability that the number of such ρ intervals is infinite.

The probability of a series of a given number of intervals,

$$n = 1 + \alpha_1 + \cdots + \alpha_k,$$

is the product of the probabilities that the first interval A_1B_1 contains α_1 points, i.e.,

$$\frac{\rho^{\alpha_1}e^{-\rho}}{\alpha_1!},$$

the second interval B_1B_2 (whose length is $e^{-\alpha_1\rho}$) contains α_2 points, that is, $(\alpha_1\rho)^{\alpha_2}e^{-\alpha_1\rho}/\alpha_2!$, and so on to the last interval that contains no points, that is, $(\alpha_k\rho)^0e^{-\alpha_k\rho}/0!$. This product is

$$\frac{\alpha_1^{\alpha_2}\alpha_2^{\alpha_3}\cdots\alpha_{k-1}^{\alpha_k}\rho^{n-1}e^{-n\rho}}{\alpha_1!\alpha_2!\cdots\alpha_k!}. \tag{14-37}$$

To obtain p_n we must consider all possible distinct decompositions of $n = 1 + \sum_{i=1}^{k} \alpha_i$. For each such decomposition, we obtain an expression such as (14-37) which, in general form, may be written as

$$p_n = a_n\rho^{n-1}e^{-n}. \tag{14-38}$$

If $\rho < 1$, then $p_\infty = 0$; convergence of the sum of the p_n gives $\sum_{n=1}^{\infty} p_n = 1$, and we have

$$\rho = \sum_{n=1}^{\infty} a_n z^n, \tag{14-39}$$

where

$$z = \rho e^{-\rho}. \tag{14-40}$$

This equation has two zeros $\rho_1 < 1$, $\rho_2 > 1$ for all z between zero and e^{-1}. Note that, as ρ increases from zero to infinity, z increases to e^{-1} at $\rho = 1$ and then decreases to zero.

Exercise: Verify the last assertions.

But

$$a_n = \frac{1}{2\pi i}\int_C \frac{\rho\,dz}{z^{n+1}} = \frac{1}{2\pi i}\int_C \frac{(1-\rho)e^{n\rho}}{\rho^n}\,d\rho = \frac{n^{n-2}}{(n-1)!}, \tag{14-41}$$

obtained by expanding the integrand in series and selecting the coefficient of ρ^{-1}. Therefore,

$$p_n = \frac{n^{n-2}}{(n-1)!}\, \rho^{n-1}e^{-n\rho}. \qquad (14\text{-}42)$$

This result also holds for $\rho \geq 1$. The relation (14-39) yields ρ_1 as a function of z. However, (14-39) is also valid for $\rho = 1$, that is, $z = e^{-1}$; hence $p_\infty = 0$. For $\rho > 1$, put $\rho = \rho_2$. In that case the sum of the probabilities is $e^{\rho_1-\rho_2}$ and $p_\infty = 1 - e^{\rho_1-\rho_2}$. For $\rho < 1$, the mean number of intervals is

$$\Sigma n p_n = \frac{1}{1-\rho}, \qquad (14\text{-}43)$$

and the probable length of a series to which a point A_i belongs is

$$\frac{\Sigma n^2 p_n}{\Sigma n p_n} = \frac{1}{(1-\rho)^2}. \qquad (14\text{-}44)$$

The mean number of series of intervals contained in a segment of length A is $A(1-\rho)$.

The result in (14-42) gives the probability of n points in a series starting with the first point of the series and A_1B_1 equal to ρ. If, instead, A_1B_1 is equal to $r\rho$, then $n = r + \alpha_1 + \cdots + \alpha_k$, and we have as before

$$p_n = r\,\frac{n^{n-r-1}}{(n-r)!}\, \rho^{n-r}e^{-n\rho}, \qquad n = r, r+1, \ldots . \qquad (14\text{-}45)$$

The reader may verify this by evaluating an expression which corresponds to (14-41).

The foregoing result has a useful application, as shown by Haight [281]. Let cars be discharged at a traffic light at a uniform time separation T. Assume Poisson input to the traffic waiting line with parameter λ, which is taken as the time unit. Let $\rho = \lambda T$. Let Z cars be in the queue at the beginning of a red light, and let X be the number of cars at the beginning of the previous green light. What is the probability of Z in queue (which is called the overflow into the red light), given X and given that the number of cars discharged in the duration α of a green light is a constant N? ρ is approximately $\alpha\lambda/N$.

Let $f(z;x) \equiv \text{Prob}\,(Z = z \mid X = x)$; then

$$f(z;x) = \frac{(\alpha\lambda)^{z-x+N}e^{-\alpha\lambda}}{(z-x+N)!}, \qquad x > N,$$

$$f(0;x) = \sum_{j=x}^{N} R(j;x), \qquad z = 0, \qquad x \leq N,$$

where $R(n;r) = r(n^{n-r-1}/(n-r)!)e^{-n\rho}\rho^{n-r}$ $\quad (n = r, r+1, \ldots)$,

$$f(z;x) = e^{\rho z}\Big[R(N+z;\,x) - \sum_{j=1}^{z-1} R(z;j)f(j;x)\Big], \qquad z > 0, \qquad x \leq N.$$

The first equation shows that for $x > N$ the overflow will depend only on the input in the "green period." The second equation is simple. The third equation is obtained as follows: For fixed $x \leq N, z > 0$, the overflow would be z, and no vehicle arrives after the light turned red in the period required to clear the overflow. This occurs with probability $f(z;x)e^{-\rho z}$, which is the probability that the queue would first vanish after $N + z$ vehicles. The unconditional probability of the latter event is $R(N + z; x)$, and thus $[R(N + z; x) - f(z;x)e^{-\rho z}]$ is equal to

$$\sum_{j=1}^{z-1} f(j;x) R(z;j),$$

the probability that the queue would vanish after $N + z$ vehicles, and the overflow is j. This yields the third equation. Haight gives tables computing values of these expressions. He shows that, for fixed x, $\sum_z f(z;x) = 1$. He also gives a brief treatment of the presence of traffic from a side street and its effect on changes in the light-signal duration.

J. C. Tanner [709] has studied the problem of interference of two queues of traffic approaching a single lane from opposite directions. He makes use of the above ideas of Borel, but his results and analysis involve a great deal of detail; because of space limitation, they cannot be fully presented here with advantage. Because of the importance of the idea of interference, we give a sketch but the results are meager.

We have a road of which a certain length AB is wide enough for only a single traffic lane and outside which there is free flow in both directions. Vehicles, referred to as V_1's, arrive at A at random (i.e., by a Poisson process), at an average rate q_1 per unit time, wishing to pass from A to B. Similarly, vehicles V_2 arrive at B at random at an average rate q_2 per unit time, wishing to pass from B to A. A vehicle V_1 will pass straight into AB on arrival at A if there is no opposing vehicle V_2 in AB and if no other V_1 has entered AB during the previous time β_1. (We give specific information on β_1 later.) Otherwise it will wait until both these conditions are satisfied. Similarly, a V_2 has entered AB within the previous time β_2. We further assume that each V_1 takes exactly a time α_1 to pass through AB and that a V_2 takes exactly α_2. It is assumed that all vehicles travel at constant speed and that starting and stopping times are negligible, although these can to some extent be allowed for by adjusting the constants α_1 and α_2. Our analysis requires that $\alpha_1 > \beta_1, \alpha_2 > \beta_2$.

A VB_1 (V_1 block) is a period when AB is controlled by V_1's which is immediately preceded and followed by a period when this is not the case. VB_2's and VB_0's are similarly defined, the latter being periods when AB is empty. $t_0, t_1,$ and t_2 are lengths of VB_0's, VB_1's, and VB_2's. Each

t_i ($i = 0$, 1, 2) has a distribution whose moment-generating function (mgf) $E(e^{t_iT})$ with respect to an argument T is $M_i(T)$. r_1 and r_2 are the numbers of V_1's and V_2's waiting at the beginning of VB_1's and VB_2's, respectively. We write $M_1(-q_1) = M_1$ and $M_2(-q_2) = M_2$.

Let n be the number of units that will be served (i.e., go through the lane) before the next time that none are being served. We refer to $n\beta_1$ as an r wave. It is greater than the time which elapses (until there is no V_1 waiting in that queue) by the amount β_1.

Let $\psi(r)$ be the expected total delay to V_1 units in the queue considered during the r wave. During the unit waves the expected total delay is $(e^{(\alpha_1-\beta_1)q_1} - 1)\psi(1)$, where α_1 is the time required for a unit from that queue to pass through the lane. During the gaps and during the final $\alpha_1 - \beta_1$ the expected total delay is zero. Thus the over-all expected delay total is the sum of these, given by

$$\psi(r) + (e^{(\alpha_1-\beta_1)q_1} - 1)\psi(1), \tag{14-46}$$

which when averaged over r gives the expected total delay.

To find $\psi(r)$ one follows Borel's procedure by breaking the r wave into generations. Thus while the initial r vehicles are departing let a vehicles arrive; while these are departing let b arrive, etc. The probability that the sizes of successive ones are r, a, b, c, . . . is

$$\frac{e^{-\rho_1 r}(\rho_1 r)^a}{a!} \frac{e^{-\rho_1 a}(\rho_1 a)^b}{b!} \frac{e^{\rho_1 b}(\rho_1 b)^c}{c!} \cdots \tag{14-47}$$

where $\rho_1 = \beta_1 q_1$.

Note that, if ρ_2 is the corresponding quantity for the other queue which is defined similarly, for an equilibrium solution to exist we must have $\rho_1 + \rho_2 < 1$.

The expected total delay from the above distribution is

$$\tfrac{1}{2}\beta_1[r(r - 1) + a(a - 1) + b(b - 1) + \cdots] \\ + \tfrac{1}{2}\beta_1[ra + ab + bc + \cdots]. \tag{14-48}$$

The terms in the first brackets arise from delays to vehicles while vehicles in their own generation are departing, whereas those in the second arise from delays while the previous generation is departing.

The analysis leads to

$$\psi(r) = \frac{1}{2}\beta_1 \left[\frac{r(2\rho_1 - 1)}{(1 - \rho_1)^2} + \frac{r^2}{1 - \rho_1} \right];$$

averaging with respect to r gives

$$\overline{\psi(r)} = \frac{1}{2}\beta_1 \left[\frac{2\rho_1 - 1}{(1 - \rho_1)^2} E(r) + \frac{E(r^2)}{1 - \rho_1} \right]$$

and still further to the following expression for the mean delay to vehicles V_1 in the queue considered:

$$W_1 = \frac{1}{2(1 - \rho_1)}\left[\beta_1\rho_1 + \frac{\rho_1 E(t_2^2)(1 - \rho_1 - \rho_2)}{\gamma + \varepsilon_1\delta_2 + \varepsilon_2\delta_1}\right], \qquad (14\text{-}49)$$

where
$$\delta_1 = q_1 + q_2 - q_1 M_1, \qquad \delta_2 = q_1 + q_2 - q_2 M_2,$$
$$\gamma = M_1 + M_2 - M_1 M_2,$$
$$\varepsilon_1 = \frac{e^{(\alpha_1 - \beta_1)q_1} - 1}{q_1},$$
$$\varepsilon_2 = \frac{e^{(\alpha_2 - \beta_2)q_1} - 1}{q_2}.$$

$M_1(\theta)$ is the moment-generating function of t_1, the time in which a block of cars from the first queue controls the lane, with respect to argument θ. The same applies for M_2 and t_2.

One also has for the mean delay to vehicles V_2

$$W_2 = \frac{1}{2(1 - \rho_2)}\left[\beta_2\rho_2 + \frac{\rho_2 E(t_1^2)(1 - \rho_1 - \rho_2)}{\gamma + \varepsilon_1\delta_2 + \varepsilon_2\delta_1}\right]. \qquad (14\text{-}50)$$

For $\beta_1 = 0$, $\beta_2 = 0$ we have

$$W_1 = \frac{e^{\alpha_1 q_2}(q_2 + q_1 e^{\alpha_2(q_1 + q_2)})}{q_2 e^{\alpha_1 q_2}(e^{(\alpha_1 + \alpha_2)q_1} - e^{\alpha_2 q_1} + 1) + q_1 e^{\alpha_2 q_1}(e^{(\alpha_1 + \alpha_2)q_2} - e^{\alpha_1 q_2} + 1)}$$
$$\frac{e^{\alpha_2 q_2} - \alpha_2 q_2 - 1}{q_2}. \qquad (14\text{-}51)$$

For W_2, replace α_1 by α_2 and q_1 by q_2, q_2 by q_1.

For $\alpha_2 = \beta_2 = 0$, we have

$$M_2 = 1, \qquad \delta_2 = q_1, \qquad \varepsilon_2 = 0, \qquad \gamma = 1,$$
$$W_1 = \frac{\rho_1^2}{2q_1(1 - \rho_1)}, \qquad (14\text{-}52)$$
$$W_2 = \frac{1}{2(1 - \rho_1)^2 q_1 q_2}\, q_2[2\rho_1 - 3\rho_1^2 + 2\rho_1^3 + 2q_1\varepsilon_1(1 - \rho_1)$$
$$- 2q_1\alpha_1(1 - \rho_1^2)]$$
$$= \frac{e^{(\alpha_1 - \beta_1)q_1}}{q_1(1 - \rho_1)} - \alpha_1 - \frac{1}{q_1} + \frac{2\rho_1^3 - \rho_1^2}{2q_1(1 - \rho_1)^2}. \qquad (14\text{-}53)$$

For $\alpha_2 = \beta_1 = \beta_2 = 0$,

$$W_1 = 0, \qquad W_2 = \frac{e^{\alpha_1 q_1} - \alpha_1 q_1 - 1}{q_1}.$$

Exercise: Verify the special cases.

These ideas are applicable to intersecting streams of pedestrians in one-way traffic, e.g., a passageway, doorway, gangway, bridge, etc. The simplification $\beta_1 = \beta_2 = 0$ may be more applicable here.

If two traffic streams intersect at right angles, the mean delays, if β_1 or β_2 or both are zero, are the same as previously given.

14-4. Vehicular Stoppages in a Tunnel

Emergency garages equipped with special tractors are provided at Port Authority tunnels of New York City for the purpose of assisting disabled vehicles and effecting their removal from the tunnel. When the Lincoln Tunnel had one tube, it was equipped with a single garage, and the present two-tube Holland and Lincoln Tunnels are equipped with two garages. Plans for construction of a third tube for the Lincoln Tunnel included a third garage. By using probability theory to determine multiple breakdowns, L. C. Edie [169] found that equipment from a third garage would be required too infrequently to justify the high cost of constructing, equipping, and operating the additional garage.

It was found that an ordinary stoppage can add 200 vehicles to the entrance backup. In 1951 the Lincoln Tunnel (two tubes) had 2714 stoppages with average durations of 5 minutes. There were seven cases of double stoppages. Assuming a Poisson distribution, the probability of two simultaneous stoppages per tube in any 5-minute interval is given by

$$P(2) = (1/2!)e^{-m}m^2 = (\tfrac{1}{2})(e^{-0.0129})(0.0129)^2 = 82.5 \times 10^{-6},$$

where m = average number of stoppages per 5-minute interval per tube. There are 105,120 five-minute intervals in a year. Therefore, the expected number of double stoppages per tube is

$$nP(2) = (105,120)(82.5 \times 10^{-6}) = 8.7,$$

and for two tubes

$$2nP(2) = 17.4.$$

This result is in error, being considerably more than the actual count of 7. The count was checked and found to be erroneous. The recheck showed a count of 31 double stoppages, still differing from the calculated value. The reason for the prediction error was revealed by the hourly distribution of stoppages. About three-fourths occurred during one-third of the day. Let N be the number of stoppages during the busiest 8 hours, n the number during the remaining 16 hours, and M the average number of stoppages per 5-minute interval during the busiest 8 hours. There are about 35,000 intervals in the 8-hour period for a year. The probability of a single stoppage being in any particular interval is

$$P = \frac{1}{35,000} = 28.6 \times 10^{-6}.$$

By count, the average number of stoppages per tube was 1357.   Thus

$$N = (\tfrac{3}{4})(1357) = 1020,$$

and
$$M = NP = (1020)(28.6 \times 10^{-6})$$
$$= 29.2 \times 10^{-3} \text{ stoppage per interval,}$$
$$P(2) = (1/2!)[e^{-.0292}(.0292)^2] = 405 \times 10^{-6}.$$

The expected number of double stoppages during the busy 8 hours is

$$(35{,}000)(405 \times 10^{-6}) = 14.2.$$

Similarly, the expected number of double stoppages for the remaining 16 hours is .8.   Thus the total for a 24-hour period for two tubes is $2(14.2 + .8) = 30$.   This agrees very well with the actual count of 31. Similar computations for the Holland Tunnel predicted 140 for an actual count of 149.

The same technique was used to predict triple stoppages.

Estimates were made for three-tube operation by the same methods. The indication was that there would be 2533 intervals with single stoppages, 57 with double stoppages, and 1 with a triple stoppage each year under three-tube operation.   Thus there would be 114 stoppages occurring in pairs.   Previous experience indicated that at the Holland Tunnel 149 were adequately handled by two garages.

The recommendation was made that a third garage not be built for the third tube.   The analysis resulted in an estimated savings of $130,000. It required about 5 man-days of work.

14-5. Applications to Aviation Traffic

A number of ideas of air traffic control were encountered in the illustrative examples of the first chapter.   We start by giving some applications of ideas from queueing theory that have found use in the landing of aircraft, as investigated by T. Pearcey [553].   Here, roughly speaking, one is generally interested in minimizing landing delays.   We first compute the probability that $n - 1$ aircraft are successively delayed.

It is assumed that aircraft approach the control zone from random directions and with random, i.e., exponential, time separations with constant mean arrival rate, which is taken as the time unit.   Hence e^{-t} is the distribution of the interarrival intervals.   An aircraft which arrives in an interval smaller than a minimum period of time required for the safe landing of a previous aircraft is delayed by the minimum amount. The ratio of the safe minimum landing interval to the mean arrival interval is denoted by T (for simplicity, taken as constant for a particular airport), and one is usually interested in the case $T < 1$. The probability that an arriving aircraft is not delayed is

$$\int_T^\infty e^{-t}\, dt = e^{-T}. \tag{14-54}$$

The probability that one aircraft is delayed is obtained by examining all single delays between two undelayed aircraft. Such a delayed aircraft must arrive in an interval of time $t_1 < T$ behind the undelayed aircraft immediately preceding it, and the undelayed aircraft immediately following it must arrive in an interval of time t greater than $2T - t_1$. Thus the desired probability that $t_1 < T$ followed by the probability that $t > 2T - t_1$ is

$$\int_0^T e^{-t_1}\, dt_1 \int_{2T-t_1}^\infty e^{-t}\, dt = Te^{-2T}.$$

The probability of pairs of delayed aircraft is similarly obtained (for pairs of delayed aircraft between undelayed ones) by computing the joint probabilities of arrival:

$t_1 < T$	for the first delayed aircraft following an undelayed aircraft,
$t_2 < 2T - t_1$	for the second delayed aircraft following the first delayed aircraft,
$t > 3T - t_1 - t_2$	for the undelayed aircraft immediately following the delayed pair.

This gives

$$\int_0^T e^{-t_1}\, dt_1 \int_0^{2T-t_1} e^{-t_2}\, dt_2 \int_{3T-t_1-t_2}^\infty e^{-t}\, dt = \frac{3T^2}{2} e^{-3T} \tag{14-55}$$

for three delayed aircraft, with $\alpha_n T^{n-1} e^{-nT}$ as the general form for the probability of $n - 1$ aircraft being delayed; α_n is a coefficient depending only on n. Obviously we must have

$$\sum_{n=1}^\infty \alpha_n T^{n-1} e^{-nT} = 1, \tag{14-56}$$

or

$$\sum_{n=1}^\infty \alpha_n u^n = T, \tag{14-57}$$

where $u \equiv Te^{-T}$ is single-valued for small T; hence T may be expressed as a function of u:

$$T(u) = \sum_{n=1}^\infty \alpha_n u^n. \tag{14-58}$$

We have

$$\alpha_n = \frac{1}{2\pi i} \oint \frac{T(u)}{u^{n+1}} \, du = \frac{1}{2\pi i} \oint \frac{e^{nT}}{T^n} (1 - T) \, dT = \frac{n^{n-1}}{(n-1)!} - \frac{n^{n-2}}{(n-2)!}$$

$$= \frac{n^{n-1}}{n!}, \tag{14-59}$$

having used the fact that the origin is a multiple pole; hence the residue may be obtained by expanding the integrand in series and choosing the coefficient of T^{-1}.

We have for the probability that $n - 1$ aircraft are successively delayed

$$\frac{n^{n-1}}{n!} T^{n-1} e^{-nT} \sim \frac{n^{n-1}}{n^n e^{-n} \sqrt{2\pi n}} T^{n-1} e^{-nT} = \frac{T^{n-1} e^{n(1-T)}}{n \sqrt{2\pi n}}, \tag{14-60}$$

having used Sterling's approximation for $n!$. Pearcey gives a set of curves for this distribution.

The average number of aircraft in the system (with an undelayed initial arrival) is given by

$$L = \sum_{n=1}^{\infty} n\alpha_n T^{n-1} e^{-nT} = \frac{1}{1 - T}, \tag{14-61}$$

which is easily obtained by differentiating (14-56) with respect to T and simplifying. Note that all aircraft are delayed if $T = 1$. Similarly, the second moment about the origin is given by $1/(1 - T)^3$.

The fraction of aircraft delayed is obtained as the ratio of the average number in the system, reduced by the aircraft that is in the landing process, to the average number, i.e.,

$$\frac{L - 1}{L} = T.$$

The distribution of landing intervals is determined by the following argument: All intervals of length $t < T$ have zero frequency; those of length $t = T$ occur with frequency T; the fraction of delayed aircraft, i.e., those of length $t > T$, occur with the fraction $1 - T$ of undelayed aircraft weighted by the probability of their arrival, that is, $e^{-(t+T)}$. We use the Heaviside step function $H(T - t)$ (which is unity for positive values of its argument and zero at the origin; its derivative is the Dirac delta function) and the Dirac delta function $\delta(T - t)$ to represent this distribution by

$$(1 - T)e^{-(t-T)}H(T - t) + T\delta(T - t).$$

The delay or waiting-time distribution should now be obtainable, using

Lindley's integral equation.   However, by means of a detailed analysis, Pearcey gives for this distribution in $mT < t < (m + 1)T$

$$Q_T(t)\, dt = (1 - T)e^{t-T}\, dt \sum_{r=0}^{\infty} \frac{[(m + r + 1)T - t]^{m+r}}{(m + r + 1)!}$$
$$\times [t + (m + r + 1)(1 - T)]e^{-(m+r)T}$$

which on integrating over $0 \leq t \leq \infty$—note that one must sum over m using the intervals $[mT, (m + 1)T]$—yields T as the fraction of aircraft delayed.   It also yields

$$\frac{T^2}{2(1 - T)}$$

for the average delay.

Note how the delay increases with increasing T.   The above distribution yields criteria for the necessary handling capacity of an airport.   A more useful measure is that only a small number of aircraft should wait longer than a specified period.

In another paper, G. E. Bell [37] has examined a variety of measurements taken for air-traffic-control study and developed the results for a Poisson-input constant-service-time queue in equilibrium for a single channel.

A paper by Galliher and Wheeler [238] emphasizes the computational aspect of an aircraft-landing problem.

The arrival rate of aircraft in the New York area was observed to be periodic with the period equal to one day.   The landing times were treated as constant throughout the day, but two different values (1 and 2 minutes) were used.   The day was divided into intervals equal in length to the constant landing time (which slowly varies with time).   In the ith of these intervals the mean arrival rate λ_i is constant, and the arrival distribution in this interval is Poisson, i.e.,

$$\frac{\lambda_i^n e^{-\lambda_i}}{n!} \equiv \alpha_n^i \qquad n = 0, 1, \ldots .$$

The service discipline is first-come–first-served, and the number of channels available at the end of the $(i - 1)$st interval for use in the ith interval is c_i.   Crommelin's formulas were used to compute the steady-state probabilities p_n^i (for n aircraft to be in the system) at the end of the ith interval n terms of those of the $(i - 1)$st interval.   Once p_n^i is known, the remaining probabilities may be computationally determined in a recursive manner for all i following this interval.   We use superscripts to indicate the number of the interval.   The equations are

$$p_0^i = \alpha_0^i P_{c_i}^{i-1},$$

$$p_1^i = \alpha_1^i P_{c_i}^{i-1} + \alpha_0^i p_{c_i+1}^{i-1},$$

$$p_n^i = \alpha_n^i P_{c_i}^{i-1} + \sum_{j=1}^{n} \alpha_{n-j}^i p_{c_i+j}^{i-1}, \qquad (14\text{-}62)$$

where
$$P_n^i = \sum_{j=0}^{n} p_j^i.$$

Since p_n^i is computed at ends of intervals (i.e., for an imbedded Markoff chain), the true queue and system probabilities are obtained by interpolation. Various measures of effectiveness may also be developed in the usual manner. It took a computer little time (7 minutes) to do the computation for a day's study, including the average number in the queue and the average for the system. These were used to compute the average number of delayed aircraft, etc.

We briefly mention the outstanding work of A. M. Lee [860a, b] and his collaborators in organizing the operation of "turnround" of aircraft at London airport. The passenger arriving at or before a prescribed time preceding his time of departure can be processed through a common check-in system. If he arrives at the last minute, he joins a restricted desk line to speed up his boarding operations.

The operation of an airport depends to a large extent on the ease and economy of information diffusion. The latter is controlled from a central room which, in London, is known as the apron control (the apron being that portion of the airport where the aircraft are serviced and loaded). The status of preparedness of the aircraft for departure is continuously reported (by two-way radio) from a mobile van, and the reporter is responsible for inspecting two or three airplanes at a time. The flight data are collected and displayed in the apron control. If any delays are reported, the apron-control room telephones the supervisor of that activity (e.g., the baggage-handling room or the catering room) to inquire about causes of trouble and the delay time. When the latter becomes known, the apron room then notifies the passenger gate in order to delay the boarding of passengers.

One of the delaying factors has been the communications problem. Previously, the van reported the completion of every necessary move in preparing the airplane for flight. This caused overloading of the system in the apron control and consequently delay in the time of departure. After a study of this problem, the vans were instructed to report to the apron control only when there is trouble. By means of queueing theory applied to messages arriving in the apron-control room and handled according to priorities, work in that room was greatly simplified. Communication difficulties have been reduced to a minimum,

A recently completed study of the desk system utilized the theory of queues and was based on the relative economics of the situation: how many staff members would be operating different combinations of the desk system? A set of decisions was formulated and then used in each airport to select the type of check-in systems that are appropriate. The system utilizing an economic number of staff and without unreasonable delays was adopted. The composite desk check-in system and the restricted final desk are the solution which worked for the London airport. However, the result is not the same for other airports. Surprisingly, in some airports the common check-in system was found to be impractical.

14-6. Finite-input-source Applications to Machine Interference

The material of this section is applied to an input population of machines. Their total number m is finite. They break down, i.e., call for service, by a Poisson distribution with parameter λ. Thus they constitute input sources, as we indicated early in the chapter. (Note that λ is characteristic of each machine. The average working time of a machine is $1/\lambda$.) Service is supplied by c repairmen. The duration of service is considered as exponentially distributed with parameter μ or is a constant-service-time distribution for all repairmen and all machines. (General service times have been considered by Takács.) All machines are assumed similar as to the average number of breakdowns that each experiences in its unit working time. Interference times are those times when a machine is stopped and waits for repair attention.

An interesting problem to consider is that of determining the optimum (based on cost of repairman and machine time) number of machines that a repairman should supervise. Frequently, in practice, when a machine breaks down, a repairman is not available because he is either repairing other machines or doing ancillary work, e.g., ensuring adequate supply of raw material for the machines to work on, as in a textile factory. Our object is not to exhaust the subject of machine interference but to indicate applications of queueing theory.

1. *Poisson Input, Arbitrary-service Distribution*

Research on the finite-population problem as applied to machine interference with Poisson input and arbitrary-service distribution by a single repairman has been done by Khintchine, Kronig, Kronig and Mondria (1943), Palm (1947), Ashcroft (1950), and Takács (1957).

The ideas are analogous to those in telephony where a call is lost if all lines are busy. The problem, already mentioned in the first section of the chapter, was investigated by Erlang (1918), Fry (1928), Palm (1943), Pollaczek (1953), Takács (1956), and J. W. Cohen (1957).

Takács [900] solves for p_n, $0 \leq n \leq m$, the probability that there are n machines at any time in the m-machine, single-repairman problem.

The probability that a machine calls for service during $(t, t + \Delta t)$ is $\lambda t + o(\Delta t)$, and the cumulative-service distribution of any machine is $B(t)$; ordered-service discipline is used. He first solves for π_n, $0 \leq n \leq m$, the probability that n machines are working immediately before the termination of service of a machine, and then obtains p_n. The argument holds in the steady state whose existence is assured since the process defines a Markoff chain which is aperiodic with a finite number of states.

Now the probability r_{ij} that there are j working machines just before the end of a service, given that there were i working machines just before the end of the previous service, is

$$r_{ij} = \int_0^\infty \binom{i+1}{j} e^{-j\lambda t}(1 - e^{-\lambda t})^{i+1-j}\, dB(t), \qquad i = 1, \ldots, m-2,$$

$$r_{m-1,j} = r_{m-2,j}, \qquad i = m-1.$$

Note that it does not make sense to talk about m working machines just before the termination of service of a machine because there are m machines altogether. Also note that $j \leq i + 1$, since it is assumed that the previous machine has been repaired and nothing need have called for service.

The integral is obtained by considering (during the service time of the machine in which j machines are waiting) the probability that a machine calls for service, that is, $p \equiv 1 - e^{-\lambda t}$, and the probability that it does not call for service, that is, $q \equiv e^{-\lambda t}$. Then the binomial formula gives $\binom{i+1}{j} e^{-j\lambda t}(1 - e^{-\lambda t})^{i+1-j}$. This holds since, after the previous machine finished repair, there are $i + 1$ working machines and to have j of these work during time t is the above expression. Now since t is given by a probability distribution, we multiply the above binomial expression by this distribution and integrate over all values of t, obtaining the foregoing expression.

We have

$$\sum_{j=0}^{m-1} \pi_j = 1$$

and

$$\pi_j = \sum_{i=j-1}^{m-1} r_{ij}\pi_i, \qquad j = 1, \ldots, m-1.$$

The second expression gives the probability that just before the previous machine finished service there were $j - 1$ working machines, that it joined the working machines and nothing called for service before the present machine finished service, or that just before the previous machine finished service there were j machines working and it joined them but one machine called for service just before the present machine finished service, etc.

To obtain the π_j from the second expression, multiply the second expression by z^j, replace r_{ij} by the expression defining it, and sum over $0 \leq j \leq m - 1$. Let

$$P(z) = \sum_{j=0}^{m-1} \pi_j z^j.$$

Then

$$P(z) = \int_0^\infty (1 - e^{-\lambda t} + ze^{-\lambda t})P(1 - e^{-\lambda t} + ze^{-\lambda t})\, dB(t)$$
$$+ (1 - z)\pi_{m-1} \int_0^\infty e^{-\lambda t}(1 - e^{-\lambda t} + ze^{-\lambda t})^{m-1}\, dB(t).$$

Write

$$P(z) = \sum_{j=0}^{m-1} B_j(z - 1)^j;$$

then

$$B_j = \frac{1}{j!} \frac{d^j P(z)}{dz^j}\bigg|_{z=1}, \qquad B_0 = 1.$$

One also has

$$B_j = \sum_{i=j}^{m-1} \binom{i}{j} \pi_i$$

as binomial moments of the π_i $\left[$and hence $\pi_j = \displaystyle\sum_{n=j}^{m-1} (-1)^{n-j} \binom{n}{j} B_j\right].$

From definitions,

$$B_j = C_j \frac{\displaystyle\sum_{i=j}^{m-1} \binom{m-1}{i}(1/C_i)}{\displaystyle\sum_{i=0}^{m-1} \binom{m-1}{i}(1/C_i)},$$

where $\quad C_0 = 1, \qquad C_i = \dfrac{\beta(\lambda)\beta(2\lambda) \,\cdots\, \beta(i\lambda)}{[1 - \beta(\lambda)][1 - \beta(2\lambda)] \,\cdots\, [1 - \beta(i\lambda)]}.$

β is the Laplace-Stieltjes transform of the service distribution.

The absolute probabilities p_n are similarly derived, using transition probabilities p_{ij}, instead of r_{ij}, defined as the latter except that the expression $\mu[1 - B(t)]\, dt$ is used in the integral instead of $dB(t)$ to indicate the remaining service time. Binomial moments for p_n are also introduced which can be related to the binomial moments of π_n. The derivation leads to the following expression:

$$p_n = \frac{m\pi_{n-1}}{n(m\lambda/\mu + \pi_{n-1})}, \qquad n = 1, \ldots, m,$$

where μ is the service rate. p_0 is obtained from

$$p_0 = 1 - \sum_{n=1}^{m} p_n.$$

The average number in the system is given by

$$\frac{1 - p_m}{\lambda/\mu} = \frac{m}{m\lambda/\mu + \pi_{m-1}}.$$

The average waiting time is given by

$$\frac{m-1}{\mu} - \frac{1 - \pi_{m-1}}{\lambda},$$

where σ^2 is the service-time variance.

2. *The Exponential-service-time Case for c Repairmen*

If p_n is the steady-state probability that there are n machines waiting and being served, we have, without difficulty, the following steady-state equations:

$$m\lambda p_0 = \mu p_1,$$
$$[(m - n)\lambda + n\mu]p_n = (m - n + 1)\lambda p_{n-1} + (n + 1)\mu p_{n+1},$$
$$1 \le n < c, \quad (14\text{-}63)$$
$$[(m - n)\lambda + c\mu]p_n = (m - n + 1)\lambda p_{n-1} + c\mu p_{n+1},$$
$$c \le n \le m,$$

or by using the first relation in the second, etc., iteratively, we have

$$(n + 1)\mu p_{n+1} = (m - n)\lambda p_n, \qquad n < c,$$
$$c\mu p_{n+1} = (m - n)\lambda p_n, \qquad c \le n \le m. \quad (14\text{-}64)$$

The following solution was given by Palm [202]:

$$p_n = p_0 (c\rho)^n \binom{m}{n}, \qquad 0 \le n < c,$$

$$p_n = p_0 (c\rho)^n \binom{m}{n}\frac{n!}{c! c^{n-c}}, \qquad c \le n \le m, \quad (14\text{-}65)$$

$$p_0 = 1 - \sum_{n=1}^{m} p_n = \left(\sum_{n=0}^{m} \frac{p_n}{p_0}\right)^{-1},$$

where $\rho = \lambda/c\mu$.

Exercise: Derive this solution.

If we define with Naor [513]

$$p(k,\xi) = \frac{\xi^k e^{-\xi}}{k!},$$

$$P(k,\xi) = \sum_{i=k}^{\infty} p(i,\xi),$$

$$S\left(m,c,\frac{\mu}{\lambda}\right) = \sum_{i=0}^{c-1} \left[\frac{c^i}{i!} - \frac{c^{i-1}}{(i-1)!}\right] [1 - P(m - i + 1, \rho^{-1})]$$

$$= \sum_{i=0}^{c-1} \frac{c^i}{i!} p(m - i, \rho^{-1}) + \frac{c^{c-1}}{(c - 1)!} [1 - P(m - c + 1, \rho^{-1})],$$

then

$$p_n = \frac{c^n}{n!} \frac{p(m - n, \rho^{-1})}{S(m,c,\mu/\lambda)}, \qquad n < c,$$

$$p_n = \frac{c^{c-1}}{(c - 1)!} \frac{p(m - n, \rho^{-1})}{S(m,c,\mu/\lambda)}, \qquad c \le n \le m, \tag{14-66}$$

which can be verified by substitution.

The average number of machines being repaired is

$$c \frac{S(m - 1, c, \mu/\lambda)}{S(m,c,\mu/\lambda)}. \tag{14-67}$$

Thus, if a is the average number of machines working, b those being serviced, and L_q those waiting in line, then

$$a + b + L_q = m, \qquad \frac{a}{b} = \frac{\mu}{\lambda},$$

$$b = \sum_{n=0}^{c-1} np_n + c \sum_{n=c}^{m} p_n = c - \sum_{n=0}^{c-1} (c - n)p_n = c \frac{S(m - 1, c, \mu/\lambda)}{S(m,c,\mu/\lambda)}.$$

The average number of unoccupied repairmen is $c - b$, the operative efficiency is b/c, the coefficient of loss due to repairs is b/m, and that due to interference (i.e., presence of other machines for repair) is L_q/m. The combined coefficient of loss is $(b + L_q)/m$. The machine efficiency or availability is a/m. Naor gives illustrative examples. The probability of waiting is given by

$$P(>0) = \sum_{n=c}^{m} p_n. \tag{14-68}$$

The average waiting time for repair is

$$W_q = \frac{1}{\lambda} \sum_{n=c+1}^{m} (n - c)p_n = \frac{\text{average number in line}}{\lambda}. \tag{14-69}$$

The expected waiting time of machines that are delayed is

$$\frac{W_q}{P(>0)}. \tag{14-70}$$

Ashcroft's solution [14] for the exponential-service-time case and a single repairman is obtained by putting $c = 1$.

In another paper, Naor [514] obtains normal approximations to facilitate the computation of p_n. One introduces the frequency function

$$p_c(j) = \frac{(c-1)!}{j!c^{c-j}}(c-j), \qquad j = 0, 1, \ldots, c-1,$$

with first and second moments, respectively:

$$E_c\{j\} = \frac{1}{p(c,c)}\{cp(c,c) - [1 - P(c,c)]\}, \tag{14-71}$$

$$E_c\{j^2\} = \frac{1}{p(c,c)}\{c^2p(c,c) - [1 - P(c,c)] - 2c[1 - P(c-1,c)]\}, \tag{14-72}$$

and forms the convolution

$$\{p(k,\xi)\} * \{p_c(j)\} = v(i,c,\xi) = \sum_{n=0}^{i} \frac{(c-1)!}{n!c^{c-n}}(c-n)p(i-n,\xi),$$

$$i \leq c - 1,$$

where the summation is taken up to $c - 1$ for $i \geq c - 1$, where $i = j + k$. If one defines $u(i,c,\zeta) = v(i,c,\xi)$, $\zeta = \xi/c$, then

$$U(i,c,\zeta) = \sum_{s=0}^{i} u(s,c,\zeta) = \sum_{n=0}^{c-1} \frac{(c-1)!}{n!c^{c-n}}(c-n)[1 - P(i-n+1,c\zeta)]$$

holds for $i \geq c - 1$, the only case of interest.

It is easy to show that

$$S(i,c,\zeta) = \frac{c^{c-1}}{(c-1)!}U(i,c,\zeta).$$

Since u is a convolution, we have for large c for the mean and variance

$$E_u\{i\} = c\zeta + E_c\{j\} \approx c\zeta + c - \tfrac{1}{2}\sqrt{2\pi c} + \tfrac{1}{3}, \tag{14-73}$$

$$V_u\{i\} = c\zeta + E_c\{j^2\} - E_c^2\{j\} \approx c\zeta + c\left(2 - \frac{\pi}{2}\right) - \frac{\sqrt{2\pi c}}{6} + \frac{2}{9}. \tag{14-74}$$

For sufficiently large $c\zeta$—the mean of $p_c(j)$—by the central-limit theorem we have the following approximation:

$$U(i,c,\zeta) \approx \frac{1}{\sqrt{2\pi}}\int_{-\infty}^{z} e^{-v^2/2}\,dy,$$

where
$$z = \frac{i + \frac{1}{2} - c\zeta - E_c\{j\}}{(c\zeta + E_c\{j^2\} - E_c^2\{j\})^{\frac{1}{2}}}, \qquad (14\text{-}75)$$

using the correction for continuity. Note that to obtain the standard form of a normal distribution one uses $y = (x - \mu)/\sigma$, where μ is the mean and σ is the standard deviation. Here we have replaced x by $i + \frac{1}{2}$, the corrected value of i. This approximation now makes the computation of p_n easier.

3. *The Constant-service-time Case for a Single Repairman*

This problem was solved by Ashcroft [14], Westgarth [756], and Benson and Cox [52].

To obtain the solution with Benson and Cox, we suppose that service is accomplished in k stages, each of which is exponentially distributed with mean $1/\mu$. The total service time will be a chi-squared distribution with $2k$ degrees of freedom and mean k/μ. If k and μ become infinite, so that k/μ remains constant, we have the solution of the constant-service-time case. We have in the chi-square service-time case for p_n, the probability that n in a total of m machines are stopped for repair,

$$p_n = \sum_{i=1}^{n} i \binom{m - n + i}{i} \nabla^{i-1} \left\{ \frac{[1 + (\lambda/\mu)(m - n)]^k - 1}{m - n} \right\} p_{n-i},$$

$$(14\text{-}76)$$

where ∇ is the backward difference operator ($\nabla f_i = f_i - f_{i-1}$).

The solution for the constant-service-time case is then given by

$$p_n = \sum_{i=1}^{n} i \binom{m - n + i}{i} \nabla^{i-1} \left\{ \frac{\exp\ [(m - n)(k\lambda/\mu)] - 1}{m - n} \right\} p_{n-i} \quad (14\text{-}77)$$

together with

$$\sum_{n=0}^{m} p_n = 1.$$

Ashcroft gives tables of the average number of machines running.

4. *Two Types of Stoppages, Single Repairman*

We now assume that there are two types of machine stoppages, each of which occurs at random with parameters λ_1 and λ_2, respectively, with a higher priority placed on repairing the first stoppage. There are m machines in all. Suppose that the corresponding repair times by a single repairman are exponential with parameters μ_1 and μ_2, respectively. We have for the steady-state probability $p_r(n_1, n_2)$ that there are n_1 stopped machines of the first kind and n_2 of the second and that the stoppage being cleared is of the rth type ($r = 1, 2$):

$$[\mu_1 + (m - n_1 - n_2)(\lambda_1 + \lambda_2)]p_1(n_1, n_2)$$
$$= (m - n_1 - n_2 + 1)\lambda_1 p_1(n_1 - 1, n_2)$$
$$+ (m - n_1 - n_2 + 1)\lambda_2 p_1(n_1, n_2 - 1)$$
$$+ \mu_1 p_1(n_1 + 1, n_2) + \mu_2 p_2(n_1, n_2 + 1), \quad n_1, n_2 \neq 0, m,$$

$$[\mu_1 + (m - n_1)(\lambda_1 + \lambda_2)]p_1(n_1, 0) = (m - n_1 + 1)\lambda_1 p_1(n_1 - 1, 0)$$
$$+ \mu_1 p_1(n_1 + 1, 0) + \mu_2 p_2(n_1, 1), \quad n_1 \neq m,$$

$$\mu_1 p_1(m, 0) = \lambda_1 p_1(m - 1, 0),$$

$$[\mu_2 + (m - n_1 - n_2)(\lambda_1 + \lambda_2)]p_2(n_1, n_2)$$
$$= (m - n_1 - n_2 + 1)\lambda_1 p_2(n_1 - 1, n_2)$$
$$+ (m - n_1 - n_2 + 1)\lambda_2 p_2(n_1, n_2 - 1), \quad n_1, n_2 \neq 0, m,$$

$$[\mu_2 + (m - n_2)(\lambda_1 + \lambda_2)]p_2(0, n_2)$$
$$= (m - n_2 + 1)\lambda_2 p_2(0, n_2 - 1) + \mu_1 p_1(1, n_2)$$
$$+ \mu_2 p_2(0, n_2 + 1), \quad n_2 \neq m,$$

$$\mu_2 p_2(0, m) = \lambda_2 p_2(0, m - 1),$$

$$m(\lambda_1 + \lambda_2)p(0, 0) = \mu_1 p_1(1, 0) + \mu_2 p_2(0, 1),$$

$$p(0, 0) + \sum_{n_1=1}^{m} \sum_{n_2=0}^{m-n_1} p_1(n_1, n_2) + \sum_{n_2=1}^{m} \sum_{n_1=0}^{m-n_2} p_2(n_1, n_2) = 1.$$

One is interested in solving these equations for p_n, the total probability that n machines are stopped. No general solution is given.

5. Each of c Repairmen Specializing in One Type of Stoppage with c Kinds of Stoppages in All

Let the total number of machines be m. Let the stoppage distribution be Poisson for c different stoppages that a machine may suffer with corresponding parameter λ_i for the ith stoppage. Let each of c repairmen specialize in one type of stoppage. Let the repair distribution for the ith stoppage be exponential with parameter μ_i. Let $p(n_1, \ldots, n_c)$ be the steady-state probability that there are n_1 machines with stoppage of type 1, n_2 of type 2, etc. We have

$$[(m - n) \sum_{i=1}^{c} \lambda_i + \sum_{i=1}^{c} \mu_i] \, p(n_1, \ldots, n_c)$$

$$= \sum_{i=1}^{c} (m - n + 1)\lambda_i p(n_1, \ldots, n_i - 1, \ldots, n_c)$$

$$+ \sum_{i=1}^{c} \mu_i p(n_1, \ldots, n_i + 1, \ldots, n_c),$$

$$n_i = 1, \ldots, m - 1, \quad n \leq m - 1$$

$$\sum_{i=1}^{c} \mu_i p(n_1, \ldots, n_c) = \sum_{i=1}^{c} \lambda_i p(n_1, \ldots, n_i - 1, \ldots, n_c),$$

$$n_i = 1, \ldots, m - 1, \, n = m,$$

$$[(m - n) \sum_{i=1}^{c} \lambda_i + \sum_j \mu_i] \, p(n_1, \ldots, n_i, \ldots, 0, \ldots)$$

$$= \sum_j (m - n + 1)\lambda_i p(n_1, \ldots, n_i - 1, \ldots, 0, \ldots)$$

$$+ \sum_j \mu_i p(n_1, \ldots, n_i + 1, \ldots, 0, \ldots)$$

$$+ \sum_{n-j} \mu_i p(n_1, \ldots, n_i, \ldots, 1, \ldots),$$

$$\mu_i p(0, \ldots, m, \ldots, 0) = \lambda_i p(0, \ldots, m - 1, \ldots, 0),$$

$$m \sum_{i=1}^{c} \mu_i p(0, \ldots, 0, \ldots, 0) = \sum_{i=1}^{c} \mu_i p(0, \ldots, 1, \ldots, 0),$$

where $n = \sum_{i=1}^{c} n_i$, $\sum_j$ is taken over all values of i such that $n_i \neq 0$, and $\sum_{n-j}$ is taken over all values of i such that $n_i = 0$.

If we write $\rho_i = \lambda_i/\mu_i$, the solution to this system is given by

$$p(n_1, \ldots, n_c) = \frac{m!}{(m - n)!} \prod_{i=1}^{c} \rho_i^{n_i} \, p(0, \ldots, 0)$$

which, when summed over the values of n_1 and those of n_2, etc., must yield unity. This gives

$$p(0, \ldots, 0) = \sum_{i=1}^{c} \frac{\rho_i^{c-1} F(\rho_i, m)}{\prod_{\substack{i=1 \\ j \neq i}}^{c} (\rho_i - \rho_j)},$$

where $\rho_i \neq \rho_j$ for $i \neq j$ and where

$$F(\rho_i, m) = \sum_{n=0}^{m} \frac{m!}{(m - n)!} \rho_i^n = 1 + m\rho_i + \cdots + m!\rho_i^m.$$

The mean number of running machines is

$$E[m - n] = m \sum_{i=1}^{c} \frac{\rho_i^{c-1} F(\rho_i, m - 1)}{\prod_{\substack{i=1 \\ j \neq i}}^{c} (\rho_i - \rho_j)} \, p(0, \ldots, 0).$$

Machine availability is given by

$$\frac{E[m - n]}{m} [p(0, \ldots, 0)]^{-1}$$

As an exercise, determine the last quantity for $c = 2$.

If a fraction L of the operative's (repairman's) time is spent in ancillary work and the remainder in machine care, it is possible to determine the number m of machines that yield unity minus this fraction and thus determine m to obtain the available efficiency. The maximum rate of production is $(1 - L)/\rho$. The minimum number of machines required to yield this value is obtained, as has just been described. Benson and Cox also give tables for this type of computation.

In yet another paper Benson [53] studies modifications of the foregoing ideas when the operative has other duties, i.e., ancillary work.

6. *Constant Repair Time by a Walking Repairman*

C. Mack, T. Murphy, and N. Webb [463] have studied the efficiency of m machines which failed at random times with parameter λ but received attention by a repairman who walked from machine to machine in an ordered fashion, returning to the first, with constant walking time t_i from the $(i - 1)$st to the ith machine, and repaired a machine in constant time R. Let p_n be the probability that n machines required repair on a tour of inspection of all machines. The probability that the machine with which he started is still running at the end of the cycle is $e^{-\lambda(mT+nR)}$, where $mT = \sum_{i=1}^{m} t_i$. The equations are

$$p_n e^{-\lambda(mT+nR)} + p_{n+1} e^{-\lambda[mT+(n+1)R]} = p_n \qquad (14\text{-}78)$$

with solution

$$p_n = (a^{n-1}b - 1)(a^{n-2}b - 1) \cdots (ab - 1)(b - 1)p_0, \qquad (14\text{-}79)$$

where $a = e^{\lambda R}$, $b = e^{\lambda mT}$, and p_0 is determined from

$$\sum_{n=0}^{m} \binom{m}{n} p_n = 1. \qquad (14\text{-}80)$$

The solution is shown to be unique. Various quantities already encountered in other examples are then computed.

14-7. Provision of Spare Machines

A study of the problem of providing spare aircraft engines has been made by Taylor and Jackson [712]. Suppose that m engines are continuously operating in aircraft and that once an engine becomes unserviceable it is instantly replaced by a spare, provided that a spare is on hand, and the unserviceable engine is placed in the repair shop at once. The m engines fail at random but at a constant average rate λ per unit time. The time required to service an engine is exponentially distributed with an average of $1/\mu$. It is possible to service c engines simultaneously. Also let n be the total number of spare engines. As long as not more than n engines are being repaired, there will be m engines operating. Once $n + 1$ engines are in the repair shop, then less than m engines will be operating. In this case either of the following occurs:

1. The engines stop operating until a replacement is available from the shop.

2. Less than m engines continue to operate.

Considering situation 1, the probability of having i unserviceable engines in the shop at any time t is $P_i(t)$. The system is described by

$$
\begin{aligned}
P_0'(t) &= -\lambda P_0(t) + \mu P_1(t), & i &= 0, \\
P_i'(t) &= \lambda P_{i-1}(t) - (\lambda + i\mu)P_i(t) + (i+1)\mu P_{i+1}(t), \\
& & 0 &< i < c, \\
P_i'(t) &= \lambda P_{i-1}(t) - (\lambda + c\mu)P_i(t) + c\mu P_{i+1}(t), \\
& & c &\leq i < n + 1, \\
P_{n+1}'(t) &= \lambda P_n(t) - c\mu P_{n+1}(t), & i &= n + 1.
\end{aligned} \tag{14-81}
$$

The steady-state solutions are of interest and are given by

$$
\begin{aligned}
p_i &= \frac{(c\rho)^i}{i!}\, p_0, & i &\leq c, \\
p_i &= \rho^{i-c}\frac{(c\rho)^c}{c!}\, p_0, & c &\leq i \leq n + 1,
\end{aligned} \tag{14-82}
$$

where $\rho = \lambda/c\mu$,

$$
p_0 = \left[e^{c\rho} - \sum_{i=c+1}^{\infty} \left(\frac{c\rho}{i!}\right)^i + \rho\left(\frac{c\rho}{c!}\right)^c \frac{1 - \rho^{n+1-c}}{1 - \rho} \right]^{-1}
$$

obtained from

$$
\sum_{i=0}^{n+1} p_i = 1.
$$

The average number of engines unserviceable is

$$
U = \sum_{i=0}^{n+1} i p_i. \tag{14-83}
$$

The average number of engines being serviced is

$$
S = \sum_{i=0}^{c} i p_i + c \sum_{c+1}^{n+1} p_i. \tag{14-84}
$$

The average number of engines waiting for service is

$$
L_q = \sum_{i=c+1}^{n+1} (i - c) p_i. \tag{14-85}
$$

Consider the case where $\rho = 1$, that is, when the potential service rate equals the rate of failure. The steady-state solution shows that the probability distribution of the number of unserviceable engines is a monotonic increasing sequence. It turns out, by an analysis similar to the one given below for $\rho < 1$, that increasing the number of spares rapidly does not materially reduce the probability of an operational emergency (p_{n+1}). Also, increasing the number of spares increases at a steady rate the number of engines waiting for service.

Consider the more practical situation where $\rho < 1$.

We have the following relation for $c \leq i \leq n + 1$:

$$p_i < \frac{1 - \rho}{1 - \rho^{n-c+2}} \rho^{i-c} \tag{14-86}$$

which is a decreasing function of n and i. Also L_q, the average number of engines waiting for service, has the following upper bound:

$$L_q < \frac{\rho}{1 - \rho}, \tag{14-87}$$

which means that the average number waiting for service has an upper bound independent of n; hence increasing the number of spare engines does not increase the number of those waiting for service but does increase the operational efficiency.

14-8. An Inventory Application

We give an example of a queueing-theory approach to an inventory problem. Note that, although inventory problems form stochastic processes, few are actually found in queueing theory proper. In inventory theory we have commodity input, its storage and reorder, and finally the sales demand. The queue may consist of demand orders waiting to be satisfied, but the real queue is often that of waiting for replenishment.

W. Karush [355] has studied a problem which is a slight elaboration on the foregoing ideas. Let the total inventory be fixed at n, of which n_0 units are in stock and $n_1 = n - n_0$ are on reorder for replenishment. Each of the latter units undergoes a replenishment cycle with a given frequency distribution (identical for all units). The demand (or customer-arrival) distribution is Poisson with parameter λ. Once a unit is sold from a stock of $n_0 > 0$, a reorder is initiated to replenish the sold unit. If $n_0 = 0$, a sale is lost, the customer is turned back, and no replenishment reorder is instituted for the lost sale. The result is a queueing problem without a queue. It is desired to compute the steady-state probability (the ratio of lost sales to total demand) of being out of stock as a function of n, the inventory level. More generally, one is interested in the conditional probability $p(n_0|n)$ of n_0 units in stock, given n, the total fixed inventory.

The reorder process may proceed in r different paths, each chosen with probability α_i ($i = 1, \ldots, r$). Each path consists of k_i ($i = 1, \ldots, r$) phases, and the time spent in each phase is exponentially distributed. Once one phase finishes its service, the next phase immediately starts its service.

The mean replenishment time from the jth phase of the ith path is $1/\mu_{ij}$. At any time, the n_1 replenishments in process are distributed among the phases in some manner where we have n_{ij} units in the jth phase of the ith path. If we define the state of the system at any time by the vector

$$N = (n_0, n_{11}, \ldots, n_{1k_1}, n_{21}, \ldots, n_{rk_r}),$$

$P(N)$ the stationary probability for state N, and single-stage transitions by

$$
\begin{aligned}
T_{ij}(N) &\to N, \\
T_{ij}(N) &\equiv (n_0, \ldots, n_{ij} + 1, n_{i,j+1} - 1, \ldots), \\
T_{ik_i}(N) &\equiv (n_0 - 1, \ldots, n_{ik_i} + 1, \ldots), \\
T_{0i}(N) &\equiv (n_0 + 1, \ldots, n_{i1} - 1, \ldots),
\end{aligned}
$$

the steady-state equations are given by

$$\left(\lambda + \sum_{i,j} \mu_{ij} n_{ij}\right) p(N) = \lambda \sum_i \alpha_i p[T_{0i}(N)]$$
$$+ \sum_{i,j} \mu_{ij}(n_{ij} + 1) p[T_{ij}(N)], \qquad n_0 > 0. \quad (14\text{-}88)$$

For $n_0 = 0$, λ is omitted on the left.

Now $p(n_0|n) = \Sigma p(N)$ taken over all vectors N for which

$$\sum_{i,j} n_{ij} = n - n_0.$$

If one defines the homogeneous polynomial of degree n_1 given by

$$P(n_0) \equiv \sum_N p(N) x_{11}^{n_{11}} x_{12}^{n_{12}} \cdots x_{ij}^{n_{ij}} \cdots, \qquad \sum_{i,j} n_{ij} = n_1, \quad (14\text{-}89)$$

the above system of equations becomes

$$\lambda P(n_0) + \sum_{i,j} \mu_{ij} x_{ij} \frac{\partial P(n_0)}{\partial x_{ij}} = \lambda \left(\sum_i \alpha_i x_{i1}\right) P(n_0 + 1)$$
$$+ \sum_{j \neq k_i} \mu_{ij} x_{i,j+1} \frac{\partial P(n_0)}{\partial x_{ij}} + \sum_i \mu_{ik_i} \frac{\partial P(n_0 - 1)}{\partial x_{ik_i}}, \qquad n_0 = 1, \ldots, n.$$
$$(14\text{-}90)$$

For $n_0 = 0$ the first term on the left and the last term on the right are omitted.

The solution (whose uniqueness is easily proved by Karush) is given by

$$P(n_0) = \frac{\lambda^{n_1}}{n_1!} \left(\frac{\sum\limits_{i,j} \alpha_i x_{ij}}{\mu_{ij}} \right)^{n_1}, \qquad n_0 = 0, 1, \ldots, n, \qquad (14\text{-}91)$$

with $P(n + 1) \equiv 0$. We then have for the probability $p(N)$ that there are n_{ij} units in the jth phase of the ith path

$$p(N) = \frac{\lambda^{n_1}}{C} \prod_{i,j} \frac{\alpha_i^{n_{ij}}}{n_{ij}! \mu_{ij}^{n_{ij}}} \qquad (14\text{-}92)$$

determined as the general coefficient of the polynomial $P(n_0)$. The normalizing factor C is given by

$$C = 1 + \rho + \frac{\rho^2}{2!} + \cdots + \frac{\rho^n}{n!}, \qquad (14\text{-}93)$$

where $\rho = \lambda/\mu$ and $1/\mu$ is the mean replenishment time. We have also, using the relation between $p(n_0|n)$ and $p(N)$,

$$p(n_0|n) = \frac{1}{C} \frac{\rho^{n_1}}{n_1!}, \qquad n = n_0 + n_1, \qquad (14\text{-}94)$$

which is the desired expression.

The probability of being out of stock is $p(0|n)$. It is obtained by replacing n_1 by n in the above formula since $n = n_0 + n_1$, where $n_0 = 0$.

The result can be generalized when λ is a function of n_0, and we have

$$p(n_0|n) = \frac{1}{Cn_1!} \frac{\lambda(n)}{\mu} \frac{\lambda(n-1)}{\mu} \frac{\lambda(n_0+1)}{\mu}, n_0 = 0, 1, \ldots, n-1, \qquad (14\text{-}95)$$

where C is determined from the fact that the sum of the probabilities yields unity.

14-9. Dams and Storage Systems

In this section we shall be mainly concerned with the formulation of models rather than with solutions. Downton [164], Gani [240–243], D. G. Kendall [370], Moran [500–503] and Prabhu [595] have investigated the theory of dams and storage systems. Some of the ideas are related to queueing theory, either in theory or in application. One is generally concerned with the distribution of the amount remaining in the reservoir after use, whether in continuous or in discrete time such as years. In the latter case the input of water into the reservoir will vary from year to year according to some distribution. It is stored (except for possible overflow) to be used during the dry season. During the season it is used at a uniform rate. The total amount used may be assumed to vary from year to year. For simplicity, one may assume this quantity to be a constant M.

Let $\{X_t\}$ be a family of serially independent and identically distributed random variables. Thus if $t_1 < t_2$, X_{t_1} is independent of X_{t_2} for every t_1 and t_2. X_t denotes the input of the water into the reservoir at time t or, if t is discrete and measures years, then X_t is the input at the beginning of the tth year. Let Z_t denote the amount of water in storage at time t but before the arrival of X_t. Let Y_t be the amount released. If t denotes the time in years, then Y_t is the amount released during the dry season of the tth year. We can examine the entire problem for discrete and for continuous probability distributions and for discrete and continuous values of time. In addition, the capacity of the dam may be assumed finite with value K or infinite.

Note that the overflow W_t is given by

$$W_t = X_t + Z_t - K \tag{14-96}$$

if the latter is positive. It is zero otherwise. Its distribution can be determined from that of X_t and Z_t, once the latter of the two has been determined.

1. *Discrete Time and Discrete Probability*

We examine the discrete-time case (measured in years). For the discrete-probability-distribution case we give the formulation but not the solution since the latter is so far obtainable only numerically. We first assume that $Y_t = M$. Thus one releases a quantity equal to min $(X_t + Z_t, M)$. Let X_t assume the values $0, 1, \ldots$ at time t with probabilities $p_0, p_1, \ldots$, and let Z_t assume the values $0, 1, \ldots$ at time t w th probabilities $P_0, \ldots, P_K$ and at time $t + 1$ with probabilities $P'_0, \ldots, P'_K$; also let $K \geq M$.

We have the following equations:

$$\begin{aligned}
P'_0 &= P_0(p_0 + p_1 + \cdots + p_M) + P_1(p_0 + \cdots + p_{M-1}) \\
&\qquad + \cdots + P_M p_0, \\
P'_1 &= P_0 p_{M+1} + P_1 p_M + \cdots + P_{M+1} p_0, \\
P'_{K-M} &= P_0(p_K + \cdots) + P_1(p_{K-1} + \cdots) + \cdots + P_{K-M}(p_M + \cdots), \\
P'_{K-M+1} &= \cdots = P'_K = 0,
\end{aligned} \tag{14-97}$$

which give the distribution of Z_{t+1} from that of Z_t and so on to Z_{t+n}, giving a Markoff chain. We have convergence as $n \to \infty$ to a steady state if $p_i > 0$ $(i = 0, \ldots, K)$.

To obtain the steady-state solution, let $P'_i = P_i$ $(i = 0, \ldots, K - M)$, and also use $\sum_{i=1}^{K-M} P_i = 1$.

The problem may be solved by various numerical, Monte Carlo, or matrix methods.

2. *Discrete Time and Continuous Probabilities*

To consider the continuous-distribution case we first suppose that in the discrete case the sum of the input and reservoir content assumes values $Z_t + X_t = i$ with probability R_i; we also assume that equilibrium is attained. It is easier to obtain the distribution of Z_t this way. Thus if we know the distributions of $Z_t + X_t$ and of X_t we can determine the distribution of Z_t. We have for the stationary distribution of R_i

$$R_0 = p_0 R_0 + p_0 R_1 + \cdots + p_0 R_M,$$
$$R_1 = p_1 R_0 + p_1 R_1 + \cdots + p_1 R_M + p_0 R_{M+1},$$
$$\cdots \cdots \cdots \cdots \cdots \cdots \cdots \cdots \cdots \cdots \cdots \cdots$$
$$R_{K-M-1} = p_{K-M-1} R_0 + \cdots + p_{K-M-1} R_M + p_{K-M-2} R_{M+1}$$
$$+ \cdots + p_0 R_{K-1},$$
$$R_{K-M} = p_{K-M} R_0 + \cdots + p_{K-M} R_M + p_{K-M-1} R_{M+1} \qquad \text{(14-98)}$$
$$+ \cdots + p_1(R_K + R_{K+1} + \cdots),$$
$$\cdots \cdots \cdots \cdots \cdots \cdots \cdots \cdots \cdots \cdots \cdots \cdots$$
$$R_{K-M+s} = p_{K-M+s} R_0 + \cdots + p_{K-M+s} R_M + p_{K-M+s-1} R_{M+1}$$
$$+ \cdots + p_s(R_K + \cdots),$$
$$\cdots \cdots \cdots \cdots \cdots \cdots \cdots \cdots \cdots \cdots \cdots \cdots$$

Note that an amount M is required for the release whenever available. The reader should have no difficulty in justifying these equations.

The steady-state form of the previous set of equations is obtained from this set by noting that

$$P_0 = R_0 + \cdots + R_M, \qquad P_{K-M} = R_K + R_{K+1} + \cdots,$$

and $P_{K-M+s} = 0$ for $s > 0$.

The continuous case is analogous to the above. If $f(x)$ is the density function of X_t for $X \geq 0$ and zero otherwise and $g(x)$ that of $X_t + Z_t$, then in the steady state we have the following equation, the three terms on the right of which correspond respectively to

$0 \leq X_t + Z_t \leq M$: The reservoir runs dry.

$M < X_t + Z_t \leq K$: No overflow, and water is stored.

$K < X_t + Z_t$: An overflow given by the difference $K - (X_t + Z_t)$.

$$g(x) = f(x) \int_0^M g(y) \, dy + \int_M^K g(y) f(x + M - y) \, dy$$
$$+ f(x - K + M) \int_K^\infty g(y) \, dy. \qquad \text{(14-99)}$$

Note that $p_0 = \int_0^M g(y)\, dy$ and $p_1 = \int_K^\infty g(y)\, dy$. From the above, Moran obtains the distribution of Z_t. It is given by

$$
\begin{array}{ll}
p_0 & \text{for } Z_t = 0, \\
g(x + M) & \text{for } Z_t = x, \quad 0 < x < K - M, \\
p_1 & \text{for } Z_t = K - M.
\end{array}
\qquad (14\text{-}100)
$$

3. *Finite-and Infinite-capacity Dams*

If the capacity of a reservoir is infinite, that is, $K \to \infty$, the corresponding problem is simplified, and the equations in the steady state are

$$
P_0 = P_0(p_0 + \cdots + p_M) + P_1(p_0 + \cdots + p_{M-1})
$$
$$
+ \cdots + P_M p_0, \qquad (14\text{-}101)
$$
$$
P_i = P_0 p_{M+1} + P_1 p_{M+i-1} + \cdots + p_{M+i} p_0, \qquad i \geq 1.
$$

The stationary distribution of $\{R_i\}$ satisfies

$$
R_0 = p_0(R_0 + \cdots + R_M),
$$
$$
R_i = p_i(R_0 + \cdots + R_M) + p_{i-1} R_{M+1} + \cdots + p_0 R_{M+i},
$$
$$
i \geq 1, \quad (14\text{-}102)
$$

$$
\sum_{i=0}^\infty R_i = 1.
$$

If we identify queue length with the reservoir content, arrivals by a Poisson process with the input and service in batches of size M, or the queue length, if the latter is less than M, with the output, then the last set of equations describe the queue length at epochs just before service begins. This gives one relation between storage dams and queues.

We have the following expression:

$$
P_0 = R_0 + \cdots + R_M, \qquad P_i = R_{M+i}, \qquad i \geq 1, \quad (14\text{-}103)
$$

and hence the two sets of equations give equivalent description of the system.

Remark: Two cases are of interest here. One is when the amount stored is very large and hence Z_t becomes zero with very small probability. The other is the case where the probability distribution of the reservoir ever being full is small. We thus have a possibility of two types of solutions.

A stationary nonzero solution exists if for the first case $\sum_{n=0}^\infty n p_n > M$ and for the second case $< M$ holds.

We also have for the continuous-distribution case and $K \to \infty$

$$
g(x) = f(x) \int_0^M g(y)\, dy + \int_M^{M+x} f(M + x - y) g(y)\, dy \quad (14\text{-}104)
$$

with
$$\sum_{i=0}^{\infty} P_i = 1.$$

For $M = 1$ we have

$$P_0 = P_0(p_0 + p_1) + P_1p_0,$$
$$P_1 = P_0p_2 \qquad + P_1p_1 + P_2p_0, \qquad (14\text{-}105)$$
. . . .

which is a familiar-type set of equations.   If we introduce the generating function

$$P(z) = \sum_{i=0}^{\infty} P_iz^i, \qquad p(z) = \sum_{i=0}^{\infty} p_iz^i,$$

then as usual we obtain

$$P(z) = p_0P_0 \frac{z - 1}{z - p(z)}. \qquad (14\text{-}106)$$

The mean, i.e., the average amount in the reservoir before the new input arrives, is given by

$$\sum_{i=0}^{\infty} iP_i = \frac{p''(1)}{2[1 - p'(1)]}. \qquad (14\text{-}107)$$

Higher moments may be similarly found.

We now note that, if $\{P_i\}$ is the solution for infinite water storage in the discrete case, we have for finite storage

$$P_i' = \frac{P_i}{P_0 + \cdots + P_{K-1}}. \qquad (14\text{-}108)$$

We can always determine $\{P_i\}$ by expanding $(z - 1)/[z - p(z)]$ for sufficiently small values of z and equating coefficients, provided that $p_0 > 0$, or, using cumulative distributions,

$$G(x) = \int_0^x G(x + M - t) \, dF(t). \qquad (14\text{-}109)$$

As an example, one may solve the last equation when

$$F(x) = \sum_i c_ie^{-\lambda_ix}, \qquad (14\text{-}110)$$

where the c_i are complex constants and Re $(\lambda_i) > 0$ by substitution, obtaining conditions which the c_i and λ_i must satisfy, particularly if the (general-type) distribution $G(x)$ is also known.   As an exercise, determine $F(x)$ if $G(x) = 1 - e^{-\mu x}$ $(\mu > 0)$.

Inputs resulting in negative exponential stationary distributions of the reservoir together with approximations to the solution of finite water storage have also been studied by Gani and Prabhu [243].

4. *Continuous Time and Continuous Probabilities*

To consider the continuous-time problem with continuous distributions and infinite capacity we divide the time interval $(t, t + 1)$ into units each of length $1/n$ (the unit amount released at the end of each subinterval). The total input during $(t, t + 1)$ is the n-fold convolution of $\{p_i\}$ and therefore has the generating function $\{p(z)\}^n$. Note that we assume that $M = 1$ and use the equations previously studied. One then varies $\{p_i\}$ as n increases so that $\{p(z)\}^n$ approaches the continuous distribution.

As an example, we use the geometric distribution

$$p_i = (1 - r)r^i, \qquad i = 0, 1, \ldots , \qquad 0 < r < 1. \quad (14\text{-}111)$$
Then
$$p(z) = (1 - r)(1 - rz)^{-1}. \quad (14\text{-}112)$$

If ξ is a variate with this distribution, its characteristic function then is

$$(1 - r)(1 - re^{i\theta})^{-1}, \quad (14\text{-}113)$$

and the variate $X = \xi/n$ has the mean $r/n(1 - r)$. As $n \to \infty$ and $r \to 1$ let the mean tend to $\mu_1 = \sum_{n=0}^{\infty} np_n$. We write $r = \mu_1 n/(1 + \mu_1 n)$. The characteristic function of X is

$$\lim_{n \to \infty} \frac{1}{\mu_1 n} \left(1 - \frac{\mu_1 n}{1 + \mu_1 n} e^{i\theta/n} \right)^{-1} = (1 - \mu_1 i\theta)^{-1}. \quad (14\text{-}114)$$

Thus X has the asymptotic distribution $e^{-x/\mu_1} dx/\mu_1$, which we take as the total input distribution during $(t, t + 1)$, that is, during a unit interval. The input during any time T then is

$$\frac{1}{\mu_1 \Gamma(T)} \left(\frac{x}{\mu_1} \right)^{T-1} e^{-x/\mu_1} dx. \quad (14\text{-}115)$$

We wish to compute the probability that no water is stored in the dam for the continuous case. We start by taking the above geometric distribution as the nth convolution of $\{p_n\}$; hence the distribution of $\{p_n\}$ itself is the nth-root convolution. Its generating function is

$$p(z) = (1 - r)^{1/n}(1 - rz)^{-1/n} \quad (14\text{-}116)$$

which may be used to compute $P(z)$ and hence $\{P_i\}$.

Now $p_0 = (1 - r)^{1/n}$, $\mu_1 = r/n(1 - r)$, and the probability that there is no water behind the dam in the discrete case is

$$\frac{1 - p'(1)}{p_0} = \frac{1 - [r/n(1 - r)]}{(1 - r)^{1/n}}. \quad (14\text{-}117)$$

As $n \to \infty$, this tends to $1 - \mu_1$ (assuming $\mu_1 < 1$), the probability that there is no water behind the dam in the continuous case. This can be rigorously shown to be the desired value from the cumulative distribution $H(z)$ of Z_t, which in turn can be shown for $Z > 0$, using the integral equation, to be

$$H(z) - \tfrac{1}{2}[H(0 + 0) - H(0 - 0)] = H(z) - \tfrac{1}{2}(1 - \mu_1)$$
$$= \frac{1}{2\pi} \int_{-\infty}^{\infty} \frac{(1 - \mu_1)(1 - e^{-i\theta z})}{i\theta + \log (1 - \mu_1 i\theta)} \, d\theta.$$
$$\text{(14-118)}$$

A method for deriving this distribution will now be developed along different lines.

Downton has examined the continuous case by considering the discrete case with arrivals in batches. In the continuous case the batch size and the interarrival intervals tend to zero. Arrivals occur in batches of size n by the cumulative distribution

$$A(t) = 1 - e^{-nt}, \qquad 0 \le t < \infty. \qquad \text{(14-119)}$$

Let the cumulative-service distribution of a batch of size n be $B_n(t)$ and the corresponding cumulative-waiting-time distribution be $C_n(t)$. Units are served on a first-come–first-served basis with cumulative-service-time distribution $B(t)$ and waiting-time distribution $C(t) \equiv \text{Prob} \ (w \le t)$ for $0 \le t < \infty$. We have for the traffic intensity

$$\rho = \frac{\int_0^\infty t \, dB(t)}{\int_0^\infty t \, dA(t)}. \qquad \text{(14-120)}$$

Thus
$$\int_0^\infty t B_n(t) = \frac{\rho}{n}.$$

If
$$\beta_n(s) \equiv \int_0^\infty e^{-st} \, dB_n(t),$$

Eq. (9-9) gives for the transform of the waiting time

$$\gamma_n(s) = \int_0^\infty e^{-st} \, dC_n(t) = (1 - \rho) \left\{ 1 - \frac{n[1 - \beta_n(s)]}{s} \right\}^{-1}. \qquad \text{(14-121)}$$

We wish to determine
$$\psi(s) = \lim_{n \to \infty} \gamma_n(s), \qquad \text{(14-122)}$$

provided that the input in a finite time remains meaningful.

In the discrete case, the probability of exactly r batches during t is given by the Poisson process $(nt)^r e^{-nt}/r!$ and the distribution of the total amount in the r batches is the r-fold convolution of $B_n(t)$. Let $\phi_n(s,t)$ be the Laplace-Stieltjes transform of $F_n(x,t)$, the probability that an amount less than x arrives in time t. Because of the foregoing,

$$\phi_n(s,t) = \frac{\sum_{r=0}^{\infty} (nt)^r e^{-nt}[\beta_n(s)]^r}{r!} = \exp\{-nt[1 - \beta_n(s)]\}. \quad (14\text{-}123)$$

For continuous input in time, suppose that $F_n(x,t) \to F(x,t)$ as $t \to \infty$. Correspondingly we have $\phi_n(s,t) \to \phi(s,t)$. The continuity condition is equivalent to

$$\lim_{n \to \infty} n[1 - \beta_n(s)] = -\frac{\log \phi(s,t)}{t}. \quad (14\text{-}124)$$

The left side does not contain t; thus also the right side does not contain t, implying the existence of an equilibrium solution for a continuous input only when the latter has an additive distribution. The last expression on the right may be substituted to determine $\psi(s)$.

Moran's example, where

$$F(x,t) = \frac{1}{\Gamma(t)\rho^t} \int_0^x e^{-y/\rho} y^{t-1} \, dy, \quad (14\text{-}125)$$

gives

$$\phi(s,t) = (1 + \rho s)^{-t} \quad (14\text{-}126)$$

and

$$\psi(s) = \frac{(1 - \rho)s}{s - \log(1 + \rho s)}. \quad (14\text{-}127)$$

This is the transform from which $H(z)$ is determined.

By writing

$$M(s) = \frac{1 - \rho}{\rho} \psi(s), \quad (14\text{-}128)$$

the term $1 - \rho$ would represent the probability concentration at the origin $(Z = 0)$; hence $M(s)$ is the conditional Laplace transform $E(e^{-sZ}|Z > 0)$, the condition being that the reservoir is not dry. D. G. Kendall has given a partial solution to find the inverse transform of $M(s)$.

The foregoing approach of passing from the discrete to the continuous case can also be used to write the distribution $H_k(z)$ for a finite-capacity dam.

14-10. Hospitals and the Demand for Medical Care

It has been possible for C. Flagle [217] of the Johns Hopkins University to use ideas from a large stochastic model for the demand of medical care. Members of the population contract illness at random and demand service from private physicians, from clinics, and from hospitals either directly or in stages, e.g., population–private physician–clinic–hospital-population.   A patient returns to the population source if cured, or he may even leave it never to return.   Babies are born in a hospital and are added to the population as possible units which might demand service. It has been observed that the demand for service and the type of service required change with age.   The object of studying the system varies. The medical school faculty has a different point of view as to the object of the operation from that of the patient or from that of the trustees responsible for the survival of the institution.

An important problem is one of nursing shortage.   To control and allocate this service properly in order to obtain greater utilization of the system, a study of arrivals and lengths of stay was made, and the variations in total daily hours of direct patient care were analyzed.   Periods (almost daily) of idleness followed by periods of intense demand for care were observed.   Seasonal variations in hospital use are noticeable almost always and in most hospitals.   The arrivals were observed to be Poisson-distributed.   Intensively ill patients require a great amount of service until the patient is able to leave his hospital bed (but remain in the hospital) when he becomes more independent and self-sufficient, requiring considerably less attention.   The infinite-channel queue applies to the cases of intensively ill patients who preempt facilities and service, as long as their number does not exceed the capacity.   For this group whose needs are urgent, which is small relative to total capacity, there are in effect an infinite number of beds available.   Evidence shows that the input is Poisson and the service is exponential except in the case of epidemics.   Then illnesses and, therefore, arrivals are dependent.

A subtle preemptive priority exists in which a very ill patient is admitted and those recuperated are encouraged to leave.   The distribution of stay of intensive-care patients is best described by a log normal distribution where the variance is large compared with the mean and approaches the negative exponential.   By integrating the staff activity for supplying care service, the variance is reduced.   This enables the hospital to admit a larger number of patients and makes better use of available service.

An analysis of various types of service required by patients indicated combinations of parallel and service channels which, when working together, provided more efficient service than a set of channels in series performing the same functions.   Thus two parallel channels, each able to provide either of two types of demanded service, may operate in series, each with two other channels each of which provides one of two other types of service.   Patients may also be segregated in different facilities according to their demand for service.

Another system studied is that of scheduled patients who arrive early or late about their time of appointment.   When any variation becomes possible, one has a Poisson process (it is the limit of disintegration of a scheduled process). Irregularities in the schedule are serially correlated.

A useful measure of effectiveness for the operation is to minimize the total cost of illness. This would include the loss in productivity. There is no comprehensive analytical model for the entire operation, but a global model of it enables the isolation of parts which are then studied analytically.

In examining statistics in hospital planning and design [23] N. T. J. Bailey has attempted to reap the best advantage from the restricted resources of limited medical skill, drugs, buildings, finance, etc. One may start by assessing the population at risk, i.e., that will give rise for demand for medical care at a given group of hospitals, by taking samples from lists of addresses of patients admitted to the group in a given period. A similar analysis is done on neighboring hospitals which are likely to interact with the group. If a patients went to the group and b to the neighboring hospitals, the group's ratio is $a/(a + b)$ and gives the number at risk for the group when multiplied by N, the population at risk. When this is summed over all districts, one has the total effective population at risk. The variance of this number is the sum over all districts of $abN^2/(a + b)^3$, where the likely fact that a and b are Poisson variables is assumed. If a and b are estimates based on a constant sampling fraction λ, then the above expression for the variance is divided by λ before taking the sum.

For theoretical purposes, the time spent waiting for different numbers of beds, if the demand and the length of stay are known, may be determined by using queueing theory. One would supply the number of beds to satisfy the amount of hospitalization demanded, plus one or two beds in order to be prepared for fluctuations in the tolerated length of period waited. The effect of varying arrival rates (which are random) on the average waiting time is not fully understood. Note that acute cases may have to be admitted without waiting, either to reserve beds or at the expense of a longer waiting time for the nonacute cases waiting to be admitted.

Similar ideas apply to outpatients. An appointment system with a tolerated average wait may be established, striking a balance as to how valuable a doctor's time may be as compared with that of the patient and accepting a certain risk of idle time. The appointment times may be staggered, but not many patients need be there when the operation begins. For example, experiments with one appointment system (an average of 5-minute consultation) have shown that for stability the consultation should begin when the third patient arrives and that appointment intervals should be constant and slightly longer than 5 minutes.

In another paper, Bailey [18] shows how the computations were made to indicate that hospitals should consider relieving bottlenecks, etc., which lengthen the service times, thus reducing the capacity and utilization when the amount of service available is limited.

In yet another paper Bailey [21] uses the average time that a patient would have to wait for one of c hospital beds to become vacant under Poisson-input (with parameter λ) exponential-stay (with parameter μ) assumptions [see (4-130)] to argue for practical purposes that a long waiting line might be largely due to an accumulated backlog. If, by emergency measures, this were relieved, short waiting times could be maintained by increasing the existing number of beds only slightly to ensure that the maximum supply is greater than the average demand. One can also use the formula to determine the reduction in the average duration of stay to achieve a reasonable waiting time. Some of Bailey's work on bulk service has been used for studying optimum methods of scheduling outpatient care.

14-11. Cafeteria Counter Design

In an interesting detailed study of the design of cafeteria counters, M. Chartrand [101] has analyzed the causes of delay in cafeteria service lines with various objectives such as reducing waiting and service times, serving large numbers before the counter, and bringing people face to face with food rapidly, which then starts service for each customer earlier than would be the case if all foods were grouped in one station. Delays were found to be generated by (1) patrons visiting in line, making selection, getting proper change; (2) poor service because of untrained and insufficient personnel; (3) poor design layout at hot food, coffee, and check-out stations ; and (4) a varied menu. Questions seeking a balance between idle facilities and idle waiting were posed; for the answers, the arrivals distribution, the number of counters available, the service and its duration, and the queue discipline were examined.

Among several useful principles considered was the arrangement of stations (e.g., dessert stations) in parallel rather than in series, since adding the latter type of station to a counter tends to increase the average service time for all persons. (In the parallel case, a customer makes up his mind what he wants and joins the appropriate queue, rather than be delayed by the customers ahead of him also in service, as in the series case.) But even in the series operation, bottlenecks should be removed, e.g., by suitable grouping of menu items. The "lead time" or service time of a single customer who is not pressed by a line behind or is held back by a line in front plays an important role and must be reduced. The time that a customer spends while in service [waiting for the line of customers ahead of him (also in service) to move] must also be reduced. This, for example, can be done (in the process of solving the problem) by reducing the number of persons ahead of him. The average service time of a customer who uses the facility regularly tends to be shorter than that of a sporadic customer.

Objectives other than those cited above include reduction of the lead and service times by making it possible for a patron to leave the line rapidly when his service is finished. Service times may be reduced by eliminating multiple stops at counter stations; making it possible to break out of line; eliminating the need for attendants' attention, e.g., by packaging, etc., some foods; and making service possible with a minimum number of motions. Self-service items should be placed parallel to the moving line, whereas service items should be placed at right angles to it. Several designs of the operation have been studied and compared.

14-12. Spare Coal Faces

F. J. Toft and H. Boothroyd [719] have studied the following problem which has a machine-interference-type model. A number m of coal faces (part of a mine from which coal is obtained) must be available for a desired coal output. To maintain this number, a maximum of s spare faces are also on hand to replace working faces under repair. Once the total number of useful faces falls below $m + s$ a team starts to repair a face. There are c teams available; hence if more than c faces require repair they must wait in a queue. The failures and repairs are Poisson with parameters λ and μ, respectively. The model assumes no delay in starting production or repair on a face. All faces are treated as equivalent, although in practice there would be differences. The number of teams available can be increased in practice. The random distributions assumed are a useful simplification. If $P_n(t)$ is the probability of n faces waiting and in repair, we have for $s \geq c$

$$P'_n(t) = -(m\lambda + n\mu)P_n(t) + m\lambda P_{n-1}(t) + (n + 1)\mu P_{n+1}(t),$$
$$0 \leq n < c,$$
$$= -(m\lambda + c\mu)P_n(t) + m\lambda P_{n-1}(t) + c\mu P_{n+1}(t),$$
$$c \leq n \leq s,$$
$$= -[(m + s - n)\lambda + c\mu]P_n(t) + (m + s - n + 1)\lambda P_{n-1}(t)$$
$$+ c\mu P_{n+1}(t), \qquad s < n \leq m + s, \quad (14\text{-}129)$$

and for $s < c$

$$P'_n(t) = -(m\lambda + n\mu)P_n(t) + m\lambda P_{n-1}(t) + (n + 1)\mu P_{n+1}(t),$$
$$0 \leq n \leq s,$$
$$= -[(m + s - n)\lambda + n\mu]P_n(t) + (m + s - n + 1)\lambda P_{n-1}(t)$$
$$+ (n + 1)\mu P_{n+1}(t), \qquad s < n < c,$$
$$= -[(m + s - n)\lambda + c\mu]P_n(t) + (m + s - n + 1)\lambda P_{n-1}(t)$$
$$+ c\mu P_{n+1}(t), \qquad c \leq n \leq m + s, \quad (14\text{-}130)$$

and
$$P_{-1}(t) = P_{m+s+1}(t) = 0.$$

The steady-state solutions are given as follows:

For $s \geq c$:

$$p_n = \frac{m^n \rho^n p_0}{n!}, \qquad\qquad n \leq c,$$

$$= \frac{m^n \rho^n p_0}{c! c^{n-c}}, \qquad\qquad c \leq n \leq s + 1,$$

$$= \frac{m! m^s \rho^n p_0}{(m + s - n)! c! c^{n-c}}, \qquad s \leq n \leq m + s. \qquad (14\text{-}131)$$

For $s < c$:

$$p_n = \frac{m^n \rho^n p_0}{n!}, \qquad\qquad n \leq s + 1,$$

$$= \frac{m! m^n \rho^n p_0}{n!(m + s - n)!}, \qquad s \leq n \leq c,$$

$$= \frac{m! m^s \rho^n p_0}{c! c^{n-c}(m + s - n)!}, \qquad c \leq n \leq m + s, \qquad (14\text{-}132)$$

and p_0 is obtained from

$$\sum_{n=0}^{m+s} p_n = 1.$$

The average number of spare faces held at one time is

$$\sum_{n=0}^{s} (s - n)p_n, \qquad\qquad (14\text{-}133)$$

and the average proportion of working time lost (i.e., when $n > s$) is

$$\frac{1}{m} \sum_{n=s+1}^{m+s} (n - s)p_n. \qquad\qquad (14\text{-}134)$$

Exercise: Verify the above solution.   Justify the last two expressions.

The optimum values of s and c can be determined to maintain a balance between keeping open spare faces and the cost of losing production time on working faces.   Graphical analysis can be used to reach answers which are generally difficult to obtain otherwise.

14-13. Semiconductor Noise as a Queueing Problem

This is a subject which will mostly interest practical scientists and engineers.   D. A. Bell [782] has studied the problem of semiconductor noise in the framework of queues.   We have already examined the idea of an autocorrelation function in a problem at the end of Chap. 4.   The power spectrum is the Fourier cosine transform of the autocorrelation function.

The outstanding problem of semiconductor noise is to account for a power spectrum of which the intensity appears to increase without limit as the frequency is decreased.   There is no simple explanation in terms of relaxation times for a spectrum which is proportional to a constant power of frequency over many decades and which is found in a wide range of different materials.   But it can be explained if the rate at which carriers leave the conduction band is treated as a queueing problem rather than a relaxation problem.   The phenomenon in question is related to the presence of current, which implies that the carriers are swept away from the neighborhood of the centers from which they originated and that the time to recombination is governed by the time taken to find a vacant center elsewhere, not by the probability of falling back to the originating center.

The semiconductor-noise problem is represented by Bell as a queueing problem by the following system of equivalences:

1. The charge carriers in the conduction band constitute the queue.

2. The members of the queue join it at random times as a result of excitation from the base levels, this excitation being generally a random process.

3. The "service outlets" are vacant base levels to which the charge carriers can return (or possibly recombination centers if there are any restrictions on the method of return of a carrier from the conduction band to a base level).

4. The duration of occupation of a service outlet is the interval from the time at which a base level is reoccupied by a carrier to the time at which it becomes vacant again as a result of excitation of the carrier out of that base level into the conduction band again.   The duration of this interval is assumed to be described statistically by an exponential distribution with mean value fixed by the excitation energy through the thermodynamic relationship between "free" and "bound" times.   (This relation gives a proportionality only, the actual time depending also on the "jump frequency" of the relaxation process.)

The model is that any small region of the semiconductor contains (1) the number of vacancies created by those excitations of carriers into the conduction band which occur in that region and (2) a number of carriers governed by the number of excitations which occur in itself and in other adjacent regions.   Although the two numbers have equal average values, they will not tally precisely at all times.   We thus have precisely the type of situation that has been studied by the telephone engineer. Although the number of service outlets matches the number of demands in terms of over-all average, there are local variations which can lead to congestion and cause some members of the resulting queue to wait for times which bear no obvious relation to the times of the service processes.

The autocorrelation function is given by

$$\psi(\tau) = \int P(\geq t)P(\geq t + \tau)\, dt,$$

where $P(\geq t)$ is obtained from Eq. (11-76)—since the number of outlets is large and the excitation is random—in which the Schläfli function is replaced by the Hankel function, using the relation between these two functions. The Hankel function exists in tabulated form. In spite of simplification of $P(\geq t)$, Bell was unable to obtain the power spectrum because of analytical difficulties.

There is always the analytic difficulty that queueing formulas are not valid for a calling rate greater than or equal to the service rate, the inequality giving a definitely divergent size of queue and the equality giving an indeterminate size, of which infinity is one of the possible values; the delay times will behave similarly. But one of the difficulties of finding an analytical interpretation of the semiconductor-noise spectrum is that it has precisely this characteristic of an indeterminate solution; the fact that no indication of a low-frequency limit has ever been substantiated experimentally, i.e., that the integral of the spectrum is divergent, means that the distribution of lifetimes is also lacking in convergence, as it would be in a queue with equal calling and service rates. See also [645a].

14-14. Other Applications

L. C. Edie [167] has used various ideas from steady-state queueing theory to analyze traffic toll booths for highway traffic. For example, he obtained an optimum schedule for operating different numbers of booths at different times of day. Use of his method permitted savings in toll-collection expenses and yielded better service.

With the aid of queueing theory, G. Brigham [75] studied the optimum number of clerks to be assigned to tool-crib counters in use in the Boeing factory area. The cribs stored a variety of tools required by mechanics in shops and assembly lines. The problem resulted from complaints by foremen who felt their mechanics were waiting too long in line. This would have led the company to assign more clerks. However, management was under pressure to reduce overhead and assign fewer clerks. The analysis gave the optimum number.

L. J. Sespaniak [655] has used the probability of delay $P(>0)$ for Poisson input with rate a per unit mean service time with an exponential-service-time distribution and multiple channels with ordered queueing to obtain the optimum number c of inspectors required to inspect a product while in process. The service included the walking time of the inspector to the service station. If $r = r_a/r_c$ is the ratio of the cost per unit time of delaying the process awaiting inspection to the cost per unit time of idle time for the inspector, the normalized cost is

$$(c - a) - \frac{raP(>0)}{c - a}, \qquad c > a.$$

A graph is developed giving the normalized cost as a function of a for different values of c and for different values of r. For example, for a fixed c and various values of r, a set of curves are generated which intersect with corresponding curves for the next value of c and the same value of r. This makes it possible to draw a straight line through the intersection points. Another line is also drawn through the points of intersection of the curves with those obtained for a smaller value of c (essentially Brigham's procedure). The zone between the two lines enables determination of the number of inspectors required for the minimum normalized cost corresponding to a given arrival rate.

D. H. Schiller and M. M. Lavin [649] have used Monte Carlo simulation methods to determine the extent of new truck-dock facilities that would be required at one consolidated warehouse to handle the volume of traffic formerly accommodated at three warehouses. The variations of morning half-hourly arrivals of trucks were nearly Poisson-distributed. The data were decomposed into 14 categories based on truck type, cargo movement, warehouse location, etc. Trucks were divided into two classes, each with a different exponential service time. The movement of trucks was simulated for facilities with 12, 15, 18, and 21 docks; waiting times and the number waiting were computed. Repeated runs of a day's operations, properly averaged, led to the desired description of the waiting-line situations for different numbers of docks.

J. Delcourt [151] has studied the problem of how much oil each of a number of pumping stations should handle, taking into consideration an accepted waiting risk of demand. As the global demand on oil increases over the years, each pump is required to operate longer; this implies longer waiting for incoming demand. The pumps do not operate all the time. Thus, within a prescribed demand schedule, it is essential to determine the length of time during which the pumps must operate, accepting some delay in satisfying the demand.

The input and service-time distributions for each pump were determined and statistically tested and simplifying assumptions made. Then elementary (Poisson-input exponential-service-times) steady-state queueing theory was used to determine the average number waiting, the average waiting time in the queue and in the system, etc., enabling the determination of how much additional load a pump can handle.

There are many other possibilities for applying queueing theory (several of the items listed below have an actual application). For variety we give a few: arrival and unloading of ships in port or trucks at platforms; the passage of ships through narrow canals, e.g., the Suez Canal; the debarkation of passengers at foreign ports and customs inspection; the many ways in which people queue up for facilities, e.g., barbers, supermarkets and bakeries, at general stores, theaters, taxis, buses, etc.; gas molecules (or ants) going through a hole; a catalyst in a chemical reaction; filtration; post office backlog; production lines; automobile inspection lines; conveyor belts; nervous reactions; telegraph communications and priorities of messages; and railroad classification yards. An important application in the United States would be the increasing backlog of cases awaiting court trials because of facility and judge shortages. If psychometric measurements were applicable to the generation and input of ideas, impulses, and goal-directed tensions into the nerve net and into memory (assumed meaningful), it would be possible to study neuroses in terms of interference of these different populations, in a bid for action (mental, psychological, or physical), in errors in priority assignment as to which ones to service first, and in delay in serving high priorities, etc.

PROBLEMS

1. Assuming car arrivals to a single traffic lane, as discussed in Tanner's work, with exponentially distributed interarrival times having means .2 and .5, respectively, use the Monte Carlo method to simulate the problem, obtaining expected waiting times for cars approaching from either side. Assume that the time required to traverse the lane by any car is 0.1 minute.

2. Referring to the summary of Brigham's and the subsequent work, plot some of the curves useful in such an analysis.

3. Let

$$p_0 = \frac{(\lambda T)^c/c!}{\sum_{k=0}^{c} (\lambda T)^k/k!}.$$

Using probability graph paper, generate a set of curves for p_0 as a function of c for values of λT ranging from .01 to 9.0 [156]. Doeh, in a letter to the editor of *Operations Research*, gives the following two problems (a part of the problem is for the reader to interpret and use the above formula):

a. In 3 years of data collection, 253 ignition distributors were removed for overhaul from the aircraft of a small South American carrier. An analysis of overhaul records revealed that the average time for overhaul was 37.6 days. Personnel familiar with and responsible for meeting service and safety requirements are confident that being out of stock of this component once every 2 years can be tolerated. What is the appropriate initial spares level? Thus show that $\lambda T = 8.7$, $p_0 = .0059$, and, from the graph, $c = 16$ spares.

b. A waitress must pay the bar at the time she picks up a filled order. She finds that she is handling orders at an average rate of 16 coffees per hour and is obliged to wait an average of 45 minutes for her customers to finish dinner and pay her. If each cup of coffee costs 60 centavos and she would like to avoid running out of working capital in more than one case per thousand, how much money must she bring to work with her? Here $\lambda T = 12$, $p_0 = .001$; hence $c = 24$, and cash reserve is 144 pesos. There are 100 centavos in a peso.

Also with the above formula, show that the number of channels corresponding to a Poisson input with parameter $\lambda = 2.24$ calls per hour and average length of call $T = .00625$ hour for $p_0 = .001$ is two channels.

4. For a loss system with c channels, Poisson arrivals with parameter λ, and exponential service with parameter μ, suppose that, once $b < c$ channels for a fixed b are occupied and an arrival occurs, all channels become occupied or blocked because of a fault of the mechanism. However, the channels become unoccupied one at a time according to the servicing of customers. Arrivals are lost once all channels are occupied or blocked. If the initial number at $t = 0$ is i, we may write [684]

$$P'_{in}(t) = -(\lambda + n\mu)P_{in}(t) + \lambda P_{i,n-1}(t) + (n+1)\mu P_{i,n+1}(t), \qquad 0 \leq n \leq b,$$
$$P'_{in}(t) = -n\lambda P_{in}(t) + (n+1)\mu P_{i,n+1}(t), \qquad b-1 \leq n \leq c-1,$$
$$P'_{ic}(t) = -\mu c P_{ic}(t) + \lambda P_{ib}(t).$$

Show that the steady-state equations have the solution

$$p_n = p_0 \frac{1}{n!} \left(\frac{\lambda}{\mu}\right)^n, \qquad n = 1, \ldots, b,$$

$$p_n = \frac{\lambda}{n\mu} p_b, \qquad n = k+1, \ldots, c,$$

and determine p_0 from $\sum_{n=0}^{c} p_n = 1$. Show that the solution becomes the Erlang formula for $b = c - 1$. Find the time congestion $\sum_{n=b+1}^{c} p_n$, that is, find the probability that a call is lost. Evaluate the calling rate for the accepted calls:

$$\lambda_0 = \sum_{n=0}^{b} \lambda p_n.$$

Note that the call (traffic) congestion is now $(\lambda - \lambda_0)/\lambda$ and the traffic offered is λ/μ, whereas the traffic carried is λ_0/μ. It is the average number of channels engaged by real calls (as distinguished from the "ghost" calls which block the system after b channels are occupied) given by

$$\sum_{n=0}^{b} np_n + cp_c. \qquad (1)$$

The traffic carried by the ghost calls is

$$\sum_{n=b+1}^{c-1} np_n \qquad (2)$$

and vanishes for $b = c - 1$. Evaluate the total average number of busy channels $\sum_{n=0}^{c} np_n$ and show that it is equal to the sum of (1) and (2).

5. In a delay system of c channels with Poisson arrivals and parameter λ and with exponential service with parameter μ, suppose there are q positions for waiting. A call is lost if all these positions are occupied. The faulty operation in this case is that an arrival, when there are $c < b$ units in the system, results in blocking all the remaining waiting positions. Thus for the states $b + 1, \ldots, c + q$ the only possible transitions are terminations.

The steady-state equations are [684]

$$\lambda p_{n-1} - (\lambda + n\mu)p_n + (n + 1)\mu p_{n+1} = 0, \qquad 0 \leq n \leq c - 1,$$
$$\lambda p_{n-1} - (\lambda + c\mu)p_n + c\mu p_{n+1} = 0, \qquad c \leq n \leq b,$$
$$-c\mu p_n + c\mu p_{n+1} = 0, \qquad b + 1 \leq n \leq c + q - 1,$$
$$\lambda p_b - c\mu p_{c+q} = 0.$$

The normalizing condition here is

$$\sum_{n=0}^{c+q} p_n = 1.$$

Obtain the solution to this system and show that for $b = c + q - 1$ it reduces to what is called the Erlang loss-delay solution for a finite queue. For $q = 1$, b can only equal c. For infinite q and finite b, show that $p_0 = 0$ and no stationary solution exists. When $b = c + q - 1$ and q is infinite, obtain the limiting "Erlang" distribution for delays for $\rho = \lambda/c\mu < 1$. Again obtain the accepted call calling rate

$$\lambda_0 = \lambda \sum_{n=0}^{b} p_n,$$

and show that the traffic carried, which is the average number of terminations of real calls, is λ_0/μ. Compute the call congestion

$$\sum_{n=b+1}^{c+q} p_n$$

and the traffic carried by ghost calls:

$$c \sum_{n=b+1}^{c+q-1} p_n.$$

The total traffic waiting is $\sum_{n=c}^{c+q} (n - c)p_n$. Show that this can be split into real waiting traffic and the waiting traffic due to ghost calls. Compute the total traffic due to ghost calls:

$$\sum_{n=b+1}^{c+q-1} np_n.$$

Note that the probability that the nth item in line waits a time greater than t is given by

$$W_n(\geq t) = \int_t^\infty \frac{(cux)^{n-c}}{(n - c)!} e^{-\mu cx} c\mu \, dx, \qquad n \geq c,$$

from which the waiting-time distribution is given by

$$P(>t) = \sum_{n=c}^{b-1} W_n(\geq t)p_n + W_{c+q-1}(>t)p_b.$$

Compute this expression for $b = c + q - 1$, which is the Erlangian loss-delay system.

If q is infinite, the exponential distribution should be obtained corresponding to the classical waiting system. Show that the probability of delay $1 - P(=0)$ is given by

$$p_c \frac{1 - \rho^{b-c+1}}{1 - \rho}.$$

Show that the mean waiting time is given by

$$\frac{\rho[1 - P(=0)] - p_{c+q}[q\rho + (b - c + 1 - q)]}{\lambda(1 - \rho)},$$

and the average number of real waiting traffic

$$\sum_{n=c}^{b} (n - c)p_n + qp_{q+c}$$

is the product of λ and the average waiting time.

6. In a single-channel ordered queue, with Poisson input and exponential services at a common control with parameter η and at the channel with parameter μ, show that the waiting-time distribution in line (not including service) is given by

$$P(>t) = \sum_{i=1}^{\infty} P(i - 1, 1) \int_t^{\infty} g_0(i,y)\, dy + \sum_{i=1}^{\infty} Q(i - 1, 0) \int_t^{\infty} g_1(i,y)\, dy,$$

where the Laplace transforms of g_0 and g_1 are, respectively,

$$g_1^*(i,s) = \sigma^i, \qquad g_0^*(i,s) = \frac{\mu}{\mu + s} \sigma^{i+1},$$

and where

$$\sigma = \frac{\mu}{\mu + s} \frac{\eta}{\eta + s}.$$

We have $g_1(i,t)$ as the probability density that a call arriving when there are $i - 1$ waiting will wait time t, and the common control is occupied. The same is true for $g_0(i,t)$ for a free common control. (HINT: The total time spent after waiting is the sum of two variates and is exponentially distributed with parameters η and μ.)

7. In a similar manner, show that for the infinite-channel case one has

$$P(>t) = e^{-(\eta-\lambda)t} - \left(1 - \frac{\lambda}{\eta}\right) e^{-\eta t}$$

with mean waiting time $(1/\eta)(\lambda)/(\eta - \lambda)$ and $P(=0) = \lambda/\eta$.

8. Suppose that arrivals to a queue with N possible positions arranged uniformly in a circle occur from Poisson distributions with parameters λ_j to the jth position and are served by a single operator, who moves around always in the same direction sequentially, by an exponential distribution with parameter μ. If the operator is busy, a call is lost. When the system is empty the operator remains in the last service position. When an arrival occurs the operator immediately moves to that position, in order to provide service. Let $P_j(t)$ be the probability that the operator is in position j at time t and that he is free, and let $Q_j(t)$ be the corresponding probability when he is busy. Let

$$Q(t) = \sum_{j=1}^{N} Q_j(t); \qquad P(t) = \sum_{j=1}^{N} P_j(t).$$

Note that $P(t) + Q(t) = 1$. Let $\lambda = \sum_{j=1}^{N} \lambda_j$. The probability that during dt the operator who is in the ith position is forced to move to the jth position and has some-one there to serve is $\lambda_j\,dt$. If there is no one to serve in the jth position, this probability is $1 - \lambda\,dt$ for $i = j$ and zero otherwise. If the operator is busy in the ith position, these probabilities are $\mu\,dt$ if there is no one to serve in position j. If there is a call in position j, it will be lost because the operator is busy; hence the operator remains in position i. Consequently the probability that the operator will move to that position is $1 - \mu\,dt$ if $i = j$ and zero otherwise. The transient equations are

$$P'_j(t) = \lambda_j Q(t) - \mu P_j(t),$$
$$Q'_j(t) = \mu P_j(t) - \lambda Q_j(t).$$

Give the solution in the steady state. Solve the transient problem by means of Laplace transforms. Obtain the steady-state solution by inverting the Laplace transforms and passing to the limit and also by evaluating $\lim_{s\to 0} sP_j^*(s)$ and $\lim_{s\to 0} sQ_j^*(s)$ of the Laplace transform of $P_j(t)$ and $Q_j(t)$ with parameters s (Syski, unpublished).

9. If one assumes that, for a total finite population of size N (discussed in the text under Engset-O'Dell) calling for service, the rate of arrivals is proportional to the number not already in the system, and if A is the number in the system, then the steady-state probability that x channels are busy is given by O'Dell and is analogous to Erlang's formula for the infinite-population case:

$$\frac{\binom{N}{x}[A/(N-A)]^x}{1 + N[A/(N-A)] + \binom{N}{2}[A/(N-A)]^2 + \cdots + \binom{N}{x}[A/(N-A)]^x}.$$

The proportion of lost calls is

$$\frac{A + (A-1)N[A/(N-A)] + (A-2)\binom{N}{2}[A/(N-2)]^2 + \cdots + (A-x)\binom{N}{x}[A/(N-A)]^x}{1 + N[A/(N-A)] + \binom{N}{2}[A/(N-A)]^2 + \cdots + \binom{N}{x}[A/(N-A)]^x}.$$

In the exponential case and for a finite number of sources N, Engset has given for the call congestion (i.e., the probability of a call being lost), where c parallel channels are operating and no waiting is possible,

$$p_N = \frac{\binom{N-1}{c}\{\rho/[1 - \rho(1 - p_N)]\}^c}{\sum_{k=0}^{c}\binom{N-1}{k}\{\rho/[1 - \rho(1 - p_N)]\}^k},$$

where $\rho = \lambda/\mu$, $\lambda\,dt$ is the calling probability for each unit of the finite population, and μ is the service rate.

The time congestion (the probability that all c channels are busy) is given by

$$\frac{\binom{N}{c}\{\rho/[1 - \rho(1 - p_N)]\}^c}{\sum_{k=0}^{c}\binom{N}{k}\{\rho/[1 - \rho(1 - p_N)]\}^k}.$$

What relation exists between the two sets of formulas? Suppose that the call-for-service rate is $(N - i)a$; determine a in terms of ρ and p_N.

10. In an elegant and detailed paper, Cohen [122] has studied the telephone problem of c trunks with identical exponential service with parameter $1/h$, Poisson input with parameter νs, and a congested call (i.e., a call that finds all trunks busy). The call may be repeated later and continue to be repeated with an exponential repetition distribution (multiplied by a constant) with parameter νs, where ν is a constant. If the traffic offered to the group of trunks is $a = sh$, the mean repetition time is $h/\mu a$. In addition, the staying (waiting) time of a congested call has an exponential distribution (multiplied by a constant) with parameter μs, where μ is a constant. After its staying time is up, a call becomes impatient and is lost. The average of this distribution is $h/\mu a$. Both ν and μ give a measure of a subscriber's mood. It will be assumed that, once a given number m of congested calls, whose staying time has not elapsed, is reached, a new call is lost and cannot be repeated. Once $P(x,y,t)$ (the conditional probability for the occurrence of the state with x busy trunks and y repeated calls at time t with initial condition $x = x_0$, $y = y_0$ at $t = t_0$) is determined, the distribution of the number of busy trunks, of the time congestion (i.e., the probability that all trunks are busy), and of the call congestion (i.e., the relative number of lost calls) can also be determined.

Show that the following equations describe the above system and note how some familiar queueing problems may be obtained by specializing the parameters of this system.

$$P(x, y, t + \Delta t) = s\,\Delta t\,P(x - 1, y, t) + \frac{(y + 1)P(x - 1, y + 1, t)\nu a\,\Delta t}{h}$$

$$+ \frac{(y + 1)P(x, y + 1, t)\mu a\,\Delta t}{h} + \frac{(x + 1)P(x + 1, y, t)\,\Delta t}{h}$$

$$+ \left[1 - s\,\Delta t - y(\mu a + \nu a)\frac{\Delta t}{h} - x\frac{\Delta t}{h} \right] P(x,y,t),$$

$$0 < x < n, \qquad 0 \leq y < m,$$

$$P(x, y, t + \Delta t) = s\,\Delta t\,P(n - 1, y, t) + \frac{(y + 1)P(n - 1, y + 1, t)\nu a\,\Delta t}{h}$$

$$+ (y + 1)P(n, y + 1, t)\frac{\mu a\,\Delta t}{h} + s\,\Delta t\,P(n, y - 1, t)$$

$$+ \left(1 - s\,\Delta t - \frac{y\mu a\,\Delta t}{h} - \frac{n\,\Delta t}{h} \right) P(n,y,t), \qquad x = n, \qquad 0 < y < m,$$

$$P(x, m, t + \Delta t) = s\,\Delta t\,P(x - 1, m, t) + \frac{(x + 1)P(x + 1, m, t)\,\Delta t}{h}$$

$$+ \left[1 - s\,\Delta t - m(\mu a + \nu a)\frac{\Delta t}{h} - \frac{x\,\Delta t}{h} \right] P(x,m,t),$$

$$0 < x < n, \qquad y = m,$$

$$P(n, m, t + \Delta t) = s\,\Delta t\,P(n - 1, m, t) + s\,\Delta t\,P(n, m - 1, t)$$

$$+ \left(1 - ma\mu\frac{\Delta t}{h} - \frac{n\,\Delta t}{h} \right) P(n,m,t), \qquad x = n, \qquad y = m.$$

Note that
$$\sum_{x=0}^{n} \sum_{y=0}^{m} P(x,y,t) = 1.$$

The equations are then reduced to a differential difference system in the usual manner. The steady-state equations are also obtained and solved by Cohen. The steady state is shown to exist as $m \to \infty$. Several special cases are examined for different values of the parameters μ, ν, and m. The well-known formulas due to Erlang, Molina, and Palm are obtained for loss and delay systems. To continue, in the special case where $m \to \infty$, μ, $\nu \to 0$, $\lim (\nu/\mu) = \beta$, the mean staying time $h/\mu a$ and the mean repetition time $h/\nu a$ become infinitely large. This implies that a congested call always stays and is repeated after a very long time interval. Thus the input distribution consisting of initiated and repeated calls is Poisson with mean $(a + \varepsilon)/h$, where ε is to be determined. Show that $f(x)$, the distribution of busy trunks, is Erlangian, i.e.,

$$f(x) = \frac{(a + \varepsilon)^x/x!}{N_n(a + \varepsilon)}$$

where

$$N_n(y) = 1 + y + \frac{1}{2!} y^2 + \cdots + \frac{1}{n!} y^n.$$

Show that the traffic handled is given by

$$a_h = \sum_{x=0}^{n} x f(x) = (a + \varepsilon)[1 - E_n(a + \varepsilon)],$$

where

$$E_n(y) = f(n).$$

If every repeated call were regarded as a fresh call, the number of lost calls per unit time would be $(a + \varepsilon)E_n(a + \varepsilon)/h$. However, a congested call is, on an average, repeated β times, and the number of lost calls per unit of time is, therefore, $[(a + \varepsilon)/h(1 + \beta)]E_n(a + \varepsilon)$. Show that the lost traffic is given by

$$\frac{a + \varepsilon}{1 + \beta} E_n(a + \varepsilon).$$

Compute the total traffic a which is the sum of lost and handled traffic and show that ε must satisfy

$$\varepsilon = \frac{\beta}{1 + \beta} (a + \varepsilon)E_n(a + \varepsilon).$$

CHAPTER 15

BASIC RENEWAL THEORY

15-1. Introduction

Renewal theory deals essentially with properties of renewals of items that encounter failures and must, therefore, be replaced. The items may be light bulbs or operating aircraft engines. When an item fails, a new item replaces it; when the latter fails, it is again replaced, etc. We are generally interested in the average number of replacements over a period of time and in its variance.

It is the purpose of this chapter to derive these quantities. It may be easy to see why they are useful. However, their development requires elaborate formalism. It would help the reader to try to fill in details omitted because of space limitations and to do some exercises.

A *renewal process* is a sequence of nonnegative random variables $\{X_i\}$ independently and identically distributed. Here X_i denotes the lifetime of the article introduced at the ith renewal. Thus, in a queueing process the interarrival times from a Poisson distribution with parameter λ form a renewal process with $F(t) \equiv 1 - e^{-\lambda t}$ as the common distribution. Recurrent input to a queue, for example, constitutes a special case of renewal processes. Most of the deeper facts of renewal theory are concerned with asymptotic behavior, since one is usually interested in the behavior of the renewal function (defined below) for large values of time.

If we write the nth partial sum $S_n = X_1 + \cdots + X_n$, then

$$S_n \ (n = 1, 2, \ldots)$$

constitutes the time at which the nth renewal is made. Sometimes an additional nonnegative random variable X_0 independent of the $\{X_i\}$ is adjoined to yield the *general renewal process* $\{X_0, X_1, \ldots\}$, where X_0 gives the residual lifetime of the initial population.

An *extended renewal process* is obtained when the nonnegativity condition on the $\{X_i\}$ is dropped. Blackwell obtains a renewal process $\{Z_n\}$ from an extended renewal process by writing Z_1 for the first nonnegative partial sum and $Z_1 + \cdots + Z_k$ for the first nonnegative S_n greater than $Z_1 + \cdots + Z_{k-1}$. This gives a set of identically distributed Z_n which are nonnegative.

Smith [666] has used N_t to denote the random variable for the largest n for which $S_n \leq t$. He points out that the renewal function is a special case of a random walk with absorbing barriers where sampling the $\{X_i\}$ continues until S_n first becomes greater than t and $N_t + 1$ is the sample size at which to stop. Thus, sequential-sampling theory carries over ideas such as the existence of finite all-order moments of N_t, but more powerful results are obtained by other means. In the continuous case we shall be dealing with the renewal function $U(t) \equiv E(N_t)$. But we start our study with the discrete case.

Many of the ideas here are based on the excellent papers of W. Smith in this field. We start out by deriving expressions for the average number of total renewals by a given time, both for the discrete and for the continuous case. The results obtained are basic in renewal theory and useful for application purposes.

15-2. The Discrete Case

Consider the problem of replacing items that encounter failure during operation, e.g., light bulbs. The initial population at time $t = 0$ obtains the first renewals at $t = X_1$ for those items that fail. These renewals are then replaced at time $t = X_1 + X_2$, etc.

We shall first consider the case where time is assumed to be divided into discrete units. Our discussion of this case, with some modification, follows Feller [202]. Let a_n be the probability that a new item that is installed at a given time will fail after exactly n time units. We may also assume that the life span is positive and hence $a_0 = 0$. The entire set of items are assumed to continue in uninterrupted operation, and once an item fails it is replaced by a new one, keeping constant the total number in use. We note that the original population of a finite number of N items need not be assumed to consist of entirely new items. Thus, if there are initially c_k items which have lived k units of time, $N = \sum_{k=0}^{\infty} c_k$. Note that k, the number of life (i.e., time) units, can take on any value (see example).

Now the replacements, or renewals, at any time consist of renewals of items that were themselves renewals of other items, etc. Let $g_{i+1}(n)$ be the expected number of the $(i + 1)$st set, or generation, of renewals made at time n. We have the following relation among the numbers of $(i + 1)$st and the ith generations of renewals:

$$g_{i+1}(n) = \sum_{k=0}^{n} g_i(n-k)a_k. \tag{15-1}$$

To see why this relationship holds, one observes that the $(i + 1)$st-generation renewals at time n constitute replacements of ith-generation renewals introduced prior to n but failing at n. The ith-generation renewals introduced at time $n - k$ for $0 \leq k \leq n$ will suffer $g_i(n - k)a_k$ failures. Thus, the $(i + 1)$st-generation renewals are obtained by summing over k; this gives the above relation.

If we define the total average renewals at time n by

$$u_n \equiv \sum_{i=1}^{n} g_i(n), \tag{15-2}$$

we have the following expression for the total average renewals at time n on summing Eq. (15-1) over i:

$$u_n - g_1(n) = \sum_{k=0}^{n} u_{n-k}a_k, \tag{15-3}$$

where we may assume that $u_0 = 0$.

Now $g_1(n)$ is the average number of renewals at time n of the original population which is the sum of the numbers c_k. The a_k were defined only for new items. For a used item, such as the initial population may have, the probability of failure at time n of an item with a life of k units already expended is conditioned by r_k, the probability that the item's life span is greater than k units, that is, $r_k = a_{k+1} + a_{k+2} + \cdots = 1 - \sum_{n=0}^{k} a_n$, if $\{a_n\}$ is a probability distribution such that $\Sigma a_n = 1$. Thus, the probability of failure at time n of an item with k units already expended is

$$\frac{a_{n+k}}{r_k}. \tag{15-4}$$

Since there are c_k units with life k already expended, the quantity

$$\frac{c_k a_{n+k}}{r_k} \tag{15-5}$$

is the expected number of those units requiring renewal at time n. If we sum over all possible values of k we obtain the expected number of renewals of the original population at time n. Thus

$$g_1(n) = \sum_{k=0}^{\infty} \frac{c_k a_{n+k}}{r_k} \tag{15-6}$$

and we finally have the expression we seek for the total average renewals at time n:

$$u_n = \sum_{k=0}^{n} u_{n-k} a_k + b_n, \qquad n \geq 1, \qquad (15\text{-}7)$$

where $b_n = g_1(n)$.

Example: Suppose that, for discrete-time renewals, one is given the binomial distribution $[m!/(m-k)!k!]p^k q^{m-k}$ which yields the probability of living k time units, where $m = 5$ life units and the probability of survival in a life unit is $p = .75$. Also suppose that the initial population consists of 100 items, of which 10 initially were new, 15 had a life of one unit, and 75 a life of three units. It is desired to compute the total average renewals after six time units have elapsed.

Note from the failure distribution that no item can live more than five life units. From the binomial distribution, we obtain a_n to two decimal places, which we arbitrarily use as exact. Thus $a_0 = 0$, $a_1 = .01$, $a_2 = .09$, $a_3 = .26$, $a_4 = .40$, and $a_5 = .24$. Clearly $a_n = 0$ for $n > 5$. Then $r_0 = 1$, $r_1 = .99$, $r_2 = .90$, $r_3 = .63$, $r_4 = .24$, and $r_5 = 0$. From the information on the population we have $c_0 = 10$, since 10 units initially had zero life. Similarly, $c_1 = 15$, $c_2 = 0$, $c_3 = 75$, $c_4 = 0$. Note that we have assumed a maximum of five life units. Hence one would not start with items with five life units on them because they are used up.

We also have

$$g_1(1) = 10(.01) + \frac{15(.09)}{.99} + \frac{75(.40)}{.63} = 49.08.$$

Similarly,

$$g_1(2) = 33.41,$$
$$g_1(3) = 8.66,$$
$$g_1(4) = 7.63,$$

and

$$g_1(5) = 2.4.$$

For $n > 5$ one has $g_1(n) = 0$. We finally have $u_0 = 0$, $u_1 = 49.08$, $u_2 = 33.90$, $u_3 = 13.40$, $u_4 = 21.98$, $u_5 = 32.29$, and $u_6 = 31.12$. The total average renewals after six time units have elapsed are 181.77.

Note that, if the average renewals are to be computed for all time, it is possible to write expressions for the generating functions $a(s)$ and $b(s)$ which give $u(s)$, from which all u_n can be obtained. If the roots of the denominator of $u(s)$ are known, then $u(s)$ can be decomposed into partial fractions, each of which can be expanded in series, and the coefficients of s^n are then grouped together. This coefficient is u_n. The total renewals are the sum of u_n obtained by letting $s \to 1$ in $u(s)$.

The first expression on the right of (15-7) is a convolution of the sequences $\{a_n\}$ and $\{u_n\}$. By analogy with the continuous case of a convolution, if we introduce the generating function $a(s) = \sum\limits_{n=0}^{\infty} a_n s^n$,

$b(s) = \sum\limits_{n=0}^{\infty} b_n s^n$, and $u(s) = \sum\limits_{n=0}^{\infty} u_n s^n$, $u_0 = 0$, $b_0 = 0$, we have, on applying

the generating function to the last equation and recalling that the generating function of a convolution is the product of the generating functions of the two components,

$$u(s) = u(s)a(s) + b(s).$$

Consequently,
$$u(s) = \frac{b(s)}{1 - a(s)} \qquad (15\text{-}8)$$

and
$$u_n \equiv \frac{1}{n!} \frac{d^n u(s)}{ds^n}\bigg|_{s=0} \qquad (15\text{-}9)$$

Note that:

1. The series $a(s)$ and $b(s)$ converge for $|s| < 1$ since the coefficients are bounded.

2. If there is a fixed finite upper bound to the life span, the generating function $a(s)$ becomes a polynomial in s and existence is assured. This is the case with most practical applications.

3. The total number of renewals for all times is given by $\sum\limits_{n=0}^{\infty} u_n$. The latter series converges if $\Sigma a_k < 1$, as can be seen from (15-8) on expanding $1/[1 - a(s)]$ in powers of $a(s)$ for values of s for which $|a(s)| < 1$. These are $|s| \leq 1$, and convergence follows since $\sum\limits_{n=0}^{\infty} b_n$ gives the first-generation renewals whose totality equals N, the number of the original population; i.e., since $b_0 = 0$, it follows that $\sum\limits_{n=1}^{\infty} b_n = \sum\limits_{n=1}^{\infty} \sum\limits_{k=0}^{\infty} c_k a_{n+k}/r_k = \sum\limits_{k=0}^{\infty} c_k = N$.

4. If $a(1) = \Sigma a_k = 1$, then as $n \to \infty$ (except for the periodic case in which all the a_k are zero except perhaps those whose subscripts are integral multiples of a given fixed subscript greater than unity) we have

$$u_n \to \frac{b(1)}{\sum\limits_{n=0}^{\infty} n a_n} = \frac{N}{\sum\limits_{n=0}^{\infty} n a_n}. \qquad (15\text{-}10)$$

This theorem essentially states that the total renewals at any time will tend to the number of the original population divided by the average life span, a fact which is in harmony with intuition. In the periodic case, $\{u_n\}$ is periodic in the limit and for $a(1) = 1$; $u_{np+m} \to p(b_m + b_{p+m} + b_{2p+m} + \cdots)/\sum\limits_{n=0}^{\infty} n a_n$ as $n \to \infty$, where p is the period.

5. At time n, the number of items of age k is denoted by $c_k(n)$, which is given by the total renewals at time $n - k$ that survive to time n. Thus,

$$c_k(n) = u_{n-k}r_k. \tag{15-11}$$

If $k \geq n$, these items must start at time zero with $k - n$ life units already expended and survive to time n, that is,

$$c_k(n) = \frac{c_k(0)r_k}{r_{k-n}}. \tag{15-12}$$

Using (15-10), one observes from (15-11) that the limiting age distribution is

$$\frac{Nr_k}{\sum\limits_{n=0}^{\infty} na_n}, \tag{15-13}$$

which is independent of the initial age distribution.

15-3. The Continuous Case

In the continuous case, if we denote by $f(x) \, dx$ the failure probability during a time interval of length dx, $f(x) = 0$, $x < 0$, we have the following relation between the number of $(i + 1)$st-generation renewals and the number of ith-generation renewals at time t:

$$g_{i+1}(t) = \int_0^t g_i(t - x)f(x) \, dx, \tag{15-14}$$

since the ith-generation renewals $g_i(t - x)$ introduced at time $t - x$ suffer $g_i(t - x)f(x) \, dx$ failures by the end of a period x.

If we introduce

$$u(t) \equiv \sum_{i=1}^{\infty} g_i(t), \tag{15-15}$$

then, as before, we have for the average renewals during $(t, t + dt)$

$$u(t) = g_1(t) + \int_0^t u(t - x)f(x) \, dx. \tag{15-16}$$

If we assume an initial population of N units, the expected number of first renewals of these during $(t, t + dt)$ is given by

$$g_1(t) = Nf(t) \, dt. \tag{15-17}$$

If we consider a population of $N = 1$ items and wish to estimate the renewals during $(t, t + dt)$, we have

$$u(t) = f(t) \, dt + \int_0^t u(t - x)f(x) \, dx. \tag{15-18}$$

This expression gives the average renewals during $(t, t + dt)$. The total average renewals by time t are given by the renewal function $U(t)$ which is obtained from the integral equation of renewal theory [obtained by integrating (15-18) from zero to t]:

$$U(t) = F(t) + \int_0^t dy \int_0^y u(y - x)f(x)\, dx$$
$$= F(t) + \int_0^t dx \int_x^t u(y - x)f(x)\, dy$$
$$= F(t) + \int_0^t U(t - x)\, dF(x). \tag{15-19}$$

Note that $U(t)$ is the first moment of N_t, that is, $U(t) = E(N_t)$, as we shall see later on.   Here

$$F(t) = \int_0^t f(x)\, dx. \tag{15-20}$$

Many investigations start with Eq. (15-19).   The existence of $U(t)$ does not require that $F(x)$ be absolutely continuous; hence $f(x)$ need not exist.   We later give a derivation for $U(t)$ which does not assume the existence of $f(x)$.   Once $U(t)$ is known, all aspects of the renewal process can be determined.

It is possible to obtain the solution of the above equation as an inverse Laplace transform.   If $F(x)$ is absolutely continuous so that $f(x) = F'(x)$, we can apply the Laplace transform to (15-18).   This gives

$$u^*(s) = \frac{f^*(s)}{1 - f^*(s)},$$

where the asterisk, in conformity with our usual notation, indicates the Laplace transform of the function.

15-4. Some Properties of the Continuous Case

Properties for the discrete case can be derived by analogy with those given here for the continuous case.

By using $f(t) = \mu e^{-\mu t}$, we find that $U(t) = \mu t$ or $U(t)/t = \mu$ for all t. One of the problems assigned (with a given failure distribution) is to show that the total renewals are a sum of incomplete gamma functions.   In the same problem, by working on the equation in $U(t)$, using the fact (from Laplace-transform theory) that $\lim_{t \to \infty} u(t) = \lim_{s \to 0} su^*(s)$, and integrating the result from zero to t to obtain $U(t)$, one has

$$\frac{U(t)}{t} \sim \text{(reciprocal of mean of failure distribution)},$$

Note that in this case we have used the fact that an asymptotic expansion may be integrated and that finite (or convergent) asymptotic expansions may be multiplied. The foregoing examples of the behavior of $U(t)$ for large t are confirmed by an important fact known as the elementary renewal theorem.

Theorem 15-1: As $t \to \infty$,

$$\frac{U(t)}{t} \to \mu_1^{-1}, \tag{15-21}$$

where μ_1 is the mean of $dF(x)$.

Thus, starting with a single item, the total renewals for a large value of the time are approximately the product of the time elapsed and the failure rate. We give a proof of this theorem at the end of the chapter.

As for the renewal density function, a necessary condition for $u(t) = U'(t)$ to tend to a limit as $t \to \infty$ is that $f(x) \to 0$ as $x \to \infty$. The *renewal-density theorem* asserts that if $f(x) \to 0$ as $x \to \infty$ and if, for some $p > 1$, $|f(x)|^p$ is integrable, then as $t \to \infty$, $u(t) \to \mu_1^{-1}$. This result does not require that $\mu_1 < \infty$.

Theorem (Blackwell) 15-2: If for a continuous renewal process $\mu_1 < \infty$ and $\alpha > 0$ is a constant, then as $t \to \infty$

$$U(t + \alpha) - U(t) \to \alpha \mu_1^{-1}. \tag{15-22}$$

This is a special case of the *key renewal theorem* due to Smith which we now give.

Theorem 15-3: As $t \to \infty$,

$$\int_0^t Q(t - x) \, dU(x) \to \mu_1^{-1} \int_0^\infty Q(y) \, dy \tag{15-23}$$

for a nonnegative nonincreasing function $Q(t)$ (with $t > 0$) which belongs to $L_1(0, \infty)$ of Lebesgue integrable functions; here μ_1 may be infinite.

If $Q(t) = 1/\alpha$ for $0 < t \le \alpha$ and zero otherwise, one has Blackwell's theorem. If we have a finite second moment μ_2 of $dF(x)$, then by substituting

$$Q(t) \equiv 1 - \mu_1^{-1} \int_0^t [1 - F(x)] \, dx$$

we obtain, as $t \to \infty$,

$$U(t) - \frac{t}{\mu_1} \to \frac{\mu_2}{2\mu_1^2} - 1. \tag{15-24}$$

15-5. The Second Moment of Renewals

In the process of deriving the second moment, we derive the renewal function in a different manner. The second moment $U_2(t) \equiv E(N_t^2)$ is determined as follows: If one writes

$$z_r = \begin{cases} 1 & \text{if } S_r \leq t, \\ 0 & \text{if } S_r > t, \end{cases}$$

then $N_t = \sum_{r=1}^{\infty} z_r$ and $E(z_r) = \text{Prob } (S_r \leq t) \equiv F_r(t)$. Also

$$U(t) = E(N_t) = \sum_{r=1}^{\infty} E(z_r) = \sum_{r=1}^{\infty} F_r(t). \tag{15-25}$$

Now the characteristic function, i.e., the Fourier-Stieltjes transform of the sum of independently distributed random variables, is the product of their characteristic functions, which are in this case identical. If we use the circle to denote this transform, we have

$$F_r^{\circ}(s) = [F^{\circ}(s)]^r, \tag{15-26}$$

$$U^{\circ}(s) = \sum_{r=1}^{\infty} [F^{\circ}(s)]^r = \frac{F^{\circ}(s)}{1 - F^{\circ}(s)},$$

where $F(x)$ is the common distribution.

Using the convolution theorem, we immediately have

$$U(t) = F(t) + \int_0^t U(t - x) \, dF(x).$$

The derivation of this step is left as an exercise.

Now from the definition of z_r we have, if $r = \max(p,q)$, $z_r = z_p z_q$. Clearly, $z_r^2 = z_r$. We use these facts to obtain

$$N_t^2 = \Big(\sum_{r=1}^{\infty} z_r\Big)^2 = z_1^2 + 2z_1 z_2 + 2z_1 z_3 + \cdots + 2z_2 z_1 + z_2^2 + 2z_2 z_3 + \cdots$$
$$+ 2z_3 z_1 + 2z_3 z_2 + z_3^2 + \cdots$$
$$+ \cdots$$
$$= \sum_{r=1}^{\infty} z_r + 2 \sum_{r=1}^{\infty} (r - 1)z_r = \sum_{r=1}^{\infty} (2r - 1)z_r.$$

And we have for the second moment $U_2(t) = E(N_t^2)$

$$U_2(t) = U(t) + 2 \int_0^t U(t - x) \, dU(x), \tag{15-27}$$

whose derivation we also leave as an exercise. Higher moments are similarly derived.

Smith obtains from his key renewal theorem for $\mu_2 < \infty$ and for large t

$$\operatorname{var} N_t \sim \frac{\mu_2 - \mu_1^2}{\mu_1^3} t. \tag{15-28}$$

If we define the coefficient of variation C by $(\sigma^2/\mu_1^2)^{\frac{1}{2}}$, where $\sigma^2 = \mu_2 - \mu_1^2$ and σ is the standard deviation, then this result on the variance of the number of renewals in time t may be written as

$$\operatorname{var} N_t \sim \frac{C^2 t}{\mu_1}. \tag{15-29}$$

A heuristic proof of this result, working with the sequence $\{X_i\}$ and assuming that $f(x)$ exists, is as follows (see Chap. 3 for the sum of random variables with random subscript):

$$E(X_1 + \cdots + X_{N_t}) = E(N_t)\mu_1. \tag{15-30}$$

Thus, the expected value of the time of the N_tth renewal of an article equals the expected number of renewals multiplied by the average lifetime of the article. Also,

$$E[(X_1 + \cdots + X_{N_t}) - N_t\mu_1]^2 = E(N_t)\sigma^2. \tag{15-31}$$

Note that this is not the variance of S_{N_t} but is the expectation of the square of the difference in time between actual time and the product of averages μ_1. The expression (15-31) follows immediately from the variance of a compound distribution. After a long time t, $X_1 + \cdots + X_{N_t} \simeq t$, where N is the number of events occurring in t.

$$E(N_t) \sim \frac{t}{\mu_1}. \tag{15-32}$$

When substituted in the second relation, the latter becomes

$$E(t - N_t\mu_1)^2 \sim \frac{\sigma^2 t}{\mu_1}. \tag{15-33}$$

But
$$\operatorname{var} N_t = E[N_t - E(N_t)]^2 \sim E\left(N_t - \frac{t}{\mu_1}\right)^2; \tag{15-34}$$

therefore
$$\operatorname{var} N_t \sim \frac{\sigma^2}{\mu_1^2} \frac{t}{\mu_1} = \frac{C^2 t}{\mu_1}. \tag{15-35}$$

If also $\mu_3 < \infty$, then as $t \to \infty$

$$\operatorname{var} N_t \sim \frac{\mu_2 - \mu_1^2}{\mu_1^3} t + \frac{5\mu_2^2}{4\mu_1^4} - \frac{2\mu_3}{3\mu_1^3} - \frac{\mu_2}{2\mu_1^2} + o(1), \tag{15-36}$$

which gives another measure of the dispersion.

15-6. Residual Life

It is clear now that for a general renewal process with

$$G(x) = \text{Prob } (X_0 \le x) \ne F(x),$$

which describes the residual life, we have, by an analysis similar to the foregoing,

$$U(t) = G(t) + \int_0^t U(t - x) \, dF(x) \tag{15-37}$$

and, using a tilde for the Laplace-Stieltjes transform, i.e.,

$$\tilde{U}(t) = \int_0^\infty e^{-st} \, dU(t),$$

we have

$$\tilde{U}(s) = \frac{\tilde{G}(s)}{1 - \tilde{F}(s)}. \tag{15-38}$$

The renewal density may be similarly obtained.

In the latter, if $f^*(s)$ denotes the Laplace transform of $f(t)$, the question of inverting $u^*(s)$ and writing it in the form

$$u(t) = \sum_{k=0}^\infty C_k \exp (s_k t), \tag{15-39}$$

where the series converges absolutely for $t \ge 0$ and where the s_k are the roots of $f^*(s) = 1$, was examined by Feller. The necessary and sufficient condition for this is that

$$u^*(s) = \sum_{k=0}^\infty \frac{C_k}{s - s_k} \tag{15-40}$$

and that $\sum_{k=0}^\infty |C_k|$ converges. In this case we have

$$C_k = \frac{-g^{*\prime}(s_k)}{f^*(s_k)}, \tag{15-41}$$

where $g(t) = G'(t)$.

The reader should distinguish the foregoing from the well-known residual-life distribution of the form

$$\frac{1}{1/\mu} \int_0^t [1 - F(x)] \, dx$$

which gives the distribution of the remaining life (i.e., the remaining life is $\le t$) at an arbitrary instant to the next renewal (or equivalently from the last renewal to an arbitrary instant of time). This distribution has been encountered in formula (9-13) in the form $\ge t$. As an exercise, the reader should derive the mean value and the Laplace transform of this distribution.

15-7. Remarks

1. A central-limit theorem for renewal processes using

$$\text{Prob } (S_n \leq t) = \text{Prob } (N_t \geq n)$$

gives for $\mu_2 < \infty$, $\sigma^2 = \mu_2^2 - \mu_1^2$ as $t \to \infty$,

$$\text{Prob } \left(N_t \geq \frac{t}{\mu_1} - \frac{\alpha\sigma}{\mu_1} \sqrt{\frac{t}{\mu_1}} \right) \to \frac{1}{\sqrt{2\pi}} \int_{-\infty}^{\alpha} e^{-v^2/2} \, dy. \qquad (15\text{-}42)$$

2. A stochastic process in which a renewal process can be imbedded is known as a regenerative process. This idea has already been used in connection with imbedded Markoff chains (regeneration points) discussed in earlier chapters. The intervals separating such points are independently and identically distributed.

3. Suppose that one has n independent renewal processes, each with its own distribution; the superposition theorem asserts that, when the means of the individual processes tend to infinity in such a way that the sum $1/m$ of their reciprocals remains constant, when $n \to \infty$, the resulting process tends to have the exponential distribution with mean m. For further remarks see Palm [546] and Cox and Smith [132].

4. A cumulative process may be obtained from a regenerative process. For example, in queueing theory the total busy time since $t = 0$, or the number of customers who encounter a queue of size greater than N, constitutes a cumulative process.

5. We now give a proof of the elementary renewal theorem. Let $S_n = X_1 + \cdots + X_n$; then the events $n = N_t + 1$ and X_{n+1}, X_{n+2}, . . . are independently and identically distributed. Clearly, by definition, we have

$$E(N_t + 1) = 1 + U(t) < \infty.$$

Let $E(X_1) = \mu_1$. Roughly speaking, the average value of the time at which the $(N_t + 1)$st renewal is made equals the average total renewals $[U(t) + 1]$ multiplied by the average lifetime μ_1. Thus

$$E(S_{N_t+1}) = E(N_t + 1)E(X_1) = [1 + U(t)]\mu_1.$$

If we write $S_{N_t+1} = t + \xi_t$, where ξ_t is the excess beyond t (the residual life in use at time t), we have from the foregoing

$$E(S_{N_t+1}) = t + E(\xi_t) = [1 + U(t)]\mu_1.$$

The fact that ξ_t is positive yields from the last equality

$$\liminf_{t \to \infty} \frac{U(t)}{t} = \frac{1}{\mu_1}$$

for then $E(\xi_t) \to 0$.

Define another renewal process by

$$Y_i = \begin{cases} X_i & \text{if } X_i \leq \Delta, \\ \Delta & \text{if } X_i > \Delta, \end{cases}$$

where Δ is some constant. The new partial sums do not exceed the corresponding previous ones; hence the new N_t is not less than the previous N_t. Thus this inequality also holds for the new and old renewal functions which we denote by $U^y(t)$ and $U^x(t)$, respectively. Hence,

$$\limsup_{t \to \infty} \frac{U^x(t)}{t} \leq \limsup_{t \to \infty} \frac{U^y(t)}{t} \leq \frac{1}{\mu_1^y}$$

where μ_1^y is the mean of the new distribution $F^y(x)$. By choosing Δ large enough, μ_1^y can be made arbitrarily close to $\mu_1^x \equiv \mu_1$, and the proof is complete.

PROBLEMS

1. Obtain the solution of the renewal equation in the continuous case by means of Laplace transforms, assuming that $f(x)\, dx = dF(x)$ exists.

2. Suppose that it is desired to estimate renewals of a population of items which are subject to n types of failure independently distributed according to

$$\mu_j e^{-\mu_j x},$$

and suppose that the proportion of items used which are subject to the jth type of failure is a_j and that the total population is N. The over-all distribution of failures is

$$\sum_{j=1}^{n} a_j \mu_j e^{-\mu_j x}, \qquad \sum_{j=1}^{n} a_j = 1.$$

Using
$$f(x) = \frac{b^a}{\Gamma(a)}\, x^{a-1} e^{-bx}, \qquad a, b > 0,$$

as an approximation to the over-all distribution of failures, determine a and b by equating first and second moments. Use the resulting distribution to compute the total number of failures at time t by the method of Laplace transforms. Show that the result is an infinite sum of incomplete gamma functions. By working with $u(t)$, show that the elementary renewal theorem holds. Also show that Blackwell's theorem and the result on $U(t)$ involving the two moments μ_1 and μ_2 also hold.

3. Does Blackwell's theorem follow from the elementary renewal theorem using ideas from asymptotic expansions? Explain.

4. Derive the renewal equation and the second-moment equation rigorously, using the functions $F_r(t)$.

5. Obtain the expansion of $u(t)$ in exponential series involving the roots s_k when $f(x)$ for $x \geq 0$ equals

a. $\mu e^{-\mu x}$.
b. $x^2 e^{-x}/2$.

6. Let $p_n(t) = F_n(t) - F_{n+1}(t)$ be the probability that $N_t = n$ for large t. Then for Re $(a) > 0$,

$$p_n(t) = \frac{1}{2\pi i} \int_{a-\infty i}^{a+\infty i} \frac{e^{st}}{s} [1 - \bar{F}(s)][\bar{F}(s)]^n \, ds,$$

and using the method of steepest descent we have

$$p_n(t) \sim \frac{e^{\alpha t}[1 - \bar{F}(\alpha)][\bar{F}(\alpha)]^n}{\alpha[2\pi n K''(\alpha)]^{1/2}}.$$

Here α is the unique real root of

$$K'(s) = -\frac{t}{n}$$

and
$$K(s) = \log \bar{F}(s).$$

Compute this expression for $p_n(t)$ for

$$f(t) = \frac{t^{n-1}e^{-t}}{\Gamma(n)}, \qquad n > 0,$$

and for
$$f(t) = \begin{cases} 0, & t < \tau \\ e^{-t+\tau}, & t \geq \tau. \end{cases}$$

Here $\bar{F}(s)$ is the Laplace-Stieltjes transform of $F(x)$, the cumulative distribution of $f(x)$.

The principle of the steepest descent gives the asymptotic approximation

$$\int_A^B f(x)e^{zg(x)} \, dx \sim \frac{f(\alpha) \sqrt{2\pi} \, e^{zg(\alpha)}}{\sqrt{-zg''(\alpha)}}$$

for
$$-\frac{\pi}{2} + \epsilon \leq \arg z \leq \frac{\pi}{2} + \epsilon$$

where $f(x)$ and $g(x)$ are analytic in a connected region of the complex plane containing A and B, and α maximizes $g(x)$ uniquely. For large z the value of the integral is only slightly changed if A and B are replaced by $-\infty$ and ∞, respectively. Note that the limits of the integral of this problem can be changed to real limits by transforming the variable of integration.

GENERAL COMMENTS, BIBLIOGRAPHY

SOME GENERAL COMMENTS

The development of queueing theory has been very rapid in the past few years. Many problems which seemed complex have been attacked and solved, e.g., the birth-death equations. It is natural to attempt to view queueing problems within a broad framework and to look for a unifying theory. Time-dependent solutions are generally preferred, particularly when a steady state does not exist. More significantly, cases of dependence require more treatment (see Ventura's work in Chap. 9). Most of the assumptions made in studying queueing problems have required independence assumptions for the arrival and service distributions. Obviously, for many problems, this is far from being the practical situation, but the analytical difficulties encountered are shared by a large number of other fields in which dependence is observed and required, and a gradual breakthrough is in the making.

Drawing on past experience, the theory, if sufficiently broad, might include different input properties with different inputs. It should include most of the phenomena discussed in Chap. 1, e.g., balking, reneging, bulk properties, etc. However, it must allow for greater variability than present piecemeal studies do. (Gnedenko's contribution is an example of greater variability.) For example the theory should account for all significant types of queue disciplines, by indicating how many of them can be accounted for by a specialization of parameters.

Queueing networks of parallel and series channels require further investigation. When variability in the number is introduced in these partly specialized cooperating channels, with different queues before them, some of which may have limitations on the allowed number of customers waiting, we have a very important problem which has in it various complexities of queues. For this type of study the question of the relation of the output distribution to an arbitrary input and service-time distributions must be answered. Recent works on dynamic priorities show another direction in which to proceed, i.e., to look for refinements in existing queueing phenomena which could have meaningful

373

applications. Collusion and jockeying are important examples of queueing problems with difficult formulation.

Of course, there are still many simple problems which must be solved, such as the arbitrary-input, general-service-distribution single-channel ordered queue in the time-dependent case. The number in the queue and the busy-period distributions, in addition to waiting times (studied by Kiefer and Wolfowitz), may be an objective. Rapid progress is being made in this direction, as the literature reveals.

Some of these problems often give rise to interesting mathematical problems. Their solvability will in part depend on rapid development in research on stochastic processes.

Simulation of queueing processes for numerical answers is useful for urgent and very important problems. However, it is frequently an inefficient process. When a computer is available and an analytical answer seems remote, simulation may be advisable. From a basic-research standpoint, the presence of a computer may provide ground for experimenting with different hypotheses and, hence, should stimulate and encourage analytical investigations if properly used as a tool.

Optimization requires a systematic introduction into queueing theory to make full use of the results. We have illustrated simple cases of optimization. As yet there is no adequate geometric language for describing queues. The theory of graphs does not have the supply of ideas required for a full representation. One continues to rely on intuition and on prior knowledge of queues to communicate ideas. This interesting problem requires urgent remedies to improve communication and explore the imagination in creating new types of queues for both theoretical and applied purposes.

BIBLIOGRAPHY

The following is a relatively comprehensive bibliography of queueing theory. The subjects of traffic, telephony, dams and storage, machine interference, and renewal and counter theories are also represented for their close relations to queues. From the latter titles a research worker may find a paper with useful ideas which are transferable to the queueing subject under investigation. Even though inventory and scheduling theories constitute a broad field, with many investigators, and are related to queues in some aspects, the number of references is too large to be given in full. It is hoped that those listed here will be useful.

In preparing this list, the bibliographies of Doig and Lunger have been of assistance. Additional references are arranged alphabetically at the end of the first set.

1. Adams, W. F.: Road Traffic Considered as a Random Series, *J. Inst. Civil Engrs.* (*London*), vol. 4, pp. 121–130, 1936.
2. Adler, H. A., and K. W. Miller: A New Approach to Probability Problems in Electrical Engineering, *Trans. Am. Inst. Elec. Engrs.*, vol. 65, pp. 630–632, 1946.
3. Adler, R. B., and S. J. Fricker: The Flow of Scheduled Air Traffic, parts I and II, *Mass. Inst. Technol., Research Lab. Electronics, Tech. Repts.* 189, 199, 1951.
4. ——— and ———: Notes on the Flow of Scheduled Air Traffic, *J. Roy. Aeronaut. Soc.*, vol. 58, pp. 475–484, 1954.
5. ——— and ———: Notes on the Flow of Scheduled Air Traffic, *IRE Trans. on Aeronaut. Navigational Electronics*, vol. AN-2, p. 22, 1955.
6. Akaike, Hirotugu: On Ergodic Property of a Tandem Type Queueing Process, *Ann. Inst. Statist. Math.* (*Tokyo*), vol. 9, pp. 13–22, 1957.
7. Albert, G. E., and L. Nelson: Contributions to the Statistical Theory of Counter Data, *Ann. Math. Statist.*, vol. 24, pp. 9–22, 1953.
8. Ancker, C. J., and A. V. Gafarian: Queueing with Multiple Poisson Inputs and Exponential Service Times, System Development Corporation, *Tech. Mem.* 503, Santa Monica, Calif., 1960.
9. Anis, A. A.: Statistical Aspects of Storage Problems, Thesis, London University, 1952.
10. Anson, C. J.: Determining the Number of Machines to be Attended by an Operator in Order to Minimize the Total Cost per Unit of Production, *Time and Motion Study J.*, vol. 6, pp. 13–27, April, 1957.
11. Arfwedson, G.: Research in Collective Risk Theory, part I, *Skand. Aktuar Tidskr.*, vol. 37, pp. 191–223, 1954.
12. Arley, N.: "On the Theory of Stochastic Processes and Their Application to the Theory of Cosmic Radiation," G.E.C. Cads Forlag, Copenhagen, 1943.
13. Arrow, K. J., S. Karlin, and H. Scarf: "Studies in the Mathematical Theory of Inventory and Production," Stanford University Press, Stanford, Calif., 1958.
14. Ashcroft, H.: The Productivity of Several Machines Under the Care of One Operator, *J. Roy. Statist. Soc., Ser. B*, vol. 12, pp. 145–151, 1950.
15. Atkinson, J.: "Telephony," vol. II, Sir Isaac Pitman & Sons Ltd., London, 1950.
16. J. H. B.: A Partial Call Queueing Scheme for Sleeve-control Exchanges, *P. O. Elec. Engrs. J.*, vol. 46, part I, April, 1953.
17. Bailey, N. T. J., and J. D. Welch: Appointment Systems in Hospital Out-patient Departments, *Lancet*, vol. 262, pp. 1105–1108, 1952.
18. ———: Study of Queues and Appointment Systems in Out-patient Departments with Special Reference to Waiting Times, *J. Roy. Statist. Soc., Ser. B*, vol. 14, pp. 185–199, 1952.
19. ———: A Continuous Time Treatment of a Simple Queue, Using Generating Functions, *J. Roy. Statist. Soc., Ser. B*, vol. 16, pp. 288–291, 1954.
20. ———: On Queueing Processes with Bulk Service, *J. Roy. Statist. Soc., Ser. B*, vol. 16, pp. 80–87, 1954.
21. ———: Queueing for Medical Care, *Appl. Statist.*, vol. 3, pp. 137–145, 1954.
22. ———: A Note on Equalizing the Mean Waiting Time of Successive Customers in a Finite Queue, *J. Roy. Statist. Soc., Ser. B*, vol. 17, no. 1, pp. 262–263, 1955.
23. ———: Statistics in Hospital Planning and Design, *Appl. Statist.*, vol. 5, no. 3, November, 1956.
24. ———: "The Mathematical Theory of Epidemics," Hafner Publishing Company, New York, 1957.
25. ———: Some Further Results in the Non-equilibrium Theory of a Simple Queue, *J. Roy. Statist. Soc., Ser. B*, vol. 19, no. 2, 1957.

26. Barrer, D. Y.: Queueing with Impatient Customers and Indifferent Clerks, *Operations Research*, vol. 5, no. 5, 1957.

27. ——: Queueing with Impatient Customers and Ordered Service, *Operations Research*, vol. 5, pp. 650–656, 1957.

28. Barry, J. Y.: A Priority Queueing Problem, *Operations Research*, vol. 4, pp. 385–386, 1956.

29. Bartlett, M. S.: "An Introduction to Stochastic Processes," Cambridge University Press, London, 1955.

30. Basharin, G. P.: Multiplex Limited Number Distribution of Busy Lines in the Second Cascade Switch Board in a Telephone System with Refusals, *Soviet Phys. Dokl.*, vol. 121, no. 3, pp. 718–721 (280–283), *Doklady Akad. Nauk U.S.S.R.*, 1958.

31. ——: An Investigation, Using Probability Theory, of a Two Stage Trunk-hunting Telephone System with Refused Calls, *Soviet Phys. Dokl.*, vol. 121, no. 3, pp. 713–717 (101–104), *Doklady Akad. Nauk U.S.S.R.*, 1958.

32. Bathgate, R. J. P.: Automatic Telephone Traffic with Reference to Production Processes, *Trans. S. African Inst. Elec. Engrs.*, vol. 42, p. 384, 1951.

33. Bech, N. I.: Calculation of the Probabilities of Possible States of Occupancy in a Simple Group with Simple Ordered Searching, *Teleteknik*, vol. 5, pp. 342–349, 1954 (Danish; English summary).

34. ——: A Method of Computing the Loss in Alternative Trunking and Grading Systems, Copenhagen Telephone Company, Copenhagen (translation of article from *Teleteknik*, vol. 5, pp. 435–448, 1954).

35. ——: Statistical Equilibrium Equations for Overflow Systems, *Teleteknik* (English ed.), vol. 1, pp. 66–71, 1957.

36. Bekessy, A.: On the Probability Distribution of Pulse Counts Made by a Faulty Counter, *Magyar Tudományos Akad. Alkalm. Mat. Int. Közleményei*, vol. 3, pp. 171–181, 1955 (Hungarian; Russian and German summaries).

37. Bell, G. E.: Theoretical Studies into Traffic Congestion, part II, *Brit. Ministry of Civil Aviation, Operations Research Sect., ORS/MCA Rept.* 3, June, 1948.

38. ——: Operational Research into Air Traffic Control, *J. Roy. Aeronaut. Soc.*, vol. 53, pp. 965–978, 1949.

39. ——, D. Barnett, M. A. Young, and D. C. T. Bennett: Air Traffic Control, *J. Inst. Navigation*, pp. 211–250, November, 1950.

40. ——: Queueing Problems in Civil Aviation, *Operational Research Quart.*, vol. 3, pp. 9–11, 1952.

41. ——, D. V. Lindley, K. D. Tocher, and J. D. Welch: Marshalling and Queueing, *Operational Research Quart.*, vol. 3, pp. 4–13, 1952.

42. Beneš, V. E.: A Continuous Time Treatment of the Waiting Time in a Queueing-system Having Poisson Arrivals, a General Distribution of Service-time and a Single Service Unit (Preliminary Report), *Ann. Math. Statist.*, vol. 27, p. 872, 1956.

43. ——: On Queues with Poisson Arrivals, *Ann. Math. Statist.*, vol. 28, pp. 670–677, 1957.

44. ——: The Joint Distribution of a Set of Sufficient Statistics for the Parameters of a Single Telephone Exchange Model, *Ann. Math. Statist.*, vol. 28, p. 525, 1957.

45. ——: A Note on Fluctuations of Telephone Traffic, *Ann. Math. Statist.*, vol. 28, 1957.

46. ——: Generalization of Palm's Loss Formula for Telephone Traffic, presented at Annual Convention of Institute of Mathematics, 1958.

47. ——: The General Queue with One Server, presented at summer meeting of American Mathematics Society, 1958.

48. ———: On Trunks with Negative Exponential Holding Times Serving a Renewal Process, *Bell System Tech. J.*, vol. 38, pp. 211–258, 1959.

49. ———: General Stochastic Processes in Traffic Systems with One Server, *Bell System Tech. J.*, vol. 39, no. 1, pp. 127–160, January, 1960.

50. ———: General Stochastic Processes in the Theories of Counters and Telephone Traffic (to be published).

51. ———: Combinatory Methods and Stochastic Kolmogorov Equations in the Theory of Queues with One Server (see Ref. 783).

52. Benson, F., and D. R. Cox: The Productivity of Machines Requiring Attention at Random Intervals, *J. Roy. Statist. Soc., Ser. B*, vol. 13, pp. 65–82, 1951.

53. ———: Further Notes on the Productivity of Machines Requiring Attention at Random Intervals, *J. Roy. Statist. Soc., Ser. B*, vol. 14, pp. 200–210, 1952.

54. ———, J. Miller, and M. W. H. Townsend: Machine Interference—A Solution of Some Workload Problems Arising When an Operator Has Charge of More than One Machine, *J. Textile Inst.*, vol. 44, pp. T619–T644, 1953.

55. Berkeley, G. S.: "Traffic and Trunking Principles in Automatic Telephony," Ernest Benn, Ltd., London, 1934 (reprinted, 1949).

56. ———: Traffic and Delay Formulae, *P. O. Elec Engrs. J.*, vol. 29, pp. 188–195, 1936.

57. Berkowitz, S. M.: Analysis of a Fixed-block Terminal Area, Air Traffic Control System, *Franklin Inst. Final Rept.* F-2164-2, 1951.

58. Bernstein, P.: How Many Automatics Should a Man Run, *Factory Management and Maintenance*, vol. 99, pp. 85–164, 1941.

59. Berry, D. S., and F. H. Green: Techniques for Measuring Overall Speed in Urban Areas, *Highway Research Board, Proc.*, vol. 29, p. 311, 1949.

60. ——— and D. M. Belmont: Distribution of Vehicle Speed and Travel Times, *Proc. Second Berkeley Symposium on Math. Statist. and Probability*, pp. 589–602, University of California Press, Berkeley, Calif., 1951.

61. ———: Field Measurement of Delay at Signalized Intersections, *Highway Research Board, Proc.*, vol. 35, pp. 505–522, 1956.

62. Beukelman, B. J., and J. W. Cohen: Call Congestion of Transposed Multiples, *Philips Telecommun. Rev.*, vol. 17, no. 4, pp. 145–154, April, 1957.

63. Bickelhaupt, C. O.: Mathematical Study of Toll Traffic, *Telephony*, vol. 77, no. 9, p. 12, 1919.

64. Bishop, D. J.: The Renewal of Aircraft, *Brit. Ministry of Aircraft Production, Aeronaut. Research Comm. Rept. and Mem.* 1907(6342), 1942.

65. Bloemena, A. R.: On Queueing Processes with a Certain Type of Bulk Service, *Math. Centrum, Statist. Afdel. Rept.*, Rept. SP-67, Amsterdam (mimeo. copy).

66. Blumstein, A.: The Operations Capacity of a Runway Used for Landings and Take-offs, Cornell Aeronautical Laboratory, Inc., Buffalo, N.Y., Feb. 11, 1960, prepared for Second International Conference on Operations Research, Aix-en-Provence, France, September, 1960.

67. Bodino, G. A., and F. Brambilla: "Teoria delle code," Istituto Editoriale Cisalpino, Milano-Varese, 1959.

68. Boiteux, M.: Coûts et tarifs en face d'une demande aléatoire, *Econométrie*, pp. 223–227; Discussion (French), pp. 227–230 (*Coll. intern. centre natl. recherche sci. (Paris)*, no. 40, 1953.

69. Boldyref, A.: Determination of the Maximal Steady State Flow of Traffic through a Railroad Network, *J. Operations Research Soc. Am.*, vol. 3, pp. 443–465, 1955.

70. Borel, E.: Sur l'emploi du théorème de Bernoulli pour faciliter le calcul d'une

infinité de coefficients, application au problème de l'attente à un guichet, *Compt. rend.*, vol. 214, p. 452, 1942.

71. Bouchman, E. N.: Problems of Congestion in Telephony, *Akad. Nauk S.S.S.R. Zhur. Priklad. Mat. i Mekhan.*, vol. 6, pp. 247–256, 1942 (Russian; English summary).

72. ———: The Problem of Waiting Times, *Akad. Nauk S.S.S.R. Zhur. Priklad. Mat. i Mekhan.*, vol. 11, pp. 475–484, 1947.

73. Bowen, E. G.: Operational Research into the Air Traffic Problem, *J. Inst. Navigation*, vol. 1, pp. 338–341, 1948.

74. ——— and T. Pearcey: Delays in Air Traffic Flow, *J. Roy. Aeronaut. Soc.*, vol. 52, pp. 251–258, 1948.

75. Brigham, G.: On a Congestion Problem in an Aircraft Factory, *J. Operations Research. Soc. Am.*, vol. 3, pp. 412–428, 1955.

76. Brisby, M. D. J., and R. T. Eddison: Train Arrivals—Handling Costs, and the Holding and Storage of Raw Materials, *J. Iron Steel Inst. (London)*, vol. 172, pt. 2, pp. 171–183, October, 1952.

77. British Ministry of Civil Aviation, Operational Research Section: Notes and Data on Traffic Flow, 1950.

78. Broadhurst, S. W., and A. T. Harmston: An Electronic Traffic Analyser, *P. O. Elec. Engrs. J.*, vol. 42, pp. 181–187, 1949.

79. ——— and ———: Studies of Telephone Traffic with the Aid of a Machine, *Proc. Inst. Elec. Engrs. (London)*, pt. I, vol. 100, pp. 259–274, 1953.

80. Brockmeyer, E., H. L. Halstrom, and A. Jensen: "The Life and Works of A. K. Erlang" (all of Erlang's papers translated into English), Copenhagen Telephone Company, Copenhagen, 1948.

81. ———: The Application of the Theory of Probability to Telephony, in J. Rybner (ed.), "Larebog in Telefonteknik," chap. 9, pp. 144–187, Copenhagen, 1949 (Danish).

82. ———: The Use of Probability Calculations in Telephone Engineering on the Basis of the Research of Erlang and Moe, *Teleteknik*, vol. 3, p. 95, 1952.

83. ———: "The Simple Overflow Problem in the Theory of Telephone Traffic," Copenhagen Telephone Company, Copenhagen (Danish); *Teleteknik*, vol. 5, pp. 361–375, 1955 (English synopsis).

84. ———: A Survey of Traffic Measuring Methods in the Copenhagen Telephone Exchange, *Teleteknik* (English ed.), vol. 1, pp. 92–105, 1957.

85. Brotman, L., and J. Minker: Computer Simulation of a Complex Communications System, *Operations Research*, vol. 5, pp. 138–139, 1957.

86. Bruce, J. A., and J. B. Rudden: Denver's Traffic Control System Electronically Moves Auto Flow, *Western City*, November, 1953.

87. Brunnschweiler, D.: Machine Interference, *J. Textile Inst.*, vol. 45, no. 12, pp. T886–T895, 1954.

88. Buch, K. R.: On a Special Use of the Erlang Methods in Industry, *Teleteknik* (English ed.), vol. 1, pp. 76–80, 1957.

89. Burke, P. J.: The Output of a Queueing System, *Operations Research*, vol. 4, pp. 699–704, 1956.

90. ———: Equilibrium Delay Distribution for One Channel with Constant Holding Time, Poisson Input and Random Service, *Bell System Tech. J.*, pp. 1021–1031, July, 1959.

91. Burrows, C.: Some Numerical Results for Waiting Times in the Queue $E_k/M/1$, *Biometrika*, vol. 47, pts. 1 and 2, p. 202, 1960.

92. Buxton, H. A.: Estimating Trunking Requirements, *Telephony*, vol. 75, no. 2, 1918.

93. Camp, G. D.: Some Approximate Solutions of the Single-step Queueing Equation, *J. Operations Research Soc. Am.*, vol. 3, p. 360, 1955.

94. ———: Bounding the Solutions of Practical Queueing Problems by Analytic Methods, in "Operations Research for Management," vol. II, pp. 307–339, Johns Hopkins Press, Baltimore, 1956.

95. Campbell, L. L.: Standard Deviation of Dead Time Correction in Counters, *Can. J. Phys.*, vol. 34, pp. 929–937, 1956.

96. Carlsson, S. G., and A. Elldin: Solving Equations of State in Telephone Traffic Theory with Digital Computers, *Ericsson Tech.*, no. 2, 1958.

97. Champernowne, D. G.: An Elementary Method of the Solution of the Queueing Problem with a Single Server and Constant Parameter, *J. Roy. Statist. Soc.*, Ser. B, vol. 18, no. 1, pp. 125–128, 1956.

98. Chandler, R. E., Robert Herman, and Elliott W. Montroll: Traffic Dynamics: Studies in Car Following, *Operations Research*, pp. 165–184, March-April, 1958.

99. Chantal, René de: Sur des probabilités relatives au trafic aérien dans les aéroports, *Travaux*, vol. 37, pp. 397–403, 1953.

100. ———: The Scientific Basis for Decision (Examples), *Travail et méthodes*, pp. 3–7, December, 1954; Review, *Operational Research Quart.*, vol. 6, p. 87, 1955.

101. Chartrand, M.: The Design of Cafeteria Counters, Norton Company Rept., Worcester, Mass., 1957.

102. Chauveau, J., and E. Vaulot: Extension de la formule d'Erlang au cas où le trafic est fonction du nombre d'abonnés occupés, *Ann. télécommun.*, nos. 8–9, 1949.

103. Chernoff, H.: The Distributions of Shadows with Applications to Traffic and Counter Problems, *Ann. Math. Statist.*, vol. 27, p. 217, 1956.

104. Chovet, A.: Determination of the Allowable Load for Telephone Operators, *Ann. télécommun.*, vol. 9, pp. 35–43, 1954.

105. Christensen, P. V.: Die Wählerzahl in Automatischen Fernsprechämtern, *Elektrotech. Z.*, vol. 34, p. 1314, 1913.

106. Churchman, C. W., R. L. Ackoff, and E. L. Arnoff: "Introduction to Operations Research," John Wiley & Sons, Inc., New York; Chapman & Hall, Ltd., London, pp. 389–416, 1957.

107. Clarke, A. B.: On the Solution of the "Telephone Problem," *Univ. Michigan Eng. Research Inst. Rept.* R-32, March, 1952.

108. ———: The Time Dependent Waiting Line Problem, *Univ. Michigan Rept.* M720-1R39, 1953.

109. ———: On Time-dependent Waiting Line Processes, *Ann. Math. Statist.*, vol. 24, pp. 491–492, 1953 (abstract).

110. ———: A Waiting Time Process of Markov Type, *Ann. Math. Statist.*, vol. 27, pp. 452–459, 1956.

111. ———: Maximum Likelihood Estimates in a Simple Queue, *Ann. Math. Statist.*, vol. 28, pp. 1036–1040, 1957.

112. Clayton, A. J. H.: Road Traffic Calculations, *J. Inst. Civil Engrs. (London)*, vol. 16, 1941.

113. Clifford, E. J.: The Measurement of Traffic Flow, *Traffic Eng.*, vol. 26, pp. 243–246, 1956.

114. Clos, C.: An Aspect of the Dialing Behavior of Subscribers and Its Effect on the Trunk Plant, *Bell System Tech. J.*, vol. 27, pp. 424–445, 1948.

115. ——— and R. I. Wilkinson: Dialing Habits of Telephone Customers, *Bell System Tech. J.*, vol. 31, pp. 32–67, 1952.

116. Cobham, A.: Priority Assignment in Waiting Line Problems, *J. Operations Research Soc. Am.*, vol. 2, pp. 70–76, 1954; A Correction, *ibid.*, vol. 3, p. 547, 1955.

117. Cohen, J. W.: The Full Availability Group of Trunks with an Arbitrary Distribution of the Inter-arrival Times and a Negative Exponential Holding Time Distribution, *Natuurk. Tÿdschr. (Ghent)*, vol. 26, no. 4, pp. 169–181, 1957.

118. ———— and P. Harkema: A Study of the Delay Encountered in Telegraph Time-relay Switching, *Commun. News*, vol. 15, p. 47, 1954.

119. ————: Berechnung der Verkehrsgrössen in Wartezeitsystem aus den Verkehrsgrössen eines Verlustsystems, *NTZ-Nachrtech. Z.*, vol. 8, p. 139, 1955.

120. ————: Das Warteproblem für das Volkommene Bündel mit einer Endlichen Quellenzahl, *NTZ-Nachrtech. Z.*, vol. 8, pp. 641–645, 1955.

121. ————: Certain Delay Problems for a Full Availability Trunk Group Loaded by Two Sources, *Commun. News*, vol. 16, pp. 105–113, 1956.

122. ————: Basic Problems of Telephone Traffic Theory and the Influence of Repeated Calls, *Philips Telecommun. Rev.*, vol. 17–18, 1956–1957.

123. ————: On the Queueing Processes of Lanes, *Philips Tech. Rept.*, 1956.

124. ————: A Survey of Queueing Problems Occurring in Telephone and Telegraph Traffic Theory, *Internatl. Conf. Operations Research*, Oxford, 1957.

125. ————: The Generalized Engset Formulae, *Philips Telecommun. Rev.*, vol. 18, no. 4, pp. 158–170, November, 1957.

126. Conolly, B. W.: A Difference Equation Technique Applied to the Simple Queue, *J. Roy. Statist. Soc., Ser. B*, vol. 20, pp. 165–167, 1958.

127. ————: A Difference Equation Technique Applied to the Simple Queue with Arbitrary Arrival Interval Distribution, *J. Roy. Statist. Soc., Ser. B*, vol. 20, pp. 167–175, 1958.

128. ————: The Busy Period in Relation to the Single-server Queueing System with General Independent Arrivals and Erlangian Service-time, *J. Roy. Statist. Soc., Ser. B*, vol. 22, no. 1, pp. 89–96, 1960.

129. ————: The Busy Period in Relation to the Queueing Process, $GI/M/1$, *Biometrika*, vol. 46, pp. 246–251, 1959.

130. Conrad, R.: Performance of Telephone Operators Relative to Traffic Level, *Nature*, no. 4548, pp. 1480, December, 1956.

131. Cox, D. R., and W. L. Smith: The Superposition of Several Strictly Periodic Sequences of Events, *Biometrika*, vol. 40, pp. 1–11, 1953.

132. ———— and ————: On the Superposition of Renewal Processes, *Biometrika*, vol. 41, 1954.

133. ————: A Table for Predicting the Production from a Group of Machines under the Care of One Operator, *J. Roy. Statist. Soc., Ser. B*, vol. 16, pp. 285–287, 1954.

134. ————: A Use of Complex Probabilities in the Theory of Stochastic Process, *Proc. Cambridge Phil. Soc.*, vol. 51, pp. 313–319, 1955.

135. ————: Some Statistical Methods Connected with Series of Events, *J. Roy. Statist. Soc., Ser. B*, vol. 17, pp. 129–164, 1955.

136. ————: The Analysis of Non-Markovian Stochastic Processes by the Inclusion of Supplementary Variables, *Proc. Cambridge Phil. Soc.*, vol. 51, pp. 433–441, 1955.

137. ————: Prévision de la production d'un groupe de machines surveillées par un exécutant, *Cahiers bur. temps elémentaires*, sér. 5, no. 503–502, p. 18, 1955 (French).

138. ————: The Statistical Analysis of Congestion, *J. Roy. Statist. Soc., Ser. A*, vol. 118, pp. 324–335, 1955.

139. Cox, R. E.: Traffic Flow in an Exponential Delay System with Priority Categories, *Proc. Inst. Elec. Engrs. (London), Ser. B*, vol. 102, pp. 815–818, 1955.

140. Cramer, H.: "The Elements of Probability Theory and Some of Its Applications," 1st Am. ed., John Wiley & Sons, Inc., New York, 1955.

141. Crane, R. T., F. B. Brown, and R. O. Blanchard: An Analysis of a Railroad Classification Yard, *J. Operations Research Soc. Am.*, vol. 3, pp. 262–271; also

"Operations Research by Management," chap. 7, Johns Hopkins Press, Baltimore, 1956.

142. Crommelin, C. D.: Delay Probability Formulae When the Holding Times Are Constant, *P. O. Elec. Engrs. J.*, vol. 25, pp. 41–50, 1932.

143. ———: Delay Probability Formulae, *P. O. Elec. Engrs. J.*, vol. 26, pp. 266–274, 1934.

144. Dantzig, G. B.: A Comment on Edie's "Traffic Delays at Toll Booths," *J. Operations Research Soc. Am.*, vol. 2, pp. 339–341, 1954.

145. Daru, E.: Nombres de machines automatiques à confier à un ouvrier en vue d'atteindre le prix de revient industriel minimum, *Rev. statist. appl.*, vol. 3, pp. 103–108, 1955 (French).

146. ———: Méthodes de resolution pour quelques problèmes de files d'attente comportant des "serveurs" d'efficacités différentes, *Rev. franç. recherche opérationnelle*, vol. 2, no. 8, pp. 137–152, 1958.

147. Davis, Harold: The Build-up Time of Waiting Lines, *Naval Research Logistics Quart.*, vol. 7, no. 2, p. 185, 1960.

148. DeBaun, R. M., and S. Katz: An Approximation to Distributions of Summed Waiting Times, *Operations Research*, vol. 7, no. 6, November–December, 1959.

149. Descamps, René: Calcul direct des probabilités d'attente dans une file, *Rev. franç. recherche opérationnelle*, vol. 3, no. 2, pp. 88–100, 1959.

150. DeGourrierec, B.: Economic Study of the Sub-division of the Lines at the Output of a Stage of Selection in Automatic Telephony, *Ann. télécommun.*, vol. 9, pp. 335–344, 1954 (French).

151. Delcourt, J.: Équipement d'une station de pompage, *Rev. statist. appl.*, vol. 1, no. 2, pp. 77–85, 1959.

152. Deutsch, K. H.: Messung und Berechnung von Wartezeiten in Unvolkommenen Bündeln bei Fernsprechwählanlagen, *Funk u. Ton*, no. 8, pp. 415–421, 469–478, 1950 (German).

153. Dietrich, H.: The Problem of the Optimum Placing of the Town Telephone Exchange, *Przeglad Telekomunikacyjny*, vol. 3, no. 12, pp. 71–77, 1952 (Polish).

154. Dobbie, J. M.: Comments on Expectancy of Multiple Vehicular Breakdowns in a Tunnel, *Operations Research*, vol. 4, pp. 609–613, 740–742, 1956.

155. Dobermann, H.: Formeln für den Mittleren Besetzteinfluss von Vielfach-Schaltungen, *FTZ-Fernmeldetech. Z.*, vol. 7, pp. 23–24, 1954.

156. Doeh, Giyora: A Graphic Tool for the No-queue Model, *Operations Research*, vol. 8, no. 1, pp. 143–145, January–February, 1960.

157. Doig, A.: A Bibliography on the Theory of Queues, *Biometrika*, vol. 44, pts. 3 and 4, pp. 490–514, December, 1957.

158. Domb, C.: The Problem of Random Intervals on a Line, *Proc. Cambridge Phil. Soc.*, vol. 43, pp. 329–341, 1947.

159. ———: Some Probability Distributions Connected with Recording Apparatus, *Proc. Cambridge Phil. Soc.*, vol. 44, pp. 335–341, 1948.

160. ———: Some Probability Distributions Connected with Recording Apparatus, part II, *Proc. Cambridge Phil. Soc.*, vol. 46, pp. 429–435, 1950.

161. Doob, J. L.: "Stochastic Processes," John Wiley & Sons, Inc., New York, 1953.

162. Downton, F.: Waiting Times in Bulk Service Queues, *J. Roy. Statist. Soc.*, Ser. B, vol. 17, pp. 256–261, 1955.

163. ———: On Limiting Distributions Arising in Bulk Service Queues, *J. Roy. Statist. Soc.*, Ser. B, vol. 18, pp. 265–274, 1956.

164. ———: A Note on Moran's Theory of Dams, *Quart. J. Math. Oxford, Ser. 2*, vol. 8, pp. 282–286, 1957.

165. Dressin, S. A.: Priority Assignment—Single Channel, *U.S. Marine Corps* (10 pages).

166. Dunn, P. F., C. F. Flagle, and P. A. Hicks: The Queuiac—An Electromechanical Analog for the Simulation of Waiting Line Problems, *Operations Research*, vol. 4, pp. 648–662, 1956.

167. Edie, L. C.: Traffic Delays at Toll Booths, *J. Operations Research Soc. Am.*, vol. 2, pp. 107–138, 1954.

168. ———: Operations Research Techniques: Waiting Problems, *Operations Research*, section of *Proc. Soc. for Advancement of Management Conf.*, January, 1954.

169. ———: Expectancy of Multiple Vehicular Breakdowns in a Tunnel, *J. Operations Research Soc. Am.*, vol. 3, pp. 513–522, 1955.

170. ———: A Reply to Comments by James M. Dobbie, *Operations Research*, vol. 4, pp. 614–619, 1956.

171. ———: Operations Research in a Public Corporation, *Operations Research*, vol. 5, pp. 111–122, 1957.

172. ———: Optimization of Traffic Delays at Toll Booths, in "Operations Research for Management," vol. 2, chap. 3, Johns Hopkins Press, Baltimore, 1956.

173. Egert, P.: The Mathematical Principles of Traffic Statistics, *Tech. u. Bolkswirt. Berichte des Wirtschafts und Verkehrsministerium Nordrhein-Westfalen*, 34 pages, 1954 (German).

174. Ekberg, S.: The Telephone Traffic Machine, *Tele.* (English ed.), vol. 3, pp. 150–155, 1953.

175. ———: Telephone Traffic Research with the Aid of the Swedish Traffic Analyser, with two Appendices. Appendix 1, The Call Generating Equipment and Its Statistical Accuracy; Appendix 2, Exact Computation of a Grade of Service for Some Types of Interconnecting Arrangements, *Tele.* (English ed.), no. 1, pp. 1–29, 1955.

176. Ekelöf, S.: Calculation of Delays in an Automatic Telephone System, *Ericsson Rev.*, nos. 7–8, 1930.

177. Elldin, A.: Traffic Measurements with Lamp Panel, *Ericsson Tech.*, vol. 10, pp. 107–187, 1954.

178. ———: On the Congestion in Gradings with Random Hunting, *Ericsson Tech.*, Ser. 11, vol. 2, pp. 33–37, 1955.

179. ——— and G. Lind: Statistical Methods for Supervision of Telephone Exchanges and Networks, *Ericsson Tech.*, no. 1, 1956.

180. Elliman, E. A., and R. W. Fraser: An Artificial Traffic Machine for Automatic Telephone Studies, *Elec. Commun.*, October, 1928.

181. Ellis, H. F.: *Punch*, May 30, 1951.

182. ———: Written in a Queue, *Punch*, vol. 232, pp. 407–408, 1957.

183. Engset, T.: Die Wahrscheinlichkeits Rechnung zur Bestimmung der Wählerzahl in Automatischen Fernsprechämtern, *Elektrotech. Z.*, vol. 31, p. 304, 1918.

184. ———: Emploi du calcul des probabilités pour la détermination du nombre de sélecteurs dans les Bureaux Téléphoniques Centreaux, *Rev. gén. elect.*, vol. 9, pp. 138–140, 1921

185. Erdélyi, A. (ed.): "Tables of Integral Transforms," vol. I, McGraw-Hill Book Company, Inc., New York, 1954.

186. Erlang, A. K.: Probability and Telephone Calls, *Nyt Tidsskr. Mat.*, Ser. B, vol. 20, pp. 33–39, 1909.

187. ———: Solution of Some Probability Problems of Significance for Automatic Telephone Exchanges, *Elektroteknikeren*, vol. 13, pp. 5–13, 1917.

188. ———: Lösung einiger Probleme der Wahrscheinlichkeitrechnung von Bedeu-

tung für die selbsttätigen Fernsprechämter, *Electrotech. Z.*, vol. 39, p. 504, 1918 (German).

189. ———: Solution of Some Problems in the Theory of Probabilities of Significance in Automatic Telephone Exchanges, *P. O. Elec. Engrs. J.*, vol. 10, p. 189, 1918.

190. ———: Telefon-Ventetider. Et Stykke Sandsynlighedsregning, *Mat. Tidsskr.*, *Ser. B*, vol. 13, pp. 25–42, 1920.

191. ———: Solution de quelques problèmes de la théorie des probabilités presentant de l'importance pour les Bureaux Téléphoniques Automatiques, *Ann. P.T.T.*, vol. 11, p. 800, 1922 (French).

192. ———: Sandsynlighedsregningens Anvendelse i Telefondrift, *Elektroteknikeren*, vol. 19, pp. 99–110, 1923.

193. ———: Application du calcul des probabilités en téléphonie, *Ann. P.T.T.*, vol. 14, p. 617, 1925 (French).

194. ———: Problem 15 in *Matematisk Tidsskrift*, vol. B, p. 36, 1929. (This is Erlang's version of the problem of the extinction of surnames.)

195. Everett, J. L.: State Probabilities in Congestion Problems Characterized by Constant Holding Times, *J. Operations Research Soc. Am.*, vol. 1, pp. 279–285, 1953.

196. ———: Seaport Operations as a Stochastic Process, *J. Operations Research Soc. Am.*, vol. 1, p. 76, 1953.

197. Fabens, A. J.: The Solution of Queueing and Inventory Models, *Stanford Univ.*, *Applied Math. and Statist. Labs. Tech. Rept.* 20, Dec. 7, 1959.

198. Fagen, R. E., and J. Riordan: Queueing Systems for Single and Multiple Operations, *J. Soc. Ind. Appl. Math.*, vol. 3, pp. 73–79, 1955.

199. Farlie, D. J.: Canteen Queues, *Incorp. Statistician*, vol. 7, pp. 73–84, 1956.

200. Feller, W.: On the Theory of Stochastic Processes, with Particular Reference to Applications, First Berkeley Symposium on Mathematical Statistics and Probability, pp. 403–432, University of California Press, Berkeley, Calif., 1949.

201. ———: Fluctuation Theory of Recurrent Events, *Trans. Am. Math. Soc.*, vol. 67, pp. 98–119, 1949.

202. ———: "An Introduction to Probability Theory and Its Applications," John Wiley & Sons, Inc., New York, 1950.

203. ———: Diffusion Processes in One Dimension, *Trans. Am. Math. Soc.*, vol. 77, pp. 1–31, 1954.

204. ———: On Boundaries and Lateral Conditions for Kolmogorov Differential Equations, *Ann. Math.*, vol. 65, no. 2, pp. 527–570, 1957.

205. ———: The Birth and Death Processes as Diffusion Processes, Princeton University, Princeton, N.J., 1958; written with support of the Air Force Office of Scientific Research of the Air R & D Command under Contract AF 18(603)24.

206. Fetter, R. B.: The Assignment of Operators to Service Automatic Machines, *J. Ind. Eng.*, vol. 6, pp. 22–30, 1955.

207. Field, J. W.: Machine Utilization and Economical Assignment, *Factory Management and Maintenance*, vol. 104, pp. 288–296, 1946.

208. Finch, P. D.: On the Transient Behavior of a Simple Queue, *J. Roy. Statist. Soc.*, vol. 22, no. 2, pp. 277–283, 1960.

209. ———: Mixed Customer Impatience in a Simple Queue, paper—London School of Economics, Research Techniques Unit (mimeo. paper).

210. ———: On Distribution of Queue Size in Queueing Problems, London School of Economics, Research Techniques Unit.

211. ———: The Effect of the Size of the Waiting Room on a Simple Queue, *J. Roy. Statist. Soc., Ser. B*, vol. 20, no. 1, pp. 182–186, 1958.

212. ———: Cyclic Queues with Feedback, *J. Roy. Statist. Soc., Ser. B*, vol. 21, no. 1, pp. 153–157, 1959.

213. ———: Balking in the Queueing System $GI/M/1$, *Acta Math. Acad. Sci. Hung.*, vol. 10, no. 1/2, pp. 241–247, 1959.

214. ———: The Output Process of the Queueing System $M/G/1$, *J. Roy. Statist. Soc., Ser. B*, vol. 21, no. 2, pp. 375–380, 1959.

215. Flagle, C. D., I. W. Gabrielson, A. Soriano, and M. M. Taylor: Analysis of Congestion in an Outpatient Clinic, Johns Hopkins University, Operations Research Division (study conducted).

216. ———: Queueing Theory and Cost Concept Applied to a Problem in Inventory Control, in "Operations Research Management," vol. II, pp. 160–177, Johns Hopkins Press, Baltimore, 1956.

217. ———: The Problem of Organization for Hospital Inpatient Care, paper presented at the Sixth Annual International Meeting of the Institute of Management Science, Paris, September, 1959; Reprint 55, Pergamon Press Ltd., London.

218. Fogel, L. J.: An Analytic Approach toward Air Traffic Control, *IRE Trans. on Aeronaut. Navigational Electronics*, vol. AN-2, no. 2, 1955.

219. Forbes, T. W.: Statistical Techniques in the Field of Traffic Engineering and Traffic Research, *Proc. Second Berkeley Symposium on Math. Statist. and Probability*, pp. 603–625, 1951.

220. Ford, L. R., and D. R. Fulkerson: A Simple Algorithm for Finding Maximal Network Flows and an Application to the Hitchcock Problem, The Rand Corporation, P-743 (alternate numbering RM 1604), September, 1955.

221. Fortet, R.: Probabilité de perte d'un appel téléphonique: Régime non-stationnaire, influence du temps d'orientaton et du groupement des lignes, *Actes colloq. du calcul des probabilités de Lyon*, June, 1948.

222. ———: Sur la probabilité de perte d'un appel téléphonique, *Compt. rend.*, vol. 226, pp. 1502–1507, 1948.

223. ———: Sur la probabilité de perte d'un appel téléphonique dans un groupe de x sélecteurs, commandés par un orienteur unique, *Compt. rend.*, vol. 226, p. 159, 1948.

224. ———: Évaluation de la probabilité de perte d'un appel téléphonique, compte tenu du temps d'orientation et du groupement des lignes, *Ann. télécommun.*, vol. 5, 1950.

225. ———: Les fonctions aléatoires en téléphonie automatique. Probabilités de perte en sélection conjuguée, *Ann. télécommun.*, vol. 11, pp. 85–88, 1956.

226. ———: Random Distributions with an Application to Telephone Engineering, Third Berkeley Symposium on Mathematical Statistics and Probability, vol. 11, 1956.

227. ——— and B. Canceill: Probabilités de perte en sélection conjuguée, *Teleteknik* (English ed.), vol. 1, pp. 41–55, 1957.

228. ———: Calcul des probabilités, in "Application des théories mathématiques," part I, Centre national recherche scientifique, Paris, 1950.

229. ——— and A. Blanc-Lapierre: "Théorie des fonctions aléatoires," Masson et Cie, Paris, 1953.

230. Foster, F. G.: On Stochastic Matrices Associated with Certain Queueing Processes, *Ann. Math. Statist.*, vol. 24, pp. 355–360, 1953.

231. ———: A Unified Theory for Stock, Storage and Queue Control, *Operational Research Quart.*, vol. 10, no. 3, September, 1959.

232. Foulkes, J. D., W. Prager, and W. H. Warner: On Bus Schedules, *Management Sci.*, vol. 1, pp. 41–48, 1954.

233. Frank, P.: Taboo Generating Functions and Other Topics in Markov Chains, *Columbia Univ. Statist. Engr. Group, Tech. Rept.* 2(N), Feb. 26, 1959.

234. Frazer, R. A., W. J. Duncan, and A. R. Collar: "Elementary Matrices," Cambridge University Press, London, 1955.

235. Frost, G. R., W. Keister, and A. E. Ritchie: A Throwdown Machine for Telephone Traffic Studies, *Bell System Tech. J.*, vol. 32, pp. 292–359, 1953.

236. Fry, T. C.: The Theory of Probability as Applied to Problems of Congestion, in "Probability and Its Engineering Uses," D. Van Nostrand Company, Inc., Princeton, N.J., 1928.

237. Fulkerson, D. R., and G. B. Dantzig: Computation of Maximum Flows in Networks, *Naval Research Logistics Quart.*, vol. 2, pp. 277–283, 1955.

238. Galliher, H. P., and R. C. Wheeler: Nonstationary Queueing Probabilities for Landing Congestion of Aircraft, *Operations Research*, vol. 6, no. 2, pp. 264–275, 1958.

239. Gander, R. S.: Operational Research on Queueing Problems, *Research*, vol. 9, pp. 295–301, 1956.

240. Gani, J.: Some Problems in the Theory of Provisioning and of Dams, *Biometrika*, vol. 42, pp. 179–200, 1955.

241. ———— and P. A. P. Moran: The Solution of Dam Equations by Monte Carlo Methods, *Australian J. Appl. Sci.*, vol. 6, pp. 267–273, 1955.

242. ————: Problems in the Probability Theory of Storage Systems, *J. Roy. Statist. Soc.*, Ser. B, vol. 19, no. 2, 1957.

243. ———— and N. V. Prabhu: Stationary Distributions of the Negative Exponential Type for the Infinite Dam, *J. Roy. Statist. Soc.*, Ser. B, vol. 19, pp. pp. 342–351, 1957.

244. Garber, N. H.: A Class of Queueing Problems, Ph.D. Thesis, Massachusetts Institute of Technology, Electrical Engineering Department, 1955.

245. Garwood, F.: Application of the Theory of Probability to the Operation of Vehicular-controlled Traffic Signals, *J. Roy. Statist. Soc.* (suppl.), vol. 7, pp. 65–77, 1940–1941.

246. Gautzsch, O.: Verlustzeitung und Entlohnung bei mehr Maschinenbedienung, *Maschinenbau*, vol. 15, pp. 179–182, 1936.

247. Gaver, D. P.: The Influence of Servicing Times in Queueing Processes, *J. Operations Research Soc. Am.*, vol. 2, pp. 139–149, 1954.

248. ————: Imbedded Markov Chain Analysis of a Waiting Line Process in Continuous Time, *Ann. Math. Statist.*, vol. 30, no. 3, pp. 698–720, September, 1959.

249. Gazis, D. G., Robert Hermann, and Renfrey B. Potts: *Operations Research*, pp. 499–505, July–August, 1959.

250. Gerlough, D. L., and A. Schuhl: Poisson and Traffic, part I, The Use of the Poisson Distribution in Highway Traffic (D.L.G.); The Calculation of Probabilities and the Distribution of Traffic on Two Lane Roads. Eno Foundation for Highway Control, Saugatuck, Conn. (Schuhl's paper is a revised translation of *Travaux*, vol. 39, p. 24, 1955).

251. ———— and J. H. Mathewson: Approaches to Operational Problems in Street and Highway Traffic—A Review, *Operations Research*, vol. 4, pp. 32–41, 1956.

252. ————: Simulation of Freeway Traffic by Electronic Computer, *Highway Research Board, Proc.*, vol. 35, pp. 543–547, 1956.

253. ————, W. W. Mosher, and C. T. Weingarten: The Optimization of Traffic Flow through Certain Signalized Intersections, *Operations Research*, vol. 5, p. 135, 1957.

254. Gilbert, E. N., and H. O. Pollak: Coincidences in Poisson Patterns, *Bell System Tech. J.*, vol. 35, no. 4, pp. 1005–1033, July, 1957.

255. Giltay, J.: Over Stationnaire en Overgangsstatisticken bij Telefoonverkeer, *Ingenieur (Utrecht)*, vol. 61, p. E45, 1949 (Dutch).

256. ———: Static and Transient Statistics in Telephone-traffic Problems, *Appl. Sci. Research, Ser. B*, vol. B1, pp. 413–419, 1950.

257. ———: Over Rangeringen bij de Automatische Telefonie, *Ingeniuer (Utrecht)*, vol. 65, pp. 107–118, 1953 (Dutch).

258. ———: Bijdrage tot de Stochastiek van het Telefoonverkehr, in het bijzonder van de Volkomen Bundel, Dissertation, *University of Delft*, 1953 (Dutch; English summary).

259. Girault, M.: "Initiation aux processus aléatoires," Dunod, Paris, 1959.

260. ———: Quelques exemples d'analyse opérationnelle des files d'attente et du stockage—méthodes statistiques et calculs probabilistes, *Rev. franç. recherche opérationnelle*, vol. 1, p. 106, 1957 (French).

261. Glanville, W. H.: Road Safety and Traffic Research in Great Britain, *J. Operations Research Soc. Am.*, vol. 3, pp. 283–299, 1955; also "Operations Research for Management," vol. II, chap. 4, Johns Hopkins Press, Baltimore, 1956.

262. Glazier, E. D.: The Variation of the Holding Times of Telephone Calls, *P. O. Elec. Engrs. J.*, vol. 30, pp. 46–47, 1937.

263. Gluss, B.: Four Streams of Traffic Converging on a Cross-road, *Ann. Math. Statist.*, vol. 27, pp. 215–216, 1956.

264. Gnedenko, B. V.: On the Theory of Geiger-Müller Counters, *Zhur. Eksptl. i Teoret. Fiz.*, vol. 11, pp. 101–106, 1941 (Russian).

265. Godfrey, A. I.: Stockyards, D.I.C. Thesis, Imperial College, London, 1953.

266. Gomm, G.: Wahrscheinlichkeitsprobleme im Fernsprechverkehr, *Mitteilungsb. Math. Statist.*, vol. 4, pp. 183–204, 1952 (German).

267. Good, I. J.: The Number of Individuals in a Cascade Process, *Proc. Cambridge Phil. Soc.*, vol. 45, pp. 360–363, 1948.

268. Goode, H. H., C. H. Pollmar, P. W. Warren, and J. B. Wright: Computer Simulation of Traffic, *Highway Research Board, Proc.*, vol. 35, pp. 548–557, 1956.

269. ——— and R. E. Machol: "Systems Engineering," chap. 23, McGraw-Hill Book Company, Inc., New York, 1957.

270. Greenfield, M. N.: Priority Effect in Queueing Theory, Methods of Operations Research, Course 41.616, American University, Washington, D.C., 1958.

271. Greenshields, B. D., and F. M. Weida: "Statistics with Applications to Highway Traffic Analyses," Eno Foundation for Traffic Control, Saugatuck, Conn., 1947 (238 pages).

272. ———, D. Schapiro, and E. L. Ericksen: Traffic Performance at Urban Street Intersections, *Yale Univ., Bur. Highway Traffic Tech. Rept. 1*, 1947.

273. Grinsted, W. H.: A Study of Telephone Traffic Problems with the Aid of the Principle of Probability, *P. O. Elec. Engrs. J.*, vol. 8, p. 33, 1915.

274. ———: The Theory of Probability in Telephone Traffic Problems, *P. O. Elec. Engrs. J.*, vol. 11, pp. 148–152, 1918.

275. Gross, M.: Leistungsvorrechnung bei mehr Maschinenbedienung, *Textil-Praxis*, vol. 4, p. 113, 1949.

276. ———: High-level Railroad Cooperation plus Operations Research Methods Equal More Efficient Railroading, *Ry. Age*, vol. 134, pp. 71–76, 1953.

277. Gumbel, Harold: Waiting Lines with Heterogeneous Servers, *Operations Research*, vol. 8, no. 4, p. 504, 1960.

278. Hahn, K.: Normalivkatolog-tiberlagerung, *Textil-u-Fasertofftech*, vol. 5, pp. 641–643, 1955.

279. Haight, F. A.: Queueing with Balking, *Biometrika*, vol. 44, pts. 3 and 4, December, 1957.

280. ———: Queueing with Reneging, *Metrika*, vol. 2, no. 3, pp. 186–197, 1959.

281. ———: Overflow at a Traffic Light, *Biometrika*, vol. 46, pts 3 and 4, December, 1959.
282. Haller, Raymond, and Brown, Inc.: Queueing Theory Applied to Military Communications Systems, State College, Pa., 1956.
283. Hammersley, J. M.: On Counters with Random Dead Time, *Proc. Cambridge Phil. Soc.*, vol. 49, pp. 623–637, 1953.
284. ———: Storage Problems, *Math. Ann.*, vol. 128, pp. 475–478, 1955.
285. Hardy, G. H.: "Orders of Infinity, The 'Infinitärcalcül' of Paul duBois-Reymond," Cambridge University Press, London, 1954.
286. Harris, T. E.: Branching Processes, *Ann. Math. Statist.*, vol. 19, p. 474, 1948.
287. ———: Some Mathematical Models for Branching Processes, *Proc. Second Berkeley Symposium on Math. Statist. and Probability*, 1951.
288. ———: Stationary Single-server Queueing Processes with a Finite Number of Sources, *Operations Research*, vol. 7, no. 4, pp. 458–467, July, 1959.
289. Harrison, S.: Queueing Theory, Booz Allen Applied Research, Bethesda, Md., paper submitted to *Scientific American*, 1960.
290. Haselton, M. L., and E. L. Schmidt: Automatic Inventory System for Air Travel Reservations, *Elec. Eng.*, pp. 641–646, July, 1954.
291. Hawkins, N. A.: Further Problems in Automatic Trunking—Some Interesting Solutions by Means of the Calculus, *P. O. Elec. Engrs. J.*, vol. 25, pp. 289–294, 1932.
292. Hayward, W. S.: The Reliability of Telephone Traffic Load Measurements by Switch Counts, *Bell System Tech. J.*, vol. 31, pp. 357–377, 1952.
293. Heathcote, C. R.: The Time Dependent Problem for a Queue with Preemptive Priorities, *Operations Research*, vol. 7, no. 5, September–October, 1959.
294. ——— and J. E. Moyal: The Random Walk (in Continuous Time) and Its Application to the Theory of Queues, *Biometrika*, vol. 46, pts 3 and 4, December, 1959.
295. Hénon, R.: Combien de machines peut-on confier à un seul agent? *Rev. statist. appl.*, vol. 3, pp. 73–81, 1955 (French).
296. Herman, R., Elliott W. Montroll, Renfrey B. Potts, and Richard W. Rothery: Traffic Dynamics: Analysis of Stability in Car Following, *Operations Research*, pp. 86–106, January–February, 1959.
297. Herne, H. J. L., and D. G. Nickols: *Proc. Radioactivity Conference*, Oxford, H.M. Stationery Office, London, 1951.
298. Herrey, E. M. J., and H. Herrey: Principles of Physics Applied to Traffic Movement and Road Conditions, *Am. J. Phys.*, vol. 13, pp. 1–14, 1945.
299. Heyveart, A. C., and A. Hurt: Inventory Management of Slow Moving Parts, *Operations Research*, vol. 4, pp. 572–580, 1956.
300. Hildebrand, F. B.: "Introduction to Numerical Analysis," McGraw-Hill Book Company, Inc., New York, 1956.
301. Hille, E.: Functional Analysis and Semi-groups, *Am. Math. Soc. Colloq. Publ.*, vol. 31, 1948.
302. ———: On the Integration of Kolmogoroff's Differential Equations, *Proc. Natl. Acad. Sci. U.S.*, vol. 40, no. 1, pp. 20–25, January, 1954.
303. Hines, J. G.: The Anticipation of Demand and Economic Selection, Provision and Layout of Plant (Telephone Systems), *J. Inst. Elec. Engrs. (London)*, vol. 67, pp. 594–618, 1929.
304. Holley, J. L.: Waiting Line Subject to Priorities, *J. Operations Research Soc. Am.*, vol. 2, pp. 341–343, 1954.
305. Holm, R.: Über die Benutzung der Wahrscheinlichkeitstheorie für Telefonverkehrsprobleme, *Arch. Elektrotech.*, vol. 8, p. 413, 1920.

306. ———: Calculation of Blocking Factors of Automatic Exchanges, *P. O. Elec. Engrs. J.*, vol. 15, pp. 22–38, 1922.

307. ———: The Validity of Erlang's Trunk Congestion Formula, *P. O. Elec. Engrs. J.*, vol. 17, pp. 318–321, 1925.

308. Homma, T.: On a Certain Queueing Process, *Rept. Statist. Application Research, Union Japanese Scientists and Engrs.*, vol. 4, no. 1, 1955.

309. ———: On the Many Server Queueing Process with a Particular Type of Queue Discipline, *Rept. Statist. Application Research, Union Japanese Scientists and Engrs.*, vol. 4, pp. 20–31, 1956.

310. ———: On the Theory of Queues with Some Types of Queue Discipline, *Yokohama Math. J.*, vol. 3, 1956.

311. Hunt, G. C.: Sequential Arrays of Waiting Lines, *Operations Research*, vol. 4, no. 6, pp. 674–683, December, 1956.

312. Hurst, H. E.: Long-term Storage Capacity of Reservoirs, *Trans. Am. Soc. Civil Engrs.*, vol. 116, p. 170, 1951.

313. Inglis, B. G.: Traffic Signal Systems, *J. Inst. Munic. Engrs.*, vol. 80, pp. 369–392, 1954.

314. Irwin ,J. O.: The Frequency Distribution of the Difference between Two Independent Variates Following the Same Poisson Distribution, *J. Roy. Statist. Soc.*, vol. 100, p. 415, 1937.

315. Jackson, J. R.: Networks of Waiting Lines, *Operations Research*, vol. 5, no. 4, August, 1957.

316. ———: A Note on Transient Phenomena in Simple Queues, *Univ. Calif., Management Sci. Research Project Discussion Paper 74*, Nov. 16, 1959.

317. ———: Some Problems in Queueing with Dynamic Priorities, *Univ. Calif., Management Sci. Research Project Paper 62*, Nov. 30, 1959.

318. Jackson, R. R. P.: Queueing Systems with Phase Type Service, *Operations Research Quart.*, vol. 5, pp. 109–120, 1954.

318a. ———: Queueing Processes with Phase-type Service, *J. Roy. Statist. Soc., Ser. B*, vol. 18, no. 1, pp. 129–132, 1956.

319. ——— and D. G. Nickols: Some Equilibrium Results for the Queueing Process $E/M/1$, *J. Roy. Statist. Soc., Ser. B*, vol. 18, pp. 275–279, 1956.

320. ———: Notes on Queueing Processes and Stochastic Stock Models, London School of Economics, *Operations Research Seminar Paper*, June 27, 1957.

321. Jacobs, W. W.: The Caterer Problem, *Naval Research Logistics Quart.*, vol. 1, pp. 154–165, 1954.

322. Jacobaeus, C.: Influence of the Size of Selectors on the Number of Selectors in Telephone Plants, *Ericsson Rev.*, vol. 24, pp. 2–9, 1947.

323. ———: Blocking Computations in Link Systems, *Ericsson Rev.*, vol. 24, p. 86, 1947.

324. ———: A Study on Congestion in Link Systems, part I, Theoretical Analysis and Results; part II, Experimental Methods and Results, *Ericsson Tech.*, vol. 8, pp. 1–68, 1950.

325. ———: Theoretical Considerations in the Design of Crossbar Exchanges, *Teleteknik*, vol. 2, p. 105, 1951.

326. Jaiswal, N. K.: Bulk-service Queueing Problem, *Operations Research*, vol. 8, no. 1, pp. 139–143, January–February, 1960.

327. ——— and H. C. Jain: A Priority Problem in Queueing Theory, *Defense Sci. J. India* (to be published in *Operations Research*, 1961).

328. Jenkins, J. L.: A Survey of Some Applications of Queueing, May 13, 1958 (mimeo. copy).

329. Jensen, A.: An Elucidation of A. K. Erlang's Statistical Works through the

Theory of Stochastic Processes, in E. Brockmeyer et al., "The Life and Works of A. K. Erlang," pp. 23–100, Copenhagen Telephone Company, Copenhagen, 1948.

330. ———: "Moe's Principle," Copenhagen Telephone Company, Copenhagen, 1950.

331. ———: Distribution Patterns Composed of a Limited Number of Exponential Distributions, Second Scandinavian Mathematics Congress, Trondheim, 1949, pp. 210–215, Johan Grundt Tanums Forlag, Oslo, 1952.

332. ———: A Basis for the Calculation of Congestion in Crossbar Systems, *Teleteknik*, vol. 3, p. 123, 1952.

333. ———: Markoff Chains as an Aid in the Study of Markoff Processes, *Skand. Aktvar Tidskr.*, vol. 36, pp. 87–91, 1953.

334. Jensen, E. L.: Elementary Queueing Theory, Elementaer Koteori, *Nord. Mat. Tidskr.*, 1958 (Danish).

335. Joffe, A., and P. E. Ney: A Convergence Problem in the Theory of Multi-server Queues, Cornell University, Department of Mathematics, Ithaca, N.Y., 1960.

336. Johannsen, F. W.: Waiting Times and Number of Calls, *P. O. Elec. Engrs. J.*, 1907; reprinted October, 1910, and January, 1911.

337. ———: Busy, *Ingeniørvidenskab. Skrifter*, 1908; Revised, *ibid.*, Scr. A, no. 32, 1932, as Appendix I of The Development of Telephonic Communication in Copenhagen, 1881–1931. Translated by A. C. Jarvis from original Danish written in 1908, pp. 150–153.

338. Johnson, M. H.: Theoretical Studies into Traffic Congestion, part IV, *British Ministry of Civil Aviation, Operations Research Sect., ORS/MCA Rept. 9*, May, 1949.

339. Jones, W. D.: A Simple Way to Figure Machine Down Time, *Factory Management and Maintenance*, vol. 104, pp. 118–121, 1946.

340. ———: Machine Interference, *J. Ind. Eng.*, vol. 1, pp. 10, 11, 19, 1949.

341. ———: Determining Inefficiencies in Multiple-machine Assignments, *Georgia Institute of Technology, Experiment Station Research in Engineering*, 1953.

342. ———: Graphical Determination of Work Loads for Multiple Machine Assignments, *J. Ind. Eng.*, vol. 4, pp. 16, 19, 26, 1953.

343. Jowett, G. H.: The Exponential Distribution and Its Applications, *Incorp. Statistician*, vol. 8, no. 2, pp. 89–96, January, 1958.

344. K. T. A. S.: The Development of Telephone Communication in Copenhagen, 1881–1931, *Ingeniørvidenskab. Skrifter, Ser. A*, no. 32, 1932.

345. Kac, M.: Random Walk in the Presence of Absorbing Barriers, *Ann. Math. Statist.*, vol. 16, 1945.

346. ———: Random Walk and the Theory of Brownian Motion, *Am. Math. Monthly*, vol. 54, pp. 369–391, 1947.

347. Karlin, S., and J. McGregor: Representation of a Class of Stochastic Processes, *Proc. Natl. Acad. Sci., U.S.*, vol. 41, no. 6, pp. 387–391, June, 1955.

348. ——— and ———: Many Server Queueing Processes with Poisson Input and Exponential Service Times, *Pacific J. Math.*, vol. 8, no. 1, pp. 87–118, 1958.

349. ——— and ———: The Classification of Birth and Death Processes, *Trans. Am. Math. Soc.*, vol. 86, no. 2, pp. 366–400, 1957.

350. ——— and ———: The Differential Equations of Birth and Death Processes and the Stieltjes Moment Problem, *Trans. Am. Math. Soc.*, vol. 85, no. 2, pp. 489–546, July, 1957.

351. ——— and ———: Linear Growth Birth and Death Process, *J. Math. Mech.*, vol. 7, pp. 643–662, 1958.

352. ———, R. C. Miller, Jr., and N. V. Prabhu: Note on a Moving Single Server Problem, *Ann. Math. Statist.*, vol. 30, pp. 243–246, 1959.

353. Karlsson, S. A.: Ein Analysator für den Fernsprechverkehr, *Teleteknik* (English ed.), vol. 1, pp. 113–120, 1957 (German).

354. Karoly, S.: Mozdonyok Varakozasi Idejerol, *Magyar Tudományas Akad.*, vol. 3, 1954.

355. Karush, W.: A Queueing Model for an Inventory Problem, *Operations Research*, vol. 5, no. 5, 1957.

356. Kawata, T.: A Problem in the Theory of Queues, *Repts. Statist. Application Research, Union Japanese Scientists and Engrs.*, vol. 3, pp. 122–129, 1955.

357. Keilson, J., and A. Kooharian: On the General Time Dependent Queue with a Single Server, *Sylvania Electronics Systems*, Applied Research Memo. 209, Waltham, Mass., Apr. 13, 1960.

358. ———: On Time Dependent Queueing Processes, *Ann. Math. Statist.*, vol. 31, no. 1, March, 1960.

359. Kellerer, H.: "Verkehrsstatistik," O. Elsner, Berlin, 1936 (German).

360. Kendall, D. G.: On Some Modes of Population Growth Leading to R. A. Fisher's Logarithmic Series, *Biometrika*, vol. 35, pp. 6–15, 1948.

361. ———: On the Role of Variable Generation Time in the Development of a Stochastic Birth Process, *Biometrika*, vol. 35, pp. 316–330, 1948.

362. ———: On the Generalized "Birth-and-Death" Process, *Ann. Math. Statist.*, vol. 19, p. 1, 1948.

363. ———: Stochastic Processes and Population Growth, *J. Roy. Statist. Soc., Ser. B*, no. 2, 1949.

364. ———: Random Fluctuations in the Age-distribution of a Population Whose Development Is Controlled by the Simple Birth-and-Death Process, *J. Roy. Statist. Soc., Ser. B*, vol. 12, p. 278, 1950.

365. ———: On Non-dissipative Markoff Chains with an Enumerable Infinity of States, *Proc. Cambridge Phil. Soc.*, vol. 47, p. 633, 1951.

366. ———: Some Problems in the Theory of Queues, *J. Roy. Statist. Soc., Ser. B*, vol. 13, no. 2, pp. 151–185, 1951.

367. ———: Stochastic Processes Occurring in the Theory of Queues and Their Analysis by the Method of the Imbedded Markov Chain, *Ann. Math. Statist.*, vol. 24, pp. 338–354, 1953.

368. ——— and G. E. H. Reuter: Some Pathological Markov Processes with a Denumerable Infinity of States and the Associated Semi-groups of Operations on *l*, *Proc. Intern. Congr. Math.*, Amsterdam, 1954.

369. ———: Some Further Pathological Examples in the Theory of Denumerable Markov Processes, *Quart. J. Math.*, Oxford Series, vol. 7, pp. 39–56, 1956.

370. ———: Some Problems in the Theory of Dams, *J. Roy. Statist. Soc., Ser. B*, vol. 19, no. 2, 1957.

371. ——— and G. E. H. Reuter: The Calculation of the Ergodic Properties for Markov Chains and Processes with a Countable Infinity of States, *Acta Math.*, vol. 97, pp. 103–144, 1957.

372. Kendall, M. G.: "The Advanced Theory of Statistics," vols. 1 and 2, Charles Griffin & Co., Ltd., London, 1948.

373. Kesten, H., and J. Th. Runnenburg: Some Elementary Proofs in Renewal Theory with Applications to Waiting Times, *Math. Centrum, Statist. Afdel. Rept.* S 203, Amsterdam, 1956.

374. ——— and ———: Priority in Waiting-line Problems, *Koninkl. Ned. Akad. Wetenschap., Proc.*, ser. A, vol. 60, pt. I, pp. 312–324, and pt. II, pp. 325–336, 1957.

375. Khintchine, A.: Mathematisches über die Erwartung vor einemöffentlichen Schalter, *Mat. Sbornik*, vol. 39, pp. 73–84, 1932 (Russian; German summary).

376. ———: Über die mittlere Däur des Stillstandes von Maschinen, *Mat. Sbornik,* vol. 40, pp. 119–123, 1933 (Russian; German summary).

377. ———: Mathematical Methods of the Theory of Mass Service, *Trudy Mat. Inst. im V. A. Steklova,* vol. 49, p. 122, 1955 (Russian).

378. ———: Über Poissonsche Folgen zufälliger Ereignissen, *Teoriya Veroyatnostei,* vol. 1, pp. 320–327, 1956.

379. ———: Nachwirkungsfreie Folgen von zufälligen Ereignissen, *Teoriya Veroyatnostei,* vol. 1, pp. 3–18, 1956.

380. Kiefer, J., and J. Wolfowitz: On the Theory of Queues with Many Servers, *Trans. Am. Math. Soc.,* vol. 78, pp. 1–18, 1955.

381. ——— and ———: On the Characteristics of the General Queueing Process with Application to Random Walks, *Ann. Math. Statist.,* vol. 27, no. 1, pp. 147–161, 1956.

382. Kinzer, J. P.: Applications of the Theory of Probability to Problems in Highway Traffic, B.C.E. Thesis, Polytechnic Institute of Brooklyn, 1933.

383. Kleinmann, H. A.: Les Phénomènes d'attente, *Univ. Paris, Inst. statist. bull. du séminaire de recherche opérationnelle,* June, 1955.

384. ———: Simultaneous Determination of the Operating Time of 450 Machines, *Rev. statist. appl.,* vol. 3, pp. 11–??, 1955 (French).

385. Knodel, W.: Mehrplatzarbeit, *Unternehmensforschung,* vol. 1, pp. 15–17, 1956 (German).

386. Knowlton, S. L.: A Simplified Approach to Waiting Lines, *J. Ind. Eng.,* vol. 10, no. 6, pp. 423–425, November–December, 1959.

387. Koenigsberg, E.: Birth, Death and Waiting in Line, *Am. Math. Monthly,* vol. 62, p. 543, 1952.

388. ———: Queueing with Special Service, *Operations Research,* vol. 4, pp. 213–220, 1956.

389. ———: Cyclic Queues, *Operational Research Quart.,* vol. 9, no. 1, 1958.

390. ———: Finite Queues and Cyclic Queues, *Operations Research,* vol. 8, no. 2, p. 246, March–April, 1960.

391. Kolmogorov, A. N.: Sur le problème d'attente, *Mat. Sbornik,* vol. 38, pp. 101–106, 1931 (French).

392. ———: Anfangsgründe der Theorie der Markoffschen Ketten mit unendlich vielen möglichen Zustanden, *Mat. Sbornik ,* n.s., vol. 1, p. 607, 1936.

393. Kometani, E.: On the Theoretical Solution of Highway Traffic Capacity under Mixed Traffic, *Mem. Fac. Eng., Kyoto Univ.,* vol. 17, pp. 79–88, 1955.

394. ———: An Abridged Table for Infinite Queues, *Operations Research,* vol. 7, no. 3, pp. 385–393, May–June, 1959.

395. ——— and T. Sasaki: A Safety Index for Traffic with Linear Spacing, *Operations Research,* vol. 7, no. 6, November–December, 1959.

396. Koop, H.: Planung von Kabelkanalen in Groszstadten, *Fernmelde-Praxis,* vol. 32, pp. 769–778, 1955.

397. Koopman, B. O.: New Mathematical Methods in Operations Research, *J. Operations Research Soc. Am.,* vol. 1, pp. 3–9, 1952.

398. Korn, F. A., and J. G. Ferguson: Number 5 Crossbar Dial Telephone Switching System, *Trans. Am. Inst. Elec. Engrs.,* vol. 69, pp. 244–254, 1950.

399. Korte, J. W., and W. Leutzbach: Time-gap Distribution in Case of Interrupted Traffic Flow, *Verkehr Technik,* vol. 9, pp. 17–18, 42–43, 1956 (German).

400. Kosten, L.: Sur les problèmes de blocage dans les multiples graduées, *Ann. P.T.T.,* vol. 26, p. 1002, 1937.

401. ———: Über Sperrungswahrscheinlichkeiten bei Staffelschaltungen, *Elek. Nachr.-Tech.,* vol. 14, pp. 5–12, 1937.

402. ———: On the Frequency Distribution of the Number of Discharges Counted by a Geiger-Müller Counter in a Constant Interval, *Physica*, vol. 10, pp. 749–756, 1943.

403. ———: Over Blokkeerings-en Wachtproblemen, Dissertation, University of Delft, 1942 (Dutch; French and German summaries).

404. ———: On the Influence of Repeated Calls in the Theory of Probabilities of Blocking, *Ingenieur (Utrecht)*, vol. 59, pp. 1–25, 1947 (Dutch).

405. ———: On the Measurement of Congestion Quantities by Means of Fictitious Traffic, *Het P.T.T. Bedrijf*, vol. 2, pp. 15–25, 1948.

406. ———: On the Validity of the Erlang and Engset Loss Formulae, *Het P.T.T. Bedrijf*, vol. 2, pp. 42–45, 1948.

407. ———, J. R. Manning, and F. Garwood: On the Accuracy of Measurements of Probabilities of Loss in Telephone Systems, *J. Roy. Statist. Soc., Ser. B*, vol. 11, pp. 54–67, 1949.

408. ———: Beschrijving van een Machine voor Kunstmatig Verkeer op Electronische Basis, *Ingenieur (Utrecht)*, vol. 63, p. 62, 1951.

409. ———: Waarschijnlijkheidsrekening in de Telecommunicatie-techniek, *Voord. Inst. Ing. Utrecht*, vol. 11, 1951.

410. ———: On the Accuracy of Measurements of Probabilities of Delay in Telecommunication Systems, Part I: Estimates of Probabilities of Delay, *Appl. Sci. Research, Ser. B*, vol. 2, pp. 108–130, 1952.

411. ———: On the Accuracy of Measurements of Probabilities of Delay in Telecommunication Systems, part II: Estimates of Average Times of Delay, *Appl. Sci. Research, Ser. B*, vol. 2, pp. 401–415, 1952.

412. ———: Tafel van de Stagnatiekans van de volledige Bundel met Wachtgelgenheid, *Centr. Lab. P.T.T. (Neth.)*, no. 69, 1954.

413. ———: The Historical Development of the Theory of Probability in Telephone Traffic Engineering in Europe, *Teleteknik* (English ed.), vol. 1, pp. 32–40, 1957.

414. ———: Application of Artificial Traffic Methods to Telephone Problems, *Teleteknik* (English ed.), vol. 1, pp. 107–110, 1957.

415. Kremer, H.: Das Verteilungsgesetz der Belegunslangen in der Wahltechnik, *Arch. Elektrotech.*, vol. 6, pp. 195–198, 1952 (German).

416. ———: The Statistics of the "Fully Available Group" as Applied to Telephone Network Theory, *Arch. Elektrotech.*, vol. 6, pp. 469–472, 1952.

417. Krepelien, H. Y.: The Influence of Telephone Rates on Local Traffic, *Ericsson Tech.*, no. 2, 1958.

418. Kroes, J. L. de: Calculation of the Number of Direct Junction Lines in Telephony, *Commun. News*, vol. 12, p. 132, 1952.

419. Kronig, R.: On Time Losses in Machinery Undergoing Interruptions, part I, *Physics*, vol. 10, pp. 215–224, 1943.

420. ——— and H. Mondria: On Time Losses in Machinery Undergoing Interruptions, part II, *Physics*, vol. 10, pp. 331–336, 1943.

421. Kruithof, J.: Telefoonsverkeersrekening, *Ingenieur (Utrecht)*, vol. 52, pp. 15–25, 1937 (Dutch).

422. ———: Rotary Traffic Machine, *Elec. Commun.*, vol. 23, p. 192, 1946.

423. Kurbatov, J. D., and H. B. Mann: Correction of Geiger-Müller Counter Data, *Phys. Rev.*, vol. 68, pp. 40–43, 1945.

424. Lavallee, R. S.: The Scheduling of Traffic Signals by Linear Programming, *Highway Research Board, Proc.*, vol. 35, pp. 534–542, 1956.

425. Lectures in Queueing Theory given by Univac Engineering Division, Remington Rand, PX 1301, St. Paul, Minn., May, 1959.

426. Ledermann, W., and G. E. Reuter: Spectral Theory for the Differential Equa-

tions of Simple Birth and Death Processes, *Phil. Trans. Roy. Soc. London*, *Ser. A*, vol. 246, pp. 321–369, 1954.

427. Lely, U. P.: Waarschijnlijkheidsrekening bij Automatische Telephonie, The Hague, 1918.

428. ———: On Ekelöf's Calculation of Delays in an Automatic Telephone System, *Ericsson Rev.*, pp. 7–12, 1930.

429. Leroy, R., and A. E. Vaulot: Sur la proportion d'appels perdus dans certains systèmes de téléphonie automatique ne permettant dans un groupe d'organes qu'une seule exploration simultanée, *Compt. rend.*, vol. 220, pp. 84–85, 1945.

430. Le Roy, J: Recherches en deux ordres inverses dans la sélection en téléphonie automatique, *Ann. P.T.T.*, vol. 5, pp. 366–374, 1950 (French).

431. ———: Formules matricielles du calcul du délai d'attente dans le cas des appels desseruis au hasard, *Ann. télécommun.*, vol. 12, 1957.

432. Lesourne, J.: "Technique économique et gestion industrielle," Dunod, Paris, 1958.

433. Levenson, B. D., and R. B. Tasker: High Density Military Traffic Control System, *Air Navigation Lab. Progress Repts.* 387 (1–6); *Final Eng. Rept.* 6, July, 1950–July, 1951.

434. Levert, C., and W. L. Scheen: Probability Fluctuations of Discharges in a Geiger-Müller Counter Produced by Cosmic Radiation, *Physica*, vol. 10, pp. 225–238, 1943.

435. Levinson, M. S.: The Sluggishness or Queueing Factor, *Traffic Eng.*, vol. 27, pp. 401–402, 1956.

436. Levy, P.: Sur la division d'un segment par des pointes choises au hasard, *Compt. rend.*, vol. 208, pp. 147–149, 1939.

437. ———: Convergence des séries aléatoires et loi normale, *Compt. rend.*, vol. 234, pp. 72422–72424, 1952.

438. Lewis, R., F. Neeland, and M. Gourary: An Inventory Control Bibliography, *Naval Research Logistics Quart.*, vol. 3, pp. 295–303, 1956.

439. Lighthill, M. J., and G. B. Witham: On Kinematic Waves—A Theory of Traffic Flow on Long Crowded Roads, *Proc. Roy. Soc. (London)*, *Ser. A*, vol. 229, pp. 281–316; Part II, pp. 317–345, 1955.

440. Lind, G.: Statistical Supervision of Telephone Plant, *Ericsson Tech.*, no. 2, 1958.

441. Lindelöf, E.: "Calcul des résidus," Chelsea Publishing Company, New York, 1947.

442. Lindley, D. V.: Mathematical Theory—Marshalling & Queueing, *Operational Research Quart.*, vol. 3, no. 1, 1952.

443. ———: The Theory of Queues with a Single Server, *Proc. Cambridge Phil. Soc.*, vol. 48, pp. 277–289, 1952.

444. Little, J. D. C.: The Use of Storage Water in a Hydroelectric System, *J. Operations Research Soc. Am.*, vol. 3, pp. 187–197, 1955.

445. Loeve, M.: "Probability Theory," D. Van Nostrand Company, Inc., Princeton, N.J., 1955.

446. Longley, H. A.: The Efficiency of Gradings, *P. O. Elec. Engrs. J.*, vol. 41, pp. 45–67, 1948–1949.

447. Lotka, A. J.: A Contribution to the Theory of Self-renewing Aggregates with Special Reference to Industrial Replacement, *Ann. Math. Statist.*, vol. 10, pp. 1–25, 1939.

448. Lotze, A.: Theorie der Gefahrzeit volkommener Bündel in Fernsprech-wahlsysteme, Dissertation, Technische. Hochschule, Stuttgart, 1953 (German).

449. ———: Gefahrzeit oder Verlust als Masz für die Betriebsgüte im Fernsprechverkehr, *FTZ-Fernmeldetech. Z.*, vol. 6, pp. 564–570, 1953.

450. ———: Berechnung der Verkehrsgröszen im Wartezeitsystem aus den Verkehrs-gröszen eines Verlustsystems, *FTZ-Fernmeldetech. Z.*, vol. 7, pp. 443–453, 1954.

451. ———: Erwiderung zur Stellungnahme von J. W. Cohen, *NTZ-Nachrichtentech. Z.*, vol. 8, pp. 139–140, 1955.

452. ——— and E. Schwiderski: Wartezeitprobleme im Fernsprechverkehr, *NTZ-Nachrichtentech. Z.*, vol. 8, pp. 646–649, 1955.

453. Luchak, G.: The Solution of the Single-channel Queuing Equation Character-ized by a Time-dependent Poisson-distributed Arrival Rate and a General Class of Holding Times, *Operations Research*, vol. 4, pp. 711–732, 1956.

454. ———: The Distribution of the Time Required to Reduce to Some Pre-assigned Level a Single-channel Queue Characterized by a Time-dependent Poisson-distributed Arrival Rate and a General Class of Holding Times, *Operations Research*, vol. 5, pp. 205–209, 1957.

455. ———: The Continuous Time Solution of the Equations of the Single Channel Queue with a General Class of Service-time Distributions by the Method of Generating Function, *J. Roy. Statist. Soc.*, *Ser. B*, vol. 20, pp. 176–181, 1958.

456. Lukaszewicz, J., and H. Steinhaus: On Determining the "Centre of Copper" of a Telephone Network, *Zastos Mat.*, vol. 1, pp. 299–307, 1954 (Polish; English and Russian summaries).

457. Lundkvist, K.: Calculation of the Grade of Service in Automatic Telephone Systems, *Ericsson Tech.*, vol. 4, pp. 75–81, 1936.

458. ———: Method of Computing the Grade of Service in a Selection Stage Com-posed of Primary and Secondary Switches, *Ericsson Rev.*, pp. 11–17, 1948.

459. ———: General Theory for Telephone Traffic, *Ericsson Tech.*, vol. 9, pp. 11–40, 1953.

460. ———: Analysis of General Theory for Telephone Traffic, *Ericsson Tech.*, vol. 11, pp. 3–32, 1955.

461. Lunger, G. F.: Bibliography on Queuing Theory, Sperry Rand Corporation, Univac Division, 1959.

462. Mack, C.: The Efficiency of N Machines Unidirectionally Patrolled by One Operator When Walking Time Is Constant and Repair Times Are Variables, *J. Roy. Statist. Soc.*, *Ser. B*, vol. 19, pp. 173–178, 1957.

463. ———, T. Murphy, and N. L. Webb: The Efficiency of N Machines Unidirec-tionally Patrolled by One Operative When Walking Time and Repair Times Are Constants, *J. Roy. Statist. Soc.*, *Ser. B*, vol. 19, no. 1, pp. 166–172, 1957.

464. Maier-Leibnitz, M.: Die Koinzidenzmethode und ihre Anwendung auf Kern-physicalische Probleme, *Phys. Z.*, vol. 43, pp. 333–362, 1942 (German).

465. Malcolm, D. C.: Queueing Theory in Organization Design, *J. Ind. Eng.*, vol. 6, pp. 19–26, 1955.

466. Mangelsdorf, T. A.: Applications of Waiting-line Theory to Manufacturing Problems, Master's Thesis, Harvard University, 1955.

467. Mann, H. B.: A Note on the Correction of Geiger-Müller Counter Data, *Quart. Appl. Math.*, vol. 4, pp. 307–309, 1946.

468. Marshall, A. W., and E. Reich: Study of the Characteristics of Queues in Tan-dem, *Operations Research*, vol. 4, p. 386, 1956.

469. Marshall, B. O., Jr.: Queueing Theory, in Joseph F. McCloskey and Florence N. Trefethen (eds.), "Operations Research for Management," vol. 1, pp. 134–148, Johns Hopkins Press, Baltimore, 1954.

470. Massé, D.: Les Réserves et la régulation de l'avenir dans la vie économique (2 vols.). Vol. 1: Avenir determiné; vol. II: Avenir aléatoire, Actualités Scien-tifiques et Industrielles, nos. 1007 and 1008, Hermann & Cie, Paris, 1954 (French).

471. Mathewson, J. H., D. L. Trautman, and D. L. Gerlough: Study of Traffic Flow

by Simulation, presented at Thirty-fourth Annual Meeting of the Highway Research Board, Washington, D.C., 1955.

472. Mayne, A. J.: Some Further Results in the Theory of Pedestrians and Road Traffic, *Biometrika*, vol. 41, pp. 375–389, 1954.

473. McCloskey, J., and F. Trefethen (eds.): "Operations Research for Management," vol. 1, Johns Hopkins Press, Baltimore, 1954.

474. ———— and J. M. Coppinger (eds.): "Operations Research for Management," vol. 2, Johns Hopkins Press, Baltimore, 1956.

475. McGuire, C. B., and C. B. Winsten: A Theoretical Study of Traffic Congestion, *J. Operations Research Soc. Am.*, vol. 1, pp. 149–150, 1953.

476. McMillan, B., and J. Riordan: A Moving Single Server Problem, *Ann. Math. Statist.*, vol. 28, 1957.

477. Mehlis, A.: Ist der Einsatz II Vorwahlern in der Vorwählstufe Gerecht Fertigt? *FTZ-Fernmeldetech. Z.*, vol. 4, p. 545, 1951 (German).

478. ————: Wahler oder Schalter als Verbindungsorgane in der Fernsprechvermittlungstechnik, *FTZ-Fernmeldetech. Z.*, vol. 5, pp. 293–296, 1952 (German).

479. Meinesz, M.: The Problem of the Gambler's Ruin, *Statist. Need.*, vol. 10, no. 2, pp. 87–97, 1956.

480. Meisling, T.: Discrete-time Queuing Theory, *Operations Research*, vol. 6, no. 1, pp. 96–105, 1958.

481. ————: All-epoch Queuing Theory, Stanford Research Institute, English Division, July 3, 1958 (mimeo. notes).

482. Mellor, S. D.: Delayed Call Formulae When Calls Are Served in Random Order, *P. O. Elec. Engrs. J.*, vol. 35, pt. 2, pp. 53–56, 1942.

483. ————: A Method of Assessing the Effect of Flexibility in Distribution Networks, *P. O. Elec. Engrs. J.*, vol. 45, pp. 125–128, 1952.

484. Mercer, A.: A Queueing Problem in Which the Arrival Times of the Customers Are Scheduled, *J. Roy. Statist. Soc., Ser. B*, vol. 22, no. 1, pp. 108–113, 1960.

485. Merker, M.: Some Notes on the Use of the Probability Theory to Determine the Number of Switches in an Automatic Telephone Exchange, *P. O. Elec. Engrs. J.*, vol. 16–17, pp. 27–50, 347–362, 1924.

486. Miller, R. G., Jr.: Priority Queues, *Ann. Math. Statist.*, vol. 31, no. 1, March, 1900.

487. ————: A Contribution to the Theory of Bulk Queues, *J. Roy. Statist. Soc.*, vol. 21, no. 2, pp. 320–337, 1959.

488. Ministry of Transportation and Civil Aviation: Theoretical Researches into Traffic Congestion, *Operations Research Memo* 18.

489. Mitchell, H.: Machine Interference, Cotton Board Work Study Course, Tutorial Notes.

490. M.I.T. Summer Short Course in Operations Research, Technology Press, M.I.T., Cambridge, Mass., 1953.

491. M.I.T. Interim Report 2, Fundamental Investigations in Methods of Operations Research, Apr. 1, 1954–Nov. 30, 1954.

492. Modee, G.: Undersökning av Kombinations Gruppernas Utnyttjande vid det Amerikanska Koordinatväljarsystemet, *Tek. Medd. Fran Kungl. Tele.*, pp. 7–9, 1943 (Swedish).

493. Molina, E. C.: The Theory of Probabilities Applied to Telephone Trunking Problems, *Bell System Tech. J.*, vol. 1, p. 69, 1922.

494. ————: Application of the Theory of Probability to Telephone Trunking Problems, *Bell System Tech. J.*, vol. 6, pp. 461–494, 1927.

495. Molnar, I.: Delay Probability Charts for Telephone Traffic Where the Holding Times Are Constant, Auto. Elec. Co., *Engrs'. Notes* 2032, 1952.

496. ———: A Study of Traffic in Automatic Toll Telephone Systems, Ph.D. Thesis, Northwestern University, 1950.

497. ———: Toll Answering Delays and Their Measurement, *Auto. Elec. Tech. J.*, vol. 2, pp. 11–21, 1951.

498. ———: On the Accuracy of Holding-time Measurements, *Auto. Elec. Tech. J.*, vol. 3, pp. 63–72, 1952.

499. ———: The Appraisal of Delays in Gate-type Operation, *Trans. Am. Inst. Elec. Engrs.*, vol. 74, pp. 475–485, 1955.

500. Moran, P. A. P.: A Probability Theory of Dams and Storage Systems, *Australian J. Appl. Sci.*, vol. 5, pp. 116–124, 1954.

501. ———: A Probability Theory of Dams and Storage Systems—Modification of the Release Rule, *Australian J. Appl. Sci.*, vol. 6, pp. 117–130, 1955.

502. ———: A Probability Theory of a Dam with a Continuous Release, *Quart. J. Math.*, Oxford Series, vol. 2, no. 7, pp. 130–137, 1956.

503. ———: "The Theory of Storage," Methuen's Monographs, Methuen & Co., Ltd., London, 1959.

504. Morlat, G.: Sur une généralisation de la loi de Poisson, *Compt. rend.*, vol. 234, p. 933, 1952.

505. Morse, P. M., and M. L. Ernst: "Waiting Lines—and Examples and Applications—from Notes from M.I.T. Summer Course on Operations Research, June 16 to July 3, 1953," pp. 93–101, 102–108, Technology Press, M.I.T., Cambridge, Mass., 1953.

506. ———, H. N. Garber, and M. L. Ernst: A Family of Queuing Problems, *J. Operations Research Soc. Am.*, vol. 2, pp. 444–445, 1954.

507. ———: Stochastic Properties of Waiting Lines, *J. Operations Research Soc. Am.*, vol. 3, pp. 255–261, 1955.

508. ———: "Queues, Inventories and Maintenance," John Wiley & Sons, Inc., New York, 1958.

509. Mortimer, W. J.: Moving Vehicle Method of Estimating Traffic Volumes and Speeds, *Traffic Eng.*, vol. 28, pp. 539–544, 1956.

510. Moskowitz, K.: Waiting for a Gap in a Traffic Stream, *Highway Research Board, Proc.*, vol. 33, pp. 385–394, 1954.

511. Muller, H. L.: Wachtjdenproblem bij Bedienind van een Groep Machines, Doctoral Thesis, University of Delft, 1951 (Dutch).

512. Murdoch, J.: Congestion Problems in Industry, *Appl. Statist.*, vol. 3, p. 200, 1954.

513. Naor, P.: On Machine Interference, *J. Roy. Statist. Soc.*, Ser. B, vol. 18, pp. 280–287, 1956.

514. ———: Normal Approximation to Machine Interference with Repair Men, *J. Roy. Statist. Soc.*, Ser. B, vol. 19, no. 2, 1957.

515. ———: Some Problems of Machine Interference, *Proc. Intern. Conf. Operations Research*, Sept. 2–6, 1957.

516. Nelson, R. T.: Waiting Time Distributions for Application to a Series of Service Centers, *Operations Research*, vol. 6, pp. 856–862, 1958.

517. ———: An Extension of Queueing Theory Results to a Series of Service Centers, *Univ. Calif. Management Sci. Research Project, Notes and Discussion Paper 68*, Apr. 29, 1958.

518. Neovius, G.: Artificial Traffic Trials Using Digital Computers, *Ericsson Tech.*, no. 2, 1955.

519. Neveu, J.: Théorie des semi groupes de Markov, *Univ. Calif., Publs. in Statist.*, vol. 2, no. 14, pp. 319–394, 1958.

520. Newell, A.: The Capacity of a Railroad Freight Yard (A Survey of the Problem—

Not a Solution), *George Washington Univ. Logistics Research Project, App. I to Quart. Progr. Rept., Logistics Papers, Issue* 3, May 16–Aug. 15, 1950.

521. Newell, G. F.: Mathematical Models for Freely-flowing Highway Traffic, *J. Operations Research Soc. Am.*, vol. 3, pp. 176–186, 1955.

522. ———: Queues for a Fixed-cycle Traffic Light, *Ann. Math. Statist.*, vol. 31, no. 3, p. 589, 1960.

523. ———: Statistical Analysis of the Flow of Highway Traffic through a Signalized Intersection, *Appl. Math.*, vol. 13, no. 4, January, 1956.

524. Newland, W. F.: A Method of Approach and Solution to Some Fundamental Traffic Problems, *P. O. Elec. Engrs. J.*, vol. 25, pp. 119–131, 1932.

525. Neyman, J.: "Proceedings of the Berkeley Symposium on Mathematical Statistics and Probability," University of California Press, Berkeley, Calif., 1949. Also "Proceedings of the Second Symposium, 1951," and "Proceedings of the Third Symposium, 1955–1956."

526. Nicoliechia, P. E.: Sui metodi contabili del trafico telefonico, *Poste e telecomun.*, vol. 8, pp. 435–451, 1955.

527. O'Brien, G. G.: The Solution of Some Queuing Problems, *J. Soc. Ind. Appl. Math.*, vol. 2, pp. 133–142, 1954.

528. O'Connor, T. F.: "Productivity and Probability— A Treatise on Time Study and the Improvement of Industrial Efficiency," chap. VI, Mech. World Monographs 65, Emmott & Co., Ltd., London, 1952.

529. O'Dell, G. F.: The Influence of Traffic in Automatic Exchange Design, *P. O. Elec. Engrs. Inst.*, no. 85, 1920.

530. ———: Theoretical Principles of the Traffic Capacity of Automatic Switches, *P. O. Elec. Engrs. J.*, vol. 13, pp. 209–223, 1920.

531. ——— and W. W. Gibson: Automatic Trunking in Theory and Practice, *P. O. Elec. Engrs. Inst., Profess. Paper* 107, 1926 (41 pp.).

532. ———: An Outline of the Trunking Aspect of Automatic Telephony, *J. Inst. Elec. Engrs. (London)*, vol. 65, p. 185, 1927.

533. Olcott, E. S.: The Influence of Vehicular Speed and Spacing of Tunnel Capacity, *J. Operations Research Soc. Am.*, vol. 3, pp. 147–167, 1955 (also "Operations Research for Management," vol. 2, chap. 3, Johns Hopkins Press, Baltimore, 1956).

534. Olshevsky, D. E.: A Study of Flight Time Estimate Errors, *Cornell Aeronaut. Lab. Rept.* JA-693-1, May, 1951.

535. Oostrum, H. R. Van: Verkeersmetingen, Verkeersprognose en Contact-bankindeligen, *Telegram en Telefoon*, vol. 57, pp. 3–23, 1956.

536. Orden, A.: The Transshipment Problem, *Management Sci.*, vol. 2, pp. 276–285, 1956.

537. Ossoskow, G. A.: Ein Grenztheorem für Folgen Gleichartiger Ereignissen, *Teoriya Veroyatnostei*, vol. 1, pp. 274–282, 1956.

538. Palasti, I., A. Renyi, T. Szentmartony, and L. Takács: Ergänzung des Lagervorrates, I, *Magyar Tudományos Akad. Alkalm. Mat. Int. Közleményei*, vol. 2, pp. 187–201, 1953 (Hungarian; German and Russian summaries). (Also see Ref. 777.)

539. Palm, C.: Calcul exact de la perte dans les groups de circuits échelonnes, *Ericsson Tech.*, vol. 4, 1936.

540. ———: Some Investigations into Waiting Times in Telephone Plants, *Tek. Medd. Från Kungl. Teleg.*, no. 7–9, 1937 (Swedish).

541. ———: Inhomogeneous Telephone Traffic in Full Availability Groups, *Ericsson Tech.*, vol. 5, pp. 3–36, 1937.

542. ———: Étude des délais d'attente, *Ericsson Tech.*, vol. 5, p. 39, 1937.

543. ———: Analysis of the Erlang Traffic Formulae for Busy-signal Arrangements, *Ericsson Tech.*, vol. 6, pp. 39–58, 1938.

544. ———: Mätnoggrannhet vid Bestämning av Trafikmängd Enligt Genomsökningsförfarandet, *Tek. Medd. Från. Kungl. Teleg.*, nos. 7–9, 1941 (Swedish).

545. ———: A Form Factor for Judging Delay Time Distribution, *Tek. Medd. Från. Kungl. Teleg.*, nos. 1–3, 1943 (Swedish).

546. ———: Intensity Fluctuations in Telephone Traffic, *Ericsson Tech.*, vol. 1, no. 44, pp. 1–18S, 1943.

547. ———: Några Anmarkningar över de Erland'ska Formlerna for Upptaget-System, Specialnummer for Teletrafikteknik, *Tek. Medd. Från. Kungl. Teleg.*, pp. 1–110, 1946 (Swedish).

548. ———: The Distribution of Repairmen in Servicing Automatic Machines, *Industritidn. Norden*, vol. 75, pp.75 –80, 90–94, 119–123, 1947 (Swedish; an account is given in W. Feller, Ref. 202).

549. ———: "Tables of the Erlang Loss Formula," Telefon Aktiebolaget, L. M. Ericsson, Stockholm, 1947 (2d ed., 1954).

550. ———: Waiting Times When Traffic Has Variable Mean Intensity, *Ericsson Rev.*, vol. 24, pp. 102–107, 1947.

551. ———: Research on Telephone Traffic Carried by Full Availability Groups, *Tele.* (English ed.), no. 1, 1957.

552. ———: Methods of Judging Annoyance Caused by Congestion, *Tele.* (English ed.), vol. 2, pp. 1–20, 1953.

553. Pearcey, T.: Delays in Landing of Air Traffic, *J. Roy. Aeronaut. Soc.*, vol. 52, pp. 799–812, 1948.

554. Peck, L. G., and R. N. Hazelwood: "Finite Queueing Tables," John Wiley & Sons, Inc., New York, 1958.

555. Phillips, Cecil R., Jr.: A Variable-channel Queueing Model with a Limited Number of Channels, Thesis, Georgia Institute of Technology, 1960.

556. Phipps, T. E., Jr.: Machine Repair as a Priority Waiting-line Problem, *Operations Research*, vol. 4, pp. 76–85, 1956. (Comments by W. R. Van Voorhis, *ibid.*, p. 86.)

557. Piesch, J.: Die Beanspruchung der Bundel im modernen Fernsprechverkehr, *Arch. Elektrotech.*, vol. 8, pp. 324–328, 353–362, 411–429, 1954 (German).

558. Pilé, G.: Étude des délais d'attente des aéronefs à l'atterissage, *Rev. statist. appl.*, vol. 3, pp. 73–84, 1955 (French).

559. Pipes, L. A.: A Proposed Dynamic Analogy of Traffic, University of California, Special Study Institute of Transport and Traffic Engineering, vol. 11, July, 1950.

560. ———: A Mathematical Analysis of Traffic Dynamics, *J. Operations Research Soc. Am.*, vol. 1, p. 151, 1953.

561. ———: An Operational Analysis of Traffic Dynamics, *J. Appl. Phys.*, vol. 24, pp. 274–281, 1953.

562. Pitt, H. R.: A Theorem on Random Functions with an Application to a Theory of Provisioning, *J. London Math. Soc.*, vol. 21, pp. 16–22, 1946.

563. Pollaczek, F.: Über eine Aufgabe der Wahrscheinlichkeitstheorie, part I, *Math. Z.*, vol. 32, pp. 64–100, 1930; part II, *ibid.*, vol. 32, pp. 729–750, 1930.

564. ———: Theorie des Warten vor Schaltern, *Telegraphen u. Fernsprechtechnik*, vol. 19, pp. 71–78, 1930.

565. ———: Über zwei Formeln aus der Theorie des Wartens von Schaltergruppen, Gesprächverluste und Wartezeiten, *Elek. Nachr.-Tech.*, vol. 8, pp. 256–268, 268–279, 1931.

566. ———: Zur Theorie des Wartens von Schaltergruppen, *Elek. Nachr.-Tech.*, vol. 9, pp. 434–454, 1932.

567. ———: Lösung eines geometrischen Wahrscheinlichkeitsproblemes, *Math. Z.*, vol. 35, pp. 230–278, 1932.

568. ———: Kurven für die Gesprächverlust in Volkommenen Leitung-Shiendel, *Elek. Nachr.-Tech.*, vol. 10, 1933.

569. ———: Über das Warteproblem, *Math. Z.*, vol. 38, pp. 492–537, 1934.

570. ———: *Elek. Nachr.-Tech.*, vol. 11, pp. 396–399, 1934.

571. ———: *Ann. P.T.T.*, vol. 24, pp. 997–1001, 1935.

572. ———: Sur quelques lois asymptotiques de la théorie de l'encombrement des réseaux téléphoniques, *Ann. univ. Lyon, Sec. A*, vol. 5, pp. 21–35, 1942.

573. ———: Résolution de certaines équations intégrales de deuxième espace, *J. math. pures appl.*, vol. 24, pp. 73–93, 1945.

574. ———: La loi d'attente des appels téléphoniques, *Compt. rend.*, vol. 222, pp. 353–355, 1946.

575. ———: Sur l'application de la théorie des fonctions au calcul de certaines probabilités continues utilisées dans la théorie des réseaux téléphoniques, *Ann. inst. Henri Poincaré*, vol. 10, pp. 1–54, 1946.

576. ———: Sur un problème de calcul des probabilités qui se rapporte à la téléphonie, *J. math. pures appl.*, vol. 25, pp. 307–334, 1946.

577 ———: Sur la probabilité de perte d'un appel téléphonique dans le cas d'un seul groupe de lignes avec blocage temporaire, *Compt. rend.*, vol. 226, pp. 2045–2047, 1948.

578. ———: Application d'opérateurs intégrocombinatoires dans la théorie des intégrales multiples de Dirichlet, *Ann. inst. Henri Poincaré*, vol. 11, pp. 113–133, 1949.

579. ———: Réductions de différents problèmes concernant la probabilité d'attente au téléphone, à la resolution de systèmes d'équations intégrales, *Ann. inst. Henri Poincaré*, vol. 11, pp. 135–173, 1949.

580. ———: Application du calcul des probabilités au phénomène de blocage temporaire des lignes téléphoniques, *Ann. télécommun.*, vol. 6, pp. 49–53, 1951.

581. ———: Problèmes de calcul des probabilités relatifs à des systèmes téléphoniques sans possibilité d'attente, *Ann. inst. Henri Poincaré*, vol. 12, pp. 57–96, 1951.

582. ———: Répartition des délais d'attente des avions arrivant à un aéroport qui possédes pistes d'atterrissage, *Compt. rend.*, vol. 232, pp. 1901–1903, 2286–2288, 1951.

583. ———: Délais d'attente des avions atterrissant selon leur ordre d'arrivée sur un aéroport à sistes, *Compt. rend.*, vol. 234, pp. 1246–1248, 1952.

584. ———: Sur la répartition des périodes d'occupation ininterrompue d'un guichet, *Compt. rend.*, vol. 234, pp. 2042–2044, 1952.

585. ———: Fonctions caractéristiques de certaines répartitions définies au moyen de la notion d'ordre application à la théorie des attentes, *Compt. rend.*, vol. 234, pp. 2334–2336, 1952.

586. ———: Sur une généralisation de la théorie des attentes, *Compt. rend.*, vol. 236, pp. 578–580, 1953.

587. ———: Généralisation de la théorie probabiliste des systèmes téléphoniques sans dispositif d'attente, *Compt. rend.*, vol. 236, pp. 1469–1470, 1953.

588. ———: Sur la théorie stochastique des compteurs électroniques, *Compt. rend.*, vol. 238, pp. 766–768, 1954.

589. ———: Développement de la théorie stochastique des lignes téléphoniques pour un état initial quelconque, *Compt. rend.*, vol. 239, pp. 1764–1766, 1954.

590. ———: "Problèmes stochastiques posés par le phénomène de formation d'une queue d'attente à un guichet et par des phénomènes apparentes," Memorial des Sciences Mathématiques, Gauthier-Villars, Paris, 1957.

591. ———: Application de la théorie des probabilités posées par l'encombrement des réseaux téléphoniques, *Ann. télécommun.*, vol. 14, nos. 7–8, pp. 165–183, 1959.

592. Pomey, F. L.: Téléphone et statistique, *Rev. gén. élec.*, vol. 9, pp. 133–138, 1921.

593. Port of New York Authority, Operations Standards Division: A Collection of Delay Probability Formulas and Curves, 1956.

594. Prabhu, N. U.: Some Results for the Queue with Poisson Arrivals, *J. Roy. Statist. Soc., Ser. B*, vol. 22, no. 1, pp. 104–107, 1960.

595. ———: Application of Storage Theory to Queues with Poisson Arrivals, *Ann. Math. Statist.*, vol. 31, no. 2, p. 475, 1960.

596. Prager, W.: Problems of Traffic and Transportation, *Proc. Symposium on Operations Research in Business and Industry*, pp. 105–113, Midwest Research Institute, Kansas City, Mo., 1954.

597. Prince, R. K.: Queueing Up, *Proc. Eng.*, pp. 188–192, November, 1955.

598. Proceedings of the First International Congress on the Application of the Theory of Probability in Telephone Engineering and Administration, Copenhagen, June 20–23, 1955 [issued as vol. 1, no. 1, of *Teleteknik* (English ed.), 1957].

599. Pyke, R.: On Renewal Processes Related to Type I and Type II Counter Models, *Ann. Math. Statist.*, vol. 29, pp. 737–754, 1958.

600. ———: Markov Renewal Processes: Definitions and Preliminary Properties [this research at Columbia University was supported by the Office of Naval Research under Contract Nonr 266(59) Proj. No. 042-205], October, 1959.

601. Rabe, F. W.: Considerations on the Correctness of the Erlang Formula for Automatic Telephone Traffic, *Ingenieur (Utrecht)*, vol. 58, p. 12, 1946.

602. ———: Variations of Telephone Traffic, *Elec. Commun.*, vol. 26, pp. 243–248, 1949.

603. Raff, M. S.: Space-time Relationships at "Stop" Intersections, *Proc. Inst. Traffic Eng.*, vol. 20, pp. 42–49, 1949.

604. ———: "A Volume Warrant for Urban Stop Signs," chap. VI, *Eno Foundation for Highway Traffic Control*, Saugatuck, Conn., 1951.

605. ———: The Distribution of Blocks in an Uncongested Stream of Automobile Traffic, *J. Am. Statist. Assoc.*, vol. 46, no. 253, March, 1951.

606. ———: Answer to Question 35—Waiting Time for Public Transportation, *Am. Statist.*, vol. 7, no. 2, pp. 26–27, April–May, 1953: also *ibid.*, vol. 7, no. 3, p. 20, 1953.

607. Ramakrishnan, A.: Some Simple Stochastic Processes, *J. Roy. Statist. Soc.*, vol. 13(B), p. 131, 1951.

608. ——— and P. M. Mathews: A Stochastic Problem Relating to Counters, *Phil. Mag.*, vol. 44, no. 7, pp. 1122–1128, 1953.

609. Ramm, B. T.: The Determination of Branch Staff Establishments in a Multiple Shop Organization, *Appl. Statist.*, vol. 4, pp. 295–298, 1955.

610. Rapp, Y.: The Economic Optimum in Urban Telephone Network Problems, *Ericsson Tech.*, no. 49, pp. 1–132, 1950.

611. Rappleye, S. C.: A Study of the Delays Encountered by Toll Operators in Obtaining an Idle Trunk, *Bell System Tech. J.*, vol. 25, pp. 53–62, 1946.

612. Rawdin, E.: A Measure of Effectiveness for Queueing Problems, *Operations Research*, vol. 8, no. 2, p. 278, March–April, 1960.

613. Ray, D.: Stable Processes with an Absorbing Barrier, *Trans. Am. Math. Soc.*, vol. 89, pp. 16–24, 1958.

614. Reed, I. S.: Some Queue Problems Associated with Buffer-storage Systems, *Mass. Inst. Technol., Lincoln Lab. Tech. Rept.* 73, Dec: 2, 1954.

615. Reich, E.: Birth-Death Process and Tandem Queues, The Rand Corporation, *Paper* P863, 1956.

616. ———: Waiting Times When Queues Are in Tandem, *Ann. Math. Statist.*, vol. 28, no. 3, p. 768, September, 1957.

617. ———: On an Integrodifferential Equation of Takács, part I, *Ann. Math. Statist.*, vol. 29, pp. 563–570, 1958.

618. Renier P.: Criteria for Maximum Economy in Planning Local Telephone Networks, *Electro-Tech.*, vol. 40, pp. 160–163, 1953 (Italian).

619. Renyi, A., and T. Szentmartony: Wahrscheinlichkeitstheoretisch Bestimmung von Reserven von Maschinenteilen unter Ausrüstongsgegenstanden, *Mat. Lapok.*, vol. 3, pp. 129–139, 1952.

620. ——— and L. Takács: Sur les processus d'événements dérivés par un processus de Poisson et sur leurs applications techniques et physiques, *Magyar Tudományos Akad. Alkalm. Mat. Inst. Közleményei*, vol. 1, pp. 139–146, 1952 (Hungarian; French and Russian summaries).

621. Renz, A.: Graphisches Verfahren für die Addition ungleichen Verkehrswerte, *Fernmelde-Praxis*, vol. 33, pp. 243–254, 1956 (German).

622. ———: Teilung und Zusammenfassung ungleichgrosser Verkehrswerte, *Fernmelde-Praxis*, vol. 34, pp. 170–173, 1957.

623. Reports presented at the Second International Course in Traffic Engineering, Switzerland, Sept. 20–25, 1954. (W. T. & A. O. and P. I. A. R. C.) London, W. T. A. O. 1954 (Methods for Measuring Distribution of Traffic Volume).

624. Rettenberger, J.: Die Bestimmung der optimale Leitungszahlen bei Fernsprechnetzen mit Überlaufverkehr, *NTZ-Nachr.-tech. Z.*, vol. 10, pp. 53–59, 1957.

625. Reuschel, A.: The Movement of a Column of Vehicles When the Leading Vehicle Is Uniformly Accelerated or Decelerated, *Z. österr. Ing. Archit. Ver.*, vol. 95, pp. 59–62, 73–77, 1950.

626. Reuter, G. E. H.: Denumerable Markov Processes and the Associated Contraction Semigroups on 1, *Acta Math.*, vol. 97–98, pp. 1–46, 1957.

627. ——— and W. Ledermann: On Differential Equations for the Transition Probabilities of Markov Processes with Enumerably Many States, *Proc. Cambridge Phil. Soc.*, vol. 49, 1953.

628. Richards, P. I.: Shock Waves on the Highway, *Operations Research*, vol. 4, pp. 42–51, 1956.

629. Riley, V.: Bibliography of Queueing Theory, Appendix A, in "Operations Research for Management," vol. 5, pp. 541–556, Johns Hopkins Press, Baltimore, 1956.

630. Riordan, J.: Telephone Traffic Time Averages, *Bell System Tech. J.*, vol. 30, p. 1129, 1951.

631. ———: Delay Curves for Calls Served at Random, *Bell System Tech. J.*, vol. 32, pp. 100–119, 1266, 1953.

632. Rios, S.: "Introduction a los metodos de la estadistica," Instituto Estadistica, Madrid, 1957.

633. Robinson, F. D., and W. E. Duckworth: An Application of Queueing Theory to the Speed of Estimating, *Proc. Intern. Conf. Operations Research*, Sept. 2–6, 1957.

634. Romani, J.: La Teoria de las colas aplicada a un problema de produccion industrial, *Trabajos de estadistica*, vol. 6, cuaderno III, 1955.

635. ———: Distribution of the Algebraic Sum of Poisson Variables, *Trabajos de estadistica*, vol. 7, no. 2, pp. 175–181, 1956.

636. ———: Un Modelo de la teoria de colas con numero variable de canales, *Trabajos de estadistica*, vol. 8, 1957.

637. ———: A Queueing Model with a Variable Number of Channels, *Trabajos de estadistica*, vol. 8, no. 3, pp. 175–189, 1957.

638. Roy, J. le: Recherche en deux ordres inverses dans la sélection en téléphonie automatique, *Ann. P.T.T.*, vol. 5, p. 366, 1950.

639. Runnenburg, J. Th.: Machines Served by a Patrolling Operator, *Math. Centrum, Statist. Afdel. Rept.* S221 (VP13), Amsterdam, 1957 (13 pp.).

640. Ryll-Nardzewski, C.: On the Non-homogeneous Poisson Process, *Studia Math.*, vol. 14, pp. 124–128, 1953.

641. Saaty, T. L.: Résumé of Useful Formulas in Queuing Theory, *Operations Research*, vol. 5, no. 2, pp. 161–200, April, 1957.

642. ————: Five Papers by Conny Palm, *Operations Research*, vol. 6, no. 3, pp. 456–460, May–June, 1958.

643. ————: "Mathematical Methods of Operations Research," McGraw-Hill Book Company, Inc., New York, 1959.

644. ————: Some Stochastic Processes with Absorbing Barriers, *J. Roy. Statist. Soc.*, October, 1961.

645. ————: Time Dependent Solution of the Many Server Poisson Queue, *Operations Research*, November–December, 1960.

645a. ————: Approximate Solution to Semiconductor Noise as a Queueing Problem, *Proc. IRE*, June, 1961.

645b. ————: On Jockeying, Collusion, Scheduling, Optimization and Graph Theoretic Queues, Office of Naval Research, Washington, 1961.

646. Sacks, J.: Ergodicity of Queues in Series, *Ann. Math. Statist.*, vol. 31, no. 3, p. 579, 1960.

647. Salveson, M. E.: A Problem in Optimal Machine Loading, *Management Sci.*, vol. 2, p. 232, 1956; also Note, *ibid.*, vol. 3, p. 114, 1957.

648. Sarkadi, K.: On the Waiting Time of Locomotives, *Magyar Tudományos Akad. Alkalm. Mat. Int. Közleményei*, vol. 3, pp. 191–194, 1954 (Hungarian; English and Russian summaries).

649. Schiller, D. H., and M. M. Lavin: The Determination of Requirements for Warehouse Dock Facilities, *Operations Research*, vol. 4, pp. 231–243, April, 1956.

650. Schneider, E., and B. Jessen: Absatz, Produktion und Lagerhaltung bei einfacher Produktion, *Arch. Math. Wert. Sozialforsch.*, vol. 4, no. 1, 1938 (German).

651. Schofield, H.: Traffic Congestion. An Assessment of the Problem and Measures for Relief, *J. Inst. Munic. Eng.*, vol. 83, no. 2, pp. 49–62, 1956; Discussion, *ibid.*, pp. 62–63, 81–82.

652. Schouten, J. P., and J. Giltay: Oplessing van een Problem uit de Waarschijnlijkheidsrekening van Belang vaar de Automatische Telephonie, *Ingenieur (Utrecht)*, vol. 55, pp. 67–75, 1940 (Dutch).

653. Schuhl, A.: Probability Calculations and Vehicle Movement on a Two Lane Road, *Ann. ponts et chaussées*, vol. 125, pp. 631–663, 1955 (French; English summary).

654. Segerdahl, C. O.: On Homogeneous Random Processes and Collective Risk Theory, Uppsala, 1939.

655. Sespaniak, L. J.: An Application of Queueing Theory for Determining Manpower Requirements for an In-line Assembly Inspection Department, *J. Ind. Eng.*, vol. 4, pp. 265–267, July–August, 1959.

656. Simond: La Distribution hyperexponentielle, *Rev. franç. recherche opérationnelle*, vol. 2, no. 9, pp. 196–215, 1958.

657. Sittig, J.: De statistische Bases van de Planning in een Machine-Fabriek, *Statist. Need.*, vol. 9, pp. 47–69, 1955.

658. Skellam, J. G.: The Frequency Distribution of the Difference between Two Poisson Variates Belonging to Different Populations, *J. Roy. Statist. Soc.*, vol. 109, p. 296, 1946.

659. Smeed, R. J., and G. Bennett: Research on Road Safety and Traffic Flow, *Inst. Civil Engrs., Road Paper* 29, 1949.

660. Smith, H. A. B.: Some Aids to Traffic Flow, *Proc. Inst. Civil Engrs.*, II, vol. 2, pp. 416–429, 1953.

661. Smith, W. L.: On the Distribution of Queueing Times, *Proc. Cambridge Phil. Soc.*, vol. 49, pp. 449–461, 1953.

662. ———: Stochastic Sequences of Events, Ph.D. Thesis, Cambridge University, 1953.

663. ———: Asymptotic Renewal Theorems, *Proc. Roy. Soc. Edinburgh, Sec. A*, vol. 64, p. 9, 1954.

664. ———: Regenerative Stochastic Processes, *Proc. Roy. Soc. (London), Ser. A*, vol. 232, p. 6, 1955.

665. ———: Renewal Theory, Counter Problems, and Quasi-Poisson Processes, *Proc. Cambridge Phil. Soc.*, vol. 53, pp. 175–193, 1957.

666. ———: Renewal Theory and Its Ramifications, *J. Roy. Statist. Soc., Ser. B*, vol. 20, no. 2, pp. 243–302, 1958.

667. ———: On the Cumulants of Renewal Processes, *Biometrika*, vol. 46, pts. 1–2, June, 1959.

668. ———: Infinitesimal Renewal Processes, *Univ. North Carolina, Inst. of Statist., Mimeo. Ser.* 237, August, 1959.

669. Sneddon, I. N.: "Elements of Partial Differential Equations," McGraw-Hill Book Company, Inc., New York, 1957.

670. Sochor, Z.: Choice of a Suitable Size of Subscribers' Distribution Boxes in Local Telephone Networks, *Slaboproudý obzor*, vol. 14, pp. 316–321, 1953 (Czech.).

671. Stadler, F.: On the Characteristics of Telecommunication Traffic, *Slaboproudý obzor*, vol. 14, 1953 (Czech.).

672. Steffensen, J. F.: Om Sandsynligheden for at Afkommet Uddor, *Mat. Tidsskr., Ser. B*, vol. 19, 1930.

673. Steffensen, N. J.: Deux problèmes du calcul des probabilités, *Ann. inst. Henri Poincaré*, vol. 3, p. 319, 1933.

674. Stephan, F. F.: Two Queues with a Priority Rule and a Single Queue with Interrupted Service, *Operations Research*, vol. 4, p. 386, 1956.

675. Storet, H. P.: Synchronization and Its Effect on Machine Efficiency, *J. Textile Inst.*, vol. 41, pp. T225–T235, 1950.

676. Störmer, H., and F. L. Bauer: Berechnung von Wartezeiten in Vermittlungseinrichtungen mit kleinen Zubringerbündeln, *Arch. Elektrotech.*, vol. 9, p. 69, 1955.

677. ———: Wartezeitlenkung in Handbedienten Vermittlungsanlagen, *Arch. Elektrotech.*, vol. 10, pp. 58–64, 1956.

678. Stribling, J. W.: Work Load Calculations for Winding Machines, *Textile Ind.*, vol. 116, pp. 135–150, 1952.

679. Swan, A. W.: A Method of Graphing Sales Figures for Executive Use, *Appl. Statist.*, vol. 4, pp. 15–21, 1955.

680. Swensson, O.: An Approach to a Class of Queuing Problems, *Operations Research*, vol. 6, no. 2, pp. 276–295, 1958.

681. Syski, R.: The Theory of Congestion in Lost-call Systems, *A.T.E. Journal*, vol. 9, pp. 182–215, 1953.

682. ———: Analogies between the Congestion and Communication Theories, *A.T.E. Journal*, vol. 11, pp. 220–243, 1955.

683. ——— and J. W. Cohen: Second List of References to Publications on the Application of Probability Theory to Telephone Trunking and Related Problems, P.T.T. publication, paper presented at the International Teletraffic Congress at the Hague, 1958.

684. ———: "Introduction to Congestion Theory in Telephone Systems," Oliver & Boyd, Ltd., Edinburgh and London, 1960.

685. Takács, L.: Probabilistic Treatment of the Simultaneous Stoppage of Machines with Considerations of the Waiting Times, *Magyar Tudományos Akad. III (Mat. Fiz.) Oszt. Közleményei*, vol. 1, pp. 228–234, 1951 (Hungarian).

686. ———: Occurrence and Coincidence Phenomena in Case of Happenings with an Arbitrary Distribution Law of Duration, *Acta Math. Acad. Sci. Hung.*, vol. 2, pp. 275–298, 1951 (revised English translation of *Magyar Tudományos Akad. III*, vol. 1, pp. 371–386, 1951).

687. ———: A New Method for Discussing Recurrent Stochastic Processes, *Magyar Tudományos Akad. Alkalm. Mat. Int. Közleményei*, vol. 2, pp. 135–151, 1953 (Hungarian; Russian and English summaries).

688. ———: Coincidence Problems Arising in the Theory of Counters, *Magyar Tudományos Akad. Alkalm. Mat. Int. Közleményei*, vol. 2, pp. 153–163, 1953 (Hungarian; Russian and English summaries).

689. ———: Investigation of Waiting Time Problems by Reduction to Markov Processes, *Acta Math. Acad. Sci. Hung.*, vol. 6, pp. 101–129, 1955.

690. ———: On Processes of Happenings Generated by Means of a Poisson Process, *Acta Math.*, vol. 6, nos. 1–2, 1955.

691. ———: On a Probability Problem Arising in Some Traffic Investigations, *Publs. Math. Inst. Acad. Sci. Hung.*, vol. 1, pp. 99–107, 1956 (Hungarian; Russian and English summaries).

692. ———: On Some Probabilistic Problems Concerning the Counting of Particles (note in paper of A. Bekessy), *Publs. Math. Inst. Acad. Sci. Hung.*, vol. 1, pp. 93–98, 1956 (Hungarian; Russian and English summaries).

693. ———: On the Sequence of Events Selected by Counter from a Recurrent Process of Events, *Teoriya Veroyatnostei*, vol. 1, pp. 90–102, 1956 (English).

694. ———: On the Probability Problem Arising in the Theory of Counters, *Proc. Cambridge Phil. Soc.*, vol. 52, pp. 488–498, 1956.

695. ———: On Limiting Distributions Concerning a Sojourn Time Problem, *Acta Math. Acad. Sci. Hung.*, vol. 8, pp. 279–294, 1957.

696. ———: Sojourn Time Problems, *Magyar Tudományos Akad. Mat. Fiz. Oszt. Közleményei*, vol. 7, pp. 371–395, 1957.

697. ———: On a Secondary Stochastic Process Generated by a Multi-dimensional Poisson Process, *Hung. Acad. Sci.*, vol. 2, pp. 1–2, 1957.

698. ———: On a Probability Problem Concerning Telephone Traffic, *Acta Math. Acad. Sci. Hung.*, vol. 8, pp. 3–4, 1957.

699. ———: On a Queueing Problem Concerning Telephone Traffic, *Acta Math. Acad. Sci. Hung.*, vol. 8, pp. 325–395, 1957.

700. ———: On a Probability Problem Arising in Some Traffic Investigations, *Publs. Math. Inst. Acad. Sci. Hung.*, vol. 1, pp. 99–107, 1956 (Hungarian; Russian and English summaries).

701. ———: On the Generalization of Erlang's Formula, *Acta Math. Acad. Sci. Hung.*, vol. 7, pp. 419–433, 1957.

702. ———: On a Coincidence Problem Concerning Telephone Traffic, *Acta Math.*, vol. 9, pp. 1–2, 1958.

703. ———: On a Combined Waiting Time and Loss Problem Concerning Telephone Traffic, *Ann. Budapest Sect. Math.*, vol. I, 1958.

704. ———: Some Probability Questions in the Theory of Telephone Traffic, *Magyar Tudományos Akad. Mat. Fiz. Oszt. Közleményei*, vol. 8, pp. 151–210, 1958 (Hungarian).

705. ———: On a General Probability Theorem and Its Applications in the Theory of

the Stochastic Processes, reprint from *Proc. Cambridge Phil. Soc.*, vol. 54, pt. 2, pp. 219–224, 1958.

706. ————: On the Transient Behaviour of Single-server Queueing Processes with Recurrent Input and Exponentially Distributed Service Time, *Operations Research*, vol. 8, no. 2, p. 231, March–April, 1960.

707. Tacklind, S.: Fourieranalytische Behandlung von Erneuerungsproblem, *Skand. Aktuar Tidskr.*, vol. 28, pp. 68–105, 1945 (German).

708. Tanner, J. C.: The Delay to Pedestrians Crossing a Road, *Biometrika*, vol. 38, pp. 383–392, 1951.

709. ————: A Problem of Interference between Two Queues, *Biometrika*, vol. 40, pp. 58–69, 1953 (summary in *Operational Research Quart.*, vol. 4, p. 30, June, 1953).

710. ————: A Simplified Model for Delays in Overtaking on a Two-lane Road, *J. Roy. Statist. Soc.*, vol. 20, no. 2, 1958.

711. Taylor, A. E. N., and others: Studies of Traffic and the Conditions Which Influence It, Ninth Meeting of Permanent International Association of Road Congresses, Lisbon, *Permanent Intern. Assoc. Road Congr. Sect.* 2, *Rept.* 37, 1951.

712. Taylor, J., and R. R. P. Jackson: An Application of the Birth and Death Process to the Provision of Spare Machines, *Operational Research Quart.*, vol. 5, pp. 95–108, 1954.

713. Taylor, R. J.: Queues with Non-Poisson Inputs and Service Times (Including Constant Holding Times), May 13, 1958 (mimeo. copy).

714. Temple, G.: The Use of Partition Functions in Problems of Traffic Congestion, *Brit. Ministry Civil Aviation, Operations Research Sect., ORS/MCA Rept.* 2, June, 1948.

715. Thorndyke, F.: Applications of Poisson's Probability Summation, *Bell System Tech. J.*, vol. 5, p. 610, 1926.

716. Titchmarsh, E. C.: "Theory of Functions," Oxford University Press, London, 1938.

717. Tocher, K. D.: *J. Roy. Statist. Soc.*, vol. 13, no. 2, p. 181, 1951.

718. ————: Some Unsolved Problems in Queueing, *Operational Research Quart.*, vol. 3, pp. 11–13, 1952.

719. Toft, F. J., and H. Boothroyd: A Queueing Model for Spare Coal Faces, *Operational Research Quart.*, vol. 10, no. 4, pp. 245–251, December, 1959.

720. Trautman, D. L., M. Davis, J. Heilfron, E. C. Ho, and A. Rosenbloom: Analysis and Simulation of Vehicular Traffic Flow, *Univ. Calif., Inst. Transportation and Traffic Eng., Research Rept.* 20, 1954 (74 pp.).

721. Truitt, C. J.: Traffic Engineering Techniques for Determining Trunk Requirements in Alternate Routing Trunk Network, *Bell System Tech. J.*, vol. 33, pp. 277–302, 1954.

722. Turner, W. O.: Estimation of Requirements in Dial Telephone Central Offices, *Proc. Operations Research Conf.*, Case Institute of Technology, Cleveland, January, 1953.

723. Van Den Burg, A. R.: Waiting for Transport, *Sigma*, vol. 1, p. 43, 1955 (Dutch).

724. Van Dobben de Bruyn, M.: The Condition for Equilibrium in Automatic Telephone Traffic, *Ingenieur (Utrecht)*, vol. 59, pp. 1–12, 1947.

725. ————: Détermination et prévision du rendement de groupes hommes-machines, *Cahiers bur. temps élémentaires*, sér. 5, no. 503-01, 1955 (9 pp.; French).

726. Van Voorhis, W. R.: Waiting Line Theory as a Management Tool, *Operations Research*, vol. 4, pp. 221–231, 1956.

727. Vaulot, A. E.: Application du calcul des probabilités a l'exploitation téléphonique, *Rev. gén. élec.*, vol. 16, pp. 411–418, 1924.

728. ———: Application du calcul des probabilités à l'exploitation téléphonique, *Ann. P.T.T.*, vol. 14, p. 138, 1925.

729. ———: Extension des formules d'Erlang au cas où les durées des conversations suivent une loi quelconque, *Rev. gén. élec.*, vol. 22, pp. 1164–1171, 1927.

730. ———: Application du calcul des probabilités à l'exploitation téléphonique, *Rev. gén. élec.*, vol. 30, pp. 173–175, 1931.

731. ———: Étude du trafic téléphonique reçu par des lignes explorées dans un ordre déterminé, *Rev. gén. élec.*, vol. 38, pp. 651–665, 1935.

732. ———: Sur l'application du calcul des probabilités à la théorie du trafic téléphonique, *Compt. rend.*, vol. 200, pp. 1815–1818, 1935.

733. ———: Délais d'attente des appels téléphoniques, traités au hasard, *Compt. rend.*, vol. 222, pp. 268–269, 1946.

734. ———: Application du calcul des probabilités à la téléphonie, École Nationale Supérieur des Télécommunication, 1947.

735. ———: Sur la proportion d'appels perdus dans la système de téléphonie automatique où chaque organe de contrôle est commun à tous les appareils d'un groupe, *Adm. franç. P.T.T. étude 9RS*, Service des recherches et du contrôle techniques, November, 1947.

736. ———: Les Formules d'Erlang et leur calcul pratique, *Ann. télécommun.*, vol. 6, pp. 279–286, 1951.

737. ———: Délais d'attente des appels téléphoniques dans l'ordre inverse de leur arrivée, *Compt. rend.*, vol. 238, pp. 1188–1189, 1954.

738. ———: Délais d'attente des appels téléphoniques traités au hasard, *Ann. télécommun.*, vol. 9, pp. 9–14, 1954.

739. ———: Valeurs asymptotiques des expressions rencontrées dans les applications du calcul des probabilités à l'exploitation téléphonique, *Adm. franç. P.T.T. étude 140RS*, Service des recherches et du contrôle techniques.

740. ———: Sur les formules d'Erlang et leurs expressions asymptotiques, *Adm. franç. P.T.T. étude 31RS*, Service des recherches et du contrôle techniques (17 pp.).

741. Vladziyevsky, A. P.: The Probability Law of Operation of Automatic Lines and Internal Storage in Them, *Automatika i Telemekhanika*, vol. 13, pp. 277–281; Obtainable from Secretaries Dept. D.S.I.R.

742. Volberg, O.: Problemè de la queue stationnaire et nonstationnaire, *Compt. rend. acad. sci. U.R.S.S.*, vol. 24, p. 657, 1939 (French).

743. ———: *Isv. Voen. Elekt. Akad.*, vol. 17, 1939 (the contents of this paper are briefly indicated at the end of the preceding paper).

744. Von Sydow, L.: Some Aspects on the Variations in Traffic Intensity, *Teleteknik*, pp. 58–64, 1958.

745. Wagner, H. N.: Economies of Telephone Relay Applications, *Bell System Tech. J.*, vol. 33, pp. 218–275, 1954.

746. Wallstrom, B.: Artificial Traffic Trials on a Two-stage Link System Using a Digital Computer, *Ericsson Tech.*, no. 2, 1958.

747. Walz, E.: Verlustzeiten und Wartezeiten bei mehr Maschinenbedienung, *Maschinenbau*, vol. 16, pp. 11–14, 1937.

748. Wardrop, J. G.: Some Theoretical Aspects of Road Traffic Research, *Proc. Inst. Civil Engrs. (London)*, part II, vol. 1, 1952.

749. ———: Traffic Capacity of Town Streets, *Roads and Eng. Construct.*, vol. 30, no. 350, pp. 39–42; no. 351, pp. 68–71, 1952.

750. ——— and G. Charlesworth: A Method of Estimating Speed and Flow of Traffic from a Moving Vehicle, *Proc. Inst. Civil Engrs. (London)*, pt. II, vol. 3, pp. 158–171, 1954.

751. ———: The Capacity of Roads, *Operational Research Quart.*, vol. 5, pp. 14–24, 1954.

752. Waugh, W. A. O'N.: An Age-dependent Birth and Death Process, *Biometrika*, vol. 42, pts. 3–4, 1955.

753. Wax, N. (ed): "Selected Papers on Noise and Stochastic Processes," Dover Publications, New York, 1954.

754. Webster, F. V.: Traffic Signal Settings and Expected Delay, presented at International Study Week in Traffic Engineering, Stresa, Italy, Oct. 1–5, 1956; published in *World Touring and Automobile Organization*, London, 1956.

755. Weir, W. F.: Figuring Most Economical Machine Assignment, *Factory Management and Maintenance*, vol. 102, pp. 100–102, 1944.

756. Westgarth, D. R.: The Effect of Congestion in the Operation of Weaving Machines, M.Sc. Thesis, part II, London University, 1948.

757. Wheeler, R. C., Jr.: Transient Sequencing Delays as Applied to Air Traffic Control, Airborne Instrument Laboratory Inc., Mineola, N.Y., 1958.

758. White, H. C., and L. S. Christie: Queuing with Preemptive Priorities or with Breakdown, *Operations Research*, vol. 6, no. 1, pp. 79–95, 1958.

759. Whittaker, E. T., and G. N. Watson: "A Course of Modern Analysis," 4th ed., reprinted by Cambridge University Press, London, 1952.

760. ——— and G. Robinson: "The Calculus of Observations," 4th ed., Blackie & Son, Ltd., Glasgow, 1940.

761. Widder, D. V.: "The Laplace Transform," Princeton University Press, Princeton, N.J., 1946.

762. Wilkins, C. A.: On Two Queues in Parallel, *Biometrika*, vol. 47, pts. 1 and 2, p. 198, 1960.

763. Wilkinson, R. I.: The Reliability of Holding-time Measurements, *Bell System Tech. J.*, vol. 20, pp. 365–404, 1941.

764. ———: Working Curves for Delayed Exponential Calls Served in Random Order, *Bell System Tech. J.*, vol. 32, pp. 360–383, 1953.

765. ———: Theories for Toll-traffic Engineering in the U.S.A. (with an Appendix by J. Riordan), *Bell System Tech. J.*, vol. 35, pp. 421–514, 1956.

766. ———: Queueing Theory and Some of Its Industrial Uses, *Natl. Conv. Trans. Am. Soc. Quality Control*, pp. 313–330, 1958.

767. Winsten, C. B.: A Stochastic Model of Traffic Congestion, The Rand Corporation, P-459A, Nov. 19, 1953.

768. ———: Geometric Distributions in the Theory of Queues, *J. Roy. Statist. Soc.*, *Ser. B*, vol. 21, no. 1, 1959.

769. Wishart, D. M. G.: "On a Queuing Problem," Princeton University, Department of Mathematics, Princeton, N.J., 1953.

770. ———: A Queueing System with χ^2 Service-time Distribution, *Ann. Math. Statist.*, vol. 27, 1956.

771. ———: A Queuing System with Service Time Distribution of Mixed Chi-squared Type, *Operations Research*, vol. 7, no. 2, pp. 174–179, March–April, 1959.

772. Wolfowitz, J., and J. Kiefer: On the Characteristics of the General Queueing Process with Applications to Random Walks, *Ann. Math. Statist.*, vol. 27, pp. 147–161, 1956.

773. Wright, E. P. G.: Behavior of Telephone Exchange Traffic Where Non-equivalent Choice Outlets Are Commoned, *Elec. Commun.*, vol. 24, pp. 42–54, 1947.

774. ——— and J. Rice: Probability Studies Applied to Telecommunications Systems with Storage, *Elec. Commun.*, vol. 34, 1956.

775. Wright, W. R., W. G. Duvall, and H. A. Freeman: Machine Interference: Two

Solutions of Problem Raised by Multiple Machine Units, *Mech. Eng.*, vol. 58, pp. 510–514, 1936.

776. Zajontchovski, E. A.: Étude sur la question du nombre des circuits dans le service automatique, C.C.I.F. Document 20, 8th C. E., 1955.

777. Ziermann, M.: Erganzung des Lagervorrates, II. Nachbestellung, *Magyar Tudományos Akad. Alkalm. Mat. Int. Közleményei*, vol. 2, pp. 203–216, 1953 (Hungarian; Russian and German summaries).

Addendum

778. Agard, J.: Problèmes d'attente dans une compagnie aérienne, *Rev. franç. recherche opérationelle*, vol. 4, no. 4, pp. 229–243, 1960.

779. Ancker, C. J., Jr., and A. V. Gafarian: Queueing with Multiple Poisson Inputs and Exponential Service Times, Systems Development Corporation, *Tech. Mem.* 503, Santa Monica, Calif., June, 1960.

780. Bähler, W. Th., J. W. Cohen, and M. M. Jung: Calculation of the Number of First Group-selectors for Telephone Systems Provided with First and Second Concentration Stages, Taking into Account the Internal Blocking, *Commun. News*, vol. 14, pp. 51–58, 1954.

781. Bauer, F. L., and H. Störmer: Calculation of Waiting Time in Switching Systems with Small Input Line Groups, *Arch. Elektrotech.*, vol. 9, pp. 69–73, 1955.

782. Bell, D. A.: Semiconductor Noise as a Queueing Problem, *Proc. Phys. Soc. (London)*, vol. 72, pp. 27–32, 1958.

783. Beneš, V. E.: Combinatory Methods and Stochastic Kolmogorov Equations in the Theory of Queues with One Server, *Trans. Am. Math. Soc.*, vol. 94, pp. 282–294, 1960.

784. ———: Transition Probabilities for Telephone Traffic, *Bell System Tech. J.*, vol. 39, no. 5, pp. 1279–1320, 1960.

785. ———: A Sufficient Set of Statistics for a Simple Telephone Exchange Model, *Bell System Tech. J.*, vol. 36, pp. 939–964, 1957.

786. ———: Fluctuations of Telephone Traffic, *Bell System Tech. J.*, vol. 36, pp. 965–973, 1957.

787. Ben-Israel, A., and P. Naor: A Problem of Delayed Service, part I, *J. Roy. Statist. Soc.*, vol. 22, no. 2, pp. 245–269, 1960.

788. ——— and ———: A Problem of Delayed Service, part II, *J. Roy. Statist. Soc.*, vol. 22, no. 2, pp. 270–276, 1960.

789. Bharucha-Reid, A. T.: "Elements of the Theory of Markov Processes and Their Applications," McGraw-Hill Book Company, Inc., New York, 1960.

790. Blackwell, D.: A Renewal Theorem, *Duke Math. J.*, vol. 15, pp. 145–150, 1948.

791. Bradley, D. P.: Overflow Trunking of Switching and Discriminating Selector Repeaters, *Telecommun. J. Australia*, vol. 9, p. 254, 1953.

792. Bretschneider, G.: The Calculation of Groups of Lines for Overflow Traffic in Automatic Telephone Networks, *NTZ-Nachrtech. Z.*, vol. 9, pp. 533–540, 1956.

792a. Brodi, S. M.: On Integro-differential Equations for a System with Waiting Time, *Ukrain. Acad. Sci., Rept.* 6, pp. 571–573, 1959.

793. Brown, A. W.: A Note on the Use of a Pearson Type III Function in Renewal Theory, *Ann. Math. Statist.*, vol. 11, pp. 448–453, 1940.

794. Campbell, G. A.: Probability Curves Showing Poisson's Exponential Summation, *Bell System Tech. J.*, vol. 2, pp. 95–113, January, 1923.

795. Capello, F., and A. Sanneris: Studio economico delle accessibilita in una centrale telefonica automatica, *Alta frequenza*, vol. 25, nos. 3–4, pp. 305–318, 1956.

796. Cherry, E. C.: The Conceivable Future of Telecommunications, *Trans. S. African Inst. Elect. Engrs.*, vol. 49, pt. 8, pp. 271–286, August, 1958.

797. Chew, W. G. N.: The Automatic Traffic Recorder, *P. O. Elec. Engrs. J.*, vol. 28, p. 1, 1935.

798. Clos, C.: A Study of Non-blocking Switching Networks, *Bell System Tech. J.*, vol. 32, pp. 406–424, 1953.

799. ———: Automatic Alternate Routing of Telephone Traffic, *Bell Labs. Record*, vol. 32, pp. 51–57, February, 1954.

800. Cohen, J. W.: Some Examples in the Use of Implication in Switching Algebra, *Commun. News*, vol. 16, no. 1, pp. 1–10, October, 1955.

801. ———: On the Queueing Process of Lanes, *Philips Tech. Rept.*, 1956.

802. ———: On the Fundamental Problem of Telephone Traffic Theory and the Influence of Repeated Calls, *Philips Telecommun. Rev.*, vol. 18, pp. 49–100, 1957.

803. ——— and B. J. Beukelman: Call Congestion of Transposed Multiples, *Philips Telecommun. Rev.*, vol. 17, no. 4, pp. 145–154, 1957.

804. Conolly, B. W.: Queueing at a Single Serving Point with Group Arrival, *J. Roy. Statist. Soc.*, vol. 22, no. 2, pp. 285–298, 1960.

805. Conrad, R., and B. A. Hille: Telephone Operators' Adaptation to Traffic Variations, *J. Inst. Elec. Engrs. (London)*, vol. 4, pp. 10–14, 1958.

806. Cox, D. R.: The Analysis of Non-Markovian Stochastic Processes by the Inclusion of Supplementary Variables, *Proc. Cambridge Phil. Soc.*, vol. 51, pt. 3, pp. 433–441, July, 1955 (see also *Math. Rev.*, p. 277, 1956).

807. Debry, R., and J. P. Dreze: Files d'attente à plusieurs priorités absolues, *Cahiers centre d'études recherche opérationnelle*, vol. 2, no. 3, pp. 201–222, 1960.

808. Dhondt, A.: Le Problème transitorie de la file finie avec source illimitée, *Cahiers centre d'études recherche opérationnelle*, vol. 2, no. 4, pp. 309–318, 1960.

809. ———: Sur le comportement transitoire du processus d'attente simple, *Cahiers centre d'études recherche opérationnelle*, vol. 2, no. 3, 1960.

810. Dietrich, H.: The Problem of the Optimum Placing of the Town Telephone Exchange, *Telakomunikacyjny*, no. 12, pp. 371–377, December, 1952 (Polish).

811. Dobben De Bruyn, M. Van.: The Condition for Equilibrium in Automatic Telephone Traffic, *Ingenieur (Utrecht)*, vol. 59, pp. El–12, 1947 (Dutch).

812. Edie, L. C.: Car Following and Steady-state Theory for Non-congested Traffic, *Operations Research*, vol. 9, no. 1, p. 66, January–February, 1961.

813. Ekelof, S.: On the Hunting of Line Finders, *P. O. Elec. Engrs. J.*, vol. 28, p. 52, April, 1935.

814. Elldin, A.: Automatic Telephone Exchanges with Cross-bar Switches (Switches Calculations; General Survey), Switch Calculation Book B11265, Telefonaktienbolaget L. M. Ericsson, Stockholm, November, 1955.

815. ———: Applications of Equations of State in the Theory of Telephone Traffic, Thesis, University of Stockholm, 1957.

816. ———: Further Studies on Gradings with Random Hunting, *Ericsson Tech.*, vol. 13, pp. 175–257, 1957.

817. ——— and I. Tange: Long-time Observations of Telephone Traffic, *Tele.* (English ed.), no. 1, 1960.

818. ———, C. Jacobaeus, and L. Von Sydow: Traffic Measurements with Lamp Panels, *Ericsson Tech.*, vol. 10, pp. 107–187, 1954.

819. Felder, H. H., and E. N. Little: Intertoll Trunk Net Loss Maintenance under Operator Distance and Direct Distance Dialling, *Bell System Tech. J.*, vol. 35, pp. 955–972, 1956.

820. Fortet, R.: "Calcul des probabilités," Centre national de la recherche scientifique, Paris, 1950.

820a. Friend, J. K.: Two Studies in Airport Congestion, *Operational Research Quart.*, vol. 9, pp. 234–253, 1958.

821. Gandais, M., and A. Sanneris: "Principi di traffico telefonico," Stet-Torino, 1956.

822. Gaver, D. P., Jr.: A Waiting Line with Interrupted Service, Including Priorities,

Westinghouse Research Labs., Sci. Paper 6-41210-2-P4, Pittsburgh, Pa., August, 1960.

823. Gilbert, E. N.: N-terminal Switching Circuits, *Bell System Tech. J.*, vol. 30, p. 668, 1951.

824. ————: Enumeration of Labelled Graphs, *Can. J. Math.*, vol. 8, no. 3, pp. 405–411, 1956.

824a. Gnedenko, B. W.: On Some Aspects in the Development of the Theory of Queues, *Mathematik Technik Wirtschaft*, vol. 7, pp. 162–166, 1960 (in German).

824b. ————: Some Remarks on Two Works of D. Barrer, *Bullinul institutuli Politehnic din Iasi*, vol. 5, no. 9, 1960 (in German).

825. Greenberg, H., and A. Daou: The Control of Traffic Flow to Increase the Flow, *Operations Research*, vol. 8, no. 4, p. 524, 1960.

826. Gumbel, H.: Waiting Lines with Heterogeneous Servers, *Operations Research*, vol. 8, no. 4, p. 504, 1960.

827. Gurk, H. M.: Single Server, Time-limited Queue, *Bull. Am. Math. Soc.*, vol. 63, p. 400, 1957 (abstract only).

828. Haight, F. A.: The Volume, Density Relation in the Theory of Road Traffic, *Operations Research*, vol. 8, no. 4, p. 572, 1960.

829. ————: Queueing with Balking, part II, *Biometrika*, vol. 47, nos. 3 and 4, pp. 285–296, 1960.

830. Harris, L. R. F.: Time Sharing as a Basis for Electronic Telephone Switching, *Proc. Inst. Elec. Engrs. (London)*, Paper 1993R, vol. 103, pt. B, 1956.

831. Harrison, H. H.: "An Introduction to the Strowger System of Automatic Telephony," pp. 29–39 (remarks on traffic), Longmans, Green & Co., Ltd., London, 1924.

832. Hartley, H. O., and E. R. Fitch: A Chart for the Complete Beta Function and the Cumulative Binomial Distribution, *Biometrika*, vol. 38, 1951.

833. Hawkins, N. A.: Problems in Automatic Trunking–Last Contact Traffic, *P.O. Elec. Engrs. J.*, vol. 23, pp. 272–281, 1930–1931.

834. Healy, T. L.: Queues with "Exponential-type" Service-time Distributions, *Operations Research*, vol. 8, no. 5, pp. 719–721, 1960.

835. Heathcote, C. R.: A Single Queue with Several Preemptive Priority Classes, *Operations Research*, vol. 8, no. 5, pp. 630–638, 1960.

836. Herniter, J. D., and J. F. Magee: Customer Behavior as a Markov Process, *Operations Research*, vol. 9, no. 1, p. 105, 1961.

837. Hettwig, V. E., and K. Rohde: New Calculation Methods for the Design of Telephone Systems, *Siemens-Z.*, vol. 30, no. 1, pp. 1–10, January, 1956 (German).

838. Inose, H.: An Artificial Call Generator Employing Random Noise Resistors, *J. Inst. Elec. Commun. Engrs. (Japan)*, vol. 36, pp. 177–181, 1951 (Japanese).

839. Jacobaeus, C.: The Employment of Cross-bar Selectors in Bypass Circuit Systems, *Ericsson Rev.*, no. 2, p. 30, 1945.

840. ————: Measurements with Lamp Panel—A New Method for Traffic Studies (International Teletraffic Congress, Copenhagen, 1955), *Teleteknik*, vol. 1, p. 91, 1957.

841. ————: Investigations with an Electronic Telephone Traffic Machine, *Teleteknik*, vol. 5, no. 2, pp. 260–262, July, 1954 (Danish).

842. Jacobsen, C.: Investigations with an Electronic Telephone Traffic Machine, *Teleteknik*, vol. 5, pp. 260–262, 1954 (Danish).

843. Jaiswal, N. K.: Time Dependent Solution of the Bulk-service Queueing Problem, *Operations Research*, vol. 8, no. 6, p. 773, 1960.

844. Jensen, A.: Calculations of Loss in Crossbar Automatic Exchanges, *Teleteknik*, vol. 4, pp. 176–200, 1952 (Danish).

845. ———: "A Distribution Model," Munksgaard, Copenhagen, 1954.
846. ———: The Applicability of Decision Theory in the Planning and Operation of Telephone Plant (International Teletraffic Congress, Copenhagen, 1955), *Teleteknik*, vol. 1, pp. 126–129, 1957.
847. Josephs, H. J., and R. A. Hastie: Operational Research in the Post Office (Part 2, Probability Models), *P. O. Elec. Engrs. J.*, vol. 50, pp. 81–85, July, 1957.
848. Karlsson, S. A.: Konstgjord Telefontrafik som Hjalpmedel vid Behandling av Telefontrafik Problem, Helsingfors, 1945.
849. ———: Forbattrad Narmeformel fur Erlang's Ideela Gradering, *Kraft och Ljus* (Helsingfors), July–August, 1949.
850. Keinonen, A. A.: Studien und Messungen unbegrenzter Wartezeiten mit Hilfe einer Speicherkunstschaltung, *Veröffentl. Geb. Nachr. Tech.*, vol. 8, pp. 665–680, 1938.
851. Keister, W., A. Ritchie, and S. Washburn: "The Design of Switching Circuits," D. Van Nostrand Company, Inc., Princeton, N.J., 1951.
852. Kendall, D. G.: Geometric Ergodicity and the Theory of Queues, in K. J. Arrow, S. Karlin, and P. Suppes (eds.), "Mathematical Methods in the Social Sciences," chap. 12, Stanford University Press, Stanford, Calif., 1959.
853. Khintchine, A.: "Mathematical Methods in the Theory of Queueing," Charles Griffin & Co., Ltd., London, 1960.
854. Knight, N. V.: Depreciation and Service Life of Telecommunication Plant, *P. O. Elec. Engrs. J.*, Paper 209, February, 1955.
855. ———: Economic Principles of Telecommunications Plant Provision, *P. O. Elec. Engrs. J.*, Paper 210, November, 1955.
856. Kroes, J. L. De: Optimum Grouping in Uniselector Trunking, *Commun. News*, vol. 14, pp. 70–76, January, 1954.
857. Kuhn, S.: "Traffic Problems in Automatic Telephony," p. 384, P.W.N., Warszawa, Poland, 1957 (Polish).
858. Lambert, F.: Les Problèmes d'attente, *Cahiers centre d'études recherche opérationnelle*, no. 2, pp. 5–28, 1959.
859. Langer, M.: Calculation of Switches Required in Automatic Exchanges, *Elektrotech. Z.*, no. 1, 1924.
860. Laning, J. H., Jr., and R. H. Battin: "Random Processes in Automatic Control," McGraw-Hill Book Company, Inc., New York, 1956.
860a. Lee, A. M.: Some Aspects of a Control and Communication System, *Operational Research Quart.*, vol. 10, no. 4, pp. 206–216, December, 1959.
860b. ——— and P. A. Longton: Queueing Processes Associated with Airline Passenger Check-in, *Operational Research Quart.*, vol. 10, no. 1, pp. 56–71, March, 1959.
861. Lee, C. Y.: Analysis of Switching Networks, *Bell System Tech. J.*, vol. 34, pp. 1287–1315, November, 1955.
862. Le Gall, P.: Étude du blocage dans les systèmes de commutation téléphonique automatique utilisant des commutateurs électroniques du type crossbar, *Ann. télécommun.*, vol. 11, pp. 159–171, 1956; pp. 180–194, 1956; p. 197 (Erratum).
863. ———: Méthode de calcul de l'encombrement dans les systèmes téléphoniques automatiques à marquage, *Ann. télécommun.*, vol. 12, pp. 374–386, 1957.
864. Le Guillant et al.: La Neurose des téléphonistes, *Presse méd.*, vol. 64, p. 274, 1956.
864a. Leibowitz, M. A.: An Approximate Method for Treating a Class of Multiqueue Problems, *IBM J. Research Develop.*, vol. 5, no. 3, pp. 204–209, 1961.
865. Leunbach, G.: Illustration of the Application of Statistical Decision Functions in a Telephone Plant (International Teletraffic Congress, Copenhagen, 1955), *Nord. Tidsk. Tek. Okonomi*, p. 146, 1955.

866. Lewis, N. H.: Notes on the Exponential Distribution in Statistics, *P. O. Elec. Engrs. J.*, vol. 41, pp. 10–12, 1948.
867. Little, J. D. C.: Approximate Expected Delays for Several Maneuvers by a Driver in Poisson Traffic, *Operations Research*, vol. 9, no. 1, p. 39, 1961.
868. Lotka, A. J.: Application of Recurrent Series in Renewal Theory, *Ann. Math. Statist.*, vol. 19, pp. 190–206, 1948.
869. Lubberger, F.: Beobachtungen, Vorschriften und Theorien der Schwankungen im Fernsprechverkehr, in F. Lubberger (ed.), "Wahrscheinlichkeiten und Schwankungen," pp. 41–65, Springer-Verlag, Berlin, 1937.
870. Lundkvist, K.: Bestamning av Sannolikheten for Sparrning vid s.k. Stel Koppling, *Tek. Medd. Från. Kungl. Teleg.*, nos. 4–6, 1942.
871. Martin, N. H.: A Note on the Theory of Probability Applied to Telephone Traffic Problems, *P. O. Elec. Engrs. J.*, vol. 16, pp. 237–341, 1923.
872. Molina, E. C.: "Poisson's Exponential Binomial Limit," D. Van Nostrand Company, Inc., Princeton, N.J., 1947.
873. Morgan, T. J.: "Telecommunication Economics," p. 452, Macdonald & Company, Ltd., London, 1958.
874. Murray, L. J.: Crossbar Systems, *A.T.E. Tech. Soc. Paper*, Nov. 8, 1948.
875. ———: Electronics in the Switching of Telephone Calls, *Engineering*, vol. 176, pp. 413–415, Sept. 25; pp. 445–447, Oct. 2, 1953.
876. ———: Development of Automatic Telephony in Great Britain, *A.T.E. Journal*, vol. 10, no. 4, pp. 271–293, 1954.
877. Myers, O.: Common Control Telephone Switching Systems, *Bell System Tech. J.*, vol. 31, pp. 1086–1120, 1952.
878. ———: Automatic Alternate Routing of Telephone Traffic, *Bell Labs. Record*, vol. 32, p. 51, 1954.
879. Ore, O.: Studies on Directed Graphs, part I, *Ann. Math.*, vol. 63, no. 3, pp. 383–406, May, 1956.
880. Pilliod, J. J.: Fundamental Plans for Toll Telephone Plant, *Bell System Tech. J.*, vol. 31, pp. 832–850, September, 1952 (also in *Trans. Am. Inst. Elec. Engrs.*, vol. 71, no. 1, p. 248, 1952).
881. Popovic, Z.: The Efficiency of Grading with Cyclic Arrangement, *Telekomunikacije*, vol. 9, pt. 4, 1955.
882. Prigogine, I., and F. C. Andrews: A Boltzmann-like Approach for Traffic Flow, *Operations Research*, vol. 8, no. 6, p. 789, 1960.
883. Prim, R. C.: Shortest Connection Networks and Some Generalizations, *Bell System Tech. J.*, vol. 36, pp. 1389–1401, 1957.
883a. Reich, P. G.: Instability in Some Flow Systems of Civil Aviation, *Operational Research Quart.*, vol. 8, pp. 126–132, 1957.
884. Riordan, J.: Derivation of Moments of Overflow Traffic, in R. I. Wilkinson, Theories for Toll Traffic Engineering in U.S.A., App. I, *Bell System Tech. J.*, vol. 35, p. 514, 1956.
885. ——— and C. E. Shannon: The Number of Two-terminal Series-Parallel Networks, *J. Math. and Phys.*, vol. 21, pp. 83–93, 1942.
886. Rodenburg, N.: Alternative Routing of Junction Traffic, *Commun. News*, vol. 10, p. 30, 1949.
887. ———: Some Problems Relating to a Telephone System Employing Nonhoming Selectors, *Commun. News*, vol. 13, pp. 69–114, May, 1953.
888. Rohde, K.: Considerations on the Significance of Random Sampling Process in Determining the Magnitude of Telephone Traffic (International Teletraffic Congress, Copenhagen, 1955), *Teleteknik*, vol. 1, pp. 86–90, 1957 (German).

889. ——— and G. Bretchneider: Economic Traffic Standards in Telephone Systems, *Frequency*, vol. 8, no. 8, pp. 233–239, August, 1954 (German).

890. ——— and H. Störmer: "Passage Probabilities" in Switching Facilities of Communication Engineering, *Mitt. Bl. Math. Statist.*, vol. 5, nos. 2–3, 1953.

891. Rowdin, E.: Correction and Addendum to a New Measure of Effectiveness for Queuing Problems, *Operations Research*, vol. 8, no. 6, p. 868, 1960.

892. Rückle, G., and F. Lubberger: "Der Fernsprechverkehr als Massenerscheinung mit Starken Schwankungen," Springer-Verlag, Berlin, 1924.

893. Sevastyanov, B. A.: An Ergodic Theorem for Markov Processes and Its Application to Telephone Systems with Refusals, *Teoriya Veroyatnostei*, vol. 2, 1957.

894. Shannon, C. E.: A Symbolic Analysis of Relay and Switching Circuits, *Trans. Am. Inst. Elec. Engrs.*, vol. 57, pp. 713–723, 1938.

895. ———: The Synthesis of Two-terminal Networks, *Bell System Tech. J.*, vol. 28, pp. 59–98, 1949.

896. ———: Memory Requirements in a Telephone Exchange, *Bell System Tech. J.*, vol. 29, pp. 343–349, July, 1950.

897. ——— and W. Weaver: "The Mathematical Theory of Communication," University of Illinois Press, Urbana, Ill., 1949.

898. Siemens Bros. & Co. Ltd.: "Traffic Calculations for Trunk and Local Telephone Exchanges," p. 103, London, 1955.

899. Syski, R.: Determination of Waiting Time in the Simple Delay System, *A.T.E. Journal*, vol. 13, pp. 281–286, 1957.

900. Takács, L.: "Stochastic Processes, Problems and Solutions," Methuen's Monographs, Methuen & Co., Ltd., London, 1960.

901. ———: Transient Behavior of Single-server Queuing Process with Erlang Input, *Office of Naval Research, Sponsored Research Rept.* CU-40-60 Nonr-266(33) MS, March, 1960.

902. ———: On the Sequence of Events, Selected by a Counter from a Recurrent Process of Events, *Theory of Probability and Its Application*, vol. 1, no. 1, pp. 81–91, 1956.

903. ———: On a Sojourn Time Problem in the Theory of Stochastic Processes, *Trans. Am. Math. Soc.*, vol. 93, no. 3, pp. 531–540, December, 1959.

904. ———: On the Limiting Distribution of the Number of Coincidences Concerning Telephone Traffic, *Ann. Math. Statist.*, vol. 30, no. 1, pp. 134–142, March, 1959.

905. ———: On Certain Stochastic Processes Arising in the Theory of Telephone Traffic, Department of Statistics, Columbia University, New York, 1959.

906. ———: On the Probability Law of the Busy Period for Two Types of Queuing Processes, *Office of Naval Research, Sponsored Research Rept.* CU-22-61-Nonr-266 (59)MS, December, 1960.

907. Tange, I.: Optimum Methods for Determining Routes and Number of Lines in a Telephone Network with Alternative Traffic Facilities, *Tele.*, vol. 1, pp. 1–21, 1957 (Swedish); see also Document 19d, 6-ième et 7-ème Commissions d'étude du C.C.I.F. Genève, 1952–1954.

908. Ventura, E.: Sur l'utilisation des intégrales de contour dans les problèmes de stocks et de délais d'attente, *Management Sci.*, vol. 6, no. 4, p. 423, 1960.

909. Wilkinson, R. I.: The Beginnings of Switching Theory in the United States (International Teletraffic Congress, Copenhagen, 1955), *Teleteknik*, vol. 1, pp. 14–31, 1957.

910. Wishart, D. M. G.: Queuing Systems in Which the Discipline Is "Last-come, First-served," *Operations Research*, vol. 8, no. 5, pp. 591–599, 1960.

INDEX

Absorbing barriers, 117, 126, 136
Absorbing Markoff chain, 70
Accidents, 13
Action and analysis, 9
Air conditioning, 15
Aircraft, 5, 171, 318
  delayed, 319
Aircraft carrier, 181, 352
Airport, 5, 322
Allocation to channels, 250
Applications, 5, 302, 350
  accidents, 13
  air conditioning, 15
  aircraft, 5, 171, 318, 322
  aircraft carrier, 181, 352
  airport, 5, 322
  appointments, at doctor's office, 4
    and reservations, 20
  apron control, 322
  aviation traffic, 318–321
  bacteria, 117
  bakeries, 352
  bank, 10
  barbers, 352
  buses, 352
  cafeteria design, 346
  car traffic, 311
  clerks, 350
  coal faces, 347
  collusion, 10, 12, 374
  communications, 322
  computers, 298, 374
  conveyor belts, 352
  courts, 352
  dams, 336
  elevator, 171
  engines, 332
  epidemics, 117
  feedback, 294

Applications, grocery stores, 3
  highway, 11, 44
  hospitals, 9, 343–345
  influx of papers to editors, 216
  inspection line, 49
  inventory, 334–336
  jockeying, 10, 23, 284, 374
  machines, 323, 331, 332
  manufacturing units, 252
  medical care, 343
  messages, 298
  moving server, 291
  nervous impulses, 352
  networks, 7, 373
  oil pumps, 351
  passenger debarkation, 344
  perishable goods, 18
  post office, 10, 298, 352
  production-line operation, 11, 49
  repairman, 326, 329
  reservoirs, 336
  restaurants, 44, 346
  semiconductor noise, 338
  ships, 10, 12, 352
  stoppages, 317, 329
  Suez Canal, 10, 12
  supermarkets, 11, 25, 352
  taxis, 78, 352
  telegraphy, 352
  telephony, 10, 20, 171, 302
  toll collection, 350
  traffic lights, 313
  truck docks, 351
  tunnels, 317
  waitress, 352
Appointments and reservations, 20
  at doctor's office, 4
Approach time, 6
Approximation, 299, 328

415

Apron control, 322
Arbitrary input, 198, 209, 219, 275
Arrivals, 14
  batches, 5, 9, 12
  bulk, 184
  constant intervals, 10, 49, 210
  distribution, 5, 9
  input, 5, 9, 198, 209, 219, 275
  interval, 7
  oriented, 280
  periodic, 321
  Poisson, 63, 172, 193, 194
  regular, 27
  retarded, 280
  scheduled, 282
  single, 12
Asymptotic behavior, 359
Asymptotic distribution, 341
Autocorrelation function, 130
Average number, of renewals, 361
  of waiting individuals, 18
Average waiting time, 42
  for all customers, 7
  for those who wait, 7
Aviation traffic, 318, 321

Backward equation, 74
Bacteria, 117
Bakeries, 352
Balking, 6, 10, 12, 23, 270
Balking distribution, 270
Balking sequence, 275
Bank, 10
Barbers, 352
Batches, 5, 9, 171, 342
Bessel function, modified, 91
Beta function, 267
Binomial distribution, 65, 181, 272
Binomial input, arbitrary service, 181
Birth-death process, 76
  backward equations, 84
  existence and uniqueness of solution, 84
  forward equations, 83
  spectral solution of equations, 145
  steady-state solution, 87
Blackwell's theorem, 366
Blocked call, 308
Blocking, 354
Boolean algebra, 55

Boolean sigma algebra, 56
Breakdown causes, 19
Bulk arrivals and arbitrary service times, 184
Bulk queues, 171
Bulk service and arrivals, 180
Bulk waiting time, 177
Bus queues, 4
Buses, 352
Busy channels, 353
Busy period, 129, 194, 223, 227
  average length, 18
  distribution of, 99, 107, 128, 218

Cafeteria design, 346
Call congestion, 356, 357
Canceled channels, 285
Capacity of reservoir, 337
Car traffic, 311
Centavos, 352–353
Central-limit theorem, 66, 67
Chance variable, 32
Channels, busy, 46, 353
  canceled, 285
  communication, 6
  cooperating, 11, 288
  fault finding, 48
  full availability, 10, 20, 303
  idle number, 125
  infinite number of, 99, 127
  limited availability, 10
  loss system, 353
  multiple (see Multiple channels)
  multiqueues, 219
  series and parallel, 13, 261
  single, 95, 96, 161, 198, 211
  special and general, 11, 13, 290
  total idle time, 7
  variable number, 13, 285
Chapman-Kolmogorov equation, 74, 76, 185
Characteristic equation, 139
Characteristic function, 57, 65
Characteristic roots, 138
Chi square, 45, 65, 211
Circling aircraft, 6
Classification (Kendall's), 25
Clerks, optimum number, 350
Coal faces, 347
Coefficient of loss, 19, 125

Coin tossing, 66
Collusion, 10, 12, 374
Combinatorial approach, 308
Common control, 355
Common-control loss system, 309
Common-control waiting system, 310
Communication, 322
    channels of, 6
Competing enterprises, 13
Completely monotone function, 78
Compound distribution, 64
Computer, 298, 374
Conditional distribution, 47
Conditional probability, 56
Conduction band, 349
Congested traffic, 10
Congestion delays and loss of creativity,
    25
Conservative process, 85
Constant arrivals, 210
Constant input and exponential service,
    49
Constant service time, 176
Continuous parameter, 199
Continuous probability, 32
Contour representation of queues, 21
Control of air traffic, 318, 321
Conveyor belts, 352
Convolution, 58, 77, 113, 218, 299
Cooperating channels, 11, 288
Correlation, 12
Cosine transform, 130, 348
Cost analysis, 19
    of additional channel, 7
    comparisons, 125
Courts, 352
Cubic equation, solution of, 144
Cumulant, 59
Cumulant generating function, 175
Cumulative probability, 32
Cumulative process, 370
Customers, 14
    adaptation, 10
    behavior, 10
    departing, 41
    habits, 32
    wait, 7
Cycle, 334, 398
Cyclic queues, 294
Cycling, 6, 11, 13, 23, 294

Dams, 336
Data, 44
Decision process, 48
Delay distribution, 320
Delayed aircraft, 319
Delayed calls, 10, 12, 305
Density function, 57
Departing customer, 41
Departure, 255
    epochs of, 171
Dependence, 68, 214
Deterministic example, 31
Dice rolling, 31
Difference methods, 52
Different rates of service, 290
Differential-difference equation, 35
Dirac delta function, 61, 108, 110, 118,
    320
Discipline, allocation to channels, 250
    first-come first-served, 11, 38
    last-come first-served, 11, 252
    ordered, 7, 11, 12, 22, 219
    priorities, 23, 231, 234, 236, 241, 304
        dynamic, 241
        nonpreemptive, 12, 30, 231, 232
        preemptive, 12, 231, 237
    random selections, 11, 12, 23, 243–250,
        278
Discrete probabilities, 31, 66
Discrete time, 336, 337
Distribution, balking, 270
    beta function, 267
    binomial, 65, 181, 272
        negative, 273
    busy period, 99, 107, 128
    chi-square, 45, 65
    compound, 64
    conditional, 47
    Engset, 304
    Erlangian, 65, 66, 68, 167, 227, 288
    exponential (see Exponential distribu-
        tion)
    extreme value, 132
    Fisher-Tippett, 132
    flat, 78
    gamma, 65
    Gaussian or normal, 65, 254
    geometric, 40, 183
    hyperexponential, 77, 371
    hypergeometric, 65
    initial number, 101

Distribution, log normal, 65
  negative binomial or Pascal, 65, 273
  O'Dell, 304, 356
  Pearson Type-III, 65
  Poisson, 65
  probability, 14
  steep, 78
  sum of independent variables, 58, 80
  uniform or rectangular, 65
Distribution function, 57
Dynamic priorities, 241

Efficiency, 125
$E_k/D/1$, 169
$E_k/M/1$, 166
  waiting time, 169
Elementary renewal theorem, 366
Elevator, 171
Emergency, 6
Engines, 332
Engset distribution, 304
Epidemic model, 117
Epochs of departure, 171
Equilibrium, 15
Ergodic analysis, 38
Ergodic equivalence class, periodic, 70
  regular, 70
Ergodic property, 69
Ergodic system, 72
Erlang, A. K., 21
Erlang loss formula, 282, 303
Erlangian distribution, 65, 164, 288
  approximation with, 66
  as filtered Poisson distribution, 66
  modified, 60
Erlangian service, 227
Erlang's model, 38
  average waiting time, 61
  equations, 39
  limited waiting room, 61
  $P(\leq t)$, 61
  steady-state solution, 39
  waiting in system, 61
  waiting times, 60
Errors, rounding, 52
  truncation, 52
$E_s/G/1$, 189
Euler's constant, 132
Events, 55
  independence of, 56

Expected number in the line, 40
Exponential distribution, 38, 53, 65, 176,
  201, 243
  forgetfulness property of, 38
  as urn problem, 52
Exponential service, 201, 243
Exponential-service times, 176
Extra service, 269
Extreme-value distribution, 132

Facility, 3, 4
  enterprise owner, 18
  specialized, 10
  time to become idle, 29
Failure, 332
  probability of, 364
Feedback, 294
  single-service, 297
  terminal, 296
Finite input source, 323
First-come first-served basis, 11, 38
First-passage time mean, 71
Fisher-Tippett distribution, 132
Flat distribution, 78
Flow, 314
Fokker-Planck equation, 254
Forward equations, 74
Frequency histoground, 14
  of different numbers waiting, 8
Frequency spectrum, 131
Full availability, 10, 20, 303
Function of bounded operator, 137

Gamma distribution, 65
Gaussian distribution, 254
General input and service, 209
General input Erlangian service, 219
General input exponential services, 219
Generalized hypergeometric function, 113
Generating function, 35
Generations of renewals, 360
Geometric distribution, 40, 183
Geometry, 374
Ghost calls, 353
$GI/G/1$, 209
Gnedenko's generalization, 206
Graphic approach, 351
Grocery store check-out counter, 3

Hankel function, 350
Heaviside step function, 320
Highway, 11, 44
History of queues, 20, 22
Holland tunnel, 317
Honest process, 85
Hospitals, 9, 343, 345
Hyperexponential distribution, 77, 371
Hypergeometric distribution, 65
Hypothesis test, 48

Idle server, 11
Imbedded Markoff chain, 171, 370
Impatient customers, 12, 122, 278
Imperfect information, 12
Incomplete gamma function, 371
Incomplete information, 10
Increase in population, 4
Infinite number of channels, 99, 127
Influx of papers to editor, 216
Information, 7, 322
  incomplete, 10
Initial customer, 28
Initial number, distribution of, 47, 101
Input, 5, 9
  arbitrary, 198, 209, 219, 275
  limited, 12, 121, 125
  unlimited, 12
Input control, 18
Input rate, 18
Input rush, 13
Input variations, 9
Inspection line, 49
Inspector, 350
Integrodifferential equation, 198, 200
Interarrival interval distribution, 12
Interdeparture, 258
Interference, 314
Interruption rate, 287
Inventory, 334–336
Iterated logarithm, law of, 67

Jockeying, 10, 23, 284
Joint probability, 224

Key renewal theorem, 366
Khintchine's argument, 186
Kronecker's delta, 28

Lag time, 298
Lagrange's expansion, 106, 134, 293
Laguerre polynomial, 119
Landing aid and time, 6
Laplace transform, 89
Last-come first-served basis, 252
Law of large number, 67, 186
Leaving service, 206
Leaving the system, 206
Leibnitz differentiation theorem, 100, 114
Level of inventory, 334
Life queue, 5
Likelihood function, 47, 78, 126
Limited availability, 10, 311
Limited input, 12, 121
  average number of customers not in
    system, 125
Limited waiting room, 61, 127
  probabilities, 61
  waiting time, 61
Lindley's integral equation, 210
Linear growth, 129
Link systems, 308
Log normal distribution, 65
Loss delay, 304, 354
Loss-delay system, 353
Lost calls, 12, 208, 281
Lost customers, 10

Machine availability, 331
Machine interference, 323
Manufacturing units, 252
Markoff chain, 68
  aperiodic, 71
  decomposable, 70, 78
  imbedded, 171, 370
  irreducible, 71
Markoff process, 74, 199
  transition matrix, 76
Markovian and non-Markovian proc-
  esses, 54
Markovian process, 74, 219
Matrices, orthogonal idempotent, 138
Matrix Nth section, 137, 145
  solution, 136
Maximum utilization, 266
  likelihood, 46
Mean, 31
Mean waiting time, 212, 233
Measure, positive regular, 86

Measure, space, 56, 57
Measures of effectiveness, 6, 17, 18
  average number waiting of those who
    wait, 124
  expected number waiting in queue, 124
  mean waiting time, for all arrivals, 124
    for those who wait, 124
  probability, of someone waiting, 124
    that all channels are occupied, 124
Medical care, 343
$M/E_k/1$, 164
Messages, 298
Minimization, 6
Minimizing delays, 318
Models, 18
Modified Bessel function, 249
Modified Erlangian, 174
Moments, 59
  generating function, 58, 175
  about origin and mean, 32
  problem, 86
Monte Carlo, 49, 351
Moving server, 291
Multiple channels, 82, 116, 153, 155, 159,
    203, 216, 243, 259, 280, 357
  busy period, 117
  steady-state treatment, 115
  transient treatment, 110
Multiqueues, 298

Negative binomial distribution, 273
Nervous impulses, 352
Networks, 373
New approach, 215
Non-Markovian, 215
Non-Poisson queues, 24, 153
Nonpreemptive priorities, 12, 30, 231,
    232
Normal distribution, 65
Normal ordinates, 273
Null hypothesis, 48
Number served in busy period, 196

O'Dell distribution, 304, 356
Oil pumps, 351
Operative efficiency, 19
Operator, 355
Optimization, 3, 374
Optimum number of channels, 7

Ordered queue, 7, 22, 219
Oriented arrivals, 280
Outlets, 349
Output, 5, 13, 255, 257

Parallel channels, 5, 13
  (See also Multiple channels)
Parameters, 44
Partially fulfilled service, 208
Pascal or negative binomial distribution,
    65, 273
Passenger debarkation, 352
Patients, 344
Pearson Type-III distribution, 65
Periodic arrivals, 321
Perishable goods, 18
Pesos, 353
Phase-type problem, 164
Phase-type service, 105, 255
Phases, 295
Poisson arrivals, bulk service, 172
Poisson distribution, 65
Poisson input, arbitrary-service-busy-
    period distribution, 194
    number served in busy period, 196
    number in system, 193
  constant-service times, $P(w)$, 63
Poisson process, 33, 36, 59, 75
  backward equation, 125
  equation of, 34
  exponential distribution of, 37
  mean number, 37
  variance, 37
Poisson queue, 83
  single-channel solution, 88
Pole, multiple, 156, 320
Pollaczek-Khintchine formula, 40, 50
Population at risk, 345
Post office, 10, 298, 352
Power series expansion, 50
Power spectrum, 348
Preemptive priorities, 12, 231
Priorities, 23, 231, 304
  continuous number of, 236
  dynamic, 241
  and machines, 236
  multiple channels, 234
  preemptive, 237
  waiting times, 233, 237
Priority, 6, 233

Priority, selection, 7
Probabilities, addition of, law, 57
  total, law, 57
Probability, conditional, 56
  conservation of, 217
  continuous, 32
  cumulative, 32
  of exactly n channels busy, 19
  of failure, 364
  of given number at any time, 8
  joint, 224
  of not waiting, 18
  of service of no longer than n, 31
Probability distribution, 14
Probability models, 31
Procedure of service, 11
  (*See also* Discipline)
Production-line operation, 11, 49
Profit, 3
Projected planning, 4
Provision, 332
Psychometric measurement, 352
$P(\leq t)$, 201
Pure birth process, 74
Pure loss, 207

Queue convention, 25
Queuee, 5
Queuer, 5
Queues, 3
  average number in, 18
  breakdown, 13
  bulk, 171
  bus, 4
  in a circle, 355
  contour representation, 21
  cyclic, 294
  descriptive analysis, 14
  deterministic, 26, 27
  diagram construction, 8
  at difficult time of day, 9
  discipline (*see* Discipline)
  effect of waiting room, 16
  fixed length, 12, 61
  and graph theory, 374
  history of, 20, 22
  interference, 11, 314
  length probabilities, 219
  malfunction, 18
  morphological approach, 17

Queues, network flow, 7
  non-Poisson, 24, 153
  notation, 25
  ordered, 7, 22, 219
  output, 12, 13, 255
  parameter estimation, 46
  Poisson, 83, 88
  pooling, 11
  properties, 12
  ramifications, 230
  remedies, staggering working hours, 19
  resistance, 270
  selection rule, 6
  in series, 24
  specialized, 12
  stacked, 5, 6
  tandem (*see* Tandem queues)
  varieties, 9

Random selection, 7, 278
  constant service times, 249
  exponential service, 243
  for service, 23
Random variables, 14
  depending on time, 32
  infinitely divisible, 79
Ratio, of average waiting time to service
    time, 43
  of mean queueing time to mean service
    time, 19
Reflecting barriers, 136
Regeneration points, 171, 172
Regeneration process, 370
Regular arrivals, 27
Regular service, 27
Reinitiating calls, 10
Remedies, 17, 19
Reneging, 6, 10, 23, 208, 270, 277
  and balking, 277
Renewal-density theorem, 366
Renewal function, 360, 365
Renewal process, 359
  central-limit theorem for, 370
Reorder, 334
Repairman, 326, 329
Repeated calls, 357
Repeated rule, 237
Replenishment cycle, 334
Replenishment time, 335
Reservoir, 336

Residual life, 369
Residues, 156
Restaurant, 171
    operation, 44
Resume rule, 237
Retarded arrivals, 280
Riemann-Stieltjes integral, 57
Risk level, 48
Rouché's theorem, 87, 89
Rounding errors, 52
Runway, 5
    turnoffs, 6

Sampling distribution, 46
Saturated case, 237
Scale-modified distribution, 164
Scheduled arrivals, 282
Scheduling of outpatient care, 346
Schläfli's integral representation, 249, 350
Selection rules, 231
Semiconductor noise, 348
Semi-Markoff process, 189
Serially independent random variables, 337
Series and parallel channels, 11, 261
    queues, 255
Service, 5, 6, 9
    constant, 176
    cyclic, 298
    different rates, 290
    exponential, 176
    extra, 269
    outlets, 349
    partially fulfilled, 208
    procedure, 11
    rate, 18
Ships, 10, 12, 352
Sigma algebra, 55
Significance level, 46, 48
Simulation, 17, 49, 374
Single channel, 198, 211
    Erlang's model, steady-state result, 95
    priorities, 234
    random-walk solution, 96
    transient solution, 91
Small barriers, 135
Spare machines, 332
Special and general channels, 11, 13
Special service, 290

Spectral form, 137
Spectral function, 147
Stacked queue, 5, 6
Standard deviation, 31
State, absorbing, 84
    equivalences classes, 70
    ergodic, 71
    null, 71
    periodic, 71
    recurrent, 71
    reflecting, 84
    of system, 69
    transient, 71
Stationary process, 67, 74
Statistical equilibrium, 40
Statistical sampling, 7
Steady state, 15, 16, 30, 164, 167, 203, 357
Steep distribution, 78
Steepest descent, 372
Stochastic matrix, 68
    regular, 70
Stochastic process, 14, 33, 66
Stoppages, 317, 329
Storage systems, 336
Suez Canal, 10, 12
Sum of independently distributed random variables, 58, 80
Supermarket, 11, 25, 352
Superposition theorem, 370

Takác's equation, 198, 202
Tandem queues, 255, 259
    ergodicity, 268
    waiting-time distribution, 263, 267
Tauberian theorem, 218
Taxis, 78, 352
Telegraph communications, 352
Telephone, 171, 302–311
    busy signal, 10
    common control, 22, 309, 355
    grading, 21, 311
    limited availability, 21, 311
    link systems, 22, 308
    subscribers, 67
Testing hypotheses, 44
Theatres, 352
Three preemptive priorities, 238
Time, to serve all items which have waited, 28

Time, total, in system, 30
Time average, 42
Time congestion, 353, 356
Time-dependent parameters, 101, 104
Toll collection, 350
Total idle-time computation, 7
Total probability, 245
Total waiting time, 6
Traffic handled, 358
Traffic intensity, 18, 302
Traffic lane, 11
Traffic lights, 313
Transient equivalence class, 70
Transient solution, 30
Transient state, 18
Transition matrix, infinitesimal, 76
    limiting, 69
Transition probability, 68
Truck docks, 351
Truncation errors, 52
Tunnels, 317
Turnaround, 322
Two channels in parallel, 126
Type-III ordinate, 273

Uniform or rectangular distribution, 65

Variable number of channels, 13, 285
Variance, 31
    ratio test, 44
Vehicular stoppages, 317

Waiting line, 3, 12
Waiting space, 7
Waiting time, 198
    bulk, 177
    distribution, 18, 108, 198, 200, 202, 205,
        207, 210, 215
    of nth customer, 16
    variance, 18
Waitress, 352
Walking repairman, 332
Walking time, 350
Warehouse, 351
Wasted time, 4
Water storage, 337
Wiener-Hopf-type integral equation, 210
Work load, 215

A CATALOGUE OF SELECTED DOVER BOOKS
IN ALL FIELDS OF INTEREST

A CATALOGUE OF SELECTED DOVER
BOOKS IN ALL FIELDS OF INTEREST

CONDITIONED REFLEXES, Ivan P. Pavlov. Full translation of most complete statement of Pavlov's work; cerebral damage, conditioned reflex, experiments with dogs, sleep, similar topics of great importance. 430pp. 5⅜ x 8½.
60614-7 Pa. $4.50

NOTES ON NURSING: WHAT IT IS, AND WHAT IT IS NOT, Florence Nightingale. Outspoken writings by founder of modern nursing. When first published (1860) it played an important role in much needed revolution in nursing. Still stimulating. 140pp. 5⅜ x 8½.
22340-X Pa. $3.00

HARTER'S PICTURE ARCHIVE FOR COLLAGE AND ILLUSTRATION, Jim Harter. Over 300 authentic, rare 19th-century engravings selected by noted collagist for artists, designers, decoupeurs, etc. Machines, people, animals, etc., printed one side of page. 25 scene plates for backgrounds. 6 collages by Harter, Satty, Singer, Evans. Introduction. 192pp. 8⅞ x 11¾.
23659-5 Pa. $5.00

MANUAL OF TRADITIONAL WOOD CARVING, edited by Paul N. Hasluck. Possibly the best book in English on the craft of wood carving. Practical instructions, along with 1,146 working drawings and photographic illustrations. Formerly titled *Cassell's Wood Carving*. 576pp. 6½ x 9¼.
23489-4 Pa. $7.95

THE PRINCIPLES AND PRACTICE OF HAND OR SIMPLE TURNING, John Jacob Holtzapffel. Full coverage of basic lathe techniques—history and development, special apparatus, softwood turning, hardwood turning, metal turning. Many projects—billiard ball, works formed within a sphere, egg cups, ash trays, vases, jardiniers, others—included. 1881 edition. 800 illustrations. 592pp. 6⅛ x 9¼.
23365-0 Clothbd. $15.00

THE JOY OF HANDWEAVING, Osma Tod. Only book you need for hand weaving. Fundamentals, threads, weaves, plus numerous projects for small board-loom, two-harness, tapestry, laid-in, four-harness weaving and more. Over 160 illustrations. 2nd revised edition. 352pp. 6½ x 9¼.
23458-4 Pa. $6.00

THE BOOK OF WOOD CARVING, Charles Marshall Sayers. Still finest book for beginning student in wood sculpture. Noted teacher, craftsman discusses fundamentals, technique; gives 34 designs, over 34 projects for panels, bookends, mirrors, etc. "Absolutely first-rate"—E. J. Tangerman. 33 photos. 118pp. 7¾ x 10⅝.
23654-4 Pa. $3.50

ART FORMS IN NATURE, Ernst Haeckel. Multitude of strangely beautiful natural forms: Radiolaria, Foraminifera, jellyfishes, fungi, turtles, bats, etc. All 100 plates of the 19th-century evolutionist's *Kunstformen der Natur* (1904). 100pp. 9⅜ x 12¼.　　　　　　22987-4 Pa. $5.00

CHILDREN: A PICTORIAL ARCHIVE FROM NINETEENTH-CENTURY SOURCES, edited by Carol Belanger Grafton. 242 rare, copyright-free wood engravings for artists and designers. Widest such selection available. All illustrations in line. 119pp. 8⅜ x 11¼.
23694-3 Pa. $4.00

WOMEN: A PICTORIAL ARCHIVE FROM NINETEENTH-CENTURY SOURCES, edited by Jim Harter. 391 copyright-free wood engravings for artists and designers selected from rare periodicals. Most extensive such collection available. All illustrations in line. 128pp. 9 x 12.
23703-6 Pa. $4.50

ARABIC ART IN COLOR, Prisse d'Avennes. From the greatest ornamentalists of all time—50 plates in color, rarely seen outside the Near East, rich in suggestion and stimulus. Includes 4 plates on covers. 46pp. 9⅜ x 12¼.　　　　　　23658-7 Pa. $6.00

AUTHENTIC ALGERIAN CARPET DESIGNS AND MOTIFS, edited by June Beveridge. Algerian carpets are world famous. Dozens of geometrical motifs are charted on grids, color-coded, for weavers, needleworkers, craftsmen, designers. 53 illustrations plus 4 in color. 48pp. 8¼ x 11. (Available in U.S. only)　　　　　　23650-1 Pa. $1.75

DICTIONARY OF AMERICAN PORTRAITS, edited by Hayward and Blanche Cirker. 4000 important Americans, earliest times to 1905, mostly in clear line. Politicians, writers, soldiers, scientists, inventors, industrialists, Indians, Blacks, women, outlaws, etc. Identificatory information. 756pp. 9¼ x 12¾.　　　　　　21823-6 Clothbd. $40.00

HOW THE OTHER HALF LIVES, Jacob A. Riis. Journalistic record of filth, degradation, upward drive in New York immigrant slums, shops, around 1900. New edition includes 100 original Riis photos, monuments of early photography. 233pp. 10 x 7⅞.　　　　　　22012-5 Pa. $7.00

NEW YORK IN THE THIRTIES, Berenice Abbott. Noted photographer's fascinating study of city shows new buildings that have become famous and old sights that have disappeared forever. Insightful commentary. 97 photographs. 97pp. 11⅜ x 10.　　　　　　22967-X Pa. $5.00

MEN AT WORK, Lewis W. Hine. Famous photographic studies of construction workers, railroad men, factory workers and coal miners. New supplement of 18 photos on Empire State building construction. New introduction by Jonathan L. Doherty. Total of 69 photos. 63pp. 8 x 10¾.
23475-4 Pa. $3.00

THE CURVES OF LIFE, Theodore A. Cook. Examination of shells, leaves, horns, human body, art, etc., in "*the* classic reference on how the golden ratio applies to spirals and helices in nature . . . ."—Martin Gardner. 426 illustrations. Total of 512pp. 5⅜ x 8½.   23701-X Pa. $5.95

AN ILLUSTRATED FLORA OF THE NORTHERN UNITED STATES AND CANADA, Nathaniel L. Britton, Addison Brown. Encyclopedic work covers 4666 species, ferns on up. Everything. Full botanical information, illustration for each. This earlier edition is preferred by many to more recent revisions. 1913 edition. Over 4000 illustrations, total of 2087pp. 6⅛ x 9¼.   22642-5, 22643-3, 22644-1 Pa., Three-vol. set $25.50

MANUAL OF THE GRASSES OF THE UNITED STATES, A. S. Hitchcock, U.S. Dept. of Agriculture. The basic study of American grasses, both indigenous and escapes, cultivated and wild. Over 1400 species. Full descriptions, information. Over 1100 maps, illustrations. Total of 1051pp. 5⅜ x 8½.   22717-0, 22718-9 Pa., Two-vol. set $15.00

THE CACTACEAE,, Nathaniel L. Britton, John N. Rose. Exhaustive, definitive. Every cactus in the world. Full botanical descriptions. Thorough statement of nomenclatures, habitat, detailed finding keys. The one book needed by every cactus enthusiast. Over 1275 illustrations. Total of 1080pp. 8 x 10¼.   21191-6, 21192-4 Clothbd., Two-vol. set $35.00

AMERICAN MEDICINAL PLANTS, Charles F. Millspaugh. Full descriptions, 180 plants covered: history; physical description; methods of preparation with all chemical constituents extracted; all claimed curative or adverse effects. 180 full-page plates. Classification table. 804pp. 6½ x 9¼.
                                                              23034-1 Pa. $12.95

A MODERN HERBAL, Margaret Grieve. Much the fullest, most exact, most useful compilation of herbal material. Gigantic alphabetical encyclopedia, from aconite to zedoary, gives botanical information, medical properties, folklore, economic uses, and much else. Indispensable to serious reader. 161 illustrations. 888pp. 6½ x 9¼. (Available in U.S. only)
                                  22798-7, 22799-5 Pa., Two-vol. set $13.00

THE HERBAL or GENERAL HISTORY OF PLANTS, John Gerard. The 1633 edition revised and enlarged by Thomas Johnson. Containing almost 2850 plant descriptions and 2705 superb illustrations, Gerard's *Herbal* is a monumental work, the book all modern English herbals are derived from, the one herbal every serious enthusiast should have in its entirety. Original editions are worth perhaps $750. 1678pp. 8½ x 12¼.
                                                              23147-X Clothbd. $50.00

MANUAL OF THE TREES OF NORTH AMERICA, Charles S. Sargent. The basic survey of every native tree and tree-like shrub, 717 species in all. Extremely full descriptions, information on habitat, growth, locales, economics, etc. Necessary to every serious tree lover. Over 100 finding keys. 783 illustrations. Total of 986pp. 5⅜ x 8½.
                                  20277-1, 20278-X Pa., Two-vol. set $11.00

GEOMETRY, RELATIVITY AND THE FOURTH DIMENSION, Rudolf Rucker. Exposition of fourth dimension, means of visualization, concepts of relativity as Flatland characters continue adventures. Popular, easily followed yet accurate, profound. 141 illustrations. 133pp. 5⅜ x 8½.
23400-2 Pa. $2.75

THE ORIGIN OF LIFE, A. I. Oparin. Modern classic in biochemistry, the first rigorous examination of possible evolution of life from nitrocarbon compounds. Non-technical, easily followed. Total of 295pp. 5⅜ x 8½.
60213-3 Pa. $4.00

PLANETS, STARS AND GALAXIES, A. E. Fanning. Comprehensive introductory survey: the sun, solar system, stars, galaxies, universe, cosmology; quasars, radio stars, etc. 24pp. of photographs. 189pp. 5⅜ x 8½. (Available in U.S. only)
21680-2 Pa. $3.75

THE THIRTEEN BOOKS OF EUCLID'S ELEMENTS, translated with introduction and commentary by Sir Thomas L. Heath. Definitive edition. Textual and linguistic notes, mathematical analysis, 2500 years of critical commentary. Do not confuse with abridged school editions. Total of 1414pp. 5⅜ x 8½.   60088-2, 60089-0, 60090-4 Pa., Three-vol. set $18.50

Prices subject to change without notice.

Available at your book dealer or write for free catalogue to Dept. GI, Dover Publications, Inc., 180 Varick St., N.Y., N.Y. 10014. Dover publishes more than 175 books each year on science, elementary and advanced mathematics, biology, music, art, literary history, social sciences and other areas.